Optical Phenomena in Semiconductor Structures of Reduced Dimensions

NATO ASI Series

Advanced Science Institutes Series

A Series presenting the results of activities sponsored by the NATO Science Committee, which aims at the dissemination of advanced scientific and technological knowledge, with a view to strengthening links between scientific communities.

The Series is published by an international board of publishers in conjunction with the NATO Scientific Affairs Division

A Life Sciences B Physics	Plenum Publishing Corporation London and New York
C Mathematical and Physical Sciences D Behavioural and Social Sciences E Applied Sciences	Kluwer Academic Publishers Dordrecht, Boston and London
F Computer and Systems Sciences G Ecological Sciences H Cell Biology I Global Environmental Change	Springer-Verlag Berlin, Heidelberg, New York, London, Paris and Tokyo

NATO-PCO-DATA BASE

The electronic index to the NATO ASI Series provides full bibliographical references (with keywords and/or abstracts) to more than 30000 contributions from international scientists published in all sections of the NATO ASI Series.
Access to the NATO-PCO-DATA BASE is possible in two ways:

– via online FILE 128 (NATO-PCO-DATA BASE) hosted by ESRIN,
Via Galileo Galilei, I-00044 Frascati, Italy.

– via CD-ROM "NATO-PCO-DATA BASE" with user-friendly retrieval software in English, French and German (© WTV GmbH and DATAWARE Technologies Inc. 1989).

The CD-ROM can be ordered through any member of the Board of Publishers or through NATO-PCO, Overijse, Belgium.

Series E: Applied Sciences - Vol. 248

Optical Phenomena in Semiconductor Structures of Reduced Dimensions

edited by

David J. Lockwood
National Research Council,
Ottawa, Ontario, Canada

and

Aron Pinczuk
AT&T Bell Laboratories,
Murray Hill, New Jersey, U.S.A.

Kluwer Academic Publishers

Dordrecht / Boston / London

Published in cooperation with NATO Scientific Affairs Division

Proceedings of the NATO Advanced Research Workshop on
Frontiers of Optical Phenomena in Semiconductor Structures of Reduced Dimensions
Yountville, California, U.S.A.
July 27–31, 1992

Library of Congress Cataloging-in-Publication Data

```
Optical phenomena in semiconductor structures of reduced dimensions /
   edited by David J. Lockwood and Aron Pinczuk.
       p.    cm. -- (NATO ASI Series. Series E, Applied sciences ; no.
   248)
      ISBN 0-7923-2512-5 (alk. paper)
      1. Semiconductors--Optical properties--Congresses.  2. Quantum
   Hall effect--Congresses.  3. Quantum wells--Congresses.
   I. Lockwood, David J.    II. Pinczuk, Aron.   III. Series: NATO ASI
   Series.  Series E, Applied sciences ; no. 248.
   QC611.6.O606626   1993
   537.6'226--dc20                                              93-31755
```

ISBN 0-7923-2512-5

Published by Kluwer Academic Publishers,
P.O. Box 17, 3300 AA Dordrecht, The Netherlands.

Kluwer Academic Publishers incorporates the publishing programmes of
D. Reidel, Martinus Nijhoff, Dr W. Junk and MTP Press.

Sold and distributed in the U.S.A. and Canada
by Kluwer Academic Publishers,
101 Philip Drive, Norwell, MA 02061, U.S.A.

In all other countries, sold and distributed
by Kluwer Academic Publishers Group,
P.O. Box 322, 3300 AH Dordrecht, The Netherlands.

Printed on acid-free paper

All Rights Reserved
© 1993 Kluwer Academic Publishers
No part of the material protected by this copyright notice may be reproduced or utilized in any form or by any means, electronic or mechanical, including photocopying, recording or by any information storage and retrieval system, without written permission from the copyright owner.

Printed in the Netherlands

This book contains the proceedings of a NATO Advanced Research Workshop held within the programme of activities of the NATO Special Programme on Nanoscale Science as part of the activities of the NATO Science Committee.

Other books previously published as a result of the activities of the Special Programme are:

NASTASI, M., PARKING, D.M. and GLEITER, H. (eds.), *Mechanical Properties and Deformation Behavior of Materials Having Ultra-Fine Microstructures.* (ASIE 233) 1993 ISBN 0-7923-2195-2

VU THIEN BINH, GARCIA, N. and DRANSFELD, K. (eds), *Nanosources and manipulation of Atoms under High Fields and Temperatures: Applications.* (E235) 1993 ISBN 0-7923-2266-5

LEBURTON, J.-P., PASCUAL J. and SOTOMAYOR TORRES, C. (eds.), *Phonons in Semiconductor Nanostructures.* (E236) 1993 ISBN 0-7923-2277-0

AVOURIS P. (ed.), *Atomic and Nanometer-Scale Modification of Materials: Fundamentals and Applications.* (E239) 1993 ISBN 0-7923-2334-3

BLÖCHL, P. E., JOACHIM, C. and FISHER, A. J. (eds.), *Computations for the Nano-Scale.* (E240) 1993 ISBN 0-7923-2360-2

POHL, D. W. and COURJON, D. (eds.), *Near Field Optics.* (E242) 1993 ISBN 0-7923-2394-7

SALEMINK, H. W. M. and PASHLEY, M. D. (eds.), *Semiconductor Interfaces at the Sub-Nanomater Scale.* (E243) 1993 ISBN 0-7923-2397-1

BENSAHEL, D. C., CANHAM, L. T. and OSSICINI, S. (eds.), *Optical Properties of Low Dimensional Silicon Structures.* (E244) 1993 ISBN 0-7923-2446-3

HERNANDO, A. (ed.), *Nanomagenetism* (E247) 1993. ISBN 0-7923-2485-4

CONTENTS

Preface xi

QUANTUM HALL EFFECT

Electrons in a Landau Level Near Half-Filling (abstract only) 1
   B.I. Halperin

Electron Spin Resonance and Nuclear Spin Relaxation
   in GaAs/AlGaAs Heterostructures 3
   A. Berg and K. von Klitzing

Collective Spin-Flip Excitations in the Quantum Hall Regime 13
   C. Kallin and J.P. Longo

Coulomb Perturbation by Photoholes of Two-Dimensional Electron Gas InGaAs-
   Heterojunctions: The Case of Fermi-Edge Singularities (abstract only) 27
   W. Chen, M. Fritze, and A.V. Nurmikko

Photoluminescence Studies of Two-Dimensional Electrons in the
   Extreme Quantum Limit 29
   I.V. Kukushkin, N.J. Pulsford, K. von Klitzing, R.J. Haug, K. Ploog, H. Buhmann,
   G. Martinez, M. Potemski, and V.B. Timofeev

Photoluminescence Spectroscopy of Incompressible Electron
   Liquid States in GaAs 45
   J.F. Ryan, A.J. Turberfield, R.A. Ford, I.N. Harris, C.T. Foxon, and J.J. Harris

Optical Spectroscopy in the Regimes of the Integer and Fractional
   Quantum Hall Effects (extended abstract) 57
   A. Pinczuk, B.S. Dennis, L.N. Pfeiffer, and K.W. West

Magneto-Optics of the Fractional Quantum Hall Effect: Theory 63
   E.I. Rashba and V.M. Apal'kov

Photoluminescence in the Fractional Quantum Hall Effect 79
   E.H. Rezayi

Exciton Unbinding in the Quasi-Two-Dimensional Electron Gas 91
G.E.W. Bauer

TIME-RESOLVED SPECTROSCOPY

Ultrafast Time-Resolved Spectroscopy of Quantum Confined
  Semiconductor Heterostructures 101
D.S. Chemla

Femtosecond Coherent Spectroscopy of Semiconductor Nanostructures........................117
J. Shah

Free Exciton Radiative Recombination in GaAs Quantum Wells......................................129
B. Deveaud, F. Clérot, B. Sermage, C. Dumas, and D.S. Katzer

Time Resolved Four Wave Mixing in GaAs/AlAs
  Quantum Well Structures 145
E.O. Göbel, M. Koch, J. Feldmann, G. von Plessen, T. Meier, A. Schulze, P. Thomas,
S. Schmitt-Rink, K. Köhler, and K. Ploog

Spin Dynamics and Dimensionality in Magnetic Semiconductor
  Quantum Structures...157
D.D. Awschalom, J.F. Smyth, and N. Samarth

Coherent Optical Phenomena of Quantum Well Excitons
  in a Magnetic Field..173
I. Bar-Joseph, S. Bar-Ad, O. Carmel, and Y. Levinson

Coherent Nonlinear Laser Spectroscopy of Excitons in Quantum Wells 187
D.G. Steel, S.T. Cundiff, and H. Wang

Theory of Spin Dynamics of Excitons and Free Carriers
  in Quantum Wells 201
L.J. Sham

QUANTUM WELLS

All-Optical Bistability in Asymmetric Quantum Wells 213
S. Scandolo and F. Tassone

Magneto-Optics of Shallow Impurities in Superlattices..221
F.M. Peeters, J.M. Shi, and J.T. Devreese

Enhanced Linear Electro-Optic Anisotropy in GaAs Quantum Wells.............................239
S.H. Kwok, H.T. Grahn, K. Ploog, and R. Merlin

QUANTUM WIRES AND DOTS

Spectroscopy on Lateral Superlattices 247
    J.P. Kotthaus

Optical Spectroscopy of Quantum Wires and Dots (extended abstract) 265
    A.S. Plaut, K. Kash, E. Kapon, B.P. Van der Gaag, A.S. Gozdz, D.M. Hwang, E. Colas, J.P. Harbison, L.T. Florez, H. Lage, P. Grambow, D. Heitmann, K. von Klitzing, and K. Ploog

Optical Properties of Semiconductor Quantum Well Wires 267
    J.M. Calleja, A.R. Goñi, J.S. Weiner, B.S. Dennis, A. Pinczuk, L.N. Pfeiffer, and K.W. West

Optical Singularities of the Quasi One-Dimensional Electron Gas 281
    C. Tejedor and F.J. Rodríguez

Acceptor Related Photoluminescence as a Probe of Many-Electron
    States in Semiconductor Nanostructures 295
    P. Hawrylak

Towards Fully Quantized Optoelectronic Semiconductor
    Heterostructures: Quantum Boxes or Quantum Microcavities? 311
    C. Weisbuch

Luminescence Properties of GaAs Quantum Wells, Wires,
    Dots, and Antidots 327
    G. Abstreiter, G. Böhm, K. Brunner, F. Hirler, R. Strenz, and G. Weimann

Optical Properties of Strain-Induced Nanometer Scale
    Quantum Wires 337
    D. Gershoni, J.S. Weiner, E.A. Fitzgerald, L.N. Pfeiffer, and N. Chand

Spectroscopy of Quantum-Dot Atoms 351
    D. Heitmann, B. Meurer, and K. Ploog

TUNNELING

Transmissions Through Barriers and Resonant Tunneling
    in a Luttinger Liquid 365
    C.L. Kane

Novel Electro-Optical Device Structures Based
    on Quantum Wells (extended abstract) 373
    E.E. Mendez

Excited State Populations of the Quantum Wells of Double Barrier
    Resonant Tunneling Structures 377
    P.D. Buckle, J.W. Cockburn, M.S. Skolnick, D.M. Whittaker, W.I.E. Tagg, R. Grey,
    G. Hill, and M.A. Pate

Quantum Well Luminescence by Resonant Tunneling
    Injection of Electrons and Holes..387
    H.B. Evans, L. Eaves, C.R.H. White, M. Henini, P.D. Buckle, T.A. Fisher, D.J. Mowbray,
    and M.S. Skolnick

NANOSCALE STRUCTURES

Electronic Excitations and Optical Properties of C_{60} Molecules 401
    E. Burstein and M.Y. Jiang

Optical Properties of Porous Silicon...409
    D.J. Lockwood

Optical Properties of Siloxene and Siloxene Derivates...427
    M. Stutzmann, M.S. Brandt, H.D. Fuchs, M. Rosenbauer, M.K. Kelly, P. Deak, J. Weber,
    and S. Finkbeiner

Group Photograph..443

Participants...445

Author Index..449

Subject Index 451

PREFACE

These proceedings are a tangible outcome of the NATO Advanced Research Workshop (ARW) on "Frontiers of Optical Phenomena in Semiconductor Structures of Reduced Dimensions" held at the Vintage Inn, Yountville, California from the 27th to the 31st of July 1992. There were, of course, many intangible benefits accruing from the ARW: the beautiful setting amidst the Napa Valley vineyards; the excellent food (including the Vintage Inn champagne breakfast) and local wine; the bright sunshine; jogging, swimming, and tennis before breakfast; the vigorous discussions during the scientific sessions; the exchange of information; and most importantly of all, the formation and strengthening of friendships and contacts between NATO country members and with participants from Russia and Israel. These inestimable scientific and social benefits are an important aspect of small concentrated meetings like ARWs and are not commonly found at larger ordinary conferences. Special social events held during the meeting included a testimonial event for Elias Burstein to mark the occasion of his forthcoming 75th birthday and rather rushed, but successful, tours of the Sterling Vineyards and Beringer wineries in the upper Napa Valley. The successful marriage of wine and science can be gauged from a snippet of conversation overheard just as glasses were being filled during the wine tasting session at the Sterling Vineyards: "... what's the filling factor?"—and later, after the wine was poured and tasted, came the answer "... one-third".

The scientific programme of the ARW showed that steady-state and time-resolved optical spectroscopies are among the most promising experimental methods currently used in studies of the novel physics of electron-electron interactions in semiconductors of reduced dimensions. The ARW provided a forum in which experimental and theoretical physicists discussed and evaluated recent developments in the optical research of the exciting new phenomena that occur in the electron systems of semiconductor nanostructures. The topics of principal interest included the integral and fractional quantum Hall effects, the non-linear optics of low dimensional electron systems, the optical responses associated with electrons in quantum wires and dots, and the optical characterization of electron tunneling in ultra-small structures. During the ARW we also had informative presentations on recent developments in the studies of optically active porous silicon and siloxene, and on the non-linear optics of fullerenes.

Optical investigations of quantum Hall phenomena and the solid phases of the ultra-high mobility two-dimensional (2D) electron gas in GaAs–AlGaAs heterostructures were the principal subjects considered during the first day of the ARW. We heard a theoretical overview of the physics of the 2D electron gas in the extreme magnetic quantum limit. This talk emphasized the relevance of the state at filling factor 1/2 in the physics of the fractional quantum Hall effect. This work will stimulate optical experiments seeking to observe elementary excitations either by absorption or inelastic scattering. The speakers also considered the singular behaviors of electron spin in the regime of the integral quantum

Hall effect. It was demonstrated that the coupling of electron and nuclear spins provides a unique venue to study the electron density of states and relaxation phenomena. Inelastic light scattering was shown to give direct insight into the enhancement of exchange interactions, while a presentation of the theoretical analysis explained that the Hartree-Fock framework gives excellent agreement with experiment in the integral quantum Hall regime. The magneto-optics of the fractional quantum Hall effect and the 2D electron solid was considered extensively by several speakers. There was a consensus that the appearance of novel quantum states in the extreme magnetic quantum limit has striking manifestations in spectra of optical emission. There were, however, different points of view in regard to the interpretation and character of the experiments. One issue that still requires full understanding is that of the responses of the quantum liquid and solid to the charged valence state that participates in the optical recombination.

There were several talks dealing with optical nonlinearities in nanostructures presented at the meeting. The majority of presentations dealt with coherent optical nonlinearities in quantum wells. The reduction of effective dimensionality was however achieved by the applications of a magnetic field. The current experimental frontier involves the time resolved four wave mixing (TRFWM) technique using femtosecond pulses. This technique provides information about dephasing of polarization. Experiments demonstrated a significant difference in coherent nonlinear phenomena involving excitons in semiconductor nanostructures versus atomic vapors. These differences have been attributed to many body effects associated with excitons and the electron-hole plasma. The understanding of these effects is slowly emerging. Some new results were presented on TRFWM experiments involving a degenerate electron gas in quantum wells. Since linear optical processes in this case involve complicated many body effects (the Fermi edge singularity), an understanding of these experiments poses a challenge for the future.

Significant progress was also reported in optical studies of quantum wires and quantum dots, although it appears that the quality of quantum wires and dots is not, as yet, sufficient to study coherent phenomena. There is a consensus that in this field the methods used in fabrication of the semiconductor nanostructures play a major role. Fabrication procedures considered by the speakers included bias-gating, reactive-ion etching, and wet-etching of sub-micron holographic gratings. The work on nanometre-scale quantum wires involved fabrication by means of electron-beam lithography and also by the novel method of cleave-edge overgrowth. It was shown that infrared absorption and inelastic light scattering provide remarkable vistas of the spectra of elementary excitations of electrons with quasi-one-dimensional and quasi-zero-dimensional behaviors. Photoluminescence and photoluminescence excitation spectroscopies reveal new phenomena related to confinement and excitonic behaviors. Carrier relaxation phenomena were considered in a conceptual presentation that highlighted the relevance of non-equilibrium distributions in the lower-dimensionality electron systems. There was agreement that disorder has remarkable effects on the states of the one-dimensional electron gas. This is a subject of great current interest in the theoretical considerations that apply to optical Fermi-edge singularities in quantum wires and to the Tomonaga-Luttinger models of the one-dimensional electron gas. At the ARW there was a consensus that these fundamental questions are going to be the subjects of further intense study.

The tremendous success of the ARW was due in a large part to the enthusiastic contributions from the hard working participants. However, the ARW would never have been possible without the financial support of the NATO Scientific Affairs Division under the auspices of the Special Programme on Nanoscale Science, the Institute for Microstructural Sciences of the National Research Council of Canada, the Office of Naval Research USA, AT&T Bell Laboratories USA, Lawrence Berkeley Laboratory USA, and the University of Pennsylvania, Philadelphia, USA. We acknowledge and are grateful for this support.

We wish to thank Linda Charbonneau and Deb Tunney of the National Research Council and Nathalie Nys of the Lawrence Berkeley Laboratory for their invaluable assistance in the preparations for the ARW; Nathalie Nys for organizational support during the running of the ARW; and Betty Legault and, in particular, Carole Munro of the National Research Council for considerable help in producing these proceedings. We are also grateful to the many people who provided valuable ideas for the programme; our manuscript reviewers; the session chair persons; Gerhard Abstreiter and Daniel S. Chemla for their overviews of future directions in the semiconductor field; and Elias Burstein and Lu J. Sham for their meeting summary, which together with material from Pawel Hawrylak formed a basis for our meeting report given above. Special thanks are due to our organizing committee members Elias Burstein, Daniel S. Chemla, and Jörg P. Kothaus for their invaluable advice, hard work, and support throughout the entire process of organizing and holding the meeting. Finally, we thank all of the participants for their valuable contributions to the ARW itself and for providing their manuscripts for these proceedings.

David J. Lockwood
Aron Pinczuk

ELECTRONS IN A LANDAU LEVEL NEAR HALF-FILLING

B.I. HALPERIN
Physics Department
Harvard University
Cambridge, MA 02138
USA

In collaboration with Patrick Lee (MIT) and Nicholas Read (Yale University) [1], the author has developed a theory of a two-dimensional electron system in an external magnetic field when the Landau-level filling factor ν is close to 1/2, which is based on a mathematical transformation to a system of fermions interacting with a Chern-Simons gauge field, such that at $\nu = 1/2$, the average effective magnetic field seen by the fermions is zero. If one ignores fluctuations in the gauge field, this implies that for a system with no impurity scattering there should be a well-defined *Fermi surface* for the fermions. When gauge fluctuations are taken into account we find that there can be infrared-divergent corrections to the quasiparticle propagator, which we interpret as a divergence in the effective mass m*, whose form depends on the nature of the assumed electron-electron interaction v(r). For short range repulsive interactions we find power-law divergences; while for Coulomb interactions we find logarithmic corrections to m*. Nevertheless, we argue that many features of the Fermi surface are likely to exist. In the presence of a weak impurity scattering potential, we predict a finite resistivity ρ_{xx} at low temperatures, whose value we estimate. We compute an anomalous contribution to surface acoustic wave propagation in qualitative agreement with recent experiments [2]. We also make predictions for the size of the energy gap in the fractional quantized Hall state at $\nu = p/(2p + 1)$, where p is an integer. The theory can also be generalized to other filling fractions with an even denominator. The talk will review some highlights of the theory and discuss possible implications for experiments involving far-infrared absorption or Raman scattering.

References

1. B.I. Halperin, P.A. Lee, and N. Read, submitted to Phys. Rev. B.
2. R.L. Willett, M.A. Paalanen, R.R. Ruel, K.W. West, L.N. Pfeiffer, and D.J. Bishop, Phys. Rev. Lett. **65**, 112 (1990); R.L. Willett, R.R. Ruel, M.A. Paalanen, K.W. West, and L.N. Pfeiffer, to be published.

ELECTRON SPIN RESONANCE AND NUCLEAR SPIN RELAXATION IN GaAs/AlGaAs HETEROSTRUCTURES

A. BERG AND K. VON KLITZING
Max-Planck-Institut für Festkörperforschung
Heisenbergstrasse 1
W-7000 Stuttgart 80
Germany

ABSTRACT. Electron spin resonance measurements (ESR) on free electrons in two-dimensional systems give information about the bandstructure parameters, the nuclear spin polarization (via the shift of the ESR line by the Overhauser shift), and the nuclear spin-lattice relaxation process which is dominated by the Korringa relaxation due to the conduction electrons. The Korringa relaxation rate depends on the product of the density of states of electrons at the Fermi energy with different spin orientations. The analysis of the relaxation rate gives information about the enhanced g-factor of electrons in two-dimensional systems whereas the ESR energy is well described within the single-electron picture for the bandstructure. The difference between the enhanced and the bare spin splitting is proportional to the exchange energy which has typical values of 2-3 meV.

1. ESR Spectroscopy on Two-Dimensional Systems

Most of the spectroscopic investigations on low dimensional electronic systems are done in the energy range of visible or near infrared light for interband transitions (Raman spectroscopy, excitonic transitions) or in the far infrared range corresponding to the energy splittings between the electric subbands and the cyclotron resonance energy in strong magnetic fields. The relatively small spin splitting of each Landau level is not directly measurable in absorption experiments since the number of electron spins ~ 10^{10} in typical two-dimensional systems (this number is even smaller in one-dimensional and zero-dimensional structures) is too small to be detected in conventional ESR measurements. However the magnetoresistance of a two-dimensional system is extremely sensitive to the ESR process as demonstrated in Fig. 1 [1]. Under microwave illumination, the resistivity ρ_{xx} shows, in addition to the well-known Shubnikov-de Haas oscillations, a small resonance peak with a halfwidth of only $\Delta B \sim 30$ mT at the magnetic field position of the ESR. The origin of this photosignal is not completely understood, but most of the signal can be qualitatively interpreted as a small increase in the electron temperature due to the absorption process. Contrary to cyclotron resonance transitions with an absorption intensity proportional to the electron concentration N_S but independent of the filling factor $\nu = N_S h/eB$, the ESR signal disappears at magnetic field values corresponding to even filling factors $\nu = 2, 4, 6, ...$ but is strongly enhanced at odd filling factors $\nu = 1, 3, 5, ...$. This oscillation of the ESR absorption as a function of the filling factor (magnetic field) originates from the fact that the product of occupied initial states with empty final states is large if the Fermi energy is located between spin split Landau levels. For example, the

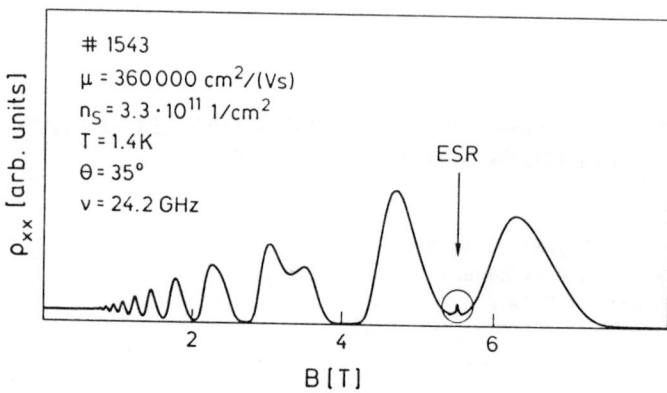

Figure 1. Shubnikov-de Haas data for a GaAs/AlGaAs heterostructure under microwave radiation. Close to filling factor i = 3 (B = 5.5 T) the microwave induced ESR signal is visible. Since the ESR absorption is strongest at odd filling factors, the magnetic field direction is tilted by θ = 35° relative to the surface normal so that the filling factor i = 3 coincides with the resonance field (which is mainly fixed by the microwave frequency of ν = 24.2 GHz).

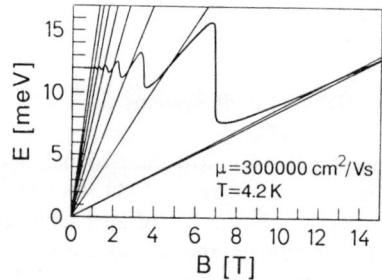

Figure 2. Energy spectrum of electrons in GaAs/AlGaAs heterostructures as a function of the magnetic field B. Each Landau level E_N is split into two spin levels with m_S = -1/2 and m_S = +1/2 (g-factor | g | = 0.4). The Fermi energy corresponds to a carrier density N_S = 3.4×10^{11} cm^{-2}.

Landau level spectrum of a GaAs/AlGaAs heterostructure with an electron concentration of about 3.4×10^{11} cm^{-2} shows clearly (see Fig. 2) that only in the magnetic field "windows" around B = 14 T (N = 0), B = 4.7 T (N = 1), B = 2.8 T (N = 2), etc. are strong ESR signals expected. The measured g-factors g = hν/μ_BB (hν: microwave energy, μ_B: Bohr magneton) change due to the nonparabolicity of the conduction band and the different slopes dE_N/dB ~ (N + 1/2) of the Landau levels E_N in a magnetic field are reflected in a similar variation of the g-factor $g_N(B)$. Experimental results are shown in Fig. 3 and the observed variation of the g-factor can be well described by [2]:

$$g_N(B) = g_0 - c\left(N + \frac{1}{2}\right)B, \tag{1}$$

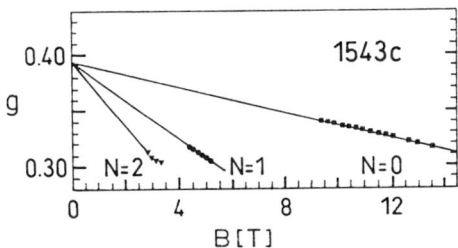

Figure 3. Experimentally determined g-factor as a function of the magnetic field for the Landau levels N = 0, N = 1, and N = 2.

g_0 is slightly smaller than the bulk-GaAs value for the electronic g-factor at the conduction band edge due to the finite electron subband confinement energy in two-dimensional systems [3]. The observed spin splitting shown in Fig. 3 can be well described on the basis of a single-electron picture for the bandstructure [4,5].

2. Influence of the Hyperfine Interaction on ESR Data

2.1. OVERHAUSER SHIFT

The position of the ESR line is not only determined by the electronic g-factor and the microwave frequency but also by the hyperfine interaction of the electrons with the nuclear spins. A nuclear-spin polarization < I > leads to an increase in the electronic spin splitting [6]:

$$\Delta E = g\mu_B B + A <I>, \qquad (2)$$

(A: hyperfine interaction constant). In thermodynamic equilibrium < I > is negligible even at liquid helium temperatures. However, the nuclear spins can be polarized dynamically [7] as a result of the flip-flop of an electronic and nuclear spin during the electron spin relaxation process of ESR excited electrons. This dynamical nuclear-spin polarization leads to a shift of the ESR line, the Overhauser shift ΔB_N:

$$\Delta E = g\mu_B[B + \Delta B_N]. \qquad (3)$$

In GaAs, the isotope ^{75}As dominates the Overhauser shift due to the relatively large probability of finding the electron at this nuclear site [8]. An Overhauser shift of up to $\Delta B_N = 0.6$ T has been observed in GaAs/AlGaAs heterostructures [9], which corresponds to a nuclear spin polarization of about $<I> / I = 10\%$.

The nuclear spin polarization relaxes with a typical time constant of several minutes so that within this time interval, the magnetic field position of the ESR line (measured with such a small microwave intensity that a change in the dynamical nuclear spin polarization is negligibly small) shifts to higher values with time. A typical result is plotted in Fig. 4. The analysis of the relaxation process of the nuclear spins gives information about the electronic properties of the two-dimensional electron gas as discussed in the following Section.

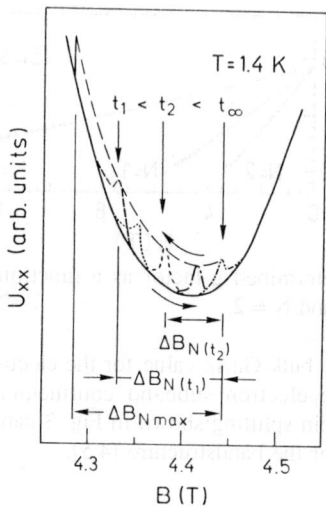

Figure 4. Time-dependent position of the ESR-line due to hyperfine interaction with the time-dependent nuclear spin polarization. The nuclear spins are dynamically polarized by pumping the ESR transition with large microwave power in a magnetic field sweep dB/dt < 0, which leads to a shift of the ESR from B = 4.44 T to B = 4.28 T. This Overhauser shift relaxes with the time constant of the nuclear-spin-lattice relaxation time T_1.

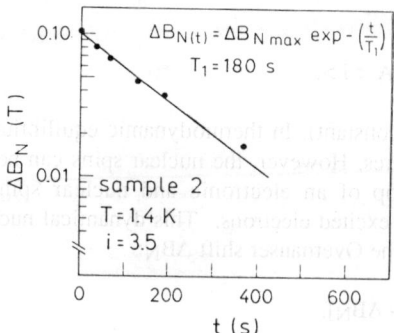

Figure 5. Relaxation of the nuclear spin lattice polarization measured via the Overhauser shift of the ESR signal (see Fig. 4). The time constant T_1 depends on the thermodynamic density of states of electrons at the Fermi energy.

2.2. KORRINGA-RELAXATION

An analysis of the time-dependent ESR peak shown in Fig. 4 demonstrates that the Overhauser shift ΔB_N and correspondingly < I > relax exponentially [10]. Small deviations from the exponential law are only visible at times much larger than the relaxation time T_1 (see Fig. 5) and originate from a longer relaxation time for the spins of the ^{69}Ga

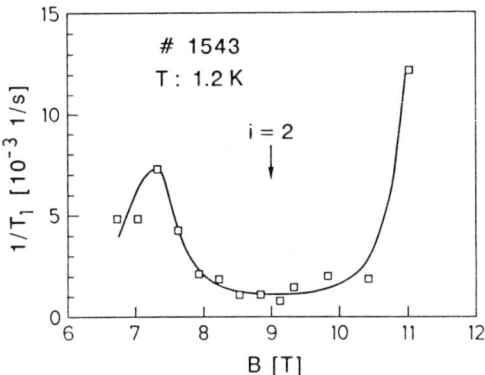

Figure 6. Nuclear-spin-lattice relaxation rate $1/T_1$ as a function of the waiting magnetic field B during the relaxation process.

and ^{71}Ga nuclei [11]. In addition, spin diffusion may lead to a non-exponential contribution. The dominating relaxation mechanism can be identified as the relaxation process of the ^{75}As nuclear-spin polarization [12].

The nuclear-spin-lattice relaxation rate $1/T_1$ varies as a function of the *waiting* magnetic field during the relaxation process. If the waiting magnetic field corresponds to an integer filling factor, the relaxation rate is slow (long relaxation time T_1), whereas a much faster relaxation is observed at half integer filling factors. Figure 6 shows a typical result. A relaxation time $T_1 \sim 1000$ s is found at filling factors around $\nu = 2$, whereas this time constant is reduced by a factor of 7 and 10 at the filling factors $\nu = 2.5$ and $\nu = 1.6$, respectively. Qualitatively, the relaxation rate for the nuclear spin polarization oscillates like the resistivity $\rho_{xx}(B)$ (which is mainly proportional to the square of the density of states at the Fermi energy if one ignores the contribution of localized states). The similarity between the magnetoresistance curves $\rho_{xx}(B)$ and the nuclear-spin-lattice relaxation rate $1/T_1(B)$ [10] leads to a two step nuclear spin relaxation model. According to this model, the first step of the nuclear spin relaxation is determined by Fermi contact interaction between the electronic and the nuclear spin system. The next step involves electronic spin transport in extended states to metallic electrodes where the spin relaxes. Within this model the nuclear spin relaxation is limited by the spin transport in extended states [13] and consequently, the nuclear spin relaxation rate is mainly determined by the conductivity. However, the assumption that the limiting step is the spin transport is not confirmed by ESR experiments which show a stronger spin relaxation within the electronic system, probably due to phonons. More accurate comparisons between the resistivity $\rho_{xx}(B)$ and the relaxation rate $1/T_1(B)$ show, that $1/T_1$ is not simply proportional to ρ_{xx}. Figure 7 and Fig. 8 shows clearly that peaks in the resistivity are not reproduced in measurements of the relaxation rate $1/T_1$.

The variation of the relaxation rate $1/T_1$ as a function of the electron density of states at the Fermi energy demonstrates clearly that the interaction with the two-dimensional electron gas dominates the nuclear-spin-lattice relaxation. This Korringa relaxation depends (if one adopts the corresponding equation developed for metals) mainly on the hyperfine structure constant A and the product of the density of states $D^{\uparrow}(E)$ for spin-up and $D^{\downarrow}(E)$ for spin-

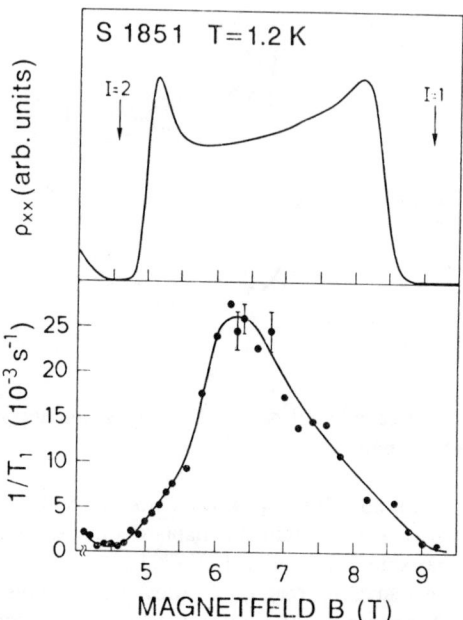

Figure 7. Comparison between the resistivity $\rho_{xx}(B)$ and the nuclear-spin-lattice relaxation rate $1/T_1(B)$ in the filling factor range $1 < i < 2$. Device parameters: $N_S = 1.8 \times 10^{11}$ cm^{-2}, $\mu = 2.1 \times 10^6$ cm^2/(Vs).

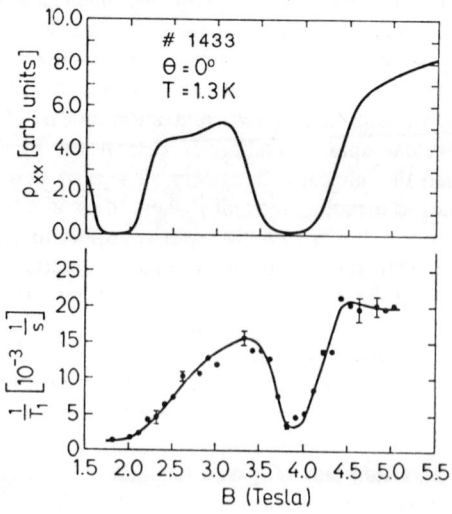

Figure 8. Similar measurements as shown in Fig. 7 for a device with $N_S = 1.0 \times 10^{11}$ cm^{-2}, $\mu = 0.3 \times 10^6$ cm^2/(Vs). Each data point for $1/T_1$ results from an analysis of a series of measurements shown in Fig. 4 and Fig. 5.

down electrons [14]:

$$\frac{1}{T_1} \sim A^2 \int_{-\infty}^{+\infty} D^\uparrow(E) D^\downarrow(E) \, f(E) \, [1-f(E)] \, dE. \tag{4}$$

The finite temperature is included in the Fermi distribution function f(E). Under the assumption that the density of states within the energy range kT around the Fermi energy E_F is constant, the integral in Eq. (4) reduces to $kTD^\uparrow(E_F)D^\downarrow(E_F)$. The experimentally determined temperature dependence of the relaxation rate $1/T_1$ is shown in Fig. 9. Strong deviations from the simple Korringa relaxation $1/T_1 \sim T$ are observed, especially for the measurements at integer filling factor i = 3.0. The relaxation rate increases more rapidly with temperature than is expected and can be explained by a temperature dependent change in the energy spectrum. It is known from transport measurements that the energy separation $E_N^\downarrow - E_N^\uparrow$ between spin split levels of the N-th Landau level increases strongly due to exchange interaction if the difference $N^\uparrow - N^\downarrow$ for the spin occupation is large [15]. This enhancement of the spin splitting can be expressed in the following form (E_{xc}: exchange energy):

$$E_N^\downarrow - E_N^\uparrow = g\mu_B B + E_{xc} \frac{N^\uparrow - N^\downarrow}{N_S} \tag{5}$$

$$\equiv g\mu_B B + E_{xc}$$

$$\equiv g^* \mu_B B.$$

At odd filling factors i the corresponding energies are

$$\begin{aligned} i=1 \quad & E_0^\downarrow - E_0^\uparrow = g\mu_B B + E_{xc} \\ i=3 \quad & E_1^\downarrow - E_1^\uparrow = g\mu_B B + \frac{1}{3} E_{xc} \\ i=5 \quad & E_2^\downarrow - E_2^\uparrow = g\mu_B B + \frac{1}{5} E_{xc}. \end{aligned} \tag{6}$$

Since the energy $g\mu_B B$ is determined by the total magnetic field B whereas the filling factor can be changed by the magnetic field component perpendicular to the two-dimensional system, the filling factor dependent contributions of the exchange energy to the spin splitting in Eq. (6) can be examined at the same total magnetic field B by tilting the sample. An analysis of the thermally activated conductivity $\rho_{xx}(T)$ at odd filling factors gives information about the energy splitting $E_N^\downarrow - E_N^\uparrow$. The experimental points in Fig. 10 show clearly, that the enhanced g-factor g* in transport measurements can be well described with an exchange energy E_{xc} = 2.8 meV [9]. This enhanced g-factor is the reason for the reduced Korringa relaxation rate $1/T_1$ at odd filling factors as shown in Figs. 6-8. The bare spin splitting $g\mu_B B$ is too small to explain the observed results on the basis of Eq. (4) when compared to the linewidth of the energy levels.

In addition, the deviation of the temperature dependent relaxation rate $1/T_1$ from the simple Korringa law $1/T_1 \sim T$ in Fig. 9 is a result of the temperature dependent variation of the enhanced g-factor. The term $N^\uparrow - N^\downarrow$ in Eq. (5) and therefore the energy separation

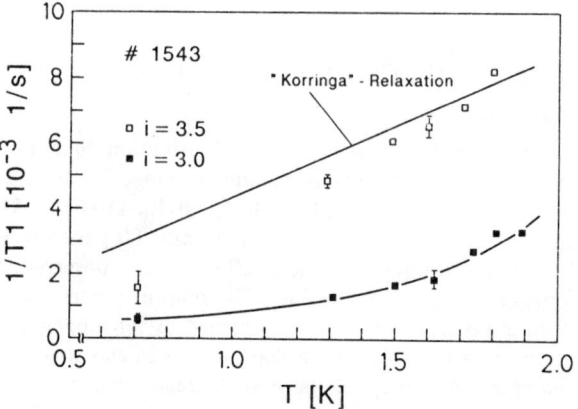

Figure 9. Temperature dependence of the nuclear-spin-lattice relaxation rate $1/T_1$ for integer and half-integer filling factors. Deviations from the Korringa-relaxation $1/T_1 \sim T$ are observed for $T > 1.5$ K and the filling factor $i = 3.0$.

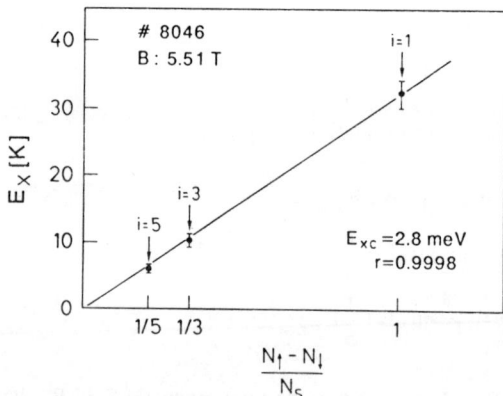

Figure 10. Contribution of the exchange energy to the spin splitting as a function of the filling factor determined from an analysis of the thermally activated resistivity (B = const. = 5.51 T, $\theta = 37°$ for $i = 1$, $\theta = 74°$ for $i = 3$, and $\theta = 81°$ for $i = 5$). Device parameter: $N_S = 1.2 \times 10^{11}$ cm^{-2}, $\mu = 7.2 \times 10^5$ cm^2/(Vs) [9].

$E_N^\downarrow - E_N^\uparrow$ becomes smaller with increasing temperature so that the combination of both the temperature dependent Fermi distribution function and the increased overlap between spin split levels leads to a superlinear dependence of the nuclear-spin-lattice relaxation rate $1/T_1$ on temperature.

A numerical fit to the measured nuclear-spin-lattice relaxation rates as a function of temperature [9] is only possible if the exchange energy is included in the spin splitting. This enhanced spin splitting is about a factor of 20 larger than the spin splitting observed in ESR experiments.

3. Conclusion

The ESR in two-dimensional systems can be detected as a microwave induced change in the resisitivity $\rho_{xx}(B)$. The measured g-values can be well explained within the one-electron picture for the band structure. It differs from the bulk value due to the energy quantization into electric subbands and Landau levels. The hyperfine interaction between the spins of electrons and nuclei leads on one hand side to a dynamical nuclear spin polarization and on the other hand side to a nuclear-spin-lattice relaxation rate which depends on the density of states of electrons at the Fermi energy. For this Korringa relaxation an enhanced spin splitting due to exchange interaction has to be included. The analysis of the relaxation rate as a function of the magnetic field shows that the density of states at the Fermi energy changes smoothly within one Landau level so that structures in $\rho_{xx}(B)$ (especially peaks in ρ_{xx} at magnetic fields close to the transition between localized and extended states at the Fermi energy) cannot be explained as a density of states contribution.

References

1. D. Stein, G. Ebert, K. v. Klitzing, and G. Weimann, Surf. Sci. **142**, 406 (1984).
2. M. Dobers, K. v. Klitzing, and G. Weimann, Phys. Rev. B **38**, 5453 (1988).
3. G. Lommer, F. Malcher, and U. Rössler, Phys. Rev. B **32**, 6965 (1985).
4. F. Malcher, G. Lommer, and U. Rössler, Superlattices and Microstructures **2**, 267 (1986).
5. G. Lommer, F. Malcher, and U. Rössler, Superlattices and Microstructures **2**, 273 (1986).
6. A. Abragam, *The Principles of Nuclear Magnetism* (Clarendon, Oxford, 1961), p. 191.
7. A.W. Overhauser, Phys. Rev. B **92**, 411 (1953).
8. D. Paget, G. Lampel, B. Sapoval, and V.S. Safarov, Phys. Rev. B **15**, 5780 (1977).
9. A. Berg, Ph.D. Thesis, Stuttgart 1992 (unpublished).
10. A. Berg, M. Dobers, R.R. Gerhardts, and K. v. Klitzing, Phys. Rev. Lett. **64**, 2563 (1990).
11. G. Kamp, H. Obloh, J. Schneider, G. Weimann, and K. Ploog, Semicond. Sci. Technol. **7**, 542 (1992).
12. M. Dobers, K. v. Klitzing, J. Schneider, G. Weimann, and K. Ploog, Phys. Rev. Lett. **61**, 1650 (1988).
13. V.I. Fal'ko, S.V. Meshkov, and I. Vagner. J. Phys.: Condens. Matter **3**, 5079 (1991).
14. C.P. Slichter, *Principles of Magnetic Resonance*, Springer Series in Solid State Sciences Vol. 1 (Springer Verlag, Berlin, 1978).
15. T. Ando and Y. Uemura, J. Phys. Soc. Jap. **37**, 1044 (1974).

COLLECTIVE SPIN-FLIP EXCITATIONS IN THE QUANTUM HALL REGIME

C. KALLIN AND J.P. LONGO
Department of Physics
McMaster University
Hamilton, Ontario L8S 4M1, Canada

ABSTRACT. In an external magnetic field, the electron inversion layer in GaAs heterojunctions supports collective charge and spin density oscillations, i.e., magnetoplasmons and spin-waves, which, at long wavelengths, can be probed by cyclotron resonance and electron spin resonance experiments. To first order, electron-electron interactions only affect these modes at finite wave vectors and, hence, their effect is not directly probed by such experiments. Recently, inelastic light scattering methods have been used with great success to probe these collective excitations at finite wave vectors. Very recently, a strong feature in these spectra, near filling factor $\nu = 1$, has been identified with a collective spin-flip excitation, which involves excitations where an electron is both promoted from the lowest Landau level to the second level and its spin is reversed. At long wavelengths, such excitations are shifted away from the sum of the cyclotron energy and the Zeeman energy, $\hbar\omega_c + g\mu B$, by a Coulomb exchange energy. At filling factor $\nu = 1$, the value of this exchange energy calculated in the time-dependent Hartree Fock approximation gives excellent agreement with the value extracted from the experimental spectra. The dispersion of collective spin-flip excitations has been calculated at integral and fractional Landau level filling within the time-dependent Hartree Fock and generalized single-mode approximations. These results are discussed, with emphasis on the features that could be probed experimentally.

1. Introduction

The effects of electron-electron interactions can be very important in two-dimensional systems such as the electron inversion layer in GaAs/AlGaAs heterojunctions [1]. This is both because of the reduced dimensionality, which tends to enhance the importance of potential energy relative to kinetic energy, and also because the effect of disorder can be relatively small in the highest mobility heterojunctions. In the presence of a large perpendicular magnetic field, the kinetic energy is further quantized—into discrete Landau levels—and, hence, the role of electron-electron interactions can be further enhanced. Because of the low electron densities in these systems, the extreme magnetic quantum limit where all the free electrons are in the lowest spin-split Landau level, or in the lowest few Landau levels, is readily accessible with available laboratory fields and temperatures. In this limit, electron-electron interactions are responsible for stabilizing such novel ground states as the incompressible quantum fluid which gives rise to the fractional quantum Hall effect [2] and possibly also electron solid phases, such as the Wigner crystal [3]. The excitations can also be strongly modified by electron-electron interactions. The most striking example is the quasiparticles in the fractional quantum Hall regime, which carry fractional charge

[4]. However, even in the integer Hall effect regime there are strong signatures of electron-electron interaction effects on the excitations. For example, there is a large exchange enhancement of the g factor which can be observed by transport measurements [5].

The collective excitations of the electron layer in the presence of a perpendicular magnetic field are also strongly perturbed by electron-electron interactions. The magnetoplasmon and spin wave excitations have been calculated in the time-dependent Hartree Fock approximation (HFA) and in the generalized single mode approximation (SMA) for various values of the Landau level filling parameter ν [6–9]. The dispersion of both of these excitations is significantly affected by electron-electron interactions at finite wavevectors [7]. However, the contributions to the energies from electron-electron interactions vanish, in the absence of disorder, at q = 0, as required by Kohn's theorem and Larmor's theorem [10]. Thus, the experimental probes of cyclotron resonance and electron spin resonance are not sensitive to correlation effects in lowest order. The presence of disorder couples the center of mass motion (which is probed by cyclotron resonance) to the relative motion (which is perturbed by interactions) and, hence, some correlation effects can be seen in cyclotron resonance experiments [11].

Inelastic light scattering has also been used to probe the collective excitations of the electron inversion layer in GaAs quantum wells [12]. Again disorder plays an important role in allowing the coupling to excitations at finite wavevectors, but these experiments probe the density of states of the collective excitations more directly than cyclotron resonance does. Inelastic light scattering has been successful in probing the dispersion due to electron-electron interactions of the magnetoplasma modes, and good agreement has been found between theory and experiment [12]. Recently, a strong feature in the light scattering spectra for filling factor near 1 has been identified as arising from an excitation in which an electron is both promoted from the lowest Landau level to the second Landau level (as for the magnetoplasmon) and its spin is reversed in the process [13]. This excitation has been referred to as a spin-flip mode, to distinquish it from the spin wave excitation, which also involves a spin reversal but does not involve a change in Landau level index.

The spin-flip mode is unusual in that no symmetry argument implies that electron-electron interactions do not affect its energy at q = 0. On the contrary, at ν = 1, the energy at q = 0 is significantly shifted from the its value in the absence of interactions, $\hbar\omega_c + |g\mu_B B|$, by Coulomb exchange effects [14]. (This fact was overlooked in Ref. [7].) The value of this energy shift calculated within the HFA gives excellent agreement with the value extracted from the experimental spectra [13]. This suggests that the HFA is a reliable approximation for integer filling and strong magnetic fields, in sharp contrast to the transport measurements of the g factor which do not agree with the value calculated within the HFA [5].

In this paper the results of calculations of the dispersion of collective spin-flip excitations a within the HFA framework, as well as within the generalized SMA, are presented [14]. The SMA is more reliable than the HFA for partially filled Landau levels, but can only be implemented if a good approximation is known for the ground state wavefunction. Hence, the SMA calculations are carried out for the fractional fillings where Laughlin's wavefunction can be used, such as for ν = 1/3, 4/3, and 7/3. All the calculations presented here are at zero temperature and in the strong magnetic field limit, which assumes that the typical Coulomb energy, $e^2/\varepsilon \ell_0$, is small compared to the cyclotron and Zeeman energies. This approximation is more applicable to the experiments than one might guess, in spite of the fact that the typical Coulomb energy is comparable to the cyclotron energy and the Zeeman energy is only about 1/60th of the cyclotron energy. This is because, first, the effective Zeeman energy for thermal excitations is strongly enhanced by exchange effects [5] and, second, the corrections to the dispersion curves even at $e^2/\varepsilon\ell_0 = \hbar\omega_c$ are only about 10–15% [15].

2. Model and Spin Density Response Function

The free electrons in the inversion layer of a GaAs/AlGaAs heterojunction, at low temperatures, behave dynamically as a two-dimensional electron gas, and in a perpendicular magnetic field may be modelled by the following Hamiltonian:

$$H = H_0 + H_{e-e}, \tag{1}$$

where

$$H_0 = \frac{1}{2m^*}\sum_i(\mathbf{p}_i - e\mathbf{A}_i/c)^2 + g\mu_B\sum_i \mathbf{B}\cdot\mathbf{S}_i, \tag{2}$$

$$H_{e-e} = \sum_{ij} v(\mathbf{r}_i - \mathbf{r}_j) - \frac{Nn_e}{2}\int d\mathbf{r}\, v(\mathbf{r}), \tag{3}$$

and $v(\mathbf{r})$ is an effective electron-electron interaction which is modified from the bare Coulomb interaction $e^2/\varepsilon r$ because of the finite thickness of the inversion layer. The effective interaction that we will use in our calculations can be written in Fourier space as [1]

$$v(q) = \frac{2\pi e^2}{\varepsilon q} F(q), \tag{4}$$

$$F(q) = \frac{1}{8}\left(1+\frac{q}{b}\right)^{-3}\left(8+9\frac{q}{b}+3\left(\frac{q}{b}\right)^2\right), \tag{5}$$

where $b^{-1} = Z_0/3$, and Z_0 is the average distance of the inversion layer electrons from the interface. A reasonable value for b is of the order of a few ℓ_0^{-1}, and in most of the following figures we have chosen the value $b\ell_0 = 3$.

The electron charge and spin density response functions have poles at the frequencies corresponding to collective excitations of the electron system. In particular, the dispersion curves for the spin flip modes are determined by calculating the following spin density response function:

$$\chi_\sigma^\pm(q,\omega) = -i\int_0^\infty dt\, e^{i\omega t}\left\langle\left[\sigma_q^\pm(t), \sigma_{-q}^\pm(0)\right]\right\rangle, \tag{6}$$

where the electron spin density operator is

$$\sigma_q^\pm(t) = e^{iHt} \sum_j e^{i\mathbf{q}\cdot\mathbf{r}_j} S_j^\pm e^{-iHt}. \tag{7}$$

In the following sections, the spin density response functions are calculated within the Hartree Fock and single mode approximations.

3. Hartree-Fock Approximation

The HFA for the density response function gives a self-consistent integral equation which can be diagonalized, leaving only a matrix equation to be solved. This calculation, which is described in detail in Ref. [7], gives the following secular equation for the excitation energies:

$$\sum_{\lambda\nu} \left[[\hbar\omega - (n_\nu - n_\lambda)\hbar\omega_c - (s_\nu - s_\lambda)|g\mu_B B|]\delta_{\alpha,\lambda}\delta_{\beta,\nu} - E_{HF}(\beta,\alpha,\nu,\lambda;q) \right] B_{\nu\lambda}(q,\omega)$$
$$= 0, \tag{8}$$

where

$$E_{HF}(\beta,\alpha,\nu,\lambda;q) = (\Sigma_\nu - \Sigma_\lambda)\delta_{\alpha,\lambda}\delta_{\beta,\nu}$$
$$+ (f_\nu - f_\lambda)[H(\beta,\alpha,\nu,\lambda;q) - X(\beta,\alpha,\nu,\lambda;q)]. \tag{9}$$

Here, f_ν is the occupation of the state with Landau level index n_ν and spin $s_\nu = \pm 1/2$,

$$H(\beta,\alpha,\nu,\lambda;q) = \delta_{s_\lambda,s_\nu} \delta_{s_\alpha,s_\beta} \frac{v(q)}{2\pi} F^*_{n_\beta n_\alpha}(q) F_{n_\nu n_\lambda}(q), \tag{10}$$

$$X(\beta,\alpha,\nu,\lambda;q) = \delta_{s_\alpha,s_\lambda} \delta_{s_\beta,s_\nu} \int \frac{d\mathbf{k}}{2\pi} v(k) e^{i(\mathbf{q}\times\mathbf{k})\cdot\hat{z}} F_{n_\alpha n_\lambda}(k) F^*_{n_\nu n_\beta}(k), \tag{11}$$

$$\Sigma_\alpha = -\sum_\beta f_\beta X(\beta,\beta,\alpha,\alpha;0), \tag{12}$$

$$F_{nm}(\mathbf{q}) = \left(\frac{m!}{n!}\right)^{1/2} \left(\frac{iq_x - q_y}{\sqrt{2}}\right)^{n-m} L_m^{n-m}\left(\frac{q^2}{2}\right) e^{-q^2/4}, \tag{13}$$

and L_m^{n-m} is a Laguerre polynomial.

The shifts in the magnetoplasma modes, $\omega(q) - \Delta n\hbar\omega_c - \Delta s|g\mu_B B|$, are given by the eigenvalues of E_{HF}. There are three different contributions to E_{HF}: (i) the direct or Hartree contributions $H(\beta,\alpha,\nu,\lambda;q)$, which are the only terms kept in the random phase approximation; (ii) the exchange or Fock contributions $X(\beta,\alpha,\nu,\lambda;q)$, which correspond to "ladder diagrams"; and (iii) the exchange self-energy corrections $\Sigma_\nu - \Sigma_\lambda$. The first term, or Hartree term, vanishes for modes where a spin flip is involved, since $\Delta s = s_\nu - s_\lambda = \pm 1$ in that case. In terms of electron and hole propagators, the second term is the direct Coulomb

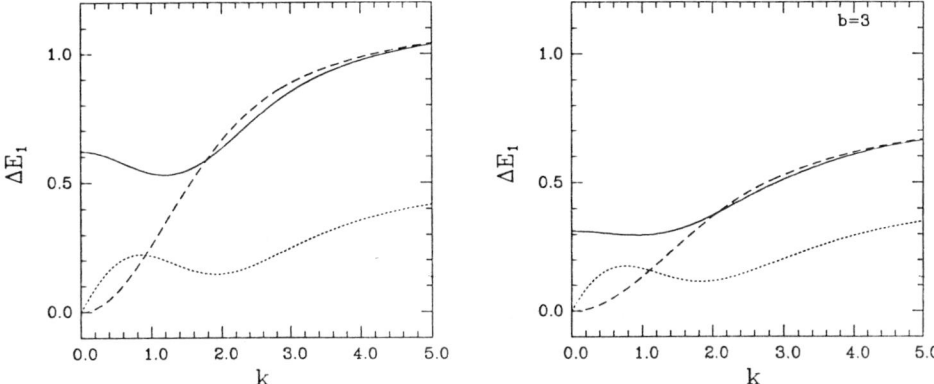

Figure 1. The dispersion of the magnetoplasmon (short dashed line), spin-wave (long dashed line), and spin-flip (solid line) modes are shown at filling factor $\nu = 1$ in the strong field limit of the Hartree Fock approximation, for (a) the strict two dimensional limit and (b) finite inversion layer width, $b = 3$. The energies are in units of $e^2/\varepsilon\ell_0$, measured with respect to the cyclotron energy, the Zeeman energy, and the sum of the cylcotron and Zeeman energies, respectively. The wave vector is given in units of ℓ_0^{-1}. Note the large exchange shift at $q = 0$ for the spin-flip mode.

interaction between the particle and the hole, and is related to the binding energy of two particles of opposite charge in a magnetic field. The third term is a constant, independent of wave vector, and represents the difference of the exchange self-energy of an electron in the excited Landau level and in the level from which the electron is removed. This is the only term which survives in the limit $q\ell_0 \to \infty$, since both H and X vanish in this limit.

The strong field approximation corresponds to restricting the sum over λ and ν in Eq. (8) to keep only the terms of order $e^2/\varepsilon\ell_0$. For example, for filling $\nu = 1$ and $n_\nu - n_\lambda = 1$, the only term which survives in the strong field approximation is $n_\beta = n_\nu = 1$, $n_\alpha = n_\lambda = 0$. If we neglect the width of the electron inversion layer so that $v(q)$ is simply the Coulomb potential, then analytic expressions (not involving integrals) can be found for all terms in E_{HF} [7]. In that case, the energy shift for the spin flip mode corresponding to $\Delta_s = 1$, at $\nu = 1$, is

$$\Delta E_{SF}^{\nu=1}(q) = \frac{e^2}{\varepsilon\ell_0}\left(\frac{\pi}{8}\right)^{1/2}\left(2 - e^{-q^2/4}\left[(1+q^2/2)I_0(q^2/4) - \frac{q^2}{2}I_1(q^2/4)\right]\right), \quad (14)$$

where I_n is a modified Bessel function of the first kind. This energy shift is shown in Fig. 1(a), together with the corresponding magnetoplasmon and spin wave energy shifts. The energy shift of the spin flip mode at $q = 0$ is $\sqrt{\pi/8}(e^2/\varepsilon\ell_0)$, which is equal to $-\Sigma_0/2$. This should be compared to the magnetoplasmon mode, $\Delta n = 1$, $\Delta s = 0$, which is not shifted away from $\hbar\omega_c$ at $q = 0$, because the change in exchange self-energy is exactly cancelled by the direct interaction between the particle and hole. The direct interaction is the same in the two cases, but the change in the self-energy is not, since the electron can no longer exchange with the electrons in the filled Landau level if its spin is flipped. It also follows

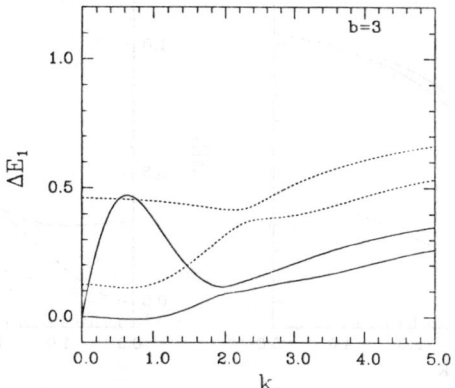

Figure 2. The dispersion of the magnetoplasmon (solid line) and spin-flip (dashed line) modes is shown for filling $\nu = 3$ and inversion layer thickness $b = 3$ within the strong field limit of the Hartree Fock approximation. The magnetoplasmon and spin-flip energies are with respect to the cyclotron energy and the sum of the cyclotron and Zeeman energies, respectively.

from Eq. (8) that the spin wave modes ($\Delta s = 1$, $\Delta n = 0$) will not be shifted away from the Zeeman energy at $q = 0$. Hence, the spin-flip modes are unusual in giving rise to large exchange shifts at zero wave vector.

For finite inversion layer width b^{-1}, the integrals in E_{HF} must be done numerically. The results for $\nu = 1$ and $b = 3$ are shown in Fig. 1(b). The finite thickness of the inversion layer reduces the effects of electron-electron interactions, as can be seen by comparing Fig. 1(a) and (b). For $\nu = 2$, the ground state is an eigenstate of total spin angular momentum with $S = 0$, and the excitations may be classified as singlets or triplets. The spin flip mode is a triplet excitation and is not shifted at $q = 0$ by exchange effects. For $\nu = 3$ an electron with majority spin can make a transition from the lowest Landau level to the second Landau level or an electron with minority spin can make a transition from the second Landau level to the third Landau level and reverse its spin in both cases. These two transitions are coupled by the Coulomb interaction and Eq. (8) gives a 2×2 matrix to diagonalize. The results are shown in Fig. 2 for $b = 3$.

The HFA may be extended to non-integral filling factors [18], although in this case there is no well-defined perturbation theory and the HFA does not give results that are exact to order $e^2/\varepsilon\ell_0$. The HFA neglects important correlations in the partially filled level. If the ground state, or a good approximation to the ground state, is known, then the single mode approximation, discussed below, does a much better job of treating these correlations. However, the HFA can still be used to give qualitative results. For example, if the filling factor is between 1 and 2, there is a second spin flip mode with $\Delta s = -1$, which will have a negative exchange shift at $q = 0$: the electron gains (negative) exchange energy when its spin is flipped because it can then exchange with the electrons in the lowest Landau level with majority spin. Within the HFA, the negative shift of this mode at $q = 0$ is exactly equal and opposite to the shift of the spin-flip mode with $\Delta s = +1$. This is shown in Fig. 3 at $\nu = 1.1$. Similar behaviour will be seen within the SMA, as discussed below.

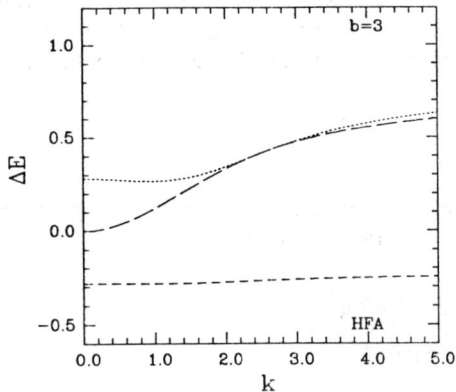

Figure 3. The dispersion of the collective spin excitations is shown for filling factor $\nu = 1.1$ and inversion layer thickness $b = 3$. The energy shifts were calculated within the strong field limit of the Hartree Fock approximation. The long dashed curve shows the spin wave energy relative to the Zeeman energy, and the other two curves show the spin-flip energies relative to $\hbar\omega_c \pm |g\mu_B B|$ with the plus sign corresponding to $\Delta s = +1$ (short dashed line) and the minus sign corresponding to $\Delta s = -1$ (medium dashed line). The energy shifts are given in units of $e^2/\varepsilon\ell_0$. Note that the exchange shifts for the spin-flip modes are equal and opposite in sign at $q = 0$.

4. Single Mode Approximation

The single mode approximation can be used to calculate the dispersion relation of excitations that are collective in nature. Typically, one assumes that the excited states are constructed by forming a density wave in the ground state,

$$|\psi_q\rangle = \rho_q |\psi_0\rangle, \tag{15}$$

where $|\psi_0\rangle$ is the ground state and ρ_q is the density operator. This can be generalized to include the case of collective spin excitations by replacing the charge density operator with the spin density operator. For example, for the spin flip excitations that we are considering, ρ_q would be replaced by $\sigma_q^\pm$.

In a magnetic field all the excitations of the non-interacting system occur at integer multiples of the cyclotron frequency, shifted by the Zeeman energy if the spin of the electron is reversed. In the interacting system, in a strong magnetic field, the excitations will still occur near integer multiples of the cyclotron frequency and they may be grouped according to the Landau levels involved. Therefore, it is useful to separate the spin density operator into contributions corresponding to transitions between different pairs of Landau levels, i.e.,

$$\sigma_q^\pm = \sum_{\nu\lambda} \sigma_q^{\nu\lambda}, \tag{16}$$

where $i = \pm$ or z and $\sigma_q^{\nu\lambda}$ is the part of the density operator which transfers an electron from the state λ to state ν. For $i = \pm$ and $\Delta s = s_\nu - s_\lambda = \pm 1$, then the ansatz excited state wave function, which is analogous to that of Eq. (15), may be written as

$$|\psi_q\rangle = \sum_{\nu\lambda} C_{\nu\lambda}(q)\sigma_q^{\nu\lambda}|\psi_0\rangle, \tag{17}$$

where for a charge density excitation, the coefficients $C_{\nu\lambda}$ are zero unless $s_\nu - s_\lambda = 0$. The excitation energy associated with this state is

$$\Delta(q) = \frac{\langle \psi_q | H - E_0 | \psi_q \rangle}{\langle \psi_q | \psi_q \rangle}, \tag{18}$$

where E_0 is the ground state energy. By using the expression for $|\psi_q\rangle$ in terms of the density operators and then minimizing with respect to the coefficients $C_{\nu\lambda}$, one finds the following secular equation for the excitation energies:

$$\sum_{\lambda\nu}\left[E(\beta,\alpha,\nu,\lambda;q) - \Delta(q)S(\beta,\alpha,\nu,\lambda;q)\right]C_{\nu\lambda}(q) = 0. \tag{19}$$

Here, the matrix elements are defined in terms of the following ground state expectation values:

$$E(\beta,\alpha,\nu,\lambda;q) = \left\langle \psi_0 \left| \sigma_{-q}^{\alpha\beta}\left[H,\sigma_q^{\nu\lambda}\right] \right| \psi_0 \right\rangle, \tag{20}$$

$$S(\beta,\alpha,\nu,\lambda;q) = \left\langle \psi_0 \left| \sigma_{-q}^{\alpha\beta}\sigma_q^{\nu\lambda} \right| \psi_0 \right\rangle. \tag{21}$$

Since the commutator of $\sigma_q^{\nu\lambda}$ with the non-interacting part of H is given by

$$\left[H_0,\sigma_q^{\nu\lambda}\right] = (\Delta n\hbar\omega_c + \Delta s|g\mu_B B|)\sigma_q^{\nu\lambda}, \tag{22}$$

it follows that

$$E(\beta,\alpha,\nu,\lambda;q) = (\Delta n\hbar\omega_c + \Delta s|g\mu_B B|)S(\beta,\alpha,\nu,\lambda;q)$$
$$+ \frac{1}{2}\int \frac{d^2k}{(2\pi)^2} v(k)\left\langle \psi_0 \left| \sigma_{-q}^{\alpha\beta}\left[\rho_{-k}\rho_k,\sigma_q^{\nu\lambda}\right] \right| \psi_0 \right\rangle. \tag{23}$$

The correlations all appear in the last term in Eq. (23). By using the definition of the partial density operators, E can be written in terms of the electron static structure factor in the ground state, $s(q)$, although the actual expression becomes increasingly complex for larger values of the Landau indices [14]. It can be shown that E reduces to E_{HF} if one uses

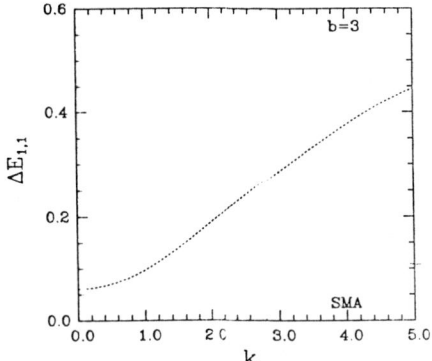

Figure 4. The energy shift of the spin-flip mode with $\Delta n = \Delta s = 1$ is shown for filling $\nu = 1/3$, in the two-dimensional limit, calculated using the generalized single mode approximation. The Hartree Fock result is the same as for $\nu = 1$, shown in Fig. 1(b), but multiplied by the filling factor 1/3. Thus, the HFA gives an exchange shift at $q = 0$ which is about 40% larger.

the uncorrelated structure factor, $s(q) = 1 - \nu_0 e^{-q^2/2}$, where ν_0 is the filling factor of the partially occupied Landau level.

If the sum over indices in Eq. (19) is restricted to $n_\nu = n_\lambda = 0$, one obtains the expression derived by Girvin, MacDonald and Platzman [8] for the low-lying magnetoroton excitations relevant to the fractional quantum Hall effect. For filling factors $\nu = 1/m$, $m = 1, 3, 5$, and 7, Laughlin's wave function is used for the ground state. The spin flip modes near the cyclotron frequency that we are considering here are obtained from the terms with $\Delta n = 1$ and $\Delta s = \pm 1$ where the values of n_ν are determined by the filling factor. At $\nu = 1/3$, we have calculated the spin-flip dispersion by using the pair correlation function derived from Laughlin's wavefunction [8]. The results are shown in Fig. 4 for the two-dimensional limit. There is a substantial exchange shift at $q = 0$, although not as large as in the HFA, which would be the same as for $\nu = 1$, but multiplied by the filling factor 1/3 (see the solid curve in Fig. 1(b)). It should be mentioned that although the SMA gives accurate results at long wavelengths (if one has a good approximation to the ground state), at larger values of q there are typically other low lying excitations that are not described by the SMA ansatz wavefunction and, hence, the dispersions obtained are probably not reliable beyond $q\ell_0 > 1$ or so.

Figure 5 shows the dispersion of the spin wave excitation ($\Delta n = 0$) and of the spin-flip excitations for filling $\nu = 4/3$. The spin-flip mode with $\Delta s = +1$ corresponds to promoting an electron from the completely filled Landau level to an empty level. For this case, the energy is the same in the HFA and in the SMA, since the electron and hole are not correlated with the partially filled Landau level. However, the spin-flip mode with $\Delta s = -1$ corresponds to exciting an electron from the partially filled level and, hence, the SMA result differs from the HFA result. At filling $\nu = 7/3$ there are three different spin-flip modes, two of which correspond to $\Delta s = +1$ and, hence, are coupled by the Coulomb interaction. As in the HFA, this leads to a 2×2 matrix to be solved. The results are shown in Fig. 6, and the oscillator weights are also shown [14]. Almost all the oscillator weight is in the more low-lying of the two $\Delta s = +1$ modes at small wave vectors.

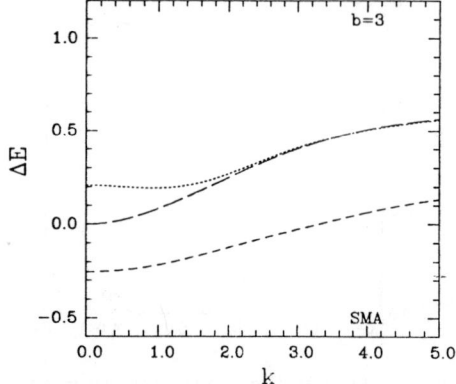

Figure 5. The energy shifts of the spin-wave (long dashed line) and spin-flip modes are shown for filling $\nu = 4/3$ and $b = 3$, within the generalized single mode approximation.

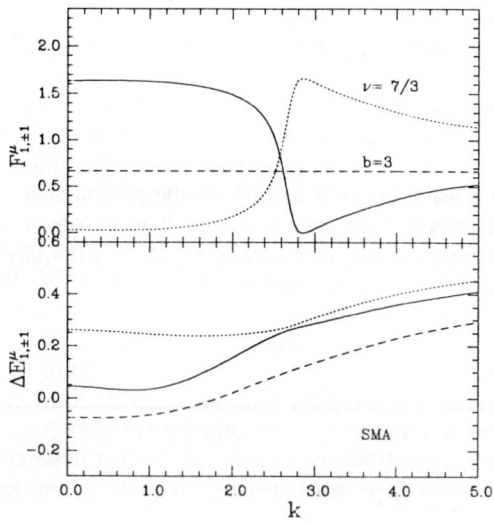

Figure 6. The energy shifts are shown for the spin flip modes at filling $\nu = 7/3$ and $b = 3$ within the generalized single mode approximation. The upper two curves correspond to $\Delta s = 1$ and the lower curve corresponds to $\Delta s = -1$. The oscillator weights associated with these three modes are shown in the upper part of the figure.

5. Discussion and Comparison with Experiments

All the calculations presented here have assumed that the spin-flip excitation is a well-defined, stable excitation. However, the spin-flip excitation can decay into a magnetoplasmon and a spin wave excitation, since the total energy of these two excitations

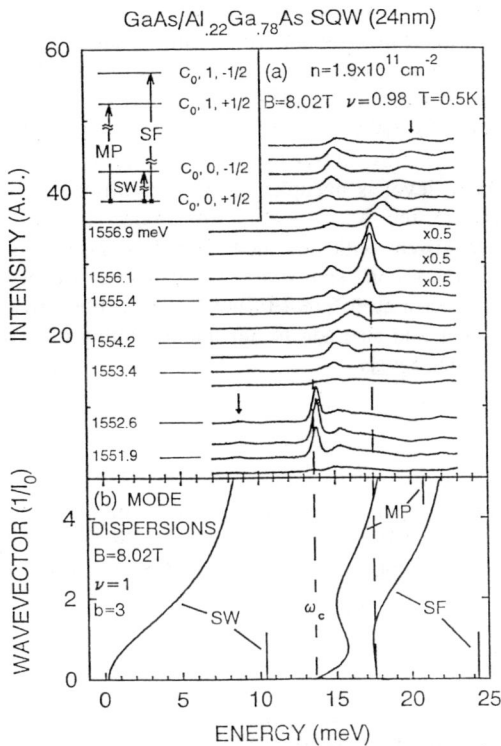

Figure 7. (a) Resonant light scattering spectra near the cyclotron frequency for filling $\nu = 1$ taken from Ref. [13]. (b) Mode dispersions calculated in the HFA for $\nu = 1$ and $b = 3$. The strong feature at 17.5 meV is believed to arise from long wavelength spin-flip excitations.

will also be $\hbar\omega_c + |g\mu_B B|$ plus interaction terms which are of order $e^2/\varepsilon \ell_0$ in the strong field limit. In general, this would be expected to alter the dispersion relation of the spin-flip mode and also to give it a finite lifetime. This decay process is not included in either the HFA or the SMA. For the case of integer filling, the HFA only includes terms where there is a single hole present at all times, in an otherwise filled Landau level. The terms relevant to the decay of a spin-flip mode involve intermediate states in which two holes are present. Calculations of the effect on the spin-flip mode, due to this decay process, are in progress.

Pinczuk and co-workers at AT&T Bell Laboratories have studied the two-dimensional electron gas in GaAs/AlGaAs heterojunctions using inelastic light scattering and Fig. 7 shows the light scattering spectra obtained for filling factor very close to 1 ($\nu = 0.98$) [13]. Near the cyclotron frequency, peaks are observed at energies corresponding to peaks in the density of states for excitations of the two-dimensional electron layer. For example, two peaks somewhat above the cyclotron frequency can be identified with the extrema in the magnetoplasmon dispersion calculated within the HFA. At higher energy, there is a large peak in the spectra which coincides with the energy of long wavelength spin-flip excitations. These experiments demonstrate that inelastic light scattering can be used to directly probe correlation effects in the two-dimensional electron gas in strong magnetic fields. The HFA, which should give reliable results in the strong magnetic field limit, gives remarkably good agreement with experiment. The results presented in this paper show how the exchange shift, which only occurs for spin-flip excitations, varies with filling factor. For example, as the filling factor is reduced below 1, one would expect the peak due to the spin-flip mode to move down in energy, closer to $\hbar\omega_c$. On the other hand, as the second spin level of the lowest Landau level is occupied, a peak significantly below $\hbar\omega_c$ should be observed, corresponding to the $\Delta s = -1$ spin flip mode. In fact, a very weak peak is observed at an energy consistent with this mode (see Fig. 7), which may be attributed to some small occupation of the second spin level. Comparisons between the results presented here and light scattering spectra at different filling factors could provide further insight into the applicability of the HFA and also add to our understanding of correlation effects in the case of partially filled Landau levels.

Acknowledgements

The work discussed in this paper was stimulated by a collaboration with Aron Pinczuk. Useful discussions with John Berlinsky, Luis Brey, and Bert Halperin are gratefully acknowledged. This work is supported by the Natural Sciences and Engineering Research Council of Canada.

References

1. T. Ando, A.B. Fowler, and F. Stern, Rev. Mod. Phys. **54**, 437 (1982).
2. R.E. Prange and S.M. Girvin, *The Quantum Hall Effect*, 2nd Ed. (Springer, New York, 1990).
3. See, for example, *Proceedings of the Ninth International Conference on Electronic Properties of Two-Dimensional Systems*, Surf. Sci. **263** (1992).
4. R.B. Laughlin, Phys. Rev. Lett. **50**, 1395 (1983).
5. A. Usher, R.J. Nicholas, J.J. Harris, and C.T. Foxon, Phys. Rev. B **41**, 1129 (1990); B.B. Goldberg, D. Heiman, and A. Pinczuk, Phys. Rev. Lett. **63**, 1102 (1989).
6. Yu.A. Bychkov, S.V. Iordanskii, and G.M. Eliashberg, Pis'ma Zh. Eksp. Teor. Fiz. **33**, 152 (1981) [JETP Lett. **33**, 143 (1981)].
7. C. Kallin and B.I. Halperin, Phys. Rev. B **30**, 5655 (1984); C. Kallin, in *Interfaces, Quantum Wells and Superlattices*, edited by C.R. Leavens and R. Taylor (Plenum Press, New York, 1988), p. 163.
8. S.M. Girvin, A.H. MacDonald, and P.M. Platzman, Phys. Rev. B **33**, 2481 (1986).
9. A.H. MacDonald, H.C.A. Oji, and S.M. Girvin, Phys. Rev. Lett. **55**, 2208 (1985); H.C.A. Oji and A.H. MacDonald, Phys. Rev. B **33**, 3810 (1986).
10. W. Kohn, Phys. Rev. **123**, 1242 (1961); D. Stein, K. v. Klitzing, and G. Weimann, Phys. Rev. Lett. **51**, 130 (1983).
11. C. Kallin, in *Interfaces, Quantum Wells and Superlattices*, edited by C.R. Leavens and R. Taylor (Plenum Press, New York, 1988), p. 175.
12. A. Pinczuk, D. Heiman, S. Schmitt-Rink, C. Kallin, B.S. Dennis, L.N. Pfeiffer, and K.W. West, in *Light Scattering in Semiconductor Structures and Superlattices*, edited by D.J.

Lockwood and J.F. Young (Plenum, New York, 1991), p. 571.
13. A. Pinczuk, B.S. Dennis, D. Heiman, C. Kallin, L. Brey, C. Tejedor, S. Schmitt-Rink, L.N. Pfeiffer, and K.W. West, Phys. Rev. Lett. **68**, 3623 (1992).
14. J.P. Longo and C. Kallin, Phys. Rev. B **47**, 4429 (1993).
15. A.H. MacDonald, J. Phys. C **18**, 1003 (1985).

COULOMB PERTURBATION BY PHOTOHOLES OF TWO-DIMENSIONAL ELECTRON GAS InGaAs-HETEROJUNCTIONS: THE CASE OF FERMI-EDGE SINGULARITIES

W. CHEN, M. FRITZE, AND A.V. NURMIKKO
Division of Engineering and Department of Physics
Brown University
Providence, RI 02912, USA

Several reports in the past two years have focused on the use of photoluminescence spectroscopy to identify optically details of behaviour of a two-dimensional (2D) electron gas in quantizing magnetic fields in the integer and fractional quantum Hall regimes, and beyond [1-4]. A critical question, of course, is the nature of the connection between such optical information and that derived from transport experiments, specifically in terms of the perturbation which the optically injected holes generate within the 2D electron system. Yet there appears to be little quantitative experimental information about the strength of this interaction to date. A second, and related question concerning the optical spectroscopy of such many electron systems is the issue of the spatial overlap between the photohole and the 2D or a lower dimensional electron gas. In the work so far, single quantum wells and heterojunctions as well as acceptor δ-doping have been employed, so that in terms of controlling the holes' spatial location considerable variable may occur. In case of the single heterojunction or a very wide quantum well, for example, this raises the question whether the exciton-like luminescence response actually originates from the heterointerface directly or indirectly.

We have studied the strength of the Coulomb perturbation by the photoholes in single GaAs (and InGaAs) quantum wells and also in wires while taking advantage of the electric subband structures in both square and parabolic well. Unoccupied subband states near the Fermi level act as virtual resonance scattering channels for enhancing the Fermi-edge singularity nature of the recombination spectrum of the 2D electrons with photoholes. This many-electron/one-hole Coulomb interaction can be continuously tuned by externally applied electric or in-plane magnetic fields. Very good agreement with a Fano resonance-type model is obtained from which a characteristic value of 0.6 meV is obtained for the intrinsic electron-hole interaction energy in "square" quantum wells of about 20.0 nm thickness. This value is large when compared with the energy scale typically associated in the FQHE of Wigner crystal states. In wide parabolic wells with shallow Fermi seas, we show how the Fermi edge singularity problem now involves virtually all the mobile electrons which Coulombically interact with holes through the many available subband channels. To gain further insight into the dynamics of the electron-hole interaction, we have employed picosecond time-resolved spectroscopy to show how the large amplitude excursions observed in steady-state luminescence amplitude in perpendicular magnetic fields, are directly related to the formation probability of the many-electron exciton complexes.

Acknowledgments

This work was carried out in collaboration with J.M. Hong and L.L. Chang (IBM), and J. Jo, M. Santos, and M. Shayegan (Princeton).

References

1. B.B. Goldberg, D. Heiman, A. Pinczuk, L. Pfeiffer, and K. West, Phys. Rev. Lett. **65**, 641 (1990).
2. H. Buhmann, W. Joss, K. von Klitzing, I.V. Kukushkin, G. Martinez, A.S. Plaut, K. Ploog, and V.B. Timofeev, Phys. Rev. Lett. **65**, 1056 (1990).
3. A.J. Turberfield, S.R. Haynes, P.A. Wright, R.A. Ford, R.G. Clark, J.F. Ryan, J.J. Harris, and C.T. Foxon, Phys. Rev. Lett. **65**, 637 (1990).
4. W. Chen, M. Fritze, A.V. Nurmikko, D. Ackley, C. Colvard, and H. Lee, Phys. Rev. Lett. **64**, 2434 (1990).

PHOTOLUMINESCENCE STUDIES OF TWO-DIMENSIONAL ELECTRONS IN THE EXTREME QUANTUM LIMIT

I.V. KUKUSHKIN,[1*] N.J. PULSFORD,[1] K. VON KLITZING,[1] R.J. HAUG,[1] K. PLOOG,[1**] H. BUHMANN,[2] G. MARTINEZ,[2] M. POTEMSKI,[2] AND V.B. TIMOFEEV[3]

[1]Max-Planck-Institut für Festkörperforschung, Heisenbergstrasse 1, D-7000 Stuttgart 80, Germany
[2]Service National des Champs Intenses, Centre National de la Recherche Scientifique, BP 166X, F-38042 Grenoble Cedex, France
[3]Institute for Solid State Physics, Russian Academy of Science, 142432 Chernogolovka, Russia

ABSTRACT. We study the radiative recombination of two-dimensional electrons in GaAs-GaAlAs heterojunctions with holes localised on acceptors in strong magnetic fields. A single magneto-optical method is applied to study the cyclotron, spin, and quasiparticle energy gaps and their dependence on temperature. Time resolved photoluminescence is used to probe the localisation of the electrons in a solid phase, and a phase boundary for the Wigner solid is derived.

1. Introduction and Acceptor Doped Structures

The high mobility two-dimensional electron gas (2DEG) realised in GaAs-GaAlAs heterojunctions and quantum wells has proven a powerful system in which to study electron-electron interactions and many body phenomena. Experimentally, milliKelvin temperatures and strong magnetic fields are required to observe the electron correlations. At odd integer filling factors, where the highest occupied Landau level is spin polarised, particle exchange strongly enhances the spin splitting (by up to a factor of 20) [1–3]. A similiar enhancement, though less dramatic (only a ~ 30% increase), is also measured in the cyclotron energy at even integer filling factors [3]. At fractional values of the filling factor, v, Coulomb interactions lead to the formation of many body ground states whose excited states are fractionally charged quasiparticles, separated from the collective ground state by a finite energy gap [4–6]. At very small filling factors and provided the disorder is weak enough, the Coulomb interactions are expected to drive the electrons into a solid phase [7,8]. Because of the strength of the correlations, activated magnetotransport in these regimes is interpreted in terms of neutral many-body excitations (magnetoplasmons, spin waves, quasiparticles) with a conserved wave vector k [9,10]. The excitation spectra can also be probed by optical spectroscopy: photoluminescence (PL) has shown the enhancement of the spin splitting at $v = 1$ (corresponding to spin wave excitations at large k) [11,12] and inelastic light scattering has demonstrated the dispersion of magnetoplasmon and spin wave modes at finite k [13]. However, in the regime of the fractional quantum Hall effect (FQHE), a number of different optical results have been reported [14–17]—it is understood

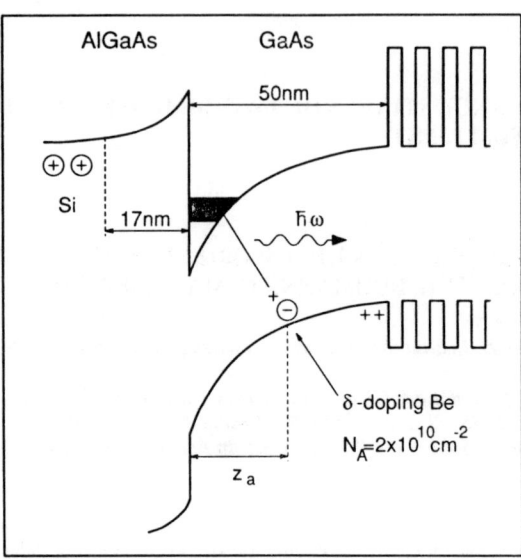

Figure 1. Schematic diagram of the acceptor doped heterostructures. The acceptor distance (z_a) is varied between 25 and 40 nm.

that the influence of the photoexcited hole on the two-dimensional (2D) electron layer is an important factor in determining the optical response in this limit.

In this paper, we review our recent measurements [18,19] for the radiative recombination of 2D electrons in GaAs-GaAlAs heterostructures with holes bound to acceptors. In Sec. 2, the relationship between the mean energy of the electrons and the first moment of the PL spectrum is developed to measure the energy gaps of the integer (IQHE) and fractional quantum Hall effects, and in Sec. 3, the condensation into a solid phase is studied by time resolved PL. In order to view the optical recombination as a probe of the correlations in the 2D electron layer, it is important to mimimize the disturbance of the electrons by the photoexcited holes. An advantage of localising the holes onto acceptors, as in the structures studied here, is that the strength of electron-hole interaction can be controlled by the acceptor position. We find, for an acceptor position z_0 = 25–40 nm from the heterointerface, the recombination time of the 2D electrons is long (τ = 0.2–1 µs) [20] and the Fermi edge singularity in the optical lineshape is small [21]. Both these indicate that the electron-hole interaction is weak: the long lifetime reflects the small electron-hole wavefunction overlap and the small singularity shows that the electron sea is only weakly perturbed by the disappearance of the valence hole. The long lifetime is also important for allowing the system to cool sufficiently under continuous illumination, and hence preserve the ground state of the 2D electrons. We estimate, from the width of the Shubnikov-de Haas oscillations, that the heating of the 2D electrons does not exceed 100 mK under our experimental conditions [22].

The samples are a series of GaAs-AlGaAs heterostructures with a delta-doped layer of Be acceptors ($N_A = 2 \times 10^{10}$ cm^{-2}) located in the GaAs buffer layer at a specified distance (25–40 nm) from the heterointerface (see Fig. 1) [25]. The weakly p-type GaAs buffer

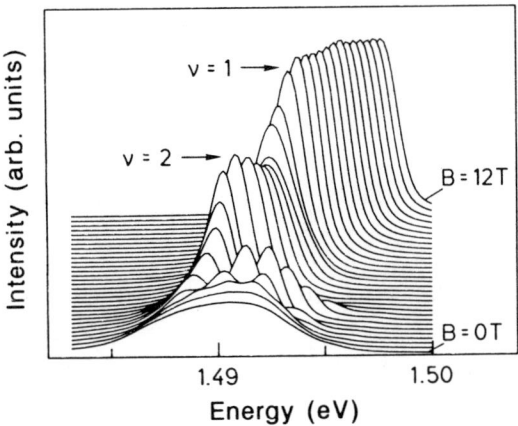

Figure 2. Photoluminescence spectra for a sample with $n_s = 1.55 \times 10^{11}$ cm^{-2} at T = 80 mK. The spectra are taken in intervals of 0.4 T and are offset for clarity.

layer produces a steep confining potential with a ~ 10 meV energy separation between the two lowest electron subbands. The PL is excited with an Ar$^+$ laser (power density 10^{-6}–10^{-3} Wcm^{-2}) and an optical fibre is used to access a dilution refrigerator. The 2D electrons at the heterointerface recombine radiatively with holes bound to the acceptors. Any photoexcited holes not trapped by the acceptors are rapidly swept by the confining potential towards the superlattice buffer layers and do not contribute to the PL. Typical spectra are shown in Fig. 2 for a heterostructure with $n_s = 1.55 \times 10^{11}$ cm^{-2} between B = 0 T and B = 12 T at a temperature T = 80 mK. In zero magnetic field, there is a broad PL spectrum reflecting the occupation of 2D electron states up to the Fermi energy E_F = 5.5 meV. The localisation of the photoexcited holes onto acceptors gives all the states up to E_F an appreciable oscillator strength (recombination spectra with free holes are dominated by states near k = 0). In a perpendicular magnetic field, the electron states are quantised into Landau levels which are resolved down to 0.8 T. For ν < 2, only a single Landau level is occupied and the PL spectra are dominated by a single peak.

2. Cyclotron, Spin and Quasiparticle Energy Gaps

The sensitivity of the PL signal to FQHE states was first observed in Si-MOSFETs [14] and subsequently in GaAs-GaAlAs heterojunctions [15–17]. An important question behind these measurements is the extent to which the physical characteristics of the FQHE states, in particular their quasiparticle energy gaps, can be derived from the optical recombination. Theoretically, two different configurations have been considered [23,24]. For the simplest case of an ideal symmetrical system in which the electrons and holes are located in the same 2D plane with the same mass, it has been shown exactly that the spectral position of the PL is not influenced by electron–electron interactions (and hence by the formation of FQHE ground states). A narrow quantum well is an approximation to such a system. In contrast, for a strongly anisotropic system in which the photoexcited holes are located a long way away from the 2D plane of the electrons, it has been shown that the first moment (M_1) of

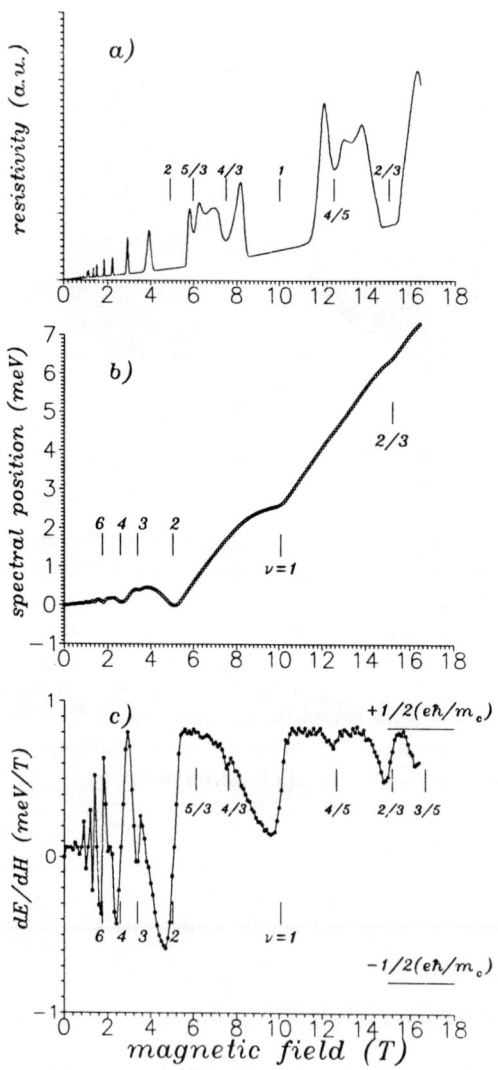

Figure 3. The correspondence between features in (a) the Shubnikov-de Haas oscillations, (b) the first moment M_1, and (c) the gradient of the first moment dM_1/dH. The positions of integer and fractional filling factors ν are indicated.

the PL line reflects the mean energy of the interacting 2D electrons [23]. The mean energy exhibits downward cusps at fractional filling factors due to the condensation into an incompressible ground state and this leads to corresponding cusps in M_1, which we have previously reported [22]. At integer filling factors, the discontinuity in the chemical potential also leads to cusps in the mean energy and we extend our optical approach, based on the behaviour of M_1, to include these gaps as well.

Figure 3 displays the Shubnikov-de Haas oscillations measured under illumination and the magnetic field dependence of M_1 for a sample with $n_s = 2.4 \times 10^{11}$ cm^{-2} at T = 100 mK. M_1 is obtained by numerical integration over the intensity of the recombination spectrum I(E):

$$M_1 = \frac{\int I(E)EdE}{\int I(E)dE}. \qquad (1)$$

Cusps are observed in the variation of M_1, aligned with Shubnikov-de Haas minima, at filling factors ν = 8,6,4,2,1,2/3. These cusps are more clearly seen by plotting (Fig. 3(c)) the derivative dM_1/dH as a function of magnetic field: discontinuities in dM_1/dH occur at filling factors ν = 10,8,6,4,2 (cyclotron gaps), ν = 5,3,1 (spin gaps) and ν = 4/5,2/3,3/5 (fractional gaps).

Following the theoretical work of Apal'kov and Rashba [23], when the acceptor is located far from the 2D electron plane ($z_0 \gg l_H$, $n_s^{-1/2}$ where l_H is the magnetic length and n_s is the 2D electron density), then in the regime of the FQHE, the first moment of the PL line (M_1) can be written as

$$M_1 = E_0 + f(\nu, \hbar\omega_c, g_e, g_h) + 2\varepsilon_{int}, \qquad (2)$$

where $E_0 + f(\nu, \hbar\omega_c, g_e, g_h)$ is the single particle energy of the electrons relative to the holes, f being an appropriate function of the g values of the electrons and holes. ε_{int} is the contribution, per particle, of correlation effects to the mean energy of the electrons. In the ideal case and for n = p/q < 1 (q odd), these authors have shown that a simple relation holds between the cusp strength in M_1 and the FQHE energy gap (Δ_q):

$$\Delta_q = \frac{1}{q}\delta\left(\frac{\partial \varepsilon_{int}}{\partial n_s}\right) = \frac{H}{2q}\delta\left(\frac{\partial M_1}{\partial H}\right). \qquad (3)$$

In general and mainly at low magnetic fields, the gap derived in this way underestimates the real value and corrections have to be applied [23,22] (and see below).

In the IQHE regime, the mean energy (ε) of the electrons is determined also by their distribution amongst different Landau levels. In particular, cusps in ε occur at even integer filling factors due to the discontinuity in the chemical potential across the cyclotron gap. In the ideal case of discrete Landau levels, without electron–electron interactions, the chemical potential μ is given by:

$$\mu = \frac{d(N\varepsilon)}{dN} = \varepsilon + N\left(\frac{\partial \varepsilon}{\partial N}\right)_{H,T}. \qquad (4)$$

Changing the variables gives:

$$\mu = \varepsilon + H\left(\frac{\partial \varepsilon}{\partial H}\right)_{\nu,T} + H\left(\frac{\partial \varepsilon}{\partial H}\right)_{N,T}. \qquad (5)$$

At even integer filling factors, both μ and $(\partial \varepsilon/\partial H)_{N,T}$ are discontinuous. Assuming $M_1 = \varepsilon$,

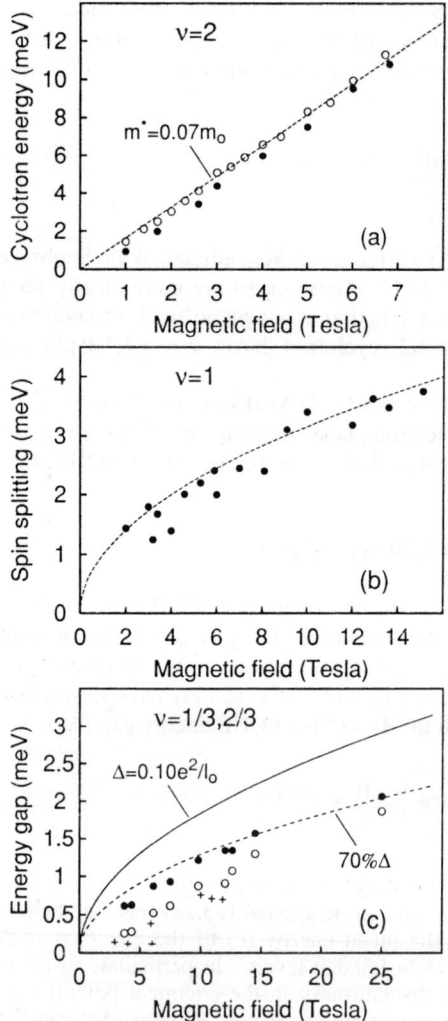

Figure 4. The gap strength derived from the optical spectra as a function of magnetic field for (a) the cyclotron gap, (b) the spin gap, and (c) the quasiparticle gap. Symbols are explained in the text.

i.e., all the Landau levels carry the same spectral weight, the cyclotron gap is related to the cusp strength by:

$$\Delta_G = H\delta\left(\frac{\partial M_1}{\partial H}\right). \tag{6}$$

Equations (3) and (6) for interacting and non-interacting electrons are very similiar; the additional coefficient "2" in Eq. (3) just reflects the pair nature of the Coulomb interaction. This similiarity was a motivation for applying the same optical method for determining the cyclotron, spin, and quasiparticle gaps in our system.

The magnetic field dependences of the various gaps are presented in Fig. 4. The cyclotron gap, measured from the derivative discontinuity at $\nu = 2$, is shown in Fig. 4(a) by the solid points. The open points correspond to the Landau level splitting measured directly from the optical spectra and the dashed line is the linear cyclotron energy for $m^* = 0.07\, m_0$. The agreement between the two sets of points confirms Eq. (6) as a measure of the cyclotron gap. The $\sim 30\%$ enhancement of the cyclotron gap at $\nu = 2$ reported in activated transport measurements [3] is not observed.

The discontinuity in the derivative at $\nu = 1$ is due to the spin splitting energy gap. However, the amplitude of this gap is not related to the normal g-factor of the electrons, but is enhanced by electron–electron interactions [1] and can be described by a spin wave dispersion [9]. Due to the low density of electrons in the upper spin state, we assign the optical gap to spin waves at large k [9], as measured in activated transport [3]. The gap is derived from Eq. (3) for interacting electrons with $q = 1$ and is shown in Fig. 4(b) (solid points). Data from a number of different heterostructures is included and the scatter in the experimental points reflects the sensitivity of the enhancement factor to the disorder. However, the spin splitting energy does follow the magnetic field dependence of the Coulomb interaction, $\Delta_s \approx e^2/l_H \approx H^{1/2}$, (dashed line) in agreement with previous experimental [2] and theoretical [1] results.

The quasiparticle gap at $\nu = 1/3, 2/3$ is presented in Fig. 4(c) for the highest mobility samples. The open points are derived from Eq. (3) directly and the solid points are with a correction factor for the final state interaction included [23,22]. The optical measurements are compared with published activated transport (crosses) [26,27] and the theoretical value of the gap strength for an ideal 2D system (solid line) [28]. The dashed line has the same field dependence as the solid line, but is scaled down to fit the solid circles—the gap reduction can be accounted for by the finite channel width and by Landau level mixing [29]. At low magnetic fields, the transport data are significantly smaller than both the solid points and the dashed line dependence, and probably reflects the Landau level broadening reducing the mobility gap.

We now turn to the temperture dependence of the gaps measured from the optical data. In Fig. 5, the behaviour of the cyclotron gap cusp at $\nu = 2$ is displayed. As the temperature is raised, the discontinuity step in dM_1/dH both decreases and broadens continuously. The resultant cyclotron energy gap derived in this way is shown in the inset and displays a low temperature quadratic correction. This temperature evolution of the cusp is expected from the Fermi-Dirac distribution function and can be reproduced with a simple model of the chemical potential around $\nu = 2$. The temperature dependence of the spin gap cusp at $\nu = 1$ is shown in Fig. 6. In contrast to the cyclotron gap, the spin gap cusp shows little change up to $T_c \approx 4$ K but then decreases abruptly. In addition, no significant broadening accompanies the cusp decrease. The different temperature behaviour of the cyclotron and spin gaps may be understood in terms of the many body character of the spin gap, which is sensitive to the population difference in the spin levels [1]: as the temperature rises, the population difference decreases and the spin splitting is reduced—this feedback effect produces the rapid decrease in the gap strength at some critical temperature T_c. However, when modelled, this simple self-consistent process does not give the observed abrupt decrease. In order to reproduce the experimental results, the exchange interaction

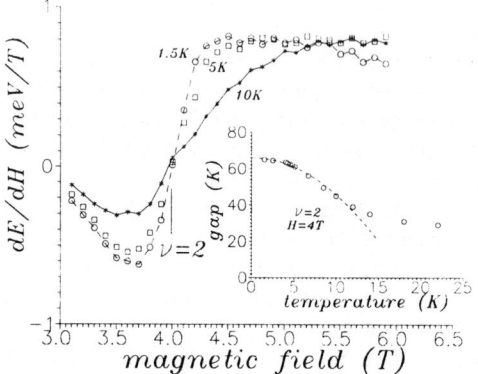

Figure 5. The behaviour of the cusp at ν = 2 for various temperatures. The inset shows the temperature dependence of the cyclotron gap measured from the cusp strength (the dashed line is a guide to the eye).

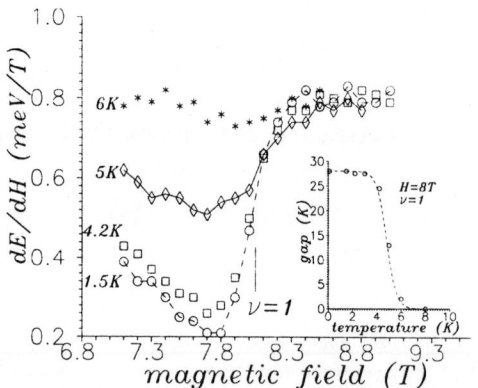

Figure 6. The behaviour of the cusp at ν = 1 for various temperatures emphasizing the rapid thermal collapse above $T_c \approx 4$ K. The corresponding temperature dependence of the spin gap is shown in the inset (the dashed line is a guide to the eye).

describing the screening, has also to decrease with temperature. We are unaware of any calculations for the temperature dependence of the enhanced spin splitting. However the similiar case of the enhanced intervalley splitting in Si-MOSFETs [30] has been calculated and the results agree qualitatively with the measured spin-gap temperature dependence shown in the inset of Fig. 3.

The temperature dependence of the M_1 cusp at ν = 2/3 is shown in Fig. 7. The strength of the cusp is constant up to a critical temperature and then drops abruptly without significant broadening. Clearly a similiar mechanism to ν = 1, i.e, a thermal excitation induced reduction of the quasiparticle energy gap takes place. In Fig. 8, we plot the

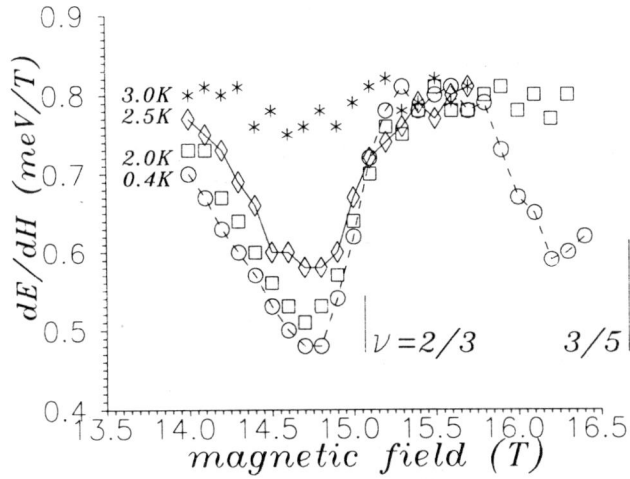

Figure 7. The behaviour of the cusp at ν = 2/3 for various temperatures.

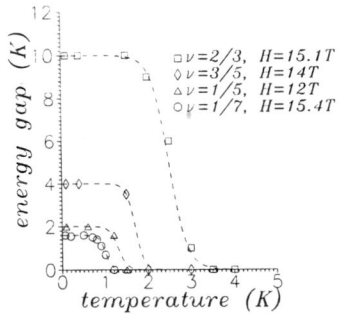

Figure 8. The temperature dependence of the energy gap at ν = 2/3, 3/5, 1/5, 1/7 measured from the cusp strength in M_1.

temperature dependence of various fractional energy gaps derived from their cusp strengths. Each fraction displays a characteristic temperature (T_c) above which it collapses. A similiar critical behaviour was also qualitatively observed in Si-MOSFETs for fractional gaps at ν = 7/3 and 8/3 [31]. For gaps originating from correlation effects, one should expect a value of the T = 0 K gap strength, $\Delta(0)$, significantly higher than the corresponding critical temperature T_c at which the gap collapses. This is observed for higher fractions and the spin gaps, but not for lower fractions. Though in the latter case the absolute value $\Delta(0)$ may be influenced by the application of the oversimplified Eq. (3), we think that there is no simple empirical relation between gaps and critical temperatures. This could be due the different quasiparticle excitation spectrum for each many body state, which can contain several gaps at different momentum values (e.g., the roton gap) [9,10]. It is probably important to include the whole quasiparticle dispersion, i.e., a number of different gaps, in discussing the thermal collapse of the fractional and spin states.

3. Time Resolved Measurements in the Electron Solid Phase

An intriguing aspect of the physics of 2D electrons is the nature of their ground state in the extreme quantum limit: the series of states of the FQHE is expected to end for low filling factors, either due to the magnetic freeze out of electrons at disorder sites or due to the formation of a Wigner solid. There have been several experiments [32–36] to observe the Wigner solid in GaAs-GaAlAs heterostructures and, although the results clearly support a pinned electron solid, no direct evidence of long range electronic ordering has been reported (for example, from Bragg reflections). A main difficulty has been to differentiate between the ordered electron solid and the extrinsic localisation of electrons due to magnetic freeze out [37].

We have previously reported the behaviour of the PL for acceptor doped samples in the regime of the Wigner crystal [35]. Below a critical filling factor $v_c = 0.27$ and below a critical temperature, (a) the integral intensity of the PL drops sharply and (b) an additional line appears in the PL spectrum. Our proposed explanation was based on the localisation of both electrons (in the pinned Wigner solid) and holes (on acceptors) in the 2D plane. With increasing magnetic field, the overlap of the electron-hole wavefunctions decreases, reducing the efficiency of radiative recombination and hence also the integrated intensity. Similiar results have been reported for the recombination with free holes [38,39], indicating that the same mechanism may be taking place, i.e., that the 'free' holes become weakly bound to disorder sites and their wavefunction in the 2D plane is then defined by the magnetic length. Additional evidence for a pinned Wigner solid phase was provided by applying a weak electric field to depin the solid domains [40]. The intensity of the additional PL line showed a threshold increase in the electric field, identifying the low radiative efficiency with the pinning of the crystal. Time resolved PL measurements presented below extend our interpretation by measuring the recombination rates of the liquid and solid phases directly.

The pulsed excitation in the time resolved measurements was achieved with an acousto-optic modulator, producing pulses with a duration of 100 ns (the rise and fall times were 15 ns) and at a frequency 1 kHz–10 MHz. The PL signal was detected by a gateable photon counting system with a gate width of 100 ns: a 100 ns time resolution was acceptable for measuring lifetimes τ of the order of microseconds. We have previously shown that in our system the electron density is dynamically reduced under continuous illumination (by up to a factor of 2–3) [41]. Thus after an excitation pulse, the electron concentration starts to increase due to the tunnelling of electrons from the AlGaAs layer into the 2D channel—in a time resolved measurement, it is important to monitor this increase. However, by studying the optical analogue of the Shubnikov-de Haas oscillations (see below), we found that the electron concentration increases by only 1% after a delay time $\Delta t = 20$ μs. We therefore restricted our measurements to $\Delta t < 2$ μs, even though it was possible to detect a signal up to $\Delta t = 50$ μs.

In Fig. 9, the PL spectra for a sample with $n = 5.3 \times 10^{10}$ cm^{-2} at $H = 16.4$ T and $T = 45$ mK are shown. The top spectrum is for continuous illumination (cw) and contains two peaks, which we previously interpreted as corresponding to the liquid (L) and solid (S) phases [35]. The time development of the PL is presented in the lower traces up to a delay time $\Delta t = 1500$ ns where Δt is defined as the time interval between the end of the laser pulse and the start of the photon counting gate. The stronger liquid line (L) dominates the spectrum for $\Delta t = 100$ ns, but decreases rapidly; only the solid line (S) is left in the

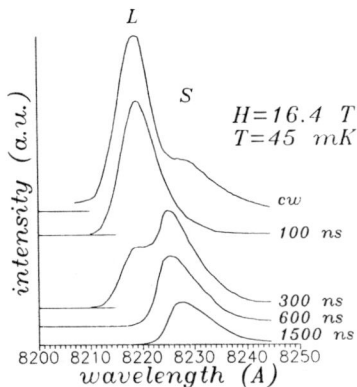

Figure 9. PL spectra for a sample with $v = 5.3 \times 10^{10}$ cm^{-2} for continuous illumination (cw) and for different delay times Δt between the excitation pulse and detection gate.

spectrum for time delays $\Delta t > 500$ ns. We interpret this behaviour in terms of a dynamic equilibrium between the two phases with some electrons recombining efficiently in the liquid phase before they have time to condense into the solid phase. From the time dependence of the integrated PL intensity, the lifetimes of the liquid and solid phases can be studied directly. For $v > 1$, the decay is exponential with a lifetime $\tau = 220$ ns corresponding to the recombination from electrons in the liquid phase. Below $v = 1$, the solid line appears, characterised by a longer decay time, which increases monotonically with the magnetic field. The longer lifetime of the solid phase is expected because the wavefunctions of both the electrons and the holes are localised in the 2D plane and their overlap, which is defined by the magnetic length, approaches zero in the limit of infinite magnetic field. The time integrated intensity of the whole PL spectrum, which is limited by the number of holes available, does not depend on the magnetic field (with an accuracy of 5%) indicating that non-radiative processes are negligible up to a time $\tau \sim 5$ µs.

Perhaps surprisingly, the solid line in the time resolved spectra appears at $v = 1$. In addition, no features are evident in the lifetime τ at fractional filling factors where the liquid phase forms the electronic ground state [7,8]. Competition between the liquid and solid phases however is reflected in the PL intensities (i.e., in the relative populations of the two phases) and not in the recombination rate τ. We can study each phase independently by setting a suitable time delay ($\Delta t = 100$ ns for the liquid and 1500 ns for the solid) and by scanning the spectrometer in accordance with the line position during a magnetic field sweep. The resultant PL intensities are plotted in Fig. 10. For the liquid phase (Fig. 10(a)), the signal is constant to $v_c = 0.26$ and then drops sharply as the filling factor decreases. Peaks in the liquid phase signal at $v = 1/5, 1/7$ are consistent with our previous results [35]. For the solid phase (Fig. 10(b)), the signal appears around $v = 1$ and shows a strong enhancement below $v_c = 0.26$. Oscillations in the intensity, similiar to transport features, are present at $v = 2/3, 2/5, 1/3, 2/7, 2/9, 1/5, 2/11, 1/7$, are used to monitor the electron concentration for different Δt (see above).

A key to understanding the behaviour of the solid line intensity is its temperature dependence, which is shown in Fig. 11. At $T = 400$ mK, the magnetic field dependence is smooth with minima at $v = 1/3, 1/5, 1/7$. At $T = 200$ mK, sharp intensity thresholds appear at

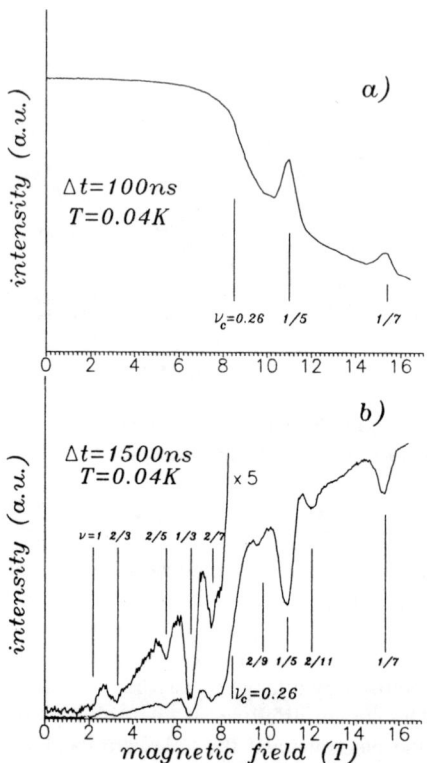

Figure 10. The intensity of (a) the liquid line (Δt = 100 ns) and (b) the solid line (Δt = 1500 ns) as a function of magnetic field. The positions of fractional filling factors are indicated.

$\nu = \nu_1, \nu_2, \nu_3$. At T = 40 mK, there is an intensity threshold at ν_c followed by minima at FQHE states. We interpret our results in the following model: starting at $\nu = 1$, a signal from localised 2D electrons due to magnetic freeze out appears. The long τ arguments of a localised phase are only appropriate below $\nu = 1$ when the magnetic length becomes comparable to or less than the interparticle distance. The proportion of localised electrons increases smoothly with field, except at fractional filling factors where the condensation of the electrons into an incompressible Fermi liquid, and absence of screening, suppresses the magnetic freeze out mechanism. At T = 200 mK, a second solid phase appears for $\nu_1 > \nu > \nu_2$ and $\nu_3 > \nu$, leading to the strong intensity enhancement. At T = 40 mK, the second phase appears over a wider field range—below $\nu_c = 0.26$ with re-entrant behaviour at $\nu = 1/5$ and 1/7. The fact that the second solid phase shows well-defined critical temperatures (see below) and filling factors demonstrates an intrinsic origin and we therefore attribute it to the Wigner solid. It contrasts with the phase due to magnetic freeze out, which shows only a smooth onset corresponding to the extrinsic process—single particle localisation.

We determine the phase boundary of the Wigner solid by measuring the intensity (Δt = 1500 ns) as a function of the temperature at fixed magnetic field. This is shown in the inset

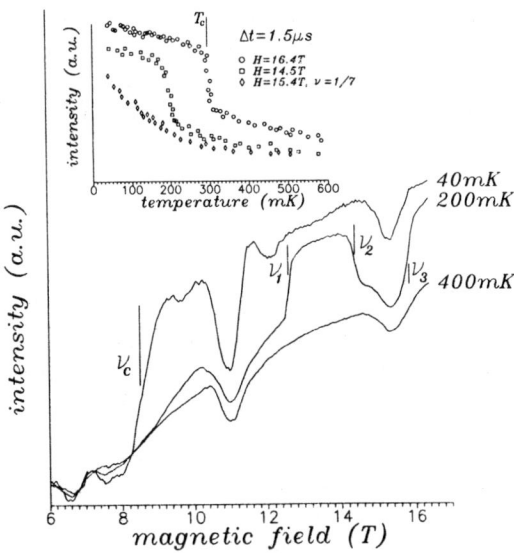

Figure 11. The intensity of the solid line for T = 40, 200, and 400 mK. The enhancement at critical filling factors is indicated. The inset shows the temperature dependence of the solid line intensity around ν = 1/7 (see text).

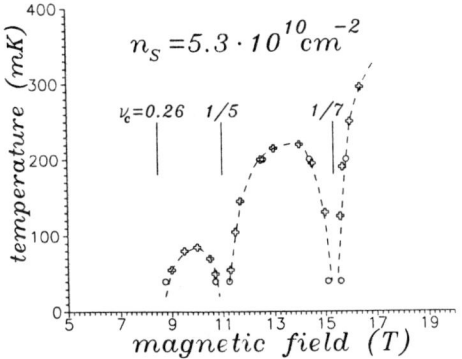

Figure 12. The phase boundary for the Wigner crystal. Crosses correspond to measurements of the transition at fixed H and circles to measurements at fixed T. The dashed line is a guide to the eye.

of Fig. 11 for magnetic fields around ν = 1/7. The sharp enhancements of the intensity for magnetic fields on either side of ν = 1/7 indicate the transition to the Wigner solid—no enhancement is observed exactly at ν = 1/7 due to the re-entrant behaviour of the liquid phase. In Fig. 12, critical temperatures (crosses) measured at various magnetic fields are plotted to form the phase boundary of the Wigner solid. The circles correspond to the values of ν_c in Fig. 11 and the dashed line is a guide to the eye. The transition temperatures all lie below the classical melting temperature T_c = 420 mK for this electron concentration and the phase boundary is similiar to previous measurements from non-linear transport and radiofrequency absorption [32–34,36] except for two differences: the onset occurs at a slightly higher filling factor ($v_c = 0.26$ rather than 0.22) and the re-entrant behaviour exists at both $v = 1/5$ and $1/7$ indicating that the liquid phase is formed at both of these fractions [8].

4. Summary

In summary, we have studied the radiative recombination of 2D electrons with holes localised on acceptors at milliKelvin temperatures and in strong magnetic fields where electron correlation effects are important. We have applied a single magneto-optical method based on the behaviour of the first moment M_1 to measure the cyclotron, spin, and quasiparticle energy gaps of the 2DEG system. The temperature dependence of these different gaps highlights the rapid collapse of any gap induced by electron–electron interactions. However, more systematic studies are necessary to identify the relation between the gap strength and the critical temperature at which the collapse occurs. In the regime of Wigner crystallisation, time-resolved PL measurements demonstrate the different lifetimes for liquid and solid phases. The long lifetime of the solid phase PL is explained in terms of localisation of both the electrons and holes in the magnetic field. The temperature dependence of the time resolved PL indicates two solid phases; one intrinsic with a sharp boundary and one extrinsic with only a smooth onset. The intrinsic phase is attributed to a pinned Wigner solid and the (H,T) phase boundary is comparable to previous transport and radiofrequency measurements.

Acknowledgments

We wish to thank Allan MacDonald and Vladimir Fal'ko for helpful discussions. NJP would like to acknowledge the Royal Society for the provision of a European Science Exchange Fellowship. The Grenoble High Magnetic Field Laboratory is a laboratory associated with the Université Joseph Fourier in Grenoble.

References

* Permanent address: Institute of Solid State Physics, Russian Academy of Science, 142432 Chernogolovka, Russia
** Present address: Paul-Drude-Insitut für Festkörperelektronik, Hausvogteiplatz 5-7, O-1096 Berlin, Germany
1. T. Ando and Y. Uemura, J. Phys. Soc. Jpn. **37**, 1044 (1974).
2. R.J. Nicholas, R.J. Haug, K. von Klitzing, and G. Weimann, Phys. Rev. B **37**, 1294 (1988).
3. A. Usher, R.J. Nicholas, J.J. Harris, and C.T. Foxon, Phys. Rev. B **41**, 1129 (1990).
4. D.C. Tsui, H.L. Störmer, and A.C. Gossard, Phys. Rev. Lett. **48**, 1559 (1982).
5. R.B. Laughlin, Phys. Rev. Lett. **50**, 1395 (1983).
6. R.G. Clark, J.R. Mallett, S.R. Haynes, J.J. Harris, and C.T. Foxon, Phys. Rev. Lett. **60**,

1747 (1988).
7. P.K. Lam and S.M. Girvin, Phys. Rev. B **30**, 473 (1984).
8. D. Lesvesque, J.J. Weis, and A.H. MacDonald, Phys. Rev. B **30**, 1056 (1984).
9. C. Kallin and B.I. Halperin, Phys. Rev. B **30**, 5655 (1984).
10. S.M. Girvin, A.H. MacDonald, and P.M. Platzman, Phys. Rev. Lett. **54**, 581 (1985).
11. B.B. Goldberg, D. Heiman, and A. Pinczuk, Phys. Rev. Lett. **63**, 1102 (1989).
12. A.S. Plaut, I.V. Kukushkin, K.von Klitzing, and K. Ploog, Phys. Rev. B **42**, 5744 (1990).
13. A. Pinczuk, B.S. Dennis, D. Heiman, C. Kallin, L. Brey, C. Tejedor, S. Schmitt-Rink, L.N. Pfeiffer, and K.W. West, Phys. Rev. Lett. **68**, 3623 (1992).
14. I.V. Kukushkin and V.B.Timofeev, JETP Lett. **44**, 228 (1986).
15. B.B. Goldberg, D. Heiman, A. Pinczuk, L. Pfeiffer, and K. West, Phys. Rev. Lett. **65**, 641 (1990).
16. A.J. Turberfield, S.R. Haynes, P.A. Wright, R.A. Ford, R.G. Clark, J.F. Ryan, J.J. Harris, and C.T. Foxon, Phys. Rev. Lett. **65**, 637 (1990).
17. H. Buhmann, W. Joss, K. von Klitzing, I.V. Kukushkin, G. Martinez, A.S. Plaut, K. Ploog, and V.B. Timofeev, Phys. Rev. Lett. **65**, 1056 (1990).
18. I.V. Kukushkin, N.J. Pulsford, K. von Klitzing, R.J. Haug, K. Ploog, H. Buhmann, M. Potemski, G. Martinez, and V.B. Timofeev, Europhys. Lett., to be published.
19. I.V. Kukushkin, N.J. Pulsford, K. von Klitzing, R.J. Haug, K. Ploog, and V.B. Timofeev, Europhys. Lett., to be published
20. A.F. Dite, I.V. Kukushkin, K. von Klitzing, V.B. Timofeev, and A.I. Filin, JETP Lett. **54**, 389 (1991).
21. P. Hawrylak, Phys. Rev. B **45**, 4237 (1992).
22. I.V. Kukushkin, N.J. Pulsford, K. von Klitzing, K. Ploog, R.J. Haug, S. Koch, and V.B. Timofeev, Europhys. Lett. **18**, 63 (1992).
23. V.M. Apal'kov and E.I. Rashba, JETP Lett. **53**, 442 (1991).
24. A.H. MacDonald, E.H. Rezayi, and D. Keller, Phys. Rev. Lett. **68**, 1939 (1992).
25. I.V. Kukushkin, K. von Klitzing, K. Ploog, and V.B. Timofeev, Phys. Rev. B **40**, 7788 (1989).
26. R.L. Willett, H.L. Störmer, D.C. Tsui, A.C. Gossard, J.H. English, and K.W. Baldwin, Surf. Sci. **196**, 257 (1988).
27. J.R. Mallett, R.G. Clark, R.J. Nicholas, R. Willett, J.J. Harris, and C.T. Foxon, Phys. Rev. B **38**, 2200 (1988).
28. G. Fano, F. Ortolani, and E. Colombo, Phys. Rev. B **34**, 2670 (1986).
29. F.C. Zhang and S. Das Sarma, Phys. Rev. B **33**, 2903 (1986).
30. H. Rauh and R. Kummel, Surf. Sci. **98**, 370 (1980).
31. I.V. Kukushkin and V.B. Timofeev, JETP Lett. **67**, 594 (1988).
32. E.Y. Andrei, G. Deville, D.C. Glattli, F.I.B. Williams, E. Paris, and B. Etienne, Phys. Rev. Lett. **60**, 2765 (1988).
33. V.J. Goldman, M. Santos, M. Shayegan, and J.E. Cunningham, Phys. Rev. Lett. **65**, 2189 (1990).
34. H.W. Jiang, R.L. Willett, H.L. Störmer, D.C. Tsui, L.N. Pfeiffer, and K.W. West, Phys. Rev. Lett. **65**, 2189 (1990).
35. H. Buhmann, W. Joss, K. von Klitzing, I.V. Kukushkin, G. Martinez, A.S. Plaut, K. Ploog, and V.B. Timofeev, Phys. Rev. Lett. **66**, 926 (1991).
36. M.A. Paalanen, R.L. Willett, R.R. Ruel, P.B. Littlewood, K.W. West, and L.N. Pfeiffer,

Phys. Rev. B **45**, 13784 (1992).
37. R.L. Willett, H.L. Störmer, D.C. Tsui, L.N. Pfeiffer, K.W. West, and K. Baldwin, Phys. Rev. B **38**, 7881 (1988).
38. B.B. Goldberg, D. Heiman, A. Pinczuk, L. Pfeiffer, and K.W. West, Surf. Sci. **263**, 9 (1992).
39. A.J. Turberfield, S.R. Haynes, P.A. Wright, R.A. Ford, R.G. Clark, J.F. Ryan, J.J. Harris, and C.T. Foxon, Surf. Sci. **263**, 1 (1992).
40. I.V. Kukushkin, N.J. Pulsford, K. von Klitzing, K. Ploog, R.J. Haug, S. Koch, and V.B. Timofeev, Phys. Rev. B **45**, 4532 (1992).
41. I.V. Kukushkin, K. von Klitzing, K. Ploog, V.E. Kirpichev, and B.N. Shepel, Phys. Rev. B **40**, 4179 (1989).

PHOTOLUMINESCENCE SPECTROSCOPY OF INCOMPRESSIBLE ELECTRON LIQUID STATES IN GaAs

J.F. RYAN,[1] A.J. TURBERFIELD,[1] R.A. FORD,[1] I.N. HARRIS,[1]
C.T. FOXON,[2] AND J.J. HARRIS[3]
[1]*University of Oxford, Clarendon Laboratory, Parks Road, Oxford OX1 3PU, UK*
[2]*Physics Department, The University, Nottingham NG7 2RD, UK*
[3]*Semiconductor Materials I.R.C., Imperial College, London SW7 2BZ, UK*

ABSTRACT. We review recent photoluminescence measurements of high-mobility electrons in GaAs in the fractional quantum Hall regime. The interaction between two-dimensional electrons and photoexcited valence band holes is fundamental in determining the photoluminescence spectrum. Measurements of a low-density electron system reveal a distinct Fermi edge singularity at zero and low magnetic fields, indicating strong multiple conduction-band electron-hole scattering at the Fermi level. The luminescence intensity is very sensitive to electron correlation (hierarchies of fractional quantum Hall states are clearly identified), whereas the luminescence energy is insensitive to electron ground state in this regime. However, spectra obtained from a higher-density electron system, where the electron-valence-hole interaction is weaker, show structure that is consistent with recombination involving quasi-particle creation and annihilation.

1. Introduction

It is now well established that incompressible liquid states, which give rise to the fractional quantum Hall effect (FQHE) in two-dimensional (2D) electron systems, have a distinctive optical response. The intensity and spectral distribution of optical transitions between the 2D electrons and photoexcited valence band holes have been shown to be sensitive probes of correlation in the electron ground state [1-3], and potentially can provide spectroscopic information which is not available from transport measurements. In order to understand the formation of and transitions between incompressible states it is important to measure directly the electronic quasi-particle excitations. The main purposes of this paper are to discuss the optical recombination processes which determine the form of the luminescence spectrum, and to evaluate just how far this potential for yielding quantitative spectroscopic data can be realised in practice. There are also important related issues. For example, recent extensions of the composite fermion (CF) model of the FQHE [4] to the half-filled Landau level (filling factor $\nu = 1/2$) suggest that a well-defined Fermi surface exists [5]: the possiblity arises of probing this novel state by optical techniques. Furthermore, the formation at low filling factor ($\nu \leq 1/5$) of a magnetic-field-induced electron solid, that is most likely pinned by defects, causes the electron system to become highly resistive; this limits the usefulness of conventional dc transport techniques and makes contactless optical techniques an important alternative probe. Tentative conclusions about the form of the liquid-solid phase diagram have already been drawn from observations of qualitative changes in the photoluminescence spectrum [6,7].

In order to obtain quantitative information from luminescence measurements the underlying physics of the recombination process itself must be established. It is necessary to understand the state of the photoexcited valence band holes and their interaction with the electron system. Clearly, this interaction must be kept relatively weak, or else the correlated electron states of interest are likely to be seriously perturbed. To this end various sample geometries have been investigated by different groups which produce significantly different initial hole states:
1. Asymmetrically doped quantum wells, in which the choice of well width controls the overlap of hole and 2D electron wavefunctions [1].
2. Modulation-doped single heterojunctions with weakly n-type GaAs, which produces a shallow confining potential for 2D electrons [2,6].
3. Acceptor-doped heterojunctions, in which the acceptor-2D electron separation determines the wavefunction overlap [3,7].

Recent theoretical work by MacDonald et al. [8] indicates that the strength of the electron-hole interaction is in fact crucial in determining the response of the holes to changes in the electron ground state. In particular, they predict that information on the many-electron states can be obtained when the electrons and holes are widely separated, but not in the limit of strong electron-hole interaction. In this paper we will review photoluminescence measurements of high mobility 2D electrons in modulation-doped GaAs/GaAlAs heterojunctions, where photoexcited hole states, being distant from the electrons, produce a relatively weak perturbation.

2. Photoluminescence from Degenerate 2D Electrons: Fermi Edge Singularity

Electron-valence-hole recombination in the presence of degenerate electrons is known to be highly sensitive to conduction-band electron-hole scattering (i.e., virtual excitations of the degenerate electrons) caused by the valence hole, and a singularity appears in the optical response at the Fermi energy [9-12]. This scattering is strongly favoured at the Fermi energy because of the availability of empty states, but is suppressed at lower energies due to the exclusion principle. For the first time in single heterojunction samples we find clear evidence for enhanced radiative recombination of electrons at the Fermi surface (Fermi edge singularity).

The two samples used here are modulation-doped GaAs/Ga$_{0.67}$Al$_{0.33}$As single heterojunctions. When illuminated the electron density in sample G648 was 3×10^{10} cm^{-2}, corresponding to a Fermi energy $E_F = 1.2$ meV (mobility at 4.2 K was 2×10^6 cm^{2}V^{-1}s^{-1}). Sample G641 has a similar structure; however, the electron density was 9.7×10^{10} cm^{-2} and mobility 9×10^6 cm^{2}V^{-1}s^{-1}. The samples were mounted in a dilution refrigerator with the temperature maintained below 100 mK. Photoluminescence was excited and detected as described previously [2,6], with the addition of filters to analyse circular polarisation of the emitted light. Well defined FQHE σ_{xx} minima were observed in simultaneous transport measurements at $\nu = 2/3$, $1/3$ and $1/5$, and at filling factors corresponding to associated hierarchical states.

Figure 1(a) shows the zero-field luminescence spectrum of G648 arising from recombination of electrons in the lowest and first excited subbands of the confining potential, E_0 and E_1. Recombination from the E_1 level, lying at 1.5137 eV, is relatively sharp. In contrast, E_0 luminescence is broad, extending over ~ 1.2 meV, i.e., the Fermi

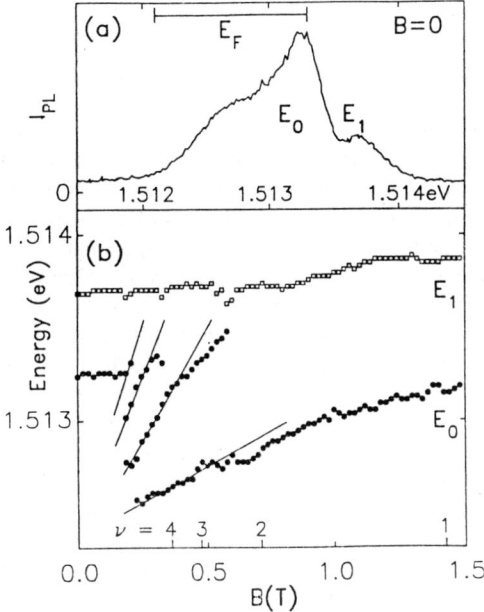

Figure 1. (a) Zero field luminescence spectrum of G648. The bar above the band of E_0 luminescence indicates the magnitude of the Fermi energy. (b) Energies of luminescence peaks as a function of magnetic field. Solid lines, indicating transitions from Landau levels 0-3, are guides to the eye.

energy, and shows strong enhancement near E_F. These results should be compared with those obtained from G641 where the many-body transition E_0 does not show a distinct Fermi edge enhancement [2].

At low applied magnetic fields perpendicular to the 2D electrons the luminescence band splits into well-resolved Landau peaks. The Fermi edge singularity persists as a strong enhancement of the intensity of recombination derived from the uppermost, partially filled E_0 Landau level; this is similar to the behaviour observed by Chen et al. [11] in GaAs quantum wells. The field-dependence of the energies of the lines is shown in Fig. 1(b) for the region $\nu > 1$. The E_1 line shows a weak diamagnetic shift with increasing field, as expected for a Wannier exciton. In contrast to this the fan of transitions originating from the many-body E_0 transition extrapolates back to a B = 0 energy corresponding to the subband edge and not to the zero-field luminescence peak at E_F. The absence of shake-up peaks in the spectrum at low energies (i.e., below the lowest Landau level transition) is consistent with the electron-valence-hole interaction being weak on the scale of the cyclotron energy [11]. Departures from linear B-dependence of the energy of the lowest E_0 line are observed at $\nu = 2, 3, 4$ and, most dramatically, in the neighbourhood of $\nu = 1$. These anomalies have been explained as arising from exchange and correlation effects [14,15], the dominant contribution being a reduction in the correlation energy of the valence band hole and the 2D electron system. The present situation, however, is more complicated than the high-temperature experiments considered previously [14,15] due to the appearance of FQHE states at filling factors $\nu > 1$.

The observation of a Fermi edge singularity in this system is consistent with the results of a mean-field calculation, incorporating a screened Coulomb electron-hole interaction, which predicts strong Fermi edge enhancement at low temperatures (100 mK) for electron densities less than 10^{11} cm^{-2} and the normal heavy-hole mass $0.3m_c$ [12]. In this theory the ground state of the degenerate electrons and a valence band hole is a many-body exciton in which the hole wavevector is at k_F. The calculated luminescence lineshape, however, does not display the lower-energy shoulder which is so pronounced for G648 in Fig. 1(a); interestingly, the theoretical lineshape is similar to that measured for G641 [2], but the latter cannot be clearly identified as a Fermi edge singularity because the dominant high-energy edge of the spectrum lies at the energy of recombination from the upper subband E_1. On the other hand, comparison of Fig. 1(a) with calculations for much higher carrier temperatures, which assume a contact electron-hole interaction [11], suggests that the hole mass in the present experiment must be significantly higher than the accepted value of the free hole mass. The need to invoke a large hole mass to explain our results is also suggested by zero-temperature calculations with a screened Coulomb electron-hole interaction which indicate that no Fermi edge enhancement is expected in GaAs with free holes because of the dynamic response of the delocalised hole [10].

Enhanced valence band hole mass might be due to in-plane localisation at a shallow defect or potential fluctuation, but this is believed to be unlikely for samples with such high electron mobility. More significantly, the interaction of holes with the 2D electron system may create a local potential minimum for holes in the vicinity of the electron layer. Preliminary calculations [16] indicate that the strength of this effect is highly dependent on the electron density, which might explain the difference in behaviour of the two samples.

Another factor that may be important in G648 is hybridization between transitions from the Fermi level and from the upper E_1 subband, allowing indirect recombination between Fermi-edge electrons and zone-centre holes [17,18]; this mechanism has been invoked by Chen et al. [13] to explain the observation of a strong Fermi edge singularity in the emission spectra of asymmetrically doped GaAs quantum wells.

3. Photoluminescence Intensities in the FQHE Regime

We now address the question of how the electron-hole interaction is modified by electron correlation that sets in at FQHE states. Figures 2 and 3 shows the magnetic field dependence of the E_0 and E_1 luminescence peak intensities for samples G648 and G641 respectively. There is a strong correlation between intensities and filling factor. E_0 and E_1 show intensity oscillations that are opposite in phase. In G648 E_1 is enhanced at $\nu = 1$ and 2/3, and there is very pronounced modulation in the $\nu = 1/3$ hierarchy as indicated. The E_0 intensity mirrors this behaviour with minima occurring at field strengths close to these filling factors. Data obtained from sample G641 in the $\nu = 2/3$ hierarchy are shown in Fig. 3. It is quite evident that the modulation is more pronounced here, and that much weaker fraction fractional states can be observed.

The intensity modulation displayed in Figs. 2 and 3 has been explained in terms of filling-factor dependent screening and correlation [2]. E_0 minima at fractional p/q filling arise from the inability of electron-valence-hole interactions to polarise the electron liquid provided that the characteristic energy gap is greater than kT. E_1 maxima occur because the two recombination processes are competing for the same photoexcited hole; also, the

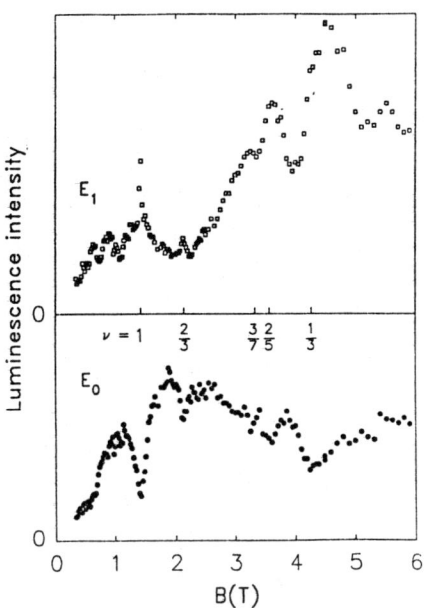

Figure 2. Magnetic field dependence of E_0 and E_1 luminescence intensities for G648 in the $\nu = 2/3$ and $1/3$ hierarchies.

spatial form of the hole wavefunction and the oscillator strength of the E_1 exciton are dependent on its interaction with E_0 electrons. For a given temperature this interaction will depend critically on the magnitude of the energy gap, which in turn depends on magnetic field strength, and also on the degree of disorder in the system. In G648 the modulation of E_0 and E_1 intensities in the 2/3 hierarchy is less pronounced than in G641, although the intensity modulation in the 1/3 hierarchy is quite distinct. Differences between the two heterojunctions may be due to the reduced magnitudes of the FQH gaps in G648 (due to the lower electron density) which causes the modulation of electron-valence-hole correlation to be less pronounced. However, the intensity oscillations in both heterojunctions are much more pronounced than in either acceptor-doped [3] or quantum well [1] structures. As a general rule it appears that the intensity modulation is strong in structures where the geometry favours weak electron-valence-hole interaction.

Figure 4 shows luminescence peak intensities for G648 in right (RCP) and left (LCP) circularly polarised spectra [19], recorded simultaneously using separate filters. It is remarkable that at $\nu = 1$ the peak heights of E_0 and E_1 are approximately independent of polarisation. This seems to belie the fact that only the lowest spin-polarised E_0 Landau level is occupied and that non-equilibrium electrons in the E_1 level should relax to the lowest (spin polarised) level from where they recombine. In fact near $\nu = 1$ a doublet with separation 0.12 meV (1.4 K) is resolved in the E_1 luminescence peak (see inset spectra); the high-energy component is unpolarised and the low-energy component is LCP. The spectrally integrated luminescence intensity is therefore dominantly LCP. There is an offset of 0.12 meV (1.4 K) between E_0 peak positions in LCP and RCP spectra; this may be due to similar behaviour although, because the E_0 line is broader, doublet structure is not well

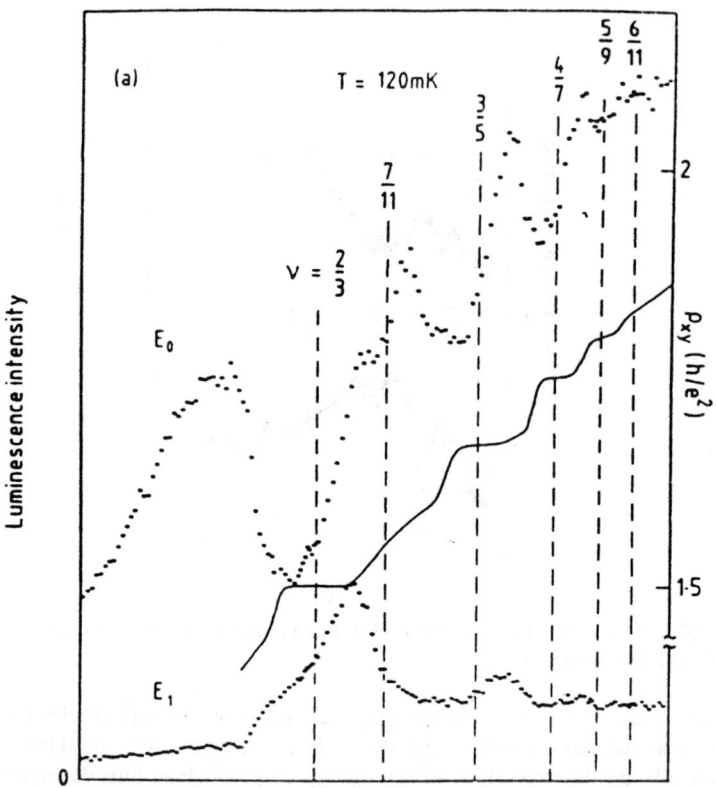

Figure 3. Magnetic field dependence of E_0 and E_1 luminescence intensities for G641 in the $\nu = 2/3$ hierarchy out to 6/11.

resolved. The measured energy separation is greater than the estimated bare electron spin splitting (0.4 K at this field strength). With increasing field E_0 and E_1 become increasingly LCP, but there is so special behaviour observed at FQHE states.

Preliminary measurements of polarised luminescence from G641 reveal a much clearer picture, with resolved energy structure (see Sec. 4) having a degree of polarisation noticeably higher than in G648. These results will be discussed in detail elsewhere [20].

The circular polarisation of luminescence from the 2D electron system is potentially a powerful probe of the spin polarisation of the electronic ground state and excitations. However, the intensity and energy of polarised luminescence depend not only on the electron spin configuration but also on the population and nature of the valence band states involved. The partial polarisation of the optical transitions observed at $\nu = 1$ and for $\nu \ll 1$, where the electron system is expected to be completely spin polarised, indicates that a quantitative interpretation of the spectra will require detailed knowledge of the valence band wavefunctions: hole wavefunctions will most certainly consist of different m_J components mixed by the electric field at the interface. A calculation of valence band states near an n-doped heterojunction in a magnetic field is clearly desirable.

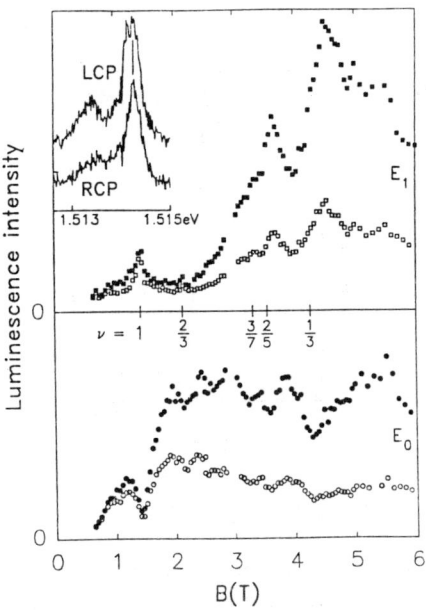

Figure 4. Luminescence peak intensities from circularly polarised spectra of G648. Open symbols, RCP; filled symbols, LCP. Spectra obtained at $\nu = 1$ are inset.

4. Photoluminescence Energy in the FQHE Regime

Various energy anomalies at FQHE states have been reported for the different sample geometries; there are differences between both magnitudes and signs of line shifts and splittings that have been measured, and the overall picture is quite confusing at the present time. This is perhaps not too surprising given the various possible initial and final states of the electron-valence-hole system. Some general statements however can be made which summarise the current situation. In the case of acceptor-doped heterojunctions the photoluminescence line lies *lower* in energy than a linear extrapolation from low magnetic field of the $l = 0$ Landau level transition; at p/q fractional filling the energy has been claimed to change discontinuously by $q\Delta$, where Δ is the energy gap for the creation of quasi-electron/quasi-hole pairs [3]. This interpretation is based on the assumption that both the initial state and the final state of the electron system is the ground state. More recent analysis of the data relates $q\Delta$ to discontinuities in the *derivative* of the luminescence energy with respect to magnetic field [21]. In the case of modulation-doped heterojunctions the luminescence line shifts are to *higher* energy, above the linearly B-dependent $l = 0$ Landau level transition. In quantum wells the shift at fractional filling is also to higher energy, and appears to be, for a given filling factor, independent of carrier density (and therefore magnetic field) [22]. In the latter two cases the luminescence spectrum may be sensitive to the different excited final states of the system.

The spectral distribution of luminescence from our single heterojunction G641 was found to be strongly filling-factor dependent; for example, in the region of $\nu = 2/3$ a

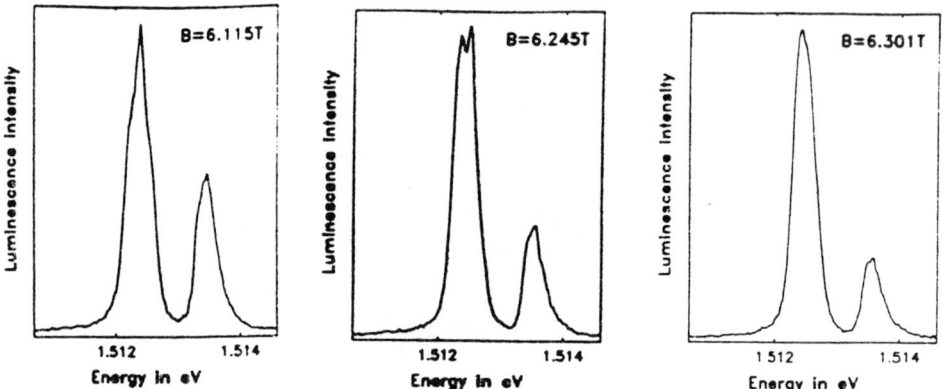

Figure 5. Luminescence spectra from G641 close to $\nu = 2/3$ (which occurs at 6 T in simultaneous ρ_{xx} measurements) showing splitting of E_0.

satellite recombination line appears, shifted up in energy by an amount somewhat smaller than Δ (see Fig 5). The surprising feature of the G648 measurements is the absence of structure in the spectra at fractional filling, although marked shifts are observed at integer filling (see Fig. 5). The difference may be due to the fact that the energy gaps for a given p/q state are smaller in G648 (as they occur at lower fields), but this would not explain the absence of structure or shifts at higher-field fractional states such as $\nu = 1/5$.

MacDonald et al. [8] have recently argued that when the electron-hole interaction is equal in magnitude to the electron-electron and hole-hole interactions (symmetric interaction), as occurs for example with coplanar electrons and holes, the photoluminescence energy is determined by the magnetoexciton energy, and not by the energies of the excited states of the many-electron system. In the case of an electron-hole interaction weak on the scale of the electron-electron interaction a new peak is predicted to appear near exact $\nu = 1/3$ filling. This satellite corresponds to annihilation of one quasielectron and creation of two quasiholes in the inter-band recombination process and is higher in energy than the main recombination channel (creation of three quasiholes) by Δ. This calculation does not support the suggestion of Ref. [3] that recombination in the presence of quasielectrons must involve annihilation of *three* of them. The prediction that the photoluminescence energy is only sensitive to the state of the many-electron system if the electron-valence hole interaction is weak may indicate that in the case of G648, the electron-hole interaction relative to the FQH gaps is significantly greater than in G641. This suggests that as the sheet density is reduced from 10^{11} cm^{-2} to 3×10^{10} cm^{-2} the ratio of electron-electron to electron-hole interaction energies falls to a level where the electron-hole interaction can no longer be considered to be weak.

Figure 7 compares the luminescence spectra from G641 at $\nu = 1/3$ and $1/5$. In the context of the composite fermion model of the FQHE these states are derived from the incompressible $\nu = 1$ integer state through the operation of the Jastrow factor D^m, where $D = \prod_{j<k} (z_j - z_k)^2$ has the effect of adding two vortices to each electron. Given the close similarity of the electron states we expect the luminescence spectra to be likewise similar.

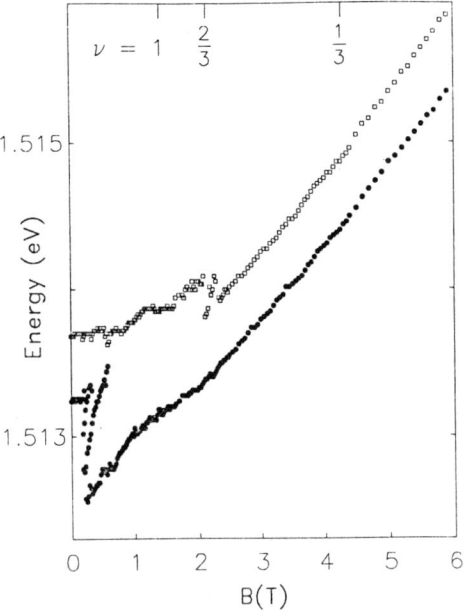

Figure 6. Energies of the E_1 and E_0 lines as a function of magnetic field for G648.

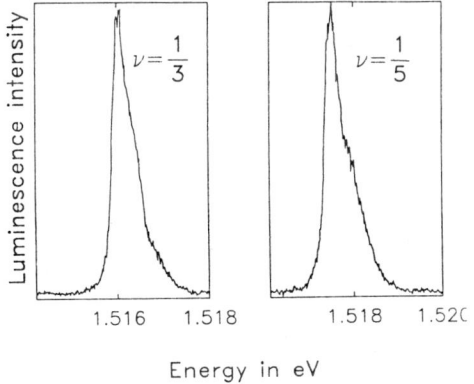

Figure 7. Spectra of G641 at $\nu = 1/3$ and $1/5$ (From Ref. [6]).

Figure 7 shows that this is indeed the case. We observe a strong E_0 peak, with a wing extending to high-energy. Following the suggestion provided by MacDonald et al. [8] we may interpret the strong low-energy component as arising from the dominant recombination process involving the generation of q quasi-holes, which corresponds to a highly-excited final state of the many-electron system. In general, recombination of electrons can proceed by n quasi-hole creation and (q-n) quasi-electron annihilation, giving

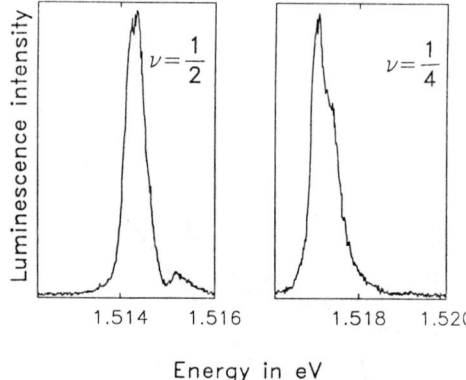

Figure 8. Spectra of G641 at the principal even-denominator fractions $\nu = 1/2$ and $1/4$ (cf. Ref. [6]).

rise to (q+1) spectral lines each separated by Δ. The relative efficiencies of these processes is not known, although processes involving annihilation of more than one quasielectron are expected to be weak due to the low probability of quasielectrons being present in the same place, i.e., the location of the valence hole. MacDonald et al. [8] find evidence for only the n = 3 and 2 processes at $\nu = 1/3$. The spectra in Fig. 7 are consistent with this interpretation, provided the recombination efficiency decreases with decreasing n.

These results provide a tantalising glimpse of the spectrum of excited states of the correlated electron system, but it is clear that improved energy resolution is required. The issue here is not the instrumental resolution, but the intrinsic width of the luminescence lines. The situation is improved, however, by detecting polarised luminescence; at $\nu = 1/3$ we find stronger evidence of resolved structure in the spectrum [20].

5. Photoluminescence at $\nu = 1/2$ and $1/4$

Halperin et al. [5] have recently extended Jain's composite fermion model of the FQHE [4] to the special cases of even-denominator fractional filling, in particular to $\nu = 1/2$. In the case that m = 1 the average field of the gauge flux quanta cancels out the applied magnetic field: the picture that results is then that of fermions in zero magnetic field, and there is predicted to be a well-defined Fermi surface. The electron system is gapless at these special values of the applied magnetic field. Experimental data for $\nu = 1/2$ and $1/4$ are rather limited: the resistivity ρ_{xx} shows a weak minimum at $\nu = 1/2$, and its absolute value is systematically smaller than that at neighbouring even-denominator fractions as predicted by the model, but the same does not appear to apply at $\nu = 1/4$ [6,23]. In the latter case it should be recalled that ρ_{xx} is found to increase very dramatically for $\nu \leq 1/4$, which may be associated with formation of an electron solid [6,23]. The discussion presented above on photoluminescence detection of FQHE gaps suggests the possibility of optically detecting this exotic quantum system.

Figure 8 presents luminescence spectra obtained from G641 at $\nu = 1/2$ and $1/4$. In the former case the strong E_0 line is remarkably symmetrical, considering the complexities of line splitting observed when the ground state is incompressible, as discussed in Sec. 4. By analogy with the true zero field situation, the full linewidth may provide a measure of the Fermi energy; the measured value is ~ 0.5 meV, which would correspond to an effective mass of approximately $0.35m_e$ (the electron density at $\nu = 1/2$ with B = 11.5 T is estimated to be 7×10^{10} cm^{-2} [6]). This result should be compared with the theoretical estimate of $0.27m_e$ at B = 10 T. However, bearing in mind the discussion in Sec. 1 on the effects of enhanced recombination at the Fermi energy this agreement is most likely fortuitous. At $\nu = 1/4$ the spectrum shows a distinct shoulder (which develops into that observed at $\nu = 1/5$ (see Fig. 7)), and it is not clear how this may be interpreted within the context of the model.

6. Conclusions

The results described above demonstrate the progress that has been achieved in optical probing of the fractional quantum Hall effect. Refinement of techniques and the establishing of systematic behaviour will undoubtedly increase our understanding of the phenomenon. However, it is clear that measurement of quasi-particle excitations of incompressible liquid states remains a severe experimental challenge.

References

1. D. Heiman, B.B. Goldberg, A. Pinczuk, C.W. Tu, A.C. Gossard, and J.H. English, Phys. Rev. Lett. **61**, 605 (1988); B.B. Goldberg, D. Heimann, A. Pinczuk, L. Pfeiffer, and K. West, Phys. Rev. Lett. **65**, 641 (1990).
2. A.J. Turberfield, S.R. Haynes, P.A. Wright, R.A. Ford, R.G. Clark, J.F. Ryan, J.J. Harris, and C.T. Foxon, Phys. Rev. Lett. **65**, 637 (1990).
3. H. Buhmann, W. Joss, K. von Klitzing, I.V. Kukushkin, G. Martinez, A.S. Plaut, K. Ploog, and V.B. Timofeev, Phys. Rev. Lett. **65**, 1056 (1990).
4. J.K. Jain, Advances in Physics **41**, 105 (1992).
5. B. I. Halperin, P.A. Lee, and N. Read, preprint.
6. R.G. Clark, R.A. Ford, S.R. Haynes, J.F. Ryan, A.J. Turberfield, P.A. Wright, C.T. Foxon, and J.J. Harris, in *High Magnetic Fields in Semiconductor Physics III*, edited by G. Landwehr (Springer-Verlag, Berlin, 1992), p. 231; A.J. Turberfield, S.R. Haynes, P.A. Wright, R.A. Ford, R.G. Clark, J.F. Ryan, J.J. Harris, and C.T. Foxon, Surface Science **263**, 1 (1992).
7. H. Buhman, W. Joss, K. von Klitzing, I.V. Kukushkin, A.S. Plaut, G. Martinez, K. Ploog, and V.B. Timofeev, Phys. Rev. Lett. **66**, 926 (1991).
8. A.H. MacDonald, E.H. Rezayi, and D. Keller, Phys. Rev. Lett. **68**, 1939 (1992).
9. M.S. Skolnick, J.M. Rorison, K.J. Nash, D.J. Mowbray, P.R. Tapster, S.J. Bass, and A.D. Pitt, Phys. Rev. Lett. **58**, 2130 (1987).
10. P. Hawrylak, Phys. Rev. B **44**, 3821 (1991).
11. T. Uenoyama and L.J. Sham, Phys. Rev. Lett. **65**, 1048 (1990).
12. G.E.W. Bauer, Phys. Rev. B **45**, 9153 (1992).
13. W. Chen, M. Fritze, A.V. Nurmikko, M. Hong, and L.L. Chang, Phys. Rev. B **43**, 14738 (1991).
14. T. Uenoyama and L.J. Sham, Phys. Rev. B **39**, 11044 (1989).

15. S. Katayama and T. Ando, Solid State Commun. **70**, 97 (1989).
16. F. Stern and S. Laux, private communication.
17. J.F. Mueller, Phys. Rev. B **42**, 11189 (1990).
18. P. Hawrylak, Phys. Rev. B **44**, 6262 (1991).
19. Left circularly polarised photons have a component of angular momentum $+\hbar$ in the direction of the magnetic field.
20. I.N. Harris, A.J. Turberfield, J.F. Ryan, C.T. Foxon, and J.J. Harris, to be published.
21. I.V. Kukushkin, N.J. Pulsford, K. von Klitzing, K. Ploog, and V.B. Timofeev, Surface Science **263**, 30 (1992).
22. B.B. Goldberg, D. Heiman, A. Pinczuk, L.N. Pfeiffer, and K.W. West, Surface Science **263**, 9 (1992).
23. H.W. Jiang, R.L. Willett, H.L. Stormer, D.C. Tsui, L.N. Pfeiffer, and K.W. West, Phys. Rev. Lett. **65**, 633 (1990).

OPTICAL SPECTROSCOPY IN THE REGIMES OF THE INTEGER AND FRACTIONAL QUANTUM HALL EFFECTS

A. PINCZUK, B.S. DENNIS, L.N. PFEIFFER, AND K.W. WEST
AT&T Bell Laboratories
Murray Hill, NJ 07974
USA

The last three years have witnessed a growing interest in optical research of the two-dimensional (2D) electron gas in the regimes of the integer and fractional quantum Hall effects. The goal of such studies is the discovery of new behaviors that are not accessible in magnetotransport experiments. The seminal photoluminescence measurements were reported in Si mosfet devices [1] and in multiple GaAs–AlGaAs quantum wells [2]. The optical recombination of the 2D electron system with acceptors in the Si devices showed anomalies at the Landau level filling factors $\nu = p/q$ of the fractional quantum Hall effect [1]. The results of intrinsic optical emission from the modulation doped quantum wells revealed temperature-dependent intensity anomalies at $\nu = 1$ and $\nu = 2/3$ [2].

The interpretation of these early experiments set up a conceptual framework that is underlying the analysis of the current optical emission research in the regime of the fractional quantum Hall effect (FQHE). In the case of acceptor recombination the goal remains the determination of values of the energy gaps associated with the emergence of the incompressible fluid of the FQHE [1,3]. In the intrinsic optical recombination the intensity anomalies are considered evidence of the paramount role played by the dynamical response of the 2D electron system to the negatively charged hole in a valence Landau level [2,4–6]. The optical anomalies also have striking manifestations in the photoluminescence intensity oscillations observed at integer values of ν [7]. The role played by the dynamical response of the electron gas in optical recombination in the FQHE regime has also been highlighted in recent theoretical work [8–10].

Inelastic light scattering was used to measure inter-Landau-level excitations at integer values of ν and also intersubband excitations at $\nu \leq 1$ [11]. In other recent work the inelastic light scattering method emerges as strikingly effective in research of quantum Hall phenomena. Studies of GaAs–AlGaAs single heterojunctions and wide quantum wells (d $\geq$ 400 Å) reveal a surprising collapse of the energies of intersubband excitations at filling factors $\nu < 1$ [12]. The collapse, as observed at incident power densities $p \geq 10^5$ W/cm^2, has been explained by a large reduction of free electron density caused by weak illumination. In narrower quantum wells (d $\leq$ 250 Å) the light scattering spectra do not show a substantial dependence on weak illumination. In these systems, at $\nu = 1$ it is possible to measure long wavelength inter-Landau-level excitations. We find the magnetoplasmon, at the cyclotron frequency ω_c, and spin-flip excitations [13]. The splitting between the long wavelength inter-Landau-level excitations is a measure of the strength of exchange enhancement in the

D. J. Lockwood and A. Pinczuk (eds.), *Optical Phenomena in Semiconductor Structures of Reduced Dimensions*, 57-61.
© 1993 *Kluwer Academic Publishers. Printed in the Netherlands.*

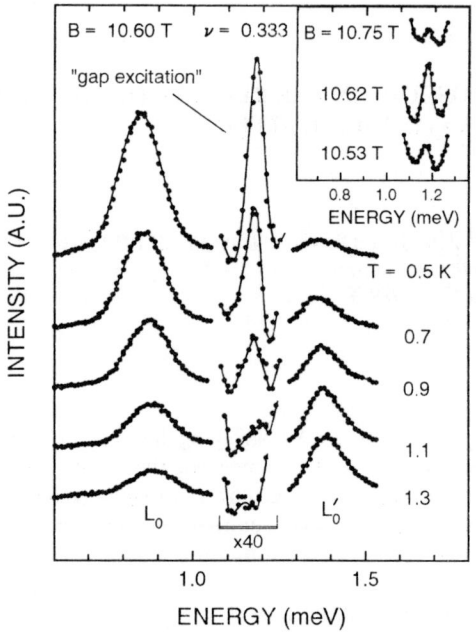

Figure 1. The sharp peak is inelastic light scattering by a low-lying excitation at $\nu = 1/3$. The bands labeled L_0 and L_0' are the characteristic doublets of intrinsic recombination. The single quantum well has density $n = 8.5 \times 10^{10}$ cm^{-2} and mobility $\mu = 2.6 \times 10^6$ cm^2/Vs at 4 K. The light scattering peak is interpreted as a $q = 0$ collective gap excitation. The temperature dependence of the L_0 and L_0' intensities is a manifestation of the photoluminescence anomaly at $\nu = 1/3$ in this sample. The inset shows the B-dependence of 0.5 K light scattering spectra. (After Ref. [16]).

spin-polarized state at $\nu = 1$.

Our most recent photoluminescence studies of the FQHE in single GaAs quantum wells have been carried out in low density and high mobility samples [14]. These asymmetric wells have $d \approx 250$ Å. Their densities are in the range 0.5–1.0×10^{11} cm^{-2} and the mobilities at 4 K are about 3×10^6 cm^2/Vs. We find that for $\nu \leq 3/5$ the ground state optical recombination consists of a well-defined doublet with a splitting of about 0.5 meV at $B = 10$ T. The high quality of these samples manifests in the sharpness of the photoluminescence peaks which have FWHM in the range of 0.1–0.2 meV. The relative intensities of the two components of the doublets show distinct anomalies at the filling factors 2/5, 1/3, and 2/7 of the fractional quantum Hall effect.

The peaks labeled L_0 and L_0' in Fig. 1 are the characteristic optical recombination doublets observed in the regime of the FQHE. The intensity anomaly at $\nu = 1/3$ is clearly seen in the figure. In the temperature range 1.3 K > T > 1.0 K the intensity of the lower energy component L_0' is largest, while in the range 1.0 K > T > 0.5 K the higher energy component L_0 dominates. Such a temperature dependence is characteristic of the phenomena of the FQHE [15]. The anomaly is similar to previous observations in higher [5] and lower [14] density systems. Photoluminescence excitation (PLE) spectra, which yield features associated with the optical absorption, show that the doublets are intrinsic. The PLE spectra also indicate that the optical excitations in the doublet are constructed from electron-hole pair states that incorporate the ground levels of the conduction and valence subbands. The holes are in the uppermost valence level. The electrons are in the lowest conduction Landau level that also houses the states occupied by the 2D electron system.

Examination of PLE spectra over a wide range of magnetic field with $\nu < 1$ indicates that a marked transfer of oscillator strength takes place between the two components of the doublet [14]. At the lower fields, close to $\nu \leq 1$, the two features in the doublet occur with similar intensities in PLE spectra. At larger fields and filling factor $\nu \leq 2/5$, the higher energy component of the doublet appears with the largest oscillator strength. In view of such results, the photoluminescence intensity anomalies in the regime of the FQHE could, at least in part, be related to the kinetics of the optical recombination. There are several important questions that require further consideration. Probably, the most significant is the issue of the many-body character of the transitions that are associated with the doublet in the ground state optical excitations. Elucidation of this question could reveal novel aspects of the incompressible fluid of the FQHE.

Our most recent light scattering work in the regime of the fractional quantum Hall effect has focused on a search of collective intra-Landau-level excitations [16]. It is well known that in the incompressible states of the FQHE the 2D electron gas should exhibit new charge-density intra-Landau-level excitations at energies fully determined by the electron-electron interactions in the condensate [17-19]. These low-lying excitations represent the gaps associated with condensation of the electron gas into incompressible states. The $q = 0$ gap excitations cannot be accessed in optical absorption experiments because intra-Landau-level transitions have vanishing optical oscillator strength [19].

Inelastic light scattering, which as a two-photon process can measure excitations that lack a net electric dipole moment, is the experimental method to yield a direct observation of collective excitations that carry the characteristic signature of the FQHE [16]. The spectra in Fig. 1 show the light scattering measurement of a low energy excitation of the FQHE state at $\nu = 1/3$. This mode, seen as a very sharp peak of FWHM = 0.04 meV, has been interpreted as a $q = 0$ gap excitation of the incompressible state. The relation of this feature to the FQHE at $\nu = 1/3$ is demonstrated in its characteristic temperature and magnetic field dependence, which is that of the FQHE [15]. As seen in Fig. 1, the mode is observed only at temperatures $T \leq 1$ K and in the narrow magnetic field range $\Delta B \approx 0.5$ T centered at $\nu = 1/3$. The measured energy, 1.18 meV at B = 10.6 T, is close to the prediction of the single-mode approximation that incorporates the finite width of the electron wavefunction along the perpendicular direction [19].

Other long wavelength collective excitations have been measured in these light scattering experiments [16]. They are the spin-wave, which is associated with the spin-flip transitions in the lowest Landau level and inter-Landau-level excitations. One striking feature of these results is the pronounced temperature and magnetic field dependences of the spin excitations, which are similar to that of the FQHE.

On the basis of these recent results, the inelastic light scattering appears to be a very important tool in studies of the remarkable behaviors of the high mobility 2D electron gas in the extreme magnetic quantum limit. Such studies would include the functional quantum Hall effect and the phenomena that a large number of publications associate with the formation of a Wigner solid. Of great current interest are the remarkable new behaviors predicted for filling factors $\nu = 1/2$ [20].

Acknowledgments

We benefited from helpful discussions with L. Brey, J.P. Eisenstein, B.B. Goldberg, Song He, D. Heiman, C. Kallin, A.H. MacDonald, P.M. Platzman, and H.L. Stormer.

References

1. I.V. Kukushkin and V.B. Timofeev, JETP Lett. **44**, 179 (1986); Surface Sci. **196**, 196 (1988).
2. B.B. Goldberg, D. Heiman, A. Pinczuk, C.W. Tu, A.C. Gossard, and J.H. English, Surface Sci. **196**, 209 (1988); D. Heiman, B.B. Goldberg, A. Pinczuk, C.W. Tu, A.C. Gossard, and J.H. English, Phys. Rev. Lett. **61**, 605 (1988).
3. H. Buhmann, W. Joss, K. von Klitzing, I.V. Kukushkin, G. Martinez, A.S. Plaut, K. Ploog, and V.B. Timofeev, Phys. Rev. Lett. **65**, 1056 (1990); I.V. Kukushkin, N.J. Pulsford, K. von Klitzing, K. Ploog, R.N. Haug, S. Koch, and V.B. Timofeev, Europhys. Lett. **18**, 63 (1992).
4. A.J. Tuberfield, S.R. Haynes, P.A. Wright, R.A. Ford, R.G. Clark, J.F. Ryan, J.J. Harris, and C.T. Foxon, Phys. Rev. Lett. **65**, 637 (1990).
5. B.B. Goldberg, D. Heiman, A. Pinczuk, L.N. Pfeiffer, and K.W. West, Phys. Rev. Lett. **65**, 641 (1990).
6. V.M. Apalkov and E.I. Rashba, JETP Lett. **53**, 442 (1991).
7. W. Chen, M. Fritze, A.V. Nurmikko, D. Ackley, C. Colvard, and H. Lee, Phys. Rev. Lett. **64**, 2434 (1990).
8. A.H. MacDonald and E.H. Rezayi, Phys. Rev. B **42**, 3224 (1990); A.H. MacDonald, E.H. Rezayi, and D. Keller, Phys. Rev. Lett. **68**, 1939 (1992).
9. B.S. Wang, J.L. Birman, and Z.-B. Su, Phys. Rev. Lett. **68**, 1605 (1992).
10. V.M. Apalkov and E.I. Rashba, Phys. Rev. B **46**, 1628 (1992).
11. A. Pinczuk, J.P. Valladares, D. Heiman, A.C. Gossard, J.H. English, C.W. Tu, L. Pfeiffer, and K.W. West, Phys. Rev. Lett. **61**, 2701 (1988); A. Pinczuk, D. Heiman, S. Schmitt-Rink, S.L. Chuang, C. Kallin, J.P. Valladares, B.S. Dennis, L.N. Pfeiffer, and K.W. West, in *Proc. of the 20th Int. Conf. on the Physics of Semiconductors*, edited by E.M. Anastassakis and J.D. Joannapoulos (World Scientific, Singapore, 1990), p. 1045.
12. D. Heiman, A. Pinczuk, B.S. Dennis, L.N. Pfeiffer, and K.W. West, Phys. Rev. B **45**, 1492 (1992); A. Pinczuk, B.S. Dennis, D. Heiman, L.N. Pfeiffer, and K.W. West, Solid State Commun. **84**, 103 (1992).
13. A. Pinczuk, B.S. Dennis, D. Heiman, C. Kallin, L. Brey, C. Tejedor, S. Schmitt-Rink, L.N. Pfeiffer, and K.W. West, Phys. Rev. Lett. **68**, 3623 (1992).
14. A. Pinczuk, D. Heiman, B.S. Dennis, L.N. Pfeiffer, and K.W. West, Bull. Am. Phys. Soc. **37**, 108 (1992).
15. D.C. Tsui, H.L. Stormer, and A.C. Gossard, Phys. Rev. Lett. **48**, 1559 (1982).
16. A. Pinczuk, B.S. Dennis, L.N. Pfeiffer, and K.W. West, submitted for publication.
17. R.B. Laughlin, in *The Quantum Hall Effect*, edited by R.E. Prange and S.M. Girvin (Springer, New York, 1990), 2nd. edition, p. 233.

18. F.D.M. Haldane and E.H. Rezayi, Phys. Rev. Lett. **54**, 237 (1985); E.H. Rezayi and F.D.M. Haldane, Phys. Rev. B **32**, 6924 (1985); F.D.M. Haldane, Ref. [17], p. 303.
19. S.M. Girvin, A.H. MacDonald, and P.M. Platzman, Phys. Rev. Lett. **54**, 581 (1985); Phys. Rev. B **33**, 2481 (1986).
20. B.I. Halperin, P.A. Lee, and N. Read, Phys. Rev. B, to be published.

MAGNETO-OPTICS OF THE FRACTIONAL QUANTUM HALL EFFECT: THEORY

E.I. RASHBA[1,2] and V.M. APAL'KOV[1]
[1]L.D. Landau Institute for Theoretical Physics, Moscow 117940, Russia
[2]Department of Physics, University of Utah, Salt Lake City, Utah 84112, USA

ABSTRACT. Collective properties of two-dimensional (2D) electrons, such as the formation of incompressible quantum liquids, the Wigner crystalization, etc., manifest themselves in the optical spectra only when a hidden symmetry inherent in the magnetospectroscopy of such systems is violated. As a result, spectra are variable, and depend on the mutual magnitudes of different perturbations. In particular, the dispersion law of magnetoexcitons strongly depends on the separation between electron and hole confinement planes. When this separation reaches some critical value, excitons become indirect, and from the magnitude of the abrupt shift of the emission band, the magnetoroton energy may be found. Analogous changes are also possible in extrinsic emission, caused by the trapping of 2D electrons by charged impurities. Unique properties are inherent in the radiative trapping of 2D electrons by neutral impurities, since the 2D system is not perturbed by such impurities in its initial quantum state. The position of the emission band shows, as a function of the filling factor, down-cusps at all the fractions corresponding to formation of incompressible liquids, the gaps for creation of charged elementary excitations may be found from the cusp strengths. A large contribution from shake-up (Auger) processes is typical for these spectra.

1. Introduction

In two-dimensional (2D) electron systems subjected to a strong magnetic field H the kinetic energy of electrons is completely quenched, hence the role of electronic interactions increases tremendously. This results in formation of new electron phases like incompressible quantum liquids (IQLs) [1], in Wigner crystallization, etc. Primarily, magnetotransport properties were the only tool for experimental investigation of this phenomena, and the discovery of the fractional quantum Hall effect (FQHE) [2], in which IQLs manifest themselves, predetermined the activity in this new field. Magnetoresistance data [3] also gave strong evidence for the existence of a re-entrant IQL–Wigner crystal phase transition.

More recently, considerable progress has been attained in the optical spectroscopy of strongly interacting 2D electrons. It was established [4,5] that the position of the extrinsic emission band shows a non-analytical behaviour as a function of filling factor ν for the same fractions for which the FQHE is observed, down to $\nu = 1/9$; the emission originates from trapping of 2D electrons by neutral impurities residing outside a heterojunction [6]. It has also been shown [7–9] that the FQHE may be evidenced from optical spectroscopy of intrinsic emission from quantum wells, and the formation of IQLs is accompanied by specific changes in the positions of the emission bands, their splitting, changes in their intensities, etc. There exists a striking difference in the patterns observed in the extrinsic and intrinsic emission from heterojunctions [10], and also a considerable dependence of intrinsic emission spectra on the shape and width of the quantum well [11]. Some controversy still exists in experimental papers concerning the procedures for treatment of experimental data, and how to extract the basic parameters of an IQL from them (for example, see Ref. [8]). Even more controversy exists as applied to spectroscopic observation of Wigner crystallization [11–13]. Apparently, this implies a strong dependence of the optical spectra on the geometry of quasi-2D systems, and it makes it clear that a reliable treatment of experimental data is only possible on the basis of a consistent theory.

In the quantum limit, $H \to \infty$, all contributions to the energy have the scale of the Coulomb energy, $\varepsilon_C = e^2/\varepsilon l$, where $l = (c/eH)^{1/2}$ is the magnetic length and $\hbar = 1$. Therefore, there is no small parameter in the problem. Hence, of special importance are both the exact results and results of numerical simulations. As applied to optical spectroscopy, the central result of this kind is the trivialization of the emission and absorption spectra for coplanar electrons and holes: all many-electron correlations are cancelled out of them, and the spectrum coincides with the spectrum of a magnetoexciton in an empty crystal [14–17]. This cancellation is caused by a hidden symmetry inherent in such systems [16,18–20]. As a result, optical spectra depend strongly on the specific way in which the hidden symmetry is violated, and the diversity of spectra, observed experimentally, acquires a natural explanation. In what follows we discuss in some detail the properties of excitons against the background of an IQL [15,16] for symmetric and nonsymmetric quantum wells, and predict abrupt changes in the position and intensity of the emission band caused by a switch-over between zero-magnetoroton and one-magnetoroton regimes in nonsymmetric systems [16,21]. A procedure that is based on this phenomenon and may permit one to find the energy of magnetorotons (MRs), neutral elementary excitations of an IQL [22], from the intrinsic emission spectra is proposed. Such experiments, if successful, will provide spectroscopic evidence of the existence of fractional charges, charged elementary excitations of an IQL [1], and also show an abrupt change in the ground state of an exciton when H changes under the conditions $v =$ constant. A theory of extrinsic emission will be also outlined. It will be shown that the singularity in the position of the emission band versus v is of the down-cusp type, i.e., of the same type as the singularity in the ground state energy [23]. However, the cusp strength strongly depends on the distance between a neutral impurity and the confinement layer. This distance is actually a degree of the asymmetry of the system, and from this dependence the gap in the spectrum of charged excitations may be found. These conclusions are in a good agreement with recent experimental findings [10]. In conclusion, we show [24] that the hidden symmetry also manifests itself in emission spectra of strongly pumped quantum wells with electron and hole populations on several Landau levels [25]; in charge symmetric systems the position of the highest frequency band does not depend on the fractional parts of electron and hole filling factors.

2. Magnetoexcitons: Hidden Symmetry

Magnetoexcitons [18,26] provide a most appropriate description of 2D electron–hole (e–h)

systems in the strong H limit. The criterion is $\varepsilon_C \ll \omega_c$, where ω_c is the cyclotron frequency. The simplest model of a e-h 2D system, which will be termed below the standard model, meets the following conditions:
1. Quantum limit. Charge carriers, electrons and holes, occupy only the ground Landau level.
2. Level mixing is neglected.
3. Charge symmetric interaction, $V_{ee}(q) = V_{hh}(q) = -V_{eh}(q)$. Here V_{ee}, V_{hh}, and V_{eh} are the potentials of e-e, h-h, and e-h interactions, respectively, and q is a 2D momentum.

In the strong magnetic field limit, optical spectroscopy of e-h systems may be completely described in terms of magnetoexcitons. The annihilation operator for a magnetoexciton with a 2D momentum $\mathbf{k}$ is

$$A(\mathbf{k}) = (2\pi/A)^{1/2} \sum_p a_{p+q_y/2} b_{p-q_y/2} \exp(ipq_x), \quad (1)$$

where a_p and b_p are electron and hole annihilation operators in the Landau gauge, and A is the normalization area. Only excitons with $\mathbf{k} = 0$ are active in optical transitions. The Hamiltonian H may be written in terms of the electron density operators $a(q)$, which are bilinear in $a_p^\dagger$ and a_p, hole density operators $b(q)$, and also operators $\hat{N}_e$ and $\hat{N}_h$ of the numbers of electrons and holes. Even more convenient than $a(q)$ and $b(q)$ are the operators

$$c(q) = a(q) - b(q),$$
$$d(q) = a(q) + b(q), \quad (2)$$

where $c(q)$ are operators of the total charge density. The following commutation rule holds:

$$[c(q), A(q)] = -2i(2\pi/A)^{1/2} \sin\frac{(\mathbf{q} \times \mathbf{k})}{2} A(q+k). \quad (3)$$

The commutation rule for $d(q)$ differs from Eq. (3) by the change of sine to cosine. Here $(\mathbf{q} \times \mathbf{k}) \equiv q_x k_y - q_y k_x$. Therefore, $A(\mathbf{k} = 0)$ commutes with all operators $c(q)$. It may be shown that for a charge symmetric interaction H does not depend on $d(q)$, and

$$[H, A^\dagger(\mathbf{k}=0)] = E_{ex}(\mathbf{k}=0) A^\dagger(\mathbf{k}=0), \quad (4)$$

where $E_{ex}(\mathbf{k})$ is the energy of a bare exciton, i.e., an exciton in the absence of the e-h background (for more detail see Refs. [15,16]).

It follows from Eq. (4) that if ψ is the wave function for an arbitrary initial, "small", system, $H\psi = \varepsilon\psi$, then

$$\Psi = A^\dagger(\mathbf{k}=0)\psi \quad (5)$$

is the exact wave function for the final, "large", system, including one additional exciton, $H\Psi = E\Psi$. The energies of these states are interrelated by the equation

$$E = \varepsilon + E_{ex}(k = 0). \tag{6}$$

Equation (6), reflecting some kind of freedom of $k = 0$ excitons, generalizes the well known result that the scattering amplitude of two excitons tends to zero when their momenta $k \to 0$ [18,27].

Functions Ψ and ψ are mutually related by Eq. (5), and when written in the configuration representation, differ in a factor equal to the wave function of a free exciton with $k = 0$, where the subsequent symmetrization of the product must be performed. Such functions as Ψ, Eq. (5), will be termed multiplicative. Despite the fact that they make up only a small portion of all the quantum states of the large system (see below), they play a very important role in optical phenomena, since, in the framework of the basic model, optical transitions are allowed only into these states or from these states and the frequency of all allowed transitions equals $E_{ex}(k = 0)$ irrespective of ν.

Despite the simplicity of Eq. (6) for the energies of multiplicative states, their electronic structure is rather complicated, and numerical methods are needed to investigate it. We used the spherical geometry [28], which is especially appropriate for spectroscopic applications, since the continuous rotation group and the associated selection rules are retained in it. In this geometry quantum states may be classified by their angular momentum L. The latter is related to the momentum k of neutral elementary excitations, in the plane geometry, by the equation $L = Rk$, where R is a sphere radius.

Figure 1 illustrates the properties of an exciton against the background of an IQL, for $\nu = 1/3$. There is a gap in the spectrum with a width of about $0.1\varepsilon_C$; it exists for all values of L and it decreases with L. Below it lie the energy levels of quantum states with a strongly screened hole. The electron density at the hole, $n(r_e = r_h)$, is close to the Fermi limit, $n_0 = 1/2\pi l^2$, in all these states. All multiplicative states belong to this part of the spectrum. In all the states belonging to the upper part of the spectrum, $n(r_e = r_h)$ is considerably less than n_0. Despite the fact that the energy shift equals zero for $k = 0$ excitons, Eq. (6), they are strongly coupled to the background. In Fig. 1(b) the distribution of electronic density is shown for all $L = 0$ states. The density distribution in the lowest multiplicative state, curve 1, differs strongly from the density distribution in the bare exciton. In the over-gap state, curve 3, the density of the screening charge is strongly reduced.

The dispersion law of a dressed exciton, $E^*_{ex}(k)$, is shown in Fig. 2(a) side by side with the dispersion law of a free exciton, $E_{ex}(k)$, in the framework of the basic model. Several specific properties of it are worth special attention:

1. There is no polaron shift of the dressed exciton energy level for $k = 0$, in agreement with Eq. (6).
2. The dressed exciton dispersion is strongly suppressed as compared to the bare exciton due to the coupling to the background. The dispersion law $E^*_{ex}(k)$ is very flat near $k = 0$. Apparently, there are no k^2 terms in $E^*_{ex}(k)$.
3. For $k \approx 1.5/l$, the curve $E^*_{ex}(k)$ and the dispersion law of magnetorotons, $\omega_{mr}(k)$, come together.

Two different situations are possible under such conditions: termination of the exciton spectrum, and formation of exciton-MR bound states [29]. The fact that $E^*_{ex}(k) < \omega_{mr}(k)$ in the region of large $k \gtrsim 1.5/l$, as well as some specific properties of exciton wave functions in this region [16], show that the second possibility is realized. The spacing between $E^*_{ex}(k)$ and $\omega_{mr}(k)$ has the meaning of an exciton-MR binding energy. The existence of bound states in the small-k region may be seen from Fig. 1(a) [15,16]. Bound

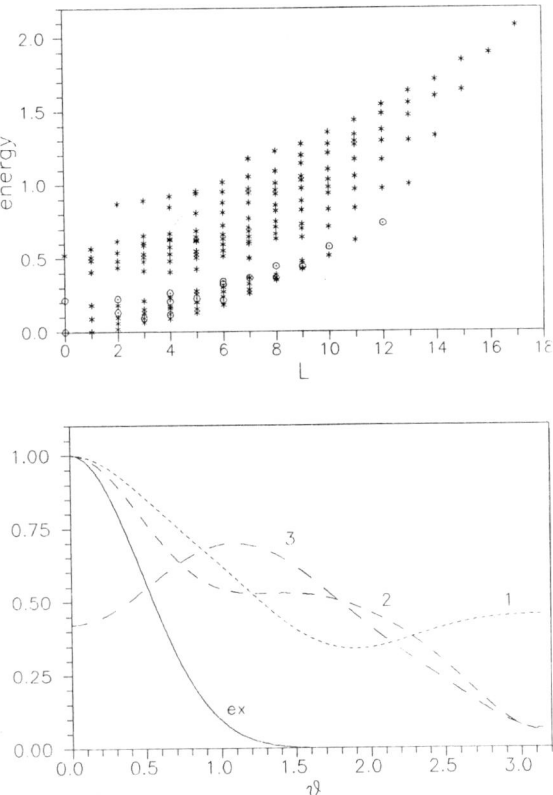

Figure 1. Exciton against the background of an IQL, $\nu = 1/3$; charge symmetric system, four electrons plus one exciton. (a) Energy spectrum, $k = \sqrt{2}L/3l$, energy in units of ε_C. The energy of a bare exciton with $k = 0$ is chosen as the origin. Multiplicative states are shown by circles. (b) Distribution of electron density around a hole. Density in units of n_0, electron-hole separation equals $r \approx 3l\vartheta/\sqrt{2}$. Curves 1, 2, and 3 correspond to three states with $L = 0$; their energies $E_1 < E_2 < E_3$. The curve "ex" represents a bare exciton.

states of this kind are well known for different strongly coupled systems, for example, three-dimensional (3D) and one-dimensional (1D) polarons and solitons [30,31].

Therefore, the optical spectra of perfect systems, which are determined by magneto-excitons with $k = 0$, are trivial in the framework of the basic model. Optical spectra of real systems are determined by the competition of different perturbations, including the effect of imperfections. One should expect a strong variability of the spectra depending on the detailed geometry of the electron and hole confinement layers.

3. Nonsymmetric Systems: Fractional Charges

In this section we consider changes in the exciton dispersion law and also in the emission

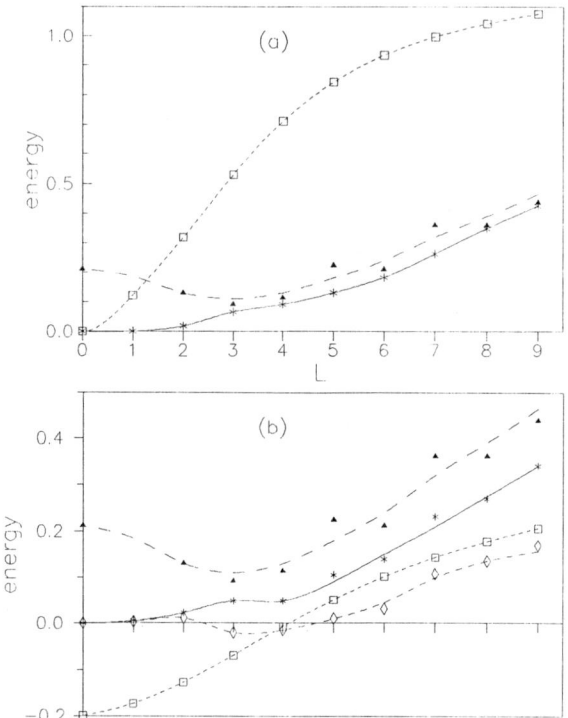

Figure 2. Dispersion law of an exciton against the background of an IQL, $\nu = 1/3$; the parameter values as in Fig. 1(a). The energy of a dressed exciton with $\mathbf{k} = 0$ is chosen as an origin. Squares: a bare exciton, $E_{ex}(L)$; asterisks: a dressed exciton, $E^*_{ex}(L)$; triangles: magnetorotons, $\omega_{mr}(L)$ (their energy does not saturate in the large L region because of a small size of the system). (a) Charge symmetric system. (b) Nonsymmetric system, $h = 0.5l$; for the curve shown by diamonds, $h = l$.

spectra, which arise due to the simplest mechanism of hidden symmetry violation [16,21]. If electrons and holes reside in different confinement planes separated by a distance h, then $V_{ee} = V_{hh} > |V_{eh}|$, and the condition of charge symmetry (see Sec. 2), is violated. The dispersion law of an exciton in such a system is shown in Fig. 2(b). It differs strikingly from Fig. 2(a):

1. The energies of bare and dressed excitons with $\mathbf{k} = 0$ differ considerably, and the energy level of a dressed exciton with $L = 0$ experiences a high energy shift. Actually, two mechanisms contribute into this shift. One contribution comes from the polaron dressing of an exciton by MRs. The other originates from the Pauli exclusion principle, since the exciton which appears against the background of an IQL sees a considerable part of the phase space already filled. The density of an electron cloud in a bare exciton coincides with the Fermi limit, n_0, for $r_e = r_h$, therefore restrictions imposed by the Pauli exclusion principle are very stringent for dressed excitons. In symmetric systems both contributions cancel out exactly at $\mathbf{k} = 0$. In nonsymmetric systems the Pauli contribution dominates, and it determines the

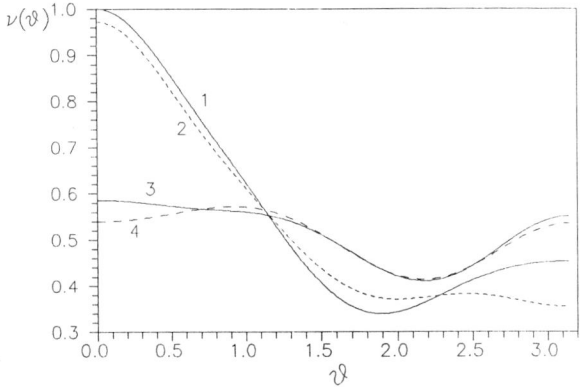

Figure 3. Distribution of the electron density, $v(\vartheta)$, in an exciton for various values of h. The electron-hole separation is $r \approx 3l\vartheta/\sqrt{2}$, ϑ is a polar angle, and the hole resides at the north pole. The background value of the filling factor actually equals $v = 0.4$, instead of 1/3, because of the final size of the system. The four curves correspond to the following values of h, of the angular momentum (L), and of the excess charge in the northern hemisphere (Q): curve 1—h = 0, L = 0, Q $\approx$ 1.06; curve 2—h = 0.845, L = 0, Q $\approx$ 1.05; curve 3—h = 0.85, L = 3, Q $\approx$ 0.67; curve 4—h = 1.0, L = 3, Q $\approx$ 0.67. All four states correspond to minimal energies at given h and L.

sign of the level shift at $\mathbf{k} = 0$.

2. In the small k region the k^2 terms have the usual order of magnitude when h ~ 1. Very interesting changes set in near $k \approx 1.5/l$. At $h \approx 0.5l$ an inflection point in the dispersion law of a dressed exciton arises, and for larger values of h an extra minimum exists. At $h = h_{cr} \approx 0.85/l$ this minimum becomes an absolute one, i.e., the bottom of the exciton band shifts from the point $\mathbf{k} = 0$ to a circle. This abrupt change in the exciton ground state has important spectroscopic implications. They will be discussed in more detail in what follows.

3. The separation of the $E^*_{ex}(\mathbf{k})$ and $\omega_{mr}(\mathbf{k})$ curves, and therefore also the exciton-MR binding energy, increases with h.

A striking difference in the distribution of the electron density in an exciton for $\mathbf{k} = 0$ and $k \approx 1.5/l$ is seen in Fig. 3. These values of k correspond to the positions of two minima on $E^*_{ex}(\mathbf{k})$ curves. Charge distributions in these states only weakly depend on h; the spacing h mainly influences their energies. It is well known [32], that the magnetoexciton momentum k is closely connected to its dipole moment, i.e., the mean separation $\bar{r}$ between an electron and a hole may be estimated as $\bar{r} \sim l^2 k$. Therefore, one can expect that in the state $k \approx 1.5/l$ a considerable part of the electron charge is removed from the central area of an exciton. It follows from Fig. 3 that the removed charge equals $Q' \approx e^*$, where $e^* = ve$ is the fractional charge of Laughlin's theory of IQLs. In these terms the change in the ground state of a magnetoexciton may be described as a transfer of one of the fractional charges entering it on a large quantum orbit.

The shifting of the minimum of $E^*_{ex}(\mathbf{k})$ should have a drastic effect on the emission spectrum at temperatures $T \approx 0$. When the absolute minimum of the dispersion spectrum is at $\mathbf{k} = 0$, both direct (zero-MR) and indirect (e.g., single-MR) transitions are allowed in the

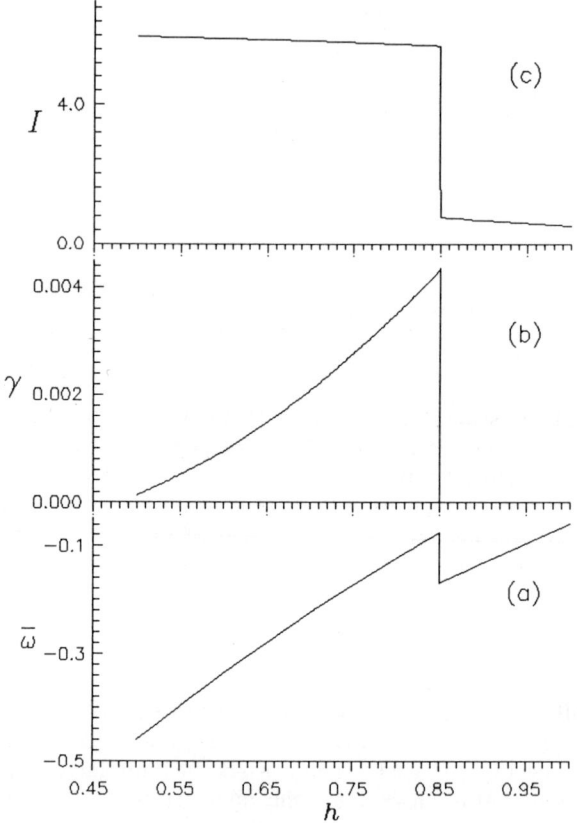

Figure 4. Basic characteristics of the exciton band as a function of h: $\bar{\omega}$, position of the band; γ, width of the band; I, intensity of the band.

emission spectrum. Since there is no small algebraical parameter in the theory, both types of transition have in many cases comparable intensities [33,34]. However, for an exciton against the background of an IQL, $\nu = 1/3$, the intensity of the direct transition dominates; it is one order of magnitude stronger than for indirect transitions. For $k \neq 0$ excitons only indirect transitions are allowed, and single-MR transitions dominate in the spectrum. Therefore, at $h = h_{cr}$, the total probability of the transition must drop abruptly, and the emission band must experience an abrupt shift that is equal to the energy of the magnetoroton, ω_{mr}. These spectacular changes in all parameters of the emission spectrum are seen in Fig. 4.

When calculating transition probabilities, I, shown in Fig. 4, one important assumption was made. Namely, the dynamics of charge carriers connected with recombination of an electron and a hole residing in different planes was neglected. For the emission probability the following equation was used:

$$I \propto |\int \Psi(r_1...r_N, r_h | r_h) \psi(r_1...r_N) dr_1...dr_N dr_h |^2 . \tag{7}$$

Here $\Psi(r_1...r_N, r_{N+1}|r_h)$ and $\psi(r_1...r_N)$ are wave functions of the large and small systems, respectively—these systems differ in a single e-h pair, r_i are electron 2D coordinates, and r_h is a hole 2D coordinate. Equation (7) is written under the assumption that an electron and a hole recombine when their 2D coordinates coincide, $r_{N+1} = r_h$.

To observe the predicted phenomenon experimentally the emission spectrum must be investigated under the conditions when the parameter h/l changes, but ν is kept constant.

Very little can be said now about the comparison of the theory with experimental data on intrinsic emission. A "blue shift" of the emission line has been observed in Refs. [7] and [11], in the vicinity of ν = 2/3. This shift depends smoothly on ν, and it does not show considerable dependence on h; it exists only with the concomitant plateaux in ρ_{xy}. It seems natural to relate this shift to the high energy shift in the energy of a dressed exciton, Fig. 3(b). Since the latter is caused by the Pauli exclusion principle and the phase space reduction, one could expect it to decrease with increasing temperature T, which would provide a blue shift of the emission line at low T. Calculations show [35] that such a blue shift of about $0.04\varepsilon_C$ can be expected at ν = 1/3 and h/l = 0.5. However, at h/l ≈ 1, when the new minimum develops in $E^*_{ex}(k)$, the shift acquires a non-monotonic temperature dependence. Calculations also predict a more sophisticated, non-monotonic, ν-dependence of the emission band position, which apparently may be attributed to the interaction of an exciton with quasi-particles [16]. To separate effects coming from different mechanisms, two sets of experiments may be especially suggestive: investigation of the spectra as a function of ν at h/l = const, and as a function of h/l at ν = const.

4. Extrinsic Emission: Energy Gaps

Depending on the electrical charge of an impurity center, emission spectra may considerably differ. Hence, in what follows they will be discussed separately. In both cases the expression for the transition probability

$$I \propto |\int \Psi(r_1...r_{N-1}r_0)\psi(r_1...r_{N-1})dr_1...dr_{N-1}|^2, \qquad (8)$$

quite analogous to Eq. (7), may be used. Here $\Psi(r_1...r_N)$ and $\psi(r_1...r_{N-1})$ are the wave functions of the initial and final states, respectively, and r_0 is the projection of r_I, the 3D coordinate of the impurity center, on the confinement plane.

In Fig. 5 emission spectra are shown for radiative trapping of 2D electrons on positive charged centers [34]. The calculations were performed using the spherical geometry [28] for different distances of the impurity from the confinement layer. The number of electrons in the initial state, N = 6, and the size of the sphere, $R = l[\nu^{-1}(N-2)/2]^{1/2} = l\sqrt{6}$, have been chosen so that an incompressible phase (a finite size analog of the ν = 1/3 phase) appeared in the final state. If r_0 is the north pole of the sphere, it follows from Eq. (8) that the following selection rule holds:

$$M = M' + S, \qquad (9)$$

where M and M' are projections of the angular momentum L on the quantization axis in the initial and final states, respectively, and $S = (R/l)^2$ is the magnetic flux through the sphere in units of the flux quantum. It is equal to the angular momentum that the

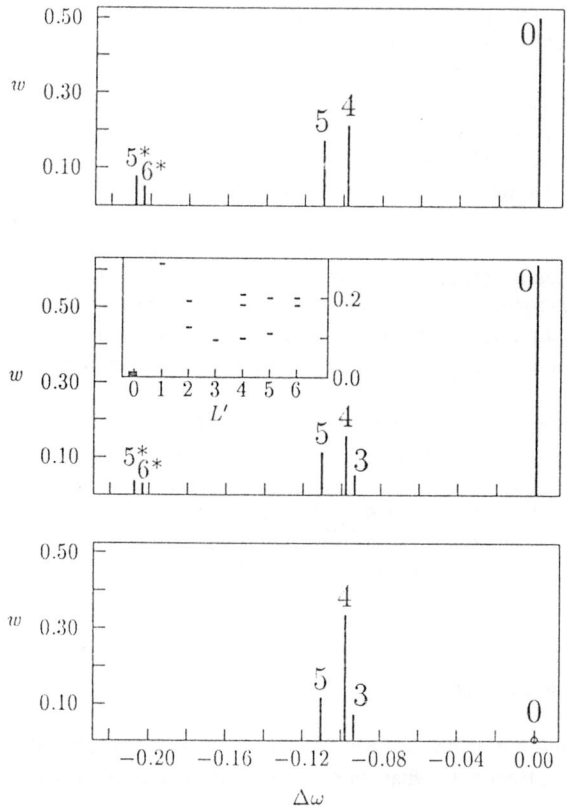

Figure 5. Emission spectrum for radiative trapping of 2D electrons on charged centers; finite system calculations with six electrons in the initial state. The energy of the zero-MR transition is chosen as an origin. Bar lengths are proportional to intensities of emission bands and the numbers by the bars indicate angular momenta (L') in final states. Data are given for three values of h: above, h = 0; in the middle, h ≈ 2l; and below, h ≈ 4l. Inset: lowest part of the spectrum of the elementary excitation in the final state.

recombining electron takes away. The final state interaction of 2D electrons with the center is neglected because of its electrical neutrality.

The spectrum consists of a zero-MR band and a single-MR satellite having a comparable intensity. The dominant contribution to it comes from the states close to the roton minimum [22]. A two-MR satellite is much weaker. A high intensity of the single-MR satellite demonstrates the importance of shake-up (Auger) processes in the quantum limit regime [33]. If h is small, the angular momentum in the ground state of the "large" system equals M = S, and all transitions are allowed. However, when h/l increases, the symmetry of the ground state is changed, and transitions to the states with small L' become forbidden by momentum conservation, Eq. (9). A rough estimate shows that it happens when h/l ≈ 2.5. As a result, the zero-MR transition disappears from the spectrum, and nearly all the intensity is concentrated in the single-MR band. The evolution of the spectrum with increasing h/l resembles the patterns described in Sec. 3, as applied to exciton emission spectra.

The important point is that the contribution of MR assisted transitions is large, and for the trapping by neutral centers it is even larger [34], since in the final state soft local modes exist in the vicinity of a Coulomb charge [36,37]. The detailed shape of the emission spectrum may be found only by numerical methods. However, the lowest moments of the spectrum may be treated by analytical methods, and such a description seems sufficient at the present time, since the fine structure of extrinsic emission spectra have not yet been resolved experimentally. Meanwhile, singularities in the v-dependence of the extrinsic emission have been observed [4,5] in gross features of the spectrum; actually, in the v-dependence of the emission band position.

The normalized first moment of the emission spectrum determines the position of its center-of-gravity. Summation over all final states results in the expression

$$\overline{\omega} = E_i - <H_f>_{av}, \qquad (10)$$

where E_i is the energy of the initial (i) state, i.e., of the ground state of N interacting particles in the absence of any external potential. H_f is the Hamiltonian of the final (f) state, i.e., the Hamiltonian of (N - 1) 2D interacting particles, $r_1...r_{N-1}$, subjected to a repulsive Coulomb potential of an impurity center residing in the point r_I. The symbol $<...>_{av}$ stands for averaging over the wave functions $\Psi_\alpha(r_1...r_N)$ of all the states α that belong to the level number i (provided that the condition $r_N = r_0$ is satisfied). The reference point is chosen in such a way that $\overline{\omega}$ includes only the Coulomb interaction energy. An exact transformation of Eq. (10), which includes averaging over r_0, permits one to express $\overline{\omega}$ in terms of the electron pair correlation function $g_v(r)$ [14]:

$$\begin{aligned}\overline{\omega} &= \int [V(|r_0 - r|) - V(|r_I - r_0|)]g_v(|r_0 - r|)dr \\ &= 2E_i/N - \int V(|r_I - r|)g_v(|r_0 - r|)dr,\end{aligned} \qquad (11)$$

where E_i is the energy of the initial level, and $V(r)$ is the e-e interaction potential. The integration is performed over a 2D layer.

For a short range potential $V(r)$, the last term in Eq. (11) may be omitted if $h = |r_0 - r_I|$ exeeds the potential radius. Therefore, for any value of $v = p/q$ at which an IQL is formed, the function $\overline{\omega}(v)$ must show singularities of the same type (down-cusps) as the ground state energy $E_i(v)$ [23], and the gap for the creation of quasi-electron–quasi-hole pairs is directly connected to the cusp strength, i.e., to the discontinuity in the first derivative of $\overline{\omega}(v), \delta\{d\overline{\omega}/dv\}$. The situation is more complicated for a Coulomb potential due to its long range behaviour. Since the non-analyticity of $E_i(v)$ originates from the non-analyticity of $g_v(r) = g_1(r) + |v - (p/q)|g_2(r)$ [38], the last term in Eq. (11) must show the singularity of the same type. Therefore, a cusp persists, but the cusp strength depends on h. It is seen from the first line of Eq. (11), that both terms on the right hand side cancel when $h \to 0$. The function $\delta\{d\overline{\omega}/dv\}$ shows h^2 behaviour at $h \ll 1$, and h^{-3} behaviour at $h \gg 1$.

In Fig. 6 the second term in Eq. (11) is shown, which was found by numerical computations in the spherical geometry. The expansion in 1/N converges slowly, hence the accuracy of the data is not especially high. However, the basic conclusions are rather

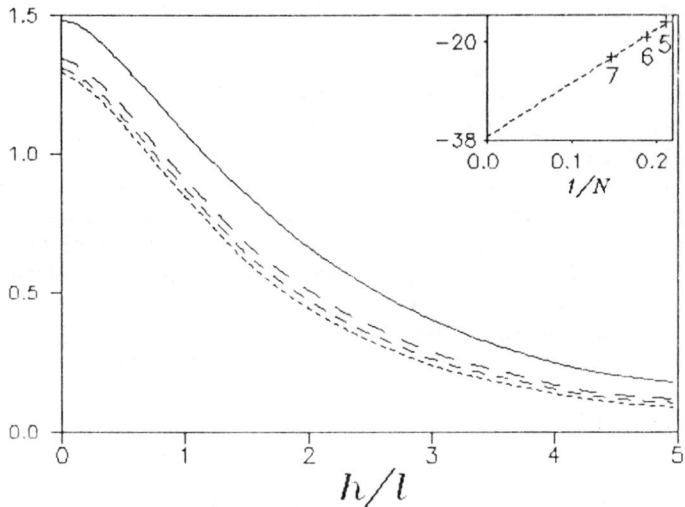

Figure 6. Extrinsic contribution to the cusp strength [second term in Eq. (11)] versus distance h, $\nu = 1/3$. The solid line is obtained by extrapolation, $1/N \to 0$, from three curves found for $N = 5, 6,$ and 7. The energy is in ε_C units. Inset: the value of the coefficient at $(l/h)^3$ in the expansion of Eq. (11).

definite: in all the region of h/l values that are of interest for the treatment of experimental data, the second term strongly depends on h/l and makes a considerable contribution to $\overline{\omega}(\nu)$.

Recent experimental data are in a good agreement with the above theoretical conclusions and their treatment based on Eq. (11), and Fig. 6 results in reasonable magnitudes of the gaps, for several fractions $\nu = p/q$ [10]. The initial, much less complete, experimental data have been discussed in terms of steps (not cusps) in the $\overline{\omega}(\nu)$ dependence [4,5]. In this connection it is natural to ask whether such a behavior is possible under any conditions, at least from the theoretical point of view. It is easy to understand that the step like behaviour may only appear in the limit when the probability of electron tunneling from a 2D layer to an impurity center tends to zero and it is the smallest parameter in the theory. Indeed, under these conditions the recombination rate is completely controlled by the tunneling probability, which reaches its maximal value when a tunneling particle takes away the largest accessible amount of energy and the 2D system is left in its ground state. In a such regime the emission probability must increase enormously when ν becomes larger than the fraction p/q, and, hence, free quasi-particles appear in the system. There are no indications of such behaviour in any available experimental papers. Therefore, we conclude that the above theory provides a satisfactory approach in a realistic range of parameter values. Of course, the 3D dynamics must result in some corrections to the model Eq. (8). Their magnitudes must strongly depend on the specific geometry.

It is worth mentioning, in conclusion, that the cancellation of cusp strengths in the $h \to 0$ limit strongly resembles the cancellation of electronic correlations in the spectra of intrinsic charge symmetric systems (Sec. 2) caused by the hidden symmetry. The distance h plays the role of an asymmetry parameter, quite analogous to intrinsic spectra.

5. Highest Landau Levels

All the above considerations were connected with electron and hole populations at the lowest Landau level, both in the conduction and in the valence bands. In this section we briefly discuss a manifestation of the hidden symmetry in the spectroscopy of the highest Landau levels. There are exciting experimental data [39–41] in which a nontrivial interaction of transient [39] and stationary [25,41] populations has revealed itself. There are also theoretical implications of the existence of some kind of hidden symmetry in the spectroscopy of excited levels [19,42–44]. Finally, it was shown that a direct generalization of Eqs. (4)–(6) exists for a definite class of systems [24]. To be concrete, let us consider a strongly pumped quantum well with a population inversion, and suppose that the interaction is charge symmetric; Landau levels 0,1...,N-1 are completely filled by charge carriers (electrons in the conduction band and holes in the valence band), level N is partially filled, and other levels are empty. The filling factors that obey the condition

$$N \leq v_e, v_h \leq N + 1 \tag{12}$$

may not be equal, $v_e \neq v_h$. If the exciton operator $A_N(k)$ for N–N transitions is defined quite analogous to Eq. (1), the following equations hold [24]:

$$[H, A_N^\dagger(\mathbf{k}=0)] = E_N^{ex}(\mathbf{k}=0) A_N^\dagger(\mathbf{k}=0), \tag{13}$$

and

$$E = \varepsilon + E_N^{ex}(\mathbf{k}=0). \tag{14}$$

Here E and ε have the same meaning as in Sec. 2, and $E_N^{ex}(\mathbf{k})$ is the exciton dispersion law when $v_e = v_h = N$. It follows from Eq. (14) that the frequencies of optical transitions do not depend on the fractional part of populations v_e and v_h. The data of Ref. [25] are in a pretty good agreement with this conclusion.

Eqs. (13) and (14) hold only for elementary excitations, excitons, and de-excitons, forming the Burstein edge [43]. The conditions of hidden symmetry are violated for excitons connected with higher, and de-excitons connected with lower, Landau levels. Therefore, their energies depend on v_e and v_h.

6. Summary

In conclusion, from extrinsic and intrinsic emission spectra two different parameters of the spectrum of elementary excitations of an IQL may be found. From the former spectra may be found the gap for creation of free charge carriers, quasi-electron–quasi-hole pairs; it coincides with the energy of MRs in the $k \to \infty$ limit. There are successful attempts of this kind. From the latter spectra the energy of MRs near the roton minimum may be found, but a special experimental procedure (v = constant regime) is needed.

Acknowledgment

The work of EIR was supported in part by NSF Grant No DMR-9116748.

References

1. R.B. Laughlin, Phys. Rev. Lett. **50**, 13 (1983).
2. D.C. Tsui, H.L. Stormer, and A.C. Gossard, Phys. Rev. Lett. **48**, 1559 (1982).
3. H.W. Jiang, R.L. Willet, H.L. Stormer, D.C. Tsui, L.H. Pfeiffer, and K.K. West, Phys. Rev. Lett. **65**, 633 (1990).
4. I.V. Kukushkin and V.B. Timofeev, Pis'ma Zh. Eksp. Teor. Fiz. **44**, 179 (1986) [JETP Lett. **44**, 228 (1986)].
5. I. Kukushkin, V.B. Timofeev, K. von Klitzing, and K. Ploog, in *Festkörperprobleme: Advances in Solid State Physics*, Vol. 28, edited by U. Rössler (Vieweg, Braunschweig, 1988), p. 21.
6. H. Buhmann, W. Joss, K. von Klitzing, I.V. Kukushkin, G. Martinez, A.S. Plaut, K. Ploog, and V.B. Timofeev, Phys. Rev. Lett. **65**, 1056 (1990).
7. D. Heiman, B.B. Goldberg, A. Pinczuk, C.W. Tu, A.C. Gossard, and J.H. English, Phys. Rev. Lett. **61**, 605 (1988).
8. A.J. Turberfield, S.R. Heynes, P.A. Wright, R.A. Ford, R.G. Clark, J.F. Ryan, J.J. Harris, and C.T. Foxon, Phys. Rev. Lett. **65**, 637 (1990).
9. B.B. Goldberg, D. Heiman, A. Pinczuk, L. Pfeiffer, and K. West, Phys. Rev. Lett. **65**, 641 (1990).
10. I.V. Kukushkin, N.J. Pulsford, K. von Klitzing, K. Ploog, and V.B. Timofeev, Surf. Sci. **263**, 30 (1992).
11. B.B. Goldberg, D. Heiman, A. Pinczuk, L. Pfeiffer, and K. West, Surf. Sci. **263**, 9 (1992).
12. H. Buhmann, W. Joss, K. von Klitzing, I.V. Kukushkin, A.S. Plaut, G. Martinez, K. Ploog, and V.B. Timofeev, Phys. Rev. Lett. **66**, 926 (1991).
13. I.V. Kukushkin, N.J. Pulsford, K. von Klitzing, K. Ploog, R.J. Haug, S. Koch, and V.B. Timofeev, Phys. Rev. B **45**, 4532 (1992).
14. V.M. Apal'kov and E.I. Rashba, Pis'ma Zh. Eksp. Teor. Fiz. **53**, 420 (1991) [JETP Lett. **53**, 442 (1991)].
15. V.M. Apal'kov and E.I. Rashba, Pis'ma Zh. Eksp. Teor. Fiz. **54**, 160 (1991) [JETP Lett. **54**, 155 (1991)].
16. V.M. Apal'kov and E.I. Rashba, Phys. Rev. B **46**, 1628 (1992).
17. A.H. MacDonald, E.H. Rezayi, and D. Keller, Phys. Rev. Lett. **68**, 1939 (1992).
18. I.V. Lerner and Y.E. Lozovik, Zh. Eksp. Teor. Fiz. **80**, 1488 (1981) [Sov. Phys. JETP **53**, 763 (1981)].
19. A.B. Dzyubenko and Y.E. Lozovik, J. Phys. A: Math. Gen. **24**, 415 (1991), and references therein.
20. A.H. MacDonald and E.H. Rezayi, Phys. Rev. B **42**, 3224 (1990).
21. V.M. Apal'kov and E.I. Rashba, Pis'ma Zh. Eksp. Teor. Fiz. **55**, 38 (1992) [JETP Lett. **55**, 37 (1992)].
22. S.M. Girvin, A.H. MacDonald, and P.M. Platzman, Phys. Rev. B **33**, 2481 (1986).
23. B.I. Halperin, Helv. Phys. Acta **53**, 75 (1983).

24. E.I. Rashba and J.L. Birman, Solid State Commun. **84**, 99 (1992).
25. L.V. Butov and V.D. Kulakovskii, Pis'ma Zh. Eksp. Teor. Fiz. **53**, 444 (1991) [JETP Lett. **53**, 466 (1991)].
26. Y.A. Bychkov, S.V. Iordanskii, and G.M. Eliashberg, Pis'ma Zh. Eksp. Teor. Fiz. **33**, 152 (1981) [JETP Lett. **33**, 143 (1981)].
27. Y.A. Bychkov, S.V. Iordanskii, and G.M. Eliashberg, Poverkhnost' [in Russian], No. 10, p. 33 (1982).
28. F.D.M. Haldane, Phys. Rev. Lett. **51**, 605 (1983).
29. I.B. Levinson and E.I. Rashba, Usp. Fiz. Nauk **111**, 683 (1973) [Sov. Phys. Usp. **16**, 892 (1974)].
30. E.I. Rashba, in *Physics of Semiconductors*, edited by B.L.H. Wilson (The Inst. of Physics, Bristol, 1979), Vol. 43, p. 1325.
31. F.J. Heeger, S. Kivelson, J.R. Schrieffer, and W.-P. Su, Rev. Mod. Phys. **60**, 781 (1988).
32. L.P. Gor'kov and I.E. Dzyaloshinskii, Zh. Eksp. Teor. Fiz. **53**, 717 (1967) [Sov. Phys. JETP **26**, 449 (1968)].
33. Y.A. Bychkov and E.I. Rashba, Zh. Eksp. Teor. Fiz. **96**, 757 (1989) [Sov. Phys. JETP **69**, 430 (1989)].
34. V.M. Apal'kov and E.I. Rashba, Pis'ma Zh. Eksp. Teor. Fiz. **53**, 46 (1991) [JETP Lett. **53**, 49 (1991)].
35. V.M. Apal'kov and E.I. Rashba, unpublished.
36. F.C. Zhang, V.C. Vulovic, Y. Guo, and S. Das Sarma, Phys. Rev. B **32**, 6920 (1985).
37. E.H. Rezayi and F.D.M. Haldane, Phys. Rev. B **32**, 6924 (1985).
38. D. Yoshioka, Phys. Rev. B **29**, 6833 (1984).
39. J.B. Stark, W.H. Knox, D.C. Chemla, W. Schäfer, S. Schmitt-Rink, and C. Stafford, Phys. Rev. Lett. **65**, 3033 (1990).
40. M. Potemskii, J.C. Maan, K. Ploog, and G. Weimann, Solid State Commun. **75**, 185 (1990).
41. L.V. Butov, V.D. Kulakovskii, A. Forchell, and D. Grutzmacher, Pis'ma Zh. Eksp. Teor. Fiz. **52**, 759 (1990) [JETP Lett. **52**, 121 (1990)].
42. C. Stafford, S. Schmitt-Rink, and W. Schäfer, Phys. Rev. **41**, 10000 (1990).
43. Y.A. Bychkov and E.I. Rashba, Pis'ma Zh. Eksp. Teor. Fiz. **52**, 1209 (1990) [JETP Lett. **52**, 624 (1990)]; Phys. Rev. B **44**, 6212 (1991).
44. L.V. Butov, V.D. Kulakovskii, and E.I. Rashba, Pis'ma Zh. Eksp. Teor. Fiz. **53**, 104 (1991) [JETP Lett. **53**, 109 (1991)].

PHOTOLUMINESCENCE IN THE FRACTIONAL QUANTUM HALL EFFECT

E.H. REZAYI
Department of Physics and Astronomy
California State University, Los Angeles
Los Angeles, CA 90032, USA

ABSTRACT. The fractional quantum Hall effect results from a correlated ground state of interacting electrons confined to a layer and subject to a strong transverse magnetic field. It exhibits dissipationaless conduction $\rho_{xx} \approx 0$ and quantization of the Hall resistance $R_H = h\nu/e^2$, with ν—the Landau level filling number—a rational fraction. This paper will address the magneto-optics of quantum Hall states. For ideal system of coplanar electrons and holes the photoluminescence (PL) spectrum occurs at the isolated magnetoexciton binding energy, independent of the underlying electron density and therefore does not reflect any of the features associated with the fractional quantum Hall effect. If, however, the electron and hole layers are far enough apart the PL spectrum coincides with the single particle spectral (tunneling) density of states $\rho(\omega)$. For quantum Hall states at primary fillings $\nu = 1/m$, $\rho(\omega)$ exhibits a single characteristic peak. In the presence of quasi-particle excitations the PL spectrum develops an additional peak with intensity proportional to the number of quasi-particles and shifted in energy from the main peak by the quasi-particle quasi-hole gap.

1. Introduction

The fractional quantum Hall effect [1] (FQHE) occurs in a clean two-dimensional electron gas in a perpendicular magnetic field. For sufficiently strong field values the Landau level degeneracies become comparable to the number of electrons and only the lowest Landau levels become occupied. In this limit, no useful perturbative expansion parameter exists and the system is intrinsically strongly correlated. The striking features seen in magneto-transport coefficients result from such a highly correlated ground state of the electrons. Laughlin's simple wavefunctions [2] for the incompressible fluid ground state and its' fractionally charged excitations provide a very accurate microscopic description of the fractional quantum Hall effect. The transport measurements have provided much valuable information on the charged excitations, but given the experimental difficulties of direct measurements of other properties of the incompressible fluid any other alternate experimental approach is highly desirable. Recent, magneto-optical measurements [3] have added a new dimension in probing this intriguing phenomenon. At present considerable confusion remains in the interpretation of these experiments. In previous publications [4,5] it has been shown that for coplanar systems of electrons and holes with symmetrical interactions the photoluminescence (PL) spectrum is completely void of structure and does not reflect any of the properties of the FQHE states. In contrast if electron-hole interactions are much weaker, than the electron-electron or hole-hole interaction the PL spectrum is related to the one-particle spectral (tunneling) density of states and should therefore contain important information on the correlated ground state of the quantum Hall effect.

2. Electrons and Holes and the Pseudo-Spin Formalism

I consider an idealized model of fully polarized electrons and holes confined to their lowest Landau level. The electrons and holes are assumed to be equivalent in other respects and in particular to have the same band envelope wavefunctions. There are many geometries that have been used to study FQHE. Haldane's [6] spherical geometry is probably the most popular one and is especially suitable for the calculation and interpretation of the PL spectrum. In what follows the relevant details of this geometry for the electron hole system are briefly discussed. In this paper, all energies are given in units of $e^2/4\pi\varepsilon_0\lambda$, where λ is the magnetic length.

In the lowest Landau level the single-particle wavefunctions of electrons and holes are given by:

$$\psi_m^e(\theta,\phi) = \sqrt{\frac{(2S+1)!}{4\pi}} \frac{u^{S+m}v^{S-m}}{\sqrt{(S+m)!(S-m)!}} \tag{1}$$

$$\psi_m^h(\theta,\phi) = \sqrt{\frac{(2S+1)!}{4\pi}} (-1)^{S+m} \frac{(u^*)^{S-m}(v^*)^{S+m}}{\sqrt{(S-m)!(S+m)!}}, \tag{2}$$

where $2S = N_\phi$ is the total flux in units of the flux quantum through the sphere and $(u,v) = (\cos\theta e^{\phi/2}, \sin\theta e^{-\phi/2})$ are the spinor coordinates. The Hamiltonian for the electron hole system is:

$$H_{e-h} = 1/2 \sum_{\{m\}} (V_{m1,m2,m3,m4}^{e-e} a_{m1}^\dagger a_{m2}^\dagger a_{m3} a_{m4} \\ + V_{m1,m2,m3,m4}^{h-h} b_{m1}^\dagger b_{m2}^\dagger b_{m3} b_{m4}) \tag{3}$$

$$+ \sum_{\{m\}} (-1)^{2S+m_2-m_3} V_{m1,m2,m3,m4}^{e-h} a_{m1}^\dagger b_{m2}^\dagger b_{m3} a_{m4}, \tag{4}$$

where a,b are the occupation space annihilation electron and hole operators respectively, and $V_{\{m\}}$ are the usual matrix elements obtained through the single-particle wave-functions for the appropriate interaction potential. The symmetric limit is defined as

$$V^{e-e}(\vec{r}) = V^{h-h}(\vec{r}) = -V^{e-h}(\vec{r}). \tag{5}$$

This will be realized if the electrons and holes are coplanar. It is unlikely that the symmetric limit can be realized experimentally. In general, the electron and hole layers are spatially separated so that $V^{e-e}(\vec{r}) \approx V^{h-h}(\vec{r}) > |V^{e-h}(\vec{r})|$. From numerical calculations results similar to those of the symmetric limit are found whenever the distance between the layers is less than a critical value. It will be seen shortly that for strongly correlated (i.e., near symmetric) electron-hole systems PL spectra is a single delta peak formed at the binding energy of an isolated exciton. This surprising result follows from the inability of the tightly bound exciton and the incompressible quantum Hall state to polarize one another. It is in fact a consequence of a hidden (pseduo-spin) symmetry present in the symmetric model [7]. To bring it out we make a canonical transformation to pseudo-spin variables $C_{m,\uparrow,\downarrow}$:

$$C_{m\uparrow} = a_m \tag{6}$$

$$C^\dagger_{m\downarrow} = (-1)^{S-m} b_{-m}. \tag{7}$$

It can be shown through these relations that the Hamiltonian in the two schemes are related by:

$$H^{e-h} = H^{sp} - E_L + N_h I \tag{8}$$

$$I = 2\frac{E_L}{N_\phi + 1}, \tag{9}$$

where E_L is the energy of the filled level which can be obtained in closed form for any size system [8]:

$$E_L = -\frac{2^{2N_\phi - 1}\{(N_\phi + 1)!\}^2}{\sqrt{\frac{N_\phi}{2}}(2N_\phi + 1)!} \tag{10}$$

and I is the binding energy of an isolated exciton. Furthermore, the azimuthal component of the total spin is related to the total number of electrons N_e and number of holes N_h:

$$S_z = \frac{N_\phi + 1 - N_e - N_h}{2}. \tag{11}$$

$$\tag{12}$$

It follows from these relations that if an arbitrary number of holes is allowed then the extreme values of the total spin are:

$$S_{max} = \frac{N_\phi + 1 - N_e}{2} \tag{13}$$

$$S_{min} = \frac{|N_\phi + 1 - N_e - 2N_h|}{2}. \tag{14}$$

This shows that each exciton pair corresponds to overturning one spin. While this mapping would translate into a model with spin-dependent forces for the non-symmetric case, it is the symmetric limit where H commutes with both total S and S_z that interesting results can be obtained immediately. It is important to distinguish two cases from the outset:

1. The ground state in the spin picture is fully polarized—the case for primary fillings $\nu = 1/m$. Then, because of pseudo-spin symmetry overturning spins, i.e., adding particle hole pairs merely adds the constant IN_h to the ground state energy of the electron-hole system, we therefore conclude:
    (a) The ground state of this system is a FQHE effect of the excess electrons and a non-interacting boson condensate of tightly bound exciton pairs. In addition, there is no residual interaction between the boson and the quantum Hall condensates.
    (b) The low lying excitations fall into branches labeled by the number of excitons (spin-wave number in the pseduo-spin language) and can be characterized by a pseduo-momentum **k**.
2. The pseduo-spin ground state is either singlet or partially polarized. This is the case for hierarchy [9] fillings $\nu = 2/5, 2/7, 4/9, 4/11,...$, then the picture of noninteracting condensates breaks down. These cases may in fact prove to be more interesting to study, particularly for partially polarized ground state where beyond a critical hole number the non-interacting behavior is recovered.

3. Photoluminescence

Turning now to the PL spectrum we write it as:

$$P(\omega) = \frac{A}{Z\hbar} \sum_{n,m} \exp(-E_i^n/k_BT) \left|\langle \psi_f^m | \hat{L} | \psi_i^n \rangle\right|^2 \delta\left(\hbar\omega - (E_i^n - E_f^m)\right), \qquad (15)$$

where i(f) refer to the initial (final) states and Z is the partition function of the initial states (exciton system), A contains the dipole matrix element [10] and overlap of the envelope functions for electrons and holes, and $\hat{L}$ is the luminescence operator:

$$\hat{L} = \int \hat{\psi}_e(\bar{\Omega}) \hat{\psi}_h(\bar{\Omega}) d^2\Omega. \qquad (16)$$

$\hat{\psi}(\bar{\Omega})$ is the appropriate field operator projected down onto the lowest Landau level. Using the single particle states we obtain

$$\hat{L} = \sum_m (-1)^{S-m} a_m b_{-m}. \qquad (17)$$

It can be recognized that $\hat{L}$ is a rotationally invariant (orbital scalar) operator. In the pseudo-spin language it corresponds to the total spin lowering operator S_-.

3.1. STRONG COUPLING

It can now be seen that the PL spectrum is strongly dependent on the geometry of the system which in the present model is controlled by the distance of the hole from the electron layer. For coplanar system the pseudo-spin symmetry is present and the initial and the final states are labeled by quantum numbers S and S_z. If the hole concentration is sufficiently dilute, which appears to be the experimental situation, then we need to retain a single hole in the initial state. In this case only two values of S and S_z are allowed:

$$S = S_{max}, \quad S = S_{max}-1 \tag{18}$$

and

$$S_z = -S_{max}, \quad S_z = S_{max}+1. \tag{19}$$

For primary fillings where the ground state is spin-polarized we immediately conclude that the only final states contributing to the PL spectra are those that are related to the initial state by the spin lowering operator:

$$|\psi_f\rangle = S_-|\psi_i\rangle \tag{20}$$

it then follows that for T = 0

$$P(\omega) = \frac{A}{\hbar} \langle \psi_i^0 | S^+ S^- | \psi_i^0 \rangle = \frac{A}{\hbar}(S_{max}(S_{max}+1) - S_z(S_z-1))\delta(\omega+I), \tag{21}$$

where $I = \sqrt{\pi/2}$ is the free exciton binding energy. Thus, for primary states, $P(\omega) = A(N_\phi + 1 - N_e)/\hbar \delta(\omega + I)$. On the other hand for hierarchy states where the ground state in the $S_z = -S_{max}+1$ manifold is not fully polarized a similar argument shows that $P(\omega) = 0$. Thus, in the symmetric limit there is no spectroscopic information on the correlated ground state contained in PL.

3.2. WEAK COUPLING

In contrast if the coupling between electrons and holes is negligible which appears to be the case if the layer separation is larger than a few magnetic lengths, then $P(\omega)$ is related to the occupied weight of the single particle density of states ($B(\omega)$ part of the spectral function A+B [11]):

$$P(\omega) = \frac{A}{Z_{N_e} \hbar (N_\phi + 1)}$$
$$\times \sum_{i,j,m} \exp(-E_i(N_e)/k_B T) |\langle \psi_j(N_e-1)|a_m|\psi_i(N_e)\rangle|^2 \delta(\hbar\omega - (E_i - E_j)), \tag{22}$$

where Z_{N_e} is the partition function of the electronic system. In this limit it follows from Eq. (22) that the total integrated PL intensity is

$$\int d\omega P(\omega) = \frac{AN_e}{N_\phi + 1}, \tag{23}$$

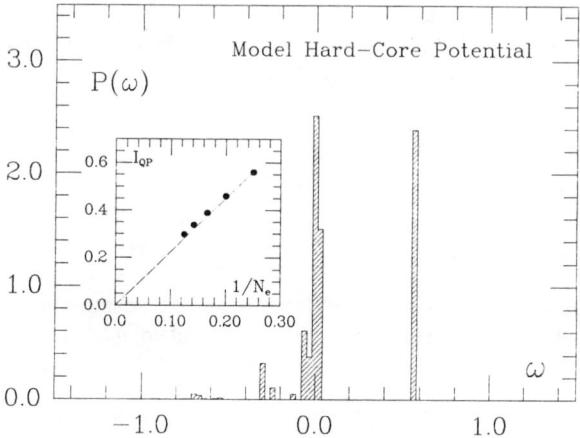

Figure 1. The spectral density for N = 8 and ν = 1/3 + a quasi-particle for the model hard core potential. The shift between two peaks is the quasi-particle quasi-hole excitation gap. The inset is the weight in the secondary peak per electron plotted versus $1/N_e$. Reproduced from Ref. [5].

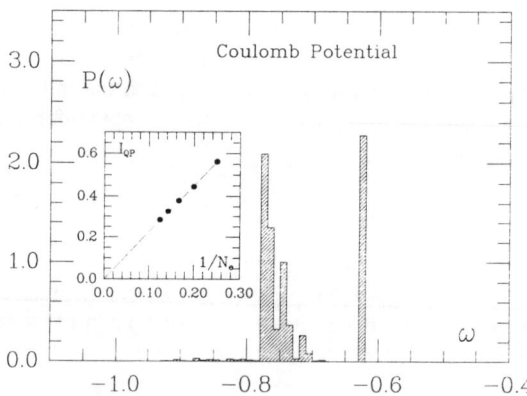

Figure 2. Same as Fig. 1 except for the Coulomb potential. Reproduced from Ref. [5].

independent of electron-electron interactions. The right hand side of Eq. (22) can be related to the imaginary part of the electronic Greens function

$$P(\omega) = \frac{A}{(N_\phi + 1)\pi} \theta(\mu/\hbar - \omega) \sum_k \mathrm{Im}\, G_{k,k}(\omega), \tag{24}$$

For the ν = 1/m sequence ρ(ω) is very strongly peaked at μ_- the chemical potential for removing an electron. In fact for the Hard-core model potential for which Laughlin's wavefunction is an exact ground state, ρ(ω) for any size is given exactly by [12]

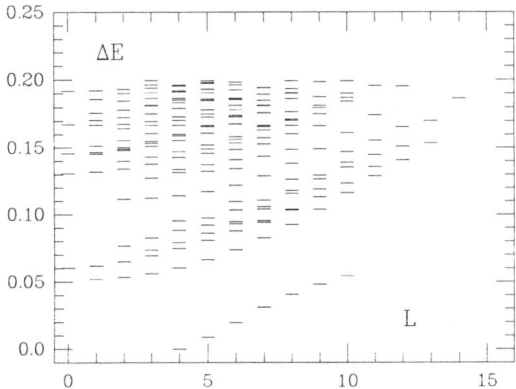

Figure 3. The low lying energy levels for six electrons plus a single hole at a distance of three magnetic lengths from the electron layer. The system is at flux $N_\phi = 14$. The series of low lying levels seen rising to the right result from the combination of two objects with angular momentum L = 7 (hole) and L = 3 (quasi-hole) which yields $4 \leq L \leq 10$. I interpret this as a fractionally charged (2e/3) object with internal dynamics.

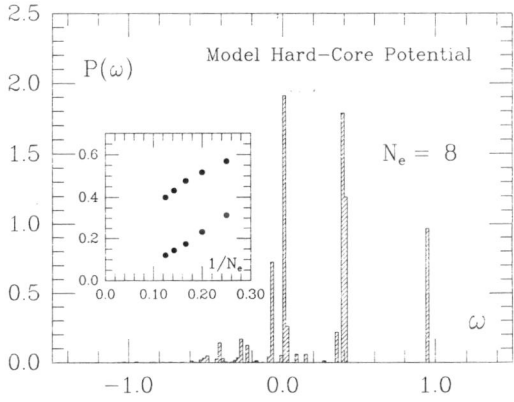

Figure 4. The spectral density for N = 8 at $\nu = 1/3$ + two quasi-particles. In addition to the previous two peaks a third high energy peak corresponding to direct annihilation of two quasi-particles by the hole is found. This process could become important if the quasi-particles can be found in sufficient numbers in the initial state.

$$\rho(\omega) = \frac{N}{N_\phi + 1} \delta(\omega) \tag{25}$$

and μ_- does not appear since it is identically zero for this model. For the more realistic Coulomb potential the peak is broadened somehow but it still remains rather sharp [12].

An important issue in interpreting PL experiments is whether the fractionally charged excitations of the incompressible ground state can produce any observable effects. These excitations are local defects in the density that heal within a distance of a few magnetic lengths. Quasi-particles (QP) are produced for $\nu > 1/3$ and quasi-holes (QH) for $\nu < 1/3$. If the number of these excitations is sufficiently small one may regard them as non-interacting and the total energy of a gas of weakly interacting defects will be

$$E = E_0 + N^+\varepsilon^+ + N^-\varepsilon^-, \tag{26}$$

where $\varepsilon^+, \varepsilon^-$ are the gap for creating the corresponding excitation. These are related to the chemical potentials $\mu^\pm$ by

$$\mu^\pm = \pm 3\varepsilon^\pm. \tag{27}$$

Thus in terms of $\Delta = \varepsilon^+ + \varepsilon^-$ the chemical potential discontinuity is

$$\Delta\mu = \mu^+ - \mu^- = 3\Delta. \tag{28}$$

Past discussion [13] of the PL predict a jump equal to $\Delta\mu$ in the main peak. Our calculations point to a different picture. We find that in the presence of quasi-particles interesting structure develops in the PL spectrum. In addition to the main peak corresponding to the process $\delta N^- = 0$, $\delta N^+ = 3$ (i.e., removing an electron from the condensate and leaving the initial quasi-particle intact, the left peak in Figs. 1 and 2), we find a second peak corresponding to $\delta N^- = -1$, $\delta N^+ = 2$ process which is a direct annihilation of the QP by the hole leaving behind two quasi-holes. Indeed for small sizes the latter process is strongly favored for finite layer separation. This can be seen from low lying spectrum for $d = 3\lambda$ (Fig. 3). There appears to be a well developed bound state of the hole and the fractionally charged quasi-particle. This suggests that one may regard the hole and the quasi-particle as a single entity. The count of levels and the value of their total angular momentum is consistent with such a composite object. The smallest possible angular momentum in the set is also consistent with the cyclotron degeneracy of a charge $2e/3$ object. The two peaks in Figs. 1 and 2 are separated by the quasi-particle quasi-hole gap Δ:

$$\varepsilon^+ + E_0(N) - 2\varepsilon^- - E_0(N-1) = \varepsilon^+ - \mu^- - 2\varepsilon^- \tag{29}$$

$$= \varepsilon^+ + \varepsilon^- = \Delta. \tag{30}$$

Figs. 1 and 2 show the PL spectrum for eight electrons for both the Coulomb and the hard core model potentials. The inset is the scaling of the QP peak per electron versus $1/N_e$. As expected this weight scales to zero almost linearly with $1/N_e$. We are thus led to an expression of the form:

$$P(\omega) = \frac{AN_e}{N_\phi + 1}\delta(\omega - \mu^-) + \frac{\gamma AN^+}{N_\phi + 1}\left[\delta(\omega - \mu^- - \Delta) - \delta(\omega - \mu^-)\right], \tag{31}$$

where γ may be considered to be a phenomenological constant whose value can be estimated from numerical calculations to be $\gamma \approx 2.1$ (see the insets to Figs. 1 and 2). The interpretation of

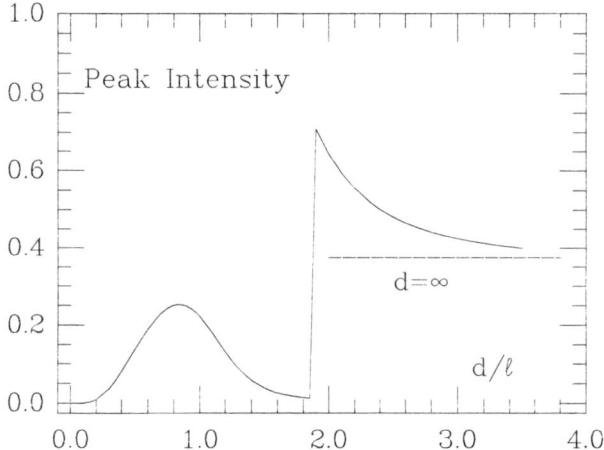

Figure 5. The PL main peak intensity for six electrons and one hole at flux $N_\phi = 16$ as a function of d. A rapid crossover from weak to strong coupling at $d_c = 1.85\lambda$ is clearly seen. The straight line is the non-interacting hole value.

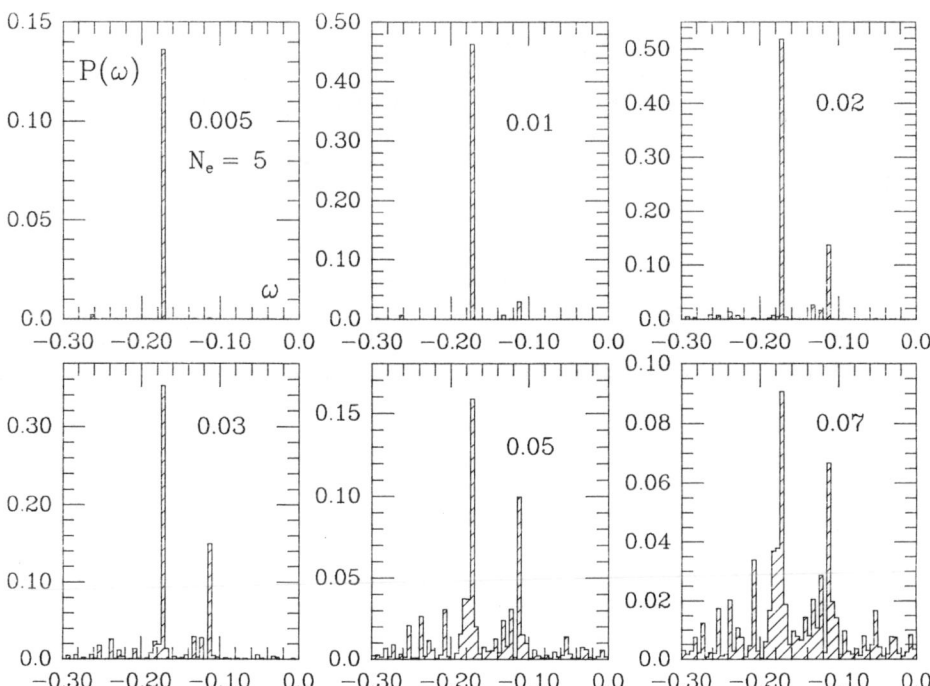

Figure 6. The PL spectrum for $N_e = 5$ and $N_h = 1$ at flux $N_\phi = 12$ and $d = 2.0\lambda$ for six different temperatures shown in units of $e^2/4\pi\varepsilon_0\lambda k_B$.

the secondary peak as direct annihilation of the quasi-particle with the hole can be confirmed in finite-size studies from the energies and angular momentum quantum numbers of the initial and final states contributing to this peak. This identification is particularly clear for the hard core model where the peak occurs exactly at the quasi-particle energy. The two quasi-hole final state belongs to a vanishing eigenvalue for this potential. That this process is allowed and survives the thermodynamic limit can be shown from a direct calculation [5] using the quasi-particle wavefunctions of Girvin and MacDonald [14]. The estimated weight from this calculation is also consistent [5] with $\gamma \approx 2.1$. It is also interesting to calculate the PL spectrum from an initial state having two quasi-particles. Fig. 4 shows $\rho(\omega)$ for the 2QP state. Here in addition to the two peaks of previous figures a third peak corresponding to the $\delta N^- = -2$ and $\delta N^+ = 1$ process appears. This results from direct annihilation of two quasi-particles by the hole leaving behind a single quasi-hole in the final states. Neglecting the repulsive energies of the quasi-particles in the initial state the shift in this peak should be approximately 2Δ. The value in the figure differs little from this.

3.3. FINITE TEMPERATURES

To investigate the crossover from weak to strong coupling behavior as the distance between the layers is varied it is convenient to choose a case where the PL intensity is zero in the symmetric limit (see earlier discussion of PL in terms of pseudo-spin quantum numbers). Fig. 5 shows the crossover to occur at $d_c = 1.85\lambda$ in a system with parameters $N_e = 6$, $N_h = 1$, and $N_\phi = 15$. The asymptote (dashed line) represents the zero coupling value. The crossover can be seen to be reasonably sharp with the weak coupling limit being attained quickly.

Figure 6 shows the PL spectrum for $N_e = 5$ and $d = 2.0\lambda$ for six temperatures where a quasi-particle is present in the initial state. The single peak seen for lower temperatures is the quasi-particle peak. It results from the tightly bound state of the hole with the quasi-particle. The tail to the left for higher temperature comes from the large angular momentum part of the levels in Fig. 3. The additional peak to the right is not easily identifiable in simple physical terms and becomes important for very high temperatures. Probably too high to be relevant to FQHE. The sharp condensate peak seen in the Green's functions which is absent here will of course be present in a thermodynamic system since there will always be excess holes or sufficient number of thermally dissociated quasi-excitons. The important observation here is to note that, in general, the QP peak will experience a finite (red) shift (relative to the $d = \infty$ gap) equal to the binding energy of the hole-quasi-particle system seen for finite d in Fig. 3.

4. Conclusion

The simple model considered here is unlikely to furnish a quantitative comparison with experiments. Nevertheless, the conclusion that FQHE features will only be seen in PL in the weak coupling limit is likely to remain valid. The picture presented here should provide a reliable framework for interpretation and design of future experiments.

Acknowledgments

Work supported by U.S. National Science Foundation under grant NSF-DMR-9113876. I would like to thank my collaborators A.H. MacDonald and D. Keller.

References

1. The history of the initial developments can be found in *The Quantum Hall Effect*, edited by R.E. Prange and S.M. Girvin (Springer-Verlag, New York, 1986).
2. R.B. Laughlin, Phys. Rev. Lett. **50**, 1395 (1983).
3. D. Heiman, B.B. Goldberg, A. Pinczuk, A.C. Gossard, and J.H. English, Phys. Rev. Lett. **61**, 605 (1988); B.B. Goldberg, D. Heiman, A. Pinczuk, L. Pfeiffer, and K. West, Phys. Rev. Lett. **65**, 641 (1990); H. Buhman, W. Joss, K. von Klitzing, I.V. Kukshin, G. Martinez, A.S. Plaut, K. Ploog, and V.B. Timofeev, Phys. Rev. Lett. **65**, 1056 (1990); A.J. Tuberfield, R.S. Haynes, P.A. Wright, R.A. Ford, R.G. Clark, J.F. Ryan, J.J. Harris, and C.T. Foxon, Phys. Rev. Lett. **65**, 637 (1990); H. Buhmann, W. Joss, K. von Klitzing, I.V. Kuksuhkin, A.S. Plaut, G. Martinez, K. Ploog, and V.B. Timofeev, Phys. Rev. Lett. **66**, 926 (1991). See also papers in these proceedings.
4. E.H. Rezayi and A.H. MacDonald, Bull. Am. Phys. Society, **36**, 915 (1991).
5. A.H. MacDonald, E.H. Rezayi, and David Keller, Phys. Rev. Lett. **68**, 1939 (1992).
6. F.D.M. Haldane, Phys. Rev. Lett. **51**, 605 (1983).
7. A.H. MacDonald and E.H. Rezayi, Phys. Rev. B **42**, 3224 (1990); see also I.V. Lerner and Yu. E. Lozovik, Zh. Eksp. Teor. Fiz. **80**, 1488 (1981) [Sov. Phys. - JETP **53**, 763 (1981)]; Y.A. Bychov and E.I. Rashba, Solid State Commun. **48**, 399 (1983); T.M. Rice, D. Paquet, and K. Ueda, Helv. Phys. Acta **58**, 410 (1985); D. Paquet, T.M. Rice, and K. Ueda, Phys. Rev. B **32**, 5208 (1985) and work quoted therein; E.I. Rashba, in these proceedings.
8. G. Fano and F. Ortolani Phys Rev. B **37**, 8179 (1988).
9. F.D.M. Haldane, Phys. Rev. Lett. **51**, 605 (1983); B.I. Halperin, Phys. Rev. Lett. **52**, 1583 (1984); R.B. Laughlin, Surf. Sci. **141**, 11 (1984).
10. For a recent review of the optical properties of semiconductor quantum wells see S. Schmitt-Rink, D.S. Chemla, and D.A.B. Miller, Advances in Physics **38**, 89 (1989).
11. A.L. Fetter and J.D. Walecka, *Quantum Theory of Many Particle Systems* (McGraw-Hill, New York, 1971).
12. E.H. Rezayi, Phys. Rev. B **35**, 3032 (1987).
13. I.V. Kikushkin and V.B. Timofeev, Pis'ma Zh. Eksp. Teor. Fiz. **44**, 179 (1986) [JETP Lett. **44**, 228 (1986)].
14. A.H. MacDonald and S.M. Girvin, Phys. Rev. B **34**, 5639 (1986); A.H. MacDonald and S.M. Girvin, Phys. Rev. B **33**, 4414 (1986).

EXCITON UNBINDING IN THE QUASI-TWO-DIMENSIONAL ELECTRON GAS

G.E.W. BAUER
Faculty of Applied Physics
Delft University of Technology
2628 CJ Delft, The Netherlands

ABSTRACT. The optical properties of the quasi-two-dimensional electron gas are investigated in the low density regime with special attention for the effects of a finite hole mass and bound excitonic states. A mean-field theory is employed which allows a consistent treatment of magnetic-field dependent screening in the integer quantum Hall regime. It is argued that higher-order scattering effects are difficult to identify by experiment. Recent luminescence experiments involving a higher subband can be explained as an equilibrium phenomenon in terms of an exciton unbinding transition proposed earlier.

1. Introduction

Optical spectroscopy of semiconductor quantum wells has become an established instrument to obtain information on the properties of the quasi-two-dimensional electron gas, which complements more conventional transport experiments [1]. Photocreated holes interact with the electrons, which tends to make optical techniques relatively invasive probes, however. This can be a nuisance if one is only interested in the electron gas itself. On the other hand the presence of the holes creates new and challenging problems even when the electron gas is allowed to be non-interacting. Mahan [2] discovered that a hole (positively charged particle) immersed into a metal can have a bound state which affects optical spectra. In his mean-field (ladder) approximation the bound state shows up as a δ-function singularity in the light absorption spectra. The optical properties are discussed here in what essentially comes down to Mahan's mean-field approximation. In Sec. 2 the phenomenon of the exciton unbinding transition will be introduced, which does not show up in the absorption, but in the luminescence spectra, which are sensitive to the ground state in the presence of the hole. Previous work by the author [3–5] is extended by explicitly including the presence of a second subband close to the Fermi energy. This situation occurs in one-sided modulation-doped heterojunctions and quantum wells in which a number of interesting excitonic effects have been observed [6–10], presumably due to the higher optical quality compared to symmetrical structures. From the theory of core hole spectra in metals it is known that the excitonic δ-function in optical spectra is asymmetrically smeared out into a power-law singularity when higher order processes (like diagrams with crossed Coulomb interaction lines) are included [11]. A number of papers has appeared [11–14] on the effect of higher order Coulombic corrections of optical spectra of the quasi-two-dimensional electron gas. In Sec. 3 the effects of these corrections

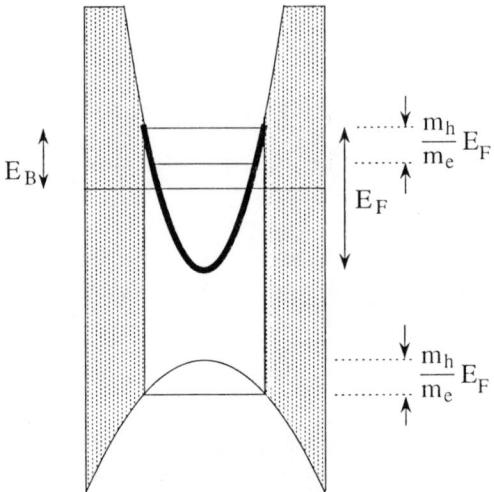

Figure 1. Schematical band picture of the Mahan exciton. The states extending from zero to the Fermi momentum are occupied by the electron gas and do not contribute to the excitonic bound state. If the exciton binding energy, as measured from the Fermi energy E_B, becomes smaller than $(m_h/m_e)E_F$ the kinetic energy of the hole becomes too large to support the bound state. The ground state in that case consists of a free hole at the top of the valence band.

are discussed, which are found to be similar to finite temperature effects, at least for the experimental conditions under which semiconductor spectra are obtained. The conclusions are discussed in Sec. 4.

2. Exciton Unbinding Transition

Let us first consider the nature of the ground state in a single-determinant approximation and at zero temperature (see Fig. 1) [3]. In two dimensions a positively charged hole will create a bound state at all densities. If the hole mass is infinite this bound state is always occupied by an electron, which means that the optical spectra should always display excitonic singularities. In the case of a finite hole mass the energy of the optically active bound state is increased because the low-momentum, low-energy pair states are excluded by the Pauli principle. Above a certain critical density ρ_c ($\approx 10^{11}$ cm^{-2} for symmetric quantum wells) a different, orthogonal state becomes energetically favorable, in which the hole is in a plane wave state at the top of the valence band. The bound state still exists, but is not occupied anymore. In luminescence the excitonic singularities are therefore absent, but they remain observable at the absorption edge. The criterion for an excitonic ground state is $|E_B| > \hbar^2 k_F^2/2m_h$, which means that the exciton binding energy E_B as measured from the absorption edge should be larger than the energy cost to excite a hole from the top of the valence band to the Fermi wave vector [3]. As pointed out in Refs. [4] and [5], this argument is easily generalized to include a second subband which is located close to the Fermi energy. For the sake of simplicity the excitons belonging to the first and second subband are assumed to be non-interacting for the moment (by the orthogonality of the subbands the form factor of the intersubband excitonic interaction vanishes at zero wave vector). In this case three wave functions compete for being the ground state, namely the free hole state and the exciton states in the first and second subbands. The exciton state associated with the second subband can be occupied in the ground state even when the second subband itself is not occupied. The reason is the higher exciton binding energy for the second subband which is caused by the larger electron-hole overlap in asymmetric structures but also by the absence (or smaller density) of electrons in the second subband which could block the states available for exciton formation.

In order to support the claim [4,5] that the results of Turberfield et al. [7,10] are caused by the exciton unbinding transition described above, a parametric study has been carried out. The results do not provide definite proof of the proposed mechanism, since the detailed electronic structure and the hole wave functions are not known, but the agreement with experiments supports the mechanism proposed here.

The details of the electron and hole structure show up in the form factors of the Coulomb interaction. The first subband is well approximated by a Gaussian wave function, for which the form factor can be evaluated analytically. The static dielectric function can therefore be calculated relatively reliably. To take into account the separation of electrons and holes, the form factors for the interaction of pair states with electrons in the first and second subband, as well as the mixing between the first and second subband, are simply approximated by multiplying the electron-electron interaction form factor f_e by constant factors f_{1111}, f_{1212}, and f_{1112}. The second subband is assumed to be located ΔE above the Fermi energy. The self energies are calculated in the quasi-static approximation, taking magnetic-field dependent screening into account [15,16]. Details of the computational method are given in Ref. [5].

In Fig. 2 the magnetoabsorption and magnetoluminescence spectra are plotted for an electron density of 10^{11} cm^{-2}, which corresponds to the sample of Ref. [7]. The Coulomb interaction between the holes and the first electron subband is chosen as $f_{1111} = 0.5$ and $\Delta E = 1.5$ meV. The subband spacing is assumed independent of magnetic field, i.e., the difference in the magnetic field dependence of the self-energy of the two subbands is disregarded. In order to avoid occupation of the second subband ΔE is chosen 0.3 meV larger for filling factor $\nu = 2$ only. This roughly models the effect of self-consistency of the Hartree problem: an occupation of the second subband is prevented by the shallowness of the Coulomb potential which would be accompanied by it. Because of the smaller electron-hole interaction in the lowest energy subbands, the Mahan excitonic peak obtained in Refs. [3]-[5] is replaced by the exciton belonging to the second subband in Fig. 2. We see that there are small cusps due to the self-energies [15,16], but the most important feature is the unbinding of the exciton when the Landau levels are only partially filled, which is caused by the strong intra-Landau-level screening. In the free ground state the luminescence originates from the band gap. The anticrossings in the absorption spectra are caused by the interexciton interaction, which does not cause any interesting effects here. These results show that the exciton unbinding mechanism is a very plausible one for the experimental results of Turberfield et al. [7] in the integer quantum Hall regime and it is concluded that non-equilibrium processes [17] are not necessarily involved. Only a few (non-)equilibrium electrons populate the second subband, which cause a weak luminescence also at non-integer occupation numbers.

Very recently the Oxford group has carried out experiments on a very low density

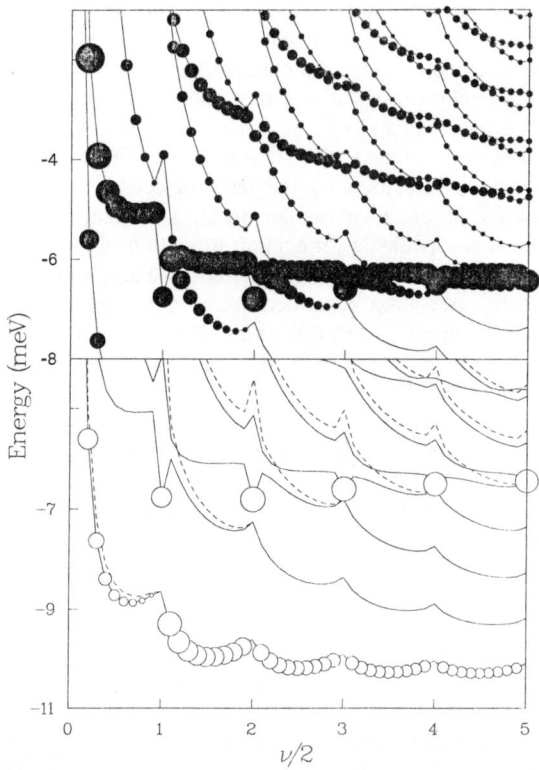

Figure 2. Absorption (upper part) and luminescence spectra (lower part) of the two-dimensional electron gas with an unoccupied second subband close to the Fermi energy as a function of the Landau level filling factor ν. The density is 10^{11}cm^{-2}, T = 100 mK, and the screening is modulated using an impurity broadening (at zero magnetic field) of $\Gamma = 0.1$ meV. The second subband is located $E_F + 1.5$ meV above the first subband at zero magnetic field. This distance is assumed to be magnetic field independent, except for $\nu = 2$, where the second subband would become occupied, where $E_F + 1.8$ meV has been adopted instead. The Coulomb interactions are modeled by form factors $f_{1212} = 2f_{1111} = 10f_{1112} = 1$ times f_e, the form factor of a 100 Å quantum well. The dashed lines are the spectral line positions without excitonic effects.

(3×10^{10} cm^{-2}) sample [10]. The second subband plays a much less prominent role, and the luminescence from the first subband displays a clear exciton at the Fermi edge that is very similar to that in [18]. Considering the high mobility of the sample, a localization of the hole is improbable. The zero-magnetic field spectrum can be explained without hole localization by the present theory since the electron density is very likely below ρ_c, even though the electrons and the hole are spatially separated. The results of a model calculation for this density are displayed in Figs. 3 and 4. The magnetic field dependence of the screening should be unimportant at very low magnetic fields and has been disregarded here. A rather high temperature has to be assumed to reproduce the low energy tail of the

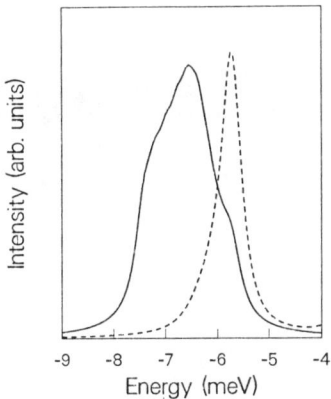

Figure 3. Absorption (dashed line) and luminescence spectra (full line) of the two-dimensional electron gas with an unoccupied second subband close to the Fermi energy, folded by a Lorentzian broadening function with a FWHM of 0.2 meV. The density is 3×10^{10} cm^{-2} and T = 2 K. The second subband is located E_F + 2.4 meV above the first subband. The form factors are $f_{1212} = 2.2 f_{1111} = 20 f_{1112} = 1$ times f_e, is the electron-electron interaction form factor of a 100 Å quantum well.

Mahan exciton as compared to the 100 mK quoted in Ref. [10]. Possibly this discrepancy is caused by higher order scattering processes discussed below, but self-energy effects like an enhanced hole mass or lifetime broadening cannot be excluded. Better agreement could have been obtained by a more complete search of parameter space. What is really called for is of course a first principles treatment of the electron and hole structure in this system.

The exciton unbinding transition can be observed only for finite hole masses at rather low densities. Still, the experiments by Chen et al. [9] have been explained similarly. The sample quality in this case is not as high than in Refs. [7] and [10] which causes partial localization of the holes. Luminescence can therefore be observed up to the Fermi energy [18]. The closeness of the second subband with larger oscillator strengths enhances the Mahan exciton, which is too weak to be observed otherwise. In contrast to the previous case, the intersubband exciton coupling is important. On the other hand, the impurity scattering appears to smear out the singular magnetic-field-dependent screening, which dominates the physics of the high-mobility samples.

The absence of excitonic effects in Si inversion layers [19] where the holes are localized to acceptors can be explained by stronger screening and weaker electron-hole interactions. In high-mobility heterojunctions and wide quantum wells where the density is higher than $\rho_c \leq 10^{11}$ cm^{-2} [20] the absence of excitons is also not surprising.

3. Mahan Exciton Versus Fermi-Edge Singularity

Ohtaka and Tanabe [11] assessed the reliability of the mean-field approximation by comparison with exact model calculations. They concluded that for an infinite hole mass the sum of ladder diagrams is a good approximation when the electron-hole interaction is sufficiently smaller than the Fermi energy and when the temperature is high enough to

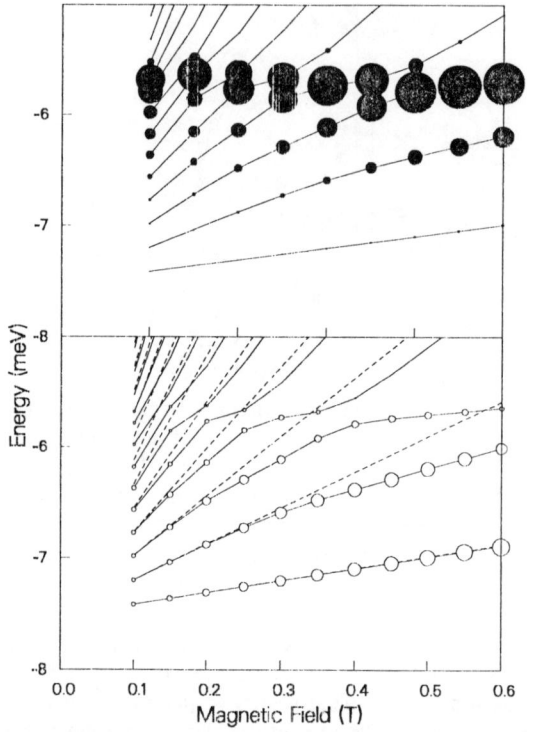

Figure 4. Absorption (upper part) and luminescence spectra (lower part) of the two-dimensional electron gas in Fig.3 at low magnetic fields. Because of the low magnetic fields involved, the magnetic field dependence of the screening is disregarded.

smear out the excitonic δ-function. In Ref. [4] mean-field spectra were found to resemble closely those calculated in Ref. [12] which included higher-order scattering effects for a hole with finite mass. Here it is argued [5] that for comparison with experiments mean-field theory should be sufficiently reliable and captures the essential physics.

The correction to the mean-field approximation can be roughly classified into two categories, namely the non-ladder diagram contributions to the vertex correection, and the excitonic corrections to the hole propagator, which are discussed in the following.

The higher order diagrams of the vertex correction reflect the effect of the sudden appearance (in absorption) or disappearance (in luminescence) of the hole, i.e., the dynamic response of the Fermi sea. Switching the hole potential non-adiabatically on or off "shakes up" the Fermi sea, which costs energy to be provided by the photons. In luminescence, we therefore obtain a low-energy tail extending from the Fermi energy to the band edge, while in absorption, oscillator strength is transferred to higher energies. In X-ray spectra the precise line shape is of considerable interest since, for example, the coefficients of the power-law singularities at the Fermi energy are experimentally accessible. In semiconductor spectra the situation is quite different, however. Firstly, the Fermi energies are much smaller. Thermal smearing causes effects which are difficult to disentangle from those caused by the non-ladder diagrams in the presence of additional broadening processes. In the case of the finite hole mass the importance of shake-up processes should be smaller, since the hole is allowed to recoil. The results obtained for an infinitely massive core hole referred to above are therefore a worst-case scenario for the mean-field approximation.

Higher-order processes also influence the single-particle propagator for the holes. These reflect the Coulomb mixing between the free state and the bound state. Since mean-field theory is not exact a finite Coulomb matrix element causes a small level repulsion between these states which has been estimated in Ref. [5]. A significant satellite in the hole spectral function is caused by excitonic corrections to the hole self energy in the density regime of the unbinding transition which are not strong enough, though, to invalidate the conclusions of mean-field theory.

Above ρ_c the excitonic state is degenerate with a continuum of finite momentum pair states. Above the unbinding transition this is an additional channel of decay for the bound state [21], which is suppressed in one-dimensional systems [22]. On the other hand, the integrated exciton oscillator strength is not diminished, and the problem is rather academic as long as the single-particle lifetime and thermal broadenings are larger than these Fano broadenings.

As long as the excitonic intersubband interaction is not large the processes discussed above have no effect on the exciton associated to the second subband.

4. Discussion

The magneto-optical properties of the uncorrelated quasi-two-dimensional electron gas can be understood within a mean-field formalism. The more spectacular effects in the integer quantum Hall regime can be semiquantitatively described as an equilibrium phenomenon. General good agreement between mean-field theory for neutral electron-hole plasmas [23] and experiment [24] has been also achieved for highly excited quantum wells.

Strong excitonic effects are also anticipated for quasi-one-dimensional systems (quantum wires) [25]. The Fano-resonance which causes a low energy tail in higher-dimensional systems has been shown to be suppressed [22], but still it is questionable if this effect is observable and/or account for the experiments of Calleja et al. [25]. Due to the higher exciton binding energy in these systems the unbinding transition is expected to occur at higher densities. Even when the holes are not localized by the impurity potential excitonic effects in quantum wires are expected to be much more robust than in two or three dimensional systems (see, however, the contribution by Rodriguez and Tejedor in these proceedings).

Is it possible to extrapolate the conclusions about the optical properties in the integer quantum Hall regime to those in the fractional quantum Hall regime? Phenomenologically, the spectra of, for example, the Oxford group [10] seem to be quite similar in both regimes. At fractional fillings, which correspond to a gap in the excitation spectrum, the higher energy feature is enhanced at the expense of the intensity of the lower energy peak in the luminescence. The competition between different ground states associated with the first and second subband has not been considered by previous theoretical approaches to the problem [26–28], but may be worthwhile investigating more closely. An explanation without non-equilibrium electrons requires that the lower state is not excitonic and the recombination process should leave the system in an excited state.

Acknowledgments

I would like to thank Paul Kelly for allowing me to use the computational facilities of the Philips Research Laboratories and Andrew Turberfield for a preprint of Ref. [10].

References

1. *Proceedings of the 9th Int. Conf. on the Electronic Properties of Two Dimensional Systems*, edited by M. Saitoh, Surf. Science **263** (1992).
2. G.D. Mahan, Phys. Rev. **153**, 882 (1967).
3. G.E.W. Bauer, Solid State Commun. **78**, 163 (1991).
4. G.E.W. Bauer, Surf. Science **263**, 482 (1992).
5. G.E.W. Bauer, Phys. Rev. B **45**, 9153 (1992).
6. W. Chen, M. Fritze, A.V. Nurmikko, D. Ackley, C. Colvard, and H. Lee, Phys. Rev. Lett. **64**, 2434 (1990).
7. A.J. Turberfield, S.R. Haynes, P.A. Wright, R.G. Clark, J.F. Ryan, J.J. Harris, and C.T. Foxon, Phys. Rev. Lett. **65**, 637 (1990).
8. S.R. Haynes, R.A. Ford, A.J. Turberfield, P.A. Wright, R.G. Clark, J.F. Ryan, J.J. Harris, and C.T. Foxon, Surf. Science **263**, 614 (1992).
9. W. Chen, M. Fritze, W. Walecki, A.V. Nurmikko, D. Ackley, J.M. Hong, and L.L. Chang, Phys. Rev. B **45**, 8464 (1992).
10. A.J. Turberfield, R.A. Ford, I.N. Harris, C.T. Foxon, and J.J. Harris, Physica Scripta T **45**, 164 (1992).
11. K. Ohtaka and Y. Tanabe, Rev. Mod. Phys. **62**, 929 (1990).
12. T. Uenoyama and L.J. Sham, Phys. Rev. Lett. **65**, 1048 (1990).
13. P. Hawrylak, Phys. Rev. B **44**, 3821 (1991).
14. P. Hawrylak, Phys. Rev. B **44**, 11236 (1991).
15. S. Katayama and T. Ando, Solid State Commun. **70**, 97 (1989).
16. T. Uenoyama and L.J. Sham, Phys. Rev. B **39**, 11044 (1989).
17. V.I. Kirchipev, I.V. Kukushkin, V.B. Timoveev, V.I. Falko, K. von Klitzing, and K. Ploog, P'isma Zh. Eksp. Teor. Fiz. **54**, 630 (1991) [JETP Lett. **54**, 636 (1991)].
18. M.S. Skolnick, J.M. Rorison, K.J. Nash, D.J. Mowbray, P.R. Tapster, S.J. Bass, and A.D. Pitt, Phys. Rev. Lett. **58**, 2130 (1987).
19. I.V. Kukushkin and V.B. Timoveev, P'isma Zh. Eksp. Teor. Fiz. **40**, 413 (1984) [JETP Lett. **40**, 1231 (1984)].
20. B.B. Goldberg, D. Heiman, A. Pinczuk, L. Pfeiffer, and K. West, Surf. Science **263**, 9 (1992).
21. A.E. Ruckenstein and S. Schmitt-Rink, Phys. Rev. B **35**, 7551 (1987).
22. T. Ogawa, A. Furusaki, and N. Nagaosa, Phys. Rev. Lett. **68**, 3638 (1992).
23. G.E.W. Bauer, in *Optics of Excitons in Confined Systems*, edited by A. D'Andrea, R. Del Sole, R. Girlanda, and A. Quattropani (IOP Publishing, Bristol, 1992) IOP Conf. Ser. 123, p. 283; Physica Scripta T **45**, 154 (1992).
24. L.V. Butov, V.D. Kulakovskii, A. Forchel, and D. Grützmacher, P'isma Zh. Eksp.

Teor. Fiz. **52**, 759 (1990) [JETP Lett. **52**, 121 (1990)].
25. J.M. Calleja, A.R. Goñi, B.S. Dennis, J.S. Weiner, A. Pinczuk, S. Schmitt-Rink, L.N. Pfeiffer, K.W. West, J.F. Mueller, and A.E. Ruckenstein, Solid State Commun. **79**, 911 (1991).
26. V.M. Apal'kov and E.I. Rashba, P'isma Zh. Eksp. Teor. Fiz. **55**, 38 (1991) [JETP Lett. **55**, 37 (1991)].
27. B.-S. Wang, J.L. Birman, and Z.-B. Su, Phys. Rev. Lett. **68**, 1605 (1992).
28. A.H. MacDonald, E.H. Rezayi, and D. Keller, Phys. Rev. Lett. **68**, 1939 (1992).

ULTRAFAST TIME-RESOLVED SPECTROSCOPY OF QUANTUM CONFINED SEMICONDUCTOR HETEROSTRUCTURES

D.S. CHEMLA
Physics Department, University of California at Berkeley
Material Sciences Division, Lawrence Berkeley Laboratory
Berkeley, CA 94720, USA

ABSTRACT. We review recent investigations of ultrafast dynamics of electronic excitations in quantum confined semiconductor heterostructures. We concentrate on two topics: the femtosecond time resolved investigations of the nonlinear optical response of semiconductor quantum wells in strong perpendicular magnetic field, and the investigations of many-body effects in the temporal lineshape of coherent emission from semiconductor quantum wells.

1. Introduction

The wealth of new physical properties and applications which has resulted from the confinement of electronic states in quasi-two dimensions (2D) in semiconductor quantum wells (QWs) [1] has triggered numerous attempts to make quasi-one-dimensional (1D) and quasi-zero-dimensional (0D) structures. These efforts have been hampered by the difficulty in obtaining defect free samples and in eliminating size fluctuations. However, by applying a large magnetic field, H, perpendicular to the QW it is possible to investigate the effects of confinement in quasi-0D on the electron-hole (e-h) states while avoiding the practical difficulty of sample preparation. The dimensionality of the QW-electronic states can be tuned continuously from quasi-2D, when the exciton (X) Bohr radius, a_0, is much smaller than the cyclotron radius, r_c, to quasi-0D, when $a_0 \gg r_c$, in materials with excellent quality and uniformity [2–6]. Further interest in magneto-exciton (MX) systems stems from the prediction of many-body theory that, at large H and thermodynamic equilibrium, an ensemble of spin polarized MXs behave like a gas of non-interacting and non-polarizable point-like Bosons [7,8].

In semiconductors, the coupling of the interband dipole, μ_k, to the electromagnetic field is renormalized by the Coulomb force [9]. In k-space the renormalized coupling is:

$$\mu_k E \rightarrow \mu_k E + \sum_{k'} V_{k,k'} \psi_{k'} \tag{1}$$

where $\mu_k E$ is the Rabi-frequency, $V_{k,k'}$ is the Coulomb potential and ψ_k is the k-component of the pair amplitude. As a consequence, polarization waves, $P = \sum_k \mu_k^* \langle \psi_k \rangle$, interact not only with E(t), but also with excitonic polarization waves [10–14]. In the case of excitation by ultrashort optical pulses, the temporal profile of the polarization waves is step-like. Hence, optical processes due to polarization waves interactions have a temporal dependence fundamentally different from that due to direct interactions with electromagnetic fields. As shown below, these processes give new information on Coulomb mediated many-body interactions.

The paper is organized as follows. In Section 2, we present femtosecond time resolved investigations of the evolution of the nonlinear optical response of GaAs QWs as their quasi-2D electronic states are further confined into quasi-0D by a strong magnetic field perpendicular to the QWs [15–18]. Section 3 present results of experiments in which this temporal-profile of the four wave mixing (FWM) signal obtained on resonance with quasi-2D X in GaAs QW has been resolved [19,20]. These experiments represent the first observations of many-body mediated polarization wave interactions and demonstrate that they can dominate the nonlinear response of semiconductors heterostructures.

2. Nonlinear Optical Response of Quasi-0D Magneto-excitons

When a magnetic field is applied perpendicular to a QW, the quasi-2D e-h pairs experience the total potential $V(r) = (\lambda r/2)^2 - 2/r]R_y$, which consists of the sum of the quadratic-potential imposed by H and the Coulomb potential. The parameter $\lambda = (a_0/r_c)^2 = \omega_c/2R_y$ measures the ratio of the magnetic and Coulomb confinement lengths and zero-point energies, (λ_c is the cyclotron frequency and R_y is the three-dimensional X-Rydberg). For $\lambda = 0$ the e-h pairs form usual quasi-2D Xs; for $\lambda \to \infty$ they tend toward e and h in pure Landau Levels (LL); and for the intermediate values, they form MXs.

The experimental linear absorption spectra of a high quality GaAs/AlGaAs-QW structure ($L_z = 84$ Å) at low temperature, T = 4 K, and in a perpendicular magnetic field, is shown in Fig. 1. The spectra were measured with σ_- polarized light as H was tuned from, $\lambda = 0 \to \sim 4$ (H = 0 $\to$ 12-T). The transition from 2D behavior at low fields to 0D behavior at high fields and the evolution of the LLs from the bound 2D-X states are nicely displayed. Figure 2 compares the 12-T linear absorption for σ_- (dashed line) and σ_+ (solid line). They are Zeeman-split in agreement with the QW selection rules at the Γ-point of the Brillouin zone. At such a high field the confinement is strong enough that the absorption strength is almost zero between the 1s and 2s MX, indicating that the MX are almost diagonal in the LL basis. These spectra are in good agreement with the theoretical calculations of Ref. [6]. It is clearly seen how the oscillator strength is now concentrated in the sharp MX-peaks which directly reflects the quasi-0D density of states (DOS).

The structure of the DOS restricts the number of available initial and final states for any transition, and therefore strongly influences relaxation. We have investigated this effect by exciting nonthermal carrier populations about 25 meV above the lowest 1shh-resonance at H = 0 T and H = 12 T at 4 K using the pump/probe technique with circular polarized light and a ~ 100 fs time resolution [15–17]. The most direct way of interpreting pump/probe experiments is to consider that the pump bean creates populations of MXs which angular momentum is determined by its polarization, and the probe measures the change of absorption induced by the presence of these populations.

MX populations affect the absorption through several mechanisms. Charge density effects (collisional broadening and Debye screening) are independent of the angular momentum, but other processes such as phase space filling (PSF), exchange (EXCH) and exciton-exciton interaction (XXI) depend critically on it [1]. It is possible, however, to separate the spin-

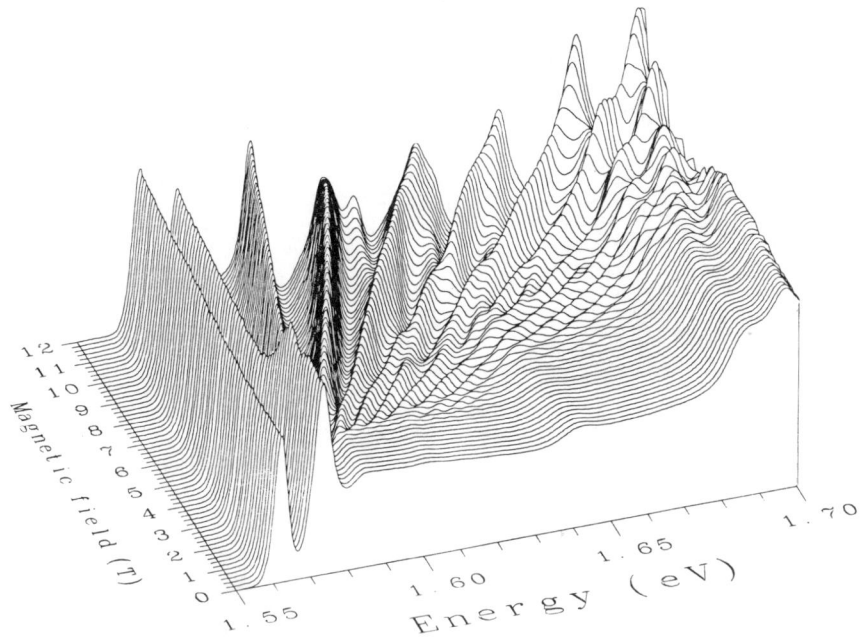

Figure 1. Linear absorption of a L_z = 84 Å GaAs/AlGaAs quantum well structure for σ_- polarized light at various magnetic fields H = 0 → 12 T.

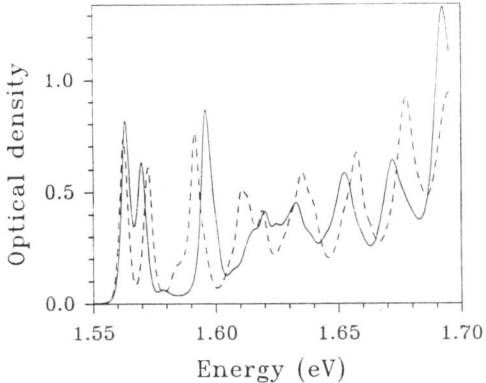

Figure 2. Comparison of the H = 12 T absorption spectra of the L_z = 84 Å GaAs/AlGaAs quantum well structure for σ_- (dashed line) and σ_+ (solid line) polarization.

dependent nonlinearities from the spin-independent ones by substracting the differential absorption spectra (DAS) measured by a σ±-probe when the sample is excited by a σ+-pump, $\Delta\alpha_\pm^+(\omega) \times L_z$, from that measured for a σ_-pump, $\Delta\alpha_\pm^-(\omega) \times L_z$ [18]. In Fig. 3 we compare the nonlinear differential circular dichroism (NDCD) spectra, $[\Delta\alpha_-^+(\omega) - \Delta\alpha_-^-(\omega)] \times L_z$, for two

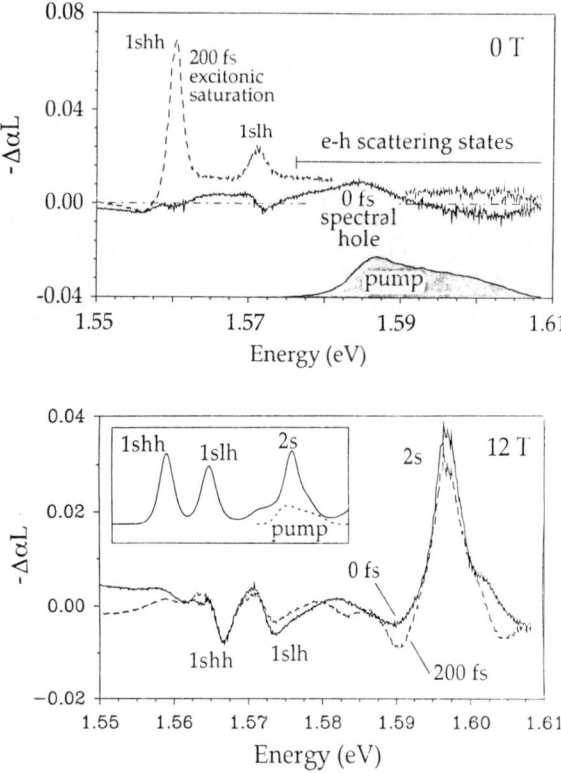

Figure 3. Circular dichroism spectra, for excitation 25 meV above the 1shh resonance and at Δt = 0 fs (solid line) and Δt = 200 fs (dashed line) for (a) H = 0 and (b) H = 12 T.

pump/probe delays, Δt = 0 fs and 200 fs, and for the two cases H = 0 (Fig. 3(a)) and H = 12 T (Fig. 3(b)). When H = 0 T one sees, in agreement with previous observations [21,22], an instantaneous, Δt = 0, spectral hole burning in the continuum at an energy slightly lower than the pump photon central energy and no response at the 1shh exciton. Hole burning in the continuum is the signature of the PSF induced by the nonthermal populations generated by the pump. At Δt = 200 fs the carriers have thermalized among themselves by carrier-carrier scattering and occupy the states at the bottom of the band, out of which the 1shh and 1slh excitons are made, blocking the transitions to these excitons. Hence the DAS reproduces the profile of these two resonances and the spectral hole has disappeared. We note that because the excess energy is smaller than the energy of optical phonons, the thermalization is internal to the plasma, i.e., the average energy remains constant as the e-h populations establish quickly a thermal distribution among themselves, at a temperature different from that of the lattice [22]. When the magnetic field is applied, the same pump excites the 2s-MX as shown in the inset of Fig. 3(b). There is immediately a strong PSF signal at the 2s-MX and a small signal due to dielectric screening at the 1shh-MX and 1slh-MX (note the different scales of the Fig. 3(a) and 3(b)). The important result is that the spectra at Δt = 0 fs and Δt = 200 fs are almost identical. The nonthermal populations are blocked 25 meV above the lowest states. They cannot emit

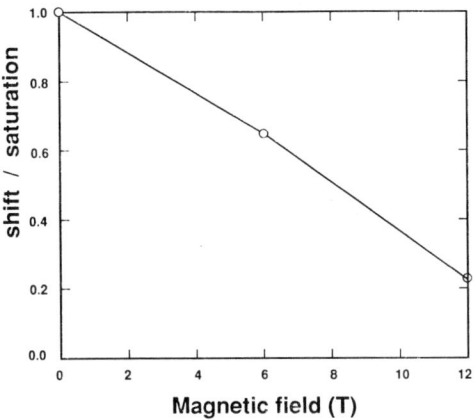

Figure 4. Evolution of the blue shift of the 1shh exciton normalized to the exciton density versus magnetic field H.

optical phonons and the carrier-carrier scatterings are quenched because of the very limited number of final states available in the quasi-0D DOS. Even if the excess energy was larger than the optical phonon energy, transitions still would be restricted by the reduced DOS as shown in Ref. [23]. Therefore the confinement in quasi 0D results in a dramatic reduction of the thermalization rates [17]. This has, of course, very important implications for device applications.

In several theoretical articles [7,8,24,25], it was shown that in the extreme magnetic limit and for a symmetric e-h system, ($V_{ee} = V_{hh} = -V_{eh}$), the energy of ground state of an MX gas is just the sum of the energies of the individual MXs. This exact result implies that MX-MX interaction disappears at very large H. XXI manifests itself directly as a blue shift of the 1shh-X-peak induced by a population of 1shh-X. This shift can be interpreted as a hard core repulsion which measures the extra energy cost necessary to create an 1shh-X in the presence of other 1shh-Xs [26]. It has been clearly resoved at H = 0 during resonant pumping of 1shh-X or subsequent to the formation of 1shh-X after excitation of e-h pairs in the continuum [27,28]. At low and moderate X-densities the blue shift is proportional to the XXI repulsive potential and to the X-density [26,28]. Therefore the variation versus H of the blue shift normalized to the X-density directly measures the interaction potential. The experimental results shown in Fig. 4 and obtained for H = 0, 6, and 12 T, clearly demonstrate that as H increases the shift and hence the XXI tend to zero.

The evolution of the nonlinear response versus H can be explained intuitively, in a way that which gives the essence of the exact thermodynamic equilibrium many-body theory [7,8,23,24], as follows. For pure parabolic bands (neglecting the Coulomb interaction), the e and h wavefunctions depend only on r_c and are therefore mass independent. As H is increased the magnetic confinement starts to dominate the Coulomb interaction, and the e and h are forced into almost identical and overlapping wavefunctions. Hence the quasi-0D MX occupy much smaller volume at high field and, for the same density, show much less PSF. Furthermore, as the magnetic potential, $(\lambda r/2)^2$, increases it restricts more and more e's and h's on the top of one another making the MXs more rigid and locally neutral and therefore much less polarizable. For exactly symmetric e-h system, $V_{ee} = V_{hh} = -V_{eh}$, this results in a perfect cancellation of the

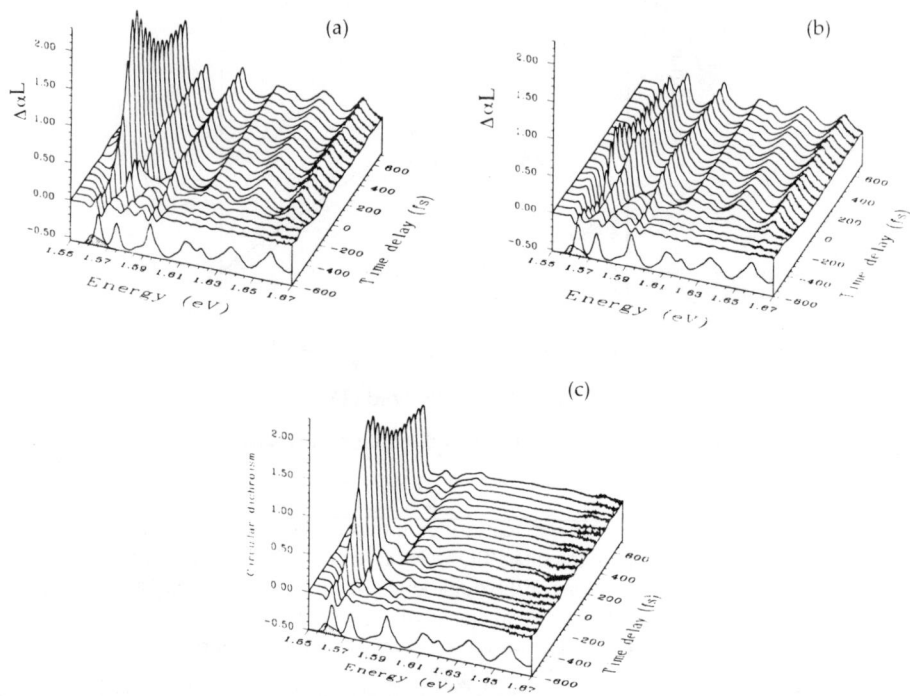

Figure 5. Differential absorption spectra, $\Delta\alpha L_z$, versus photon energy and time delay, Δt, for σ_--probe and (a) σ_--pump and (b) σ_+-pump. The nonlinear differential circular dichroism spectra (c) = (a) - (b), separates the spin dependent interaction between magneto-excitons.

MX-MX interaction. To the best of our knowledge our experiments represent the first confirmation of the remarkable, exact many-body theory result.

High density nonlinear effects are shown in Fig. 5(a) and (b), where we present the DAS, seen by a σ_--probe for excitation resonant with the lowest 1s-hh MX ($\omega \approx 1.56$ eV) with respectively σ_--pump and σ_+-pump, of the same intensity. Very strong responses are observed up to photon energies as high as $\omega = 1.67$ eV. As previously explained the spin-dependent nonlinearities can be separated from the spin-independent ones by subtracting the two sets of spectra. All the response above $\sigma \approx 1.59$ eV disappears in the NDCD spectra, Fig. 5(c), demonstrating that the high energy MXs (2s and above) are only sensitive to the charge density effects (collisional broadening and dielectric screening) induced by the resonantly-excited 1shh-σ_- or 1shh-σ_+ MXs. The same figure clearly shows that the 1s-MXs ($\omega < 1.59$ eV) on the contrary, are very sensitive to the spin of the MXs created by the pump. This behavior is detailed in Fig. 6, where we compare an expanded part of the lower portion of the absorption spectra seen by a σ_--probe at H = 0 T and H = 12 T, for Δt = -660, 0, and +660 fs after excitation by a σ_--pump and a σ_+-pump. The Δt = -660 fs the spectra are essentially that of the unexcited sample. For pump and probe both σ_- polarized, the 1sh-MX response is an instantaneous gain and the blue shift at Δt = 0, which evolves toward a strong saturation and a smaller blue shift at Δt = 660 fs. This is due to phase space filling (PSF) and exchange (EXCH)

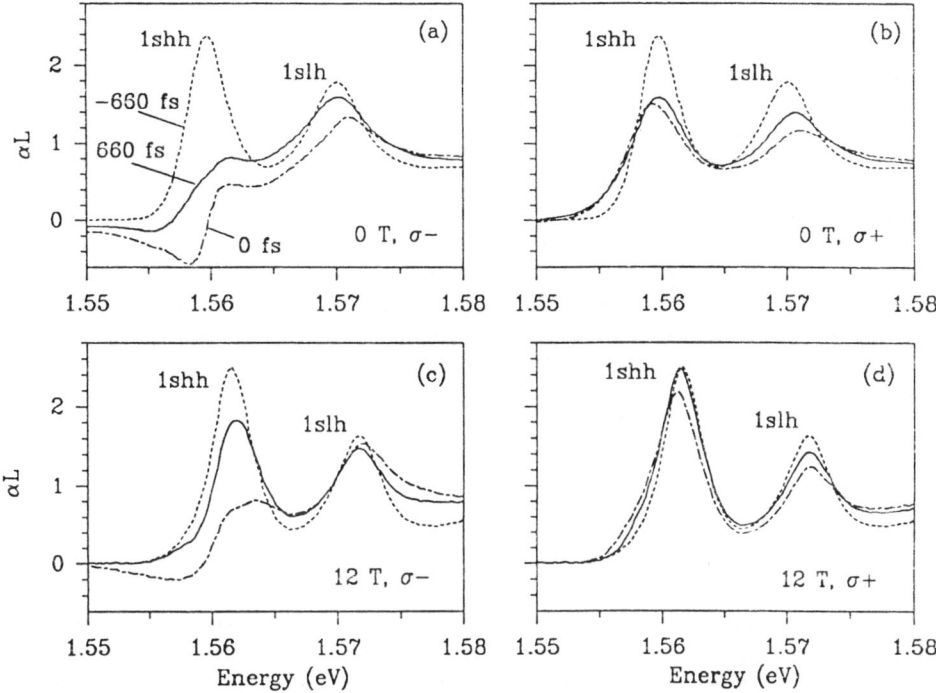

Figure 6. Detail of the absorption spectra seen by a σ_--probe near the 1s exciton resonances for Δt = -660, 0, and +660 fs and for: H = 0 T (a) σ_--pump and (b) σ_+-pump; H = 12 T, (c) σ_--pump and (d) σ_+-pump.

by the coherent ($\Delta t = 0$) and then the relaxed ($\Delta t = 660$ fs) 1shh-σ_--MXs created by the pump. The difference between the H = 0 T and the H = 12 T cases is only qualitative. The high field effects are similar to the low field ones, but significantly attenuated. For example consistent with the moderate density results discussed in the previous paragraph, the instantaneous blue shift changes from 1.9 meV → 0.3 meV. Since the 1shh and 1slh originate from distinct e and h states, the 1slh response is not due to PSF and EXCH produced by the real 1shh MXs. It comes from the pump induced virtual-populations of 1slh, i.e., AC-Stark Effect [1]. When the pump polarization is reversed to σ_+, the 1shh-σ_--MX exhibits only a small red shift and almost no saturation at H = 12 T, whereas at H = 0 it still saturates although much less than for σ_--pump. The 1slh-σ_--MX experiences a small but distinct saturation and blue shift with, however, the H = 12 T response attenuated as compared to that at H = 0 T.

Two sets of effects have to be distinguished: (i) the effects of the polarization of the MXs created by the pump (i.e., the difference between the DAS spectra for σ_- and σ_+ excitations), and (ii) the effects of magnetic confinement (i.e., the difference between the DAS spectra for H = 0 and 12 T). The polarization effects can be understood intuitively by considering [15,18]: (i) Pauli exclusion and the symmetry of the e and h states out of which the MXs are built, and (ii) the "molecular" exciton-exciton interaction potential which is attractive in the singlet state and repulsive in the triplet state. PSF and EXCH are strongly active when the pump and probe

MXs involve the same e and/or h states. The sign of the shift experienced by the probe MXs is derives from the attractive or repulsive character of the MX-MX interaction. The evolution of the nonlinear response versus H was discussed above. The combination of these two effects explains that: (i) The MX-MX interaction is reduced at high field and the shifts of the resonances, while keeping the same sign as for H = 0 T, decrease as H increases, and (ii) The inter-MX dielectric screening (similar to the band gap renormalization induced by e-h plasma), itself is strongly attenuated. This last remark explains why the significant saturation of the 1shh-σ_--MX seen at H = 0 T for σ_+-pump, which originates from this dielectric screening, almost disappears at at H = 12 T.

To go beyond these qualitative arguments a time dependent many-body theory of MX-nonlinearities was developed [16,18]. It follows the unrestricted Hartree-Fock theory, recently introduced by Schmitt-Rink et al. [9], which has been very successful in describing the near band gap nonlinear optical effects in semiconductors [29–33]. The Hartree-Fock theories are more appropriate in the case of MXs since they become exact in the extreme magnetic limit, $\lambda \rightarrow \infty$ [7,8]. In this approach, the MX wavefunctions are expanded on the LL orbitals, ϕ_ν. The optically connected conduction and valence band LLs form a set of two-level systems labeled by ν. The density matrix of the semiconductor breaks into 2×2 blocks;

$$\hat{n}_\nu(t) = \begin{bmatrix} n_{c,\nu}(t) & \psi_\nu(t) \\ \psi_\nu^*(t) & n_{v,\nu}(t) \end{bmatrix}, \tag{2}$$

where $n_{(c,v),\nu}(t)$ are the components of the populations on the conduction and the valence band LL and $\psi_\nu(t)$ are the components of the pair amplitude. The density matrix obeys the Liouville equation. The difference between this model and a collection of independent two-LL systems arises from the coupling between the LL by the Coulomb interaction, $V_{\nu,\nu'}$. This coupling affects the physics in two ways. Firstly, the conduction and valence band energies are renormalized by the excited populations in direct analogy with the mechanisms responsible for band gap renormalization,

$$\varepsilon_{j,\nu}(t) \rightarrow \varepsilon_{j,\nu} - \sum_{\nu'} V_{\nu,\nu'} \, n_{j,\nu}(t) \tag{3a}$$

(j = e or h). Secondly the coupling with the electromagnetic field, which is expressed by the Rabi frequency is modified according to:

$$\mu E(t) \rightarrow \Delta_\nu(t) = \mu E(t) + \sum_{\nu'} V_{\nu,\nu'} \psi_{\nu'}(t). \tag{3b}$$

This expresses the fact that the optically connected LLs at ν do not experience the applied field, $\mu E(t)$, rather they see the self-consistent "local field", $\Delta_\nu(t)$, which is the sum of the applied field and the "molecular" field, due to all the other e-h LLs [15,18]. This is the LL-representation of the renormalization which we have expressed in k-space by Eq. (1). The interaction between MXs appears in the Liouville equation as an "exchange" term; $\sum_{\nu'} V_{\nu,\nu'} [\psi_{\nu'} n_\nu - \psi_\nu n_{\nu'}]$. This term vanishes indentically on the diagonal, $\nu = \nu'$, giving the clue which explains the observed difference in the interaction between MXs in the same state and in different states at low and high magnetic field. For small fields, MX are composed of many LL and, therefore, there is a strong interaction between all the MXs whenever one MX-state is photoexcited. At high field the MXs become more and more diagonal in the LL basis, the interaction within a MX state vanishes and yet persists between different MX states. The residual interaction is mostly due to the self-energy corrections, which explains the high field disappearance of the 1s-MX blue-shift found experimentally. The attractive inter-MX self-energy correction explains the experimental red shift of the 2s-MX induced by a population of photoexcited 1s-MS. At high field MXs behave like two-LL systems and their nonlinear optical response become dominated by Pauli exclusion. Numerical solutions of the Liouville equation reproduce fairly well the experimental results [15,18], and in particular they show that indeed Coulomb correlation is completely quenched in the extreme magnetic limit.

3. Collective Efects in Free Induction Decay of Quasi 2D Excitons

In the Hartree-Fock description of semiconductor nonlinearities the density matrix and the Liouville equation it satisfies, are usually expressed in the plane wave basis of k-space [9,29–33]. In order to describe more directly the XXI effects seen for resonant excitation, it is convenient to first Fourier transform r-space and then to project onto the basis of solutions of the Wannier equation (i.e., to use as basis the umperturbed exciton states). It is found that the exciton amplitude, $\psi(t)$, satisfies the nonlinear Schrödinger equation [11,12]

$$\frac{\partial}{\partial t}\psi(t) = -i(\Omega - i\Gamma)\psi(t) + i\mu E(t) - i\mu E(t)\frac{|\psi(t)|^2}{\psi_s^2} - iV|\psi'(t)|^2\psi(t), \qquad (4)$$

which is valid up to third order for resonant excitation and assumes that the interband dipole element, μ, is k-independent. Ω and $\Gamma (= T_2^{-1})$ are, respectively, the exciton energy ($\hbar = 1$) and half-width at half maximum. The first line describes the linear response, while the two terms on the second line account the nonlinear response. The first nonlinear term originates from the Pauli-exclusion reduction of the exciton coupling with the electromagnetic field, $\mu E(t) \rightarrow [1 - (|\psi(t)|/\psi_s)^2]\mu E(t)$. The saturation parameter, ψ_s, is a material parameter that depends on the exciton characteristics [26]. This term is, of course, always present and accounts for the isolated-level-type nonlinearities [34,35]. The second nonlinear term is specific to dense media semiconductors and molecular crystals [11,12]. In semiconductors, it describes the Coulomb mediated exciton-exciton interaction which causes the interaction between polarization waves within the medium. Its physical interpretation is that, via the potential V, excitons at one site, $\psi(t)$, interact with excitons at other sites, $\psi'(t)$. Consistently with the previous paragraph, we call the two nonlinear terms the phase space filling (PSF) and the exciton-exciton interaction (XXI) term, respectively.

It is important, however, to note that in the mean field approximation, and for an unexcited system (E(t) → 0) in the steady state ($\partial/\partial t \rightarrow 0$), Eq. (4) has the form of a Ginzburg-Landau equation for an "order parameter" ψ. This similarity permits another interpretation to the XXI term: it describes how coherent polarization waves are driven in the medium and therefore, it has direct a analogy with the mechanism that drives the transition toward a superconducting state. Thus investigations of excitonic systems coherently driven by a ultrashort pulse, E(t), provided a unique opportunity to study the dynamics of a Ginzburg-Landau mechanism.

The simplest coherent wave-mixing configuration is that of two beam four-wave-mixing (FWM). In such a configuration two ultrashort lasers, labeled pulse-2 and pulse-1, separated by a time delay $\Delta t = t_2 - t_1$, and propagating in the directions k_2 and k_1, interfere in a sample to generate a transient grating, which diffracts photons into the background-free direction $k_S = 2k_2 - k_1$. In the case of homogeneously broadened two level system the FWM-signal is emitted immediately after the second pulse. It originates from the natural decay of the component of the nonlinear polarization P(t), which emits in the direction k_S, and corresponds to free induction decay (FID) [35]. For inhomogeneously broadened lines, the FWM-signal is delayed by Δt after the second pulse and corresponds to a "photon echo". The signal is usually measured as a function of Δt by using a slow detector to integrate the light emitted in the direction k_S. To observe the dynamics of the establishment of the polarization wave by XXI, however, it is necessary to measure the temporal lineshape of the FWM-signal itself.

Excitonic resonances in semiconductors are usually inhomogeneously broadened at low temperatures. In QW-structures, however, the quantum confinement in ultrathin layers, narrower than the bulk Bohr radius, stabilizes the quasi-2D excitons up to room temperature [1]. Collisions with the large population of thermal-phonons homogenize the resonances and shorten their dephasing time. We have investigated a sample consisting of 47 periods of 98 Å GaAs QWs and 96 Å $Al_{0.3}Ga_{0.7}As$ barrier layers. The output of a mode-locked Ti:Sapphire laser, delivering extremely stable ~ 70–100 fs transform-limited Gaussian pulses at 88 MHz, was tuned to the exciton resonance and split into three beams. Two of these beams were used to generate the FWM-signal, which, if desired, could be detected directly as a function of Δt using a slow detector in the conventional way. The signal obtained in such a measurement is called the time-integrated-signal (TIS). In order to time-resolve the FWM-signal, for each Δt the light emitted in the direction k_S was cross-correlated with the third laser beam by sum frequency generation in a highly transparent nonlinear crystal. This cross-correlation determined the temporal-profile of the FWM-signal versus the absolute time t [36–39]. The signal obtained in such a measurement is called the time-resolved-signal (TRS). The overall time resolution of the experiment is better than 100 fs.

In Fig. 7 we present TRS versus absolute time t for various values of the time delay Δt, measured with a laser intensity such that the total (generated by both pulses) exciton density is $N_x \approx 3 \times 10^{11}$ cm^{-2}. The laser pulse duration was 78 ± 3 fs. The weaker pulse-1 acts at $t = 0$ and the stronger pulse-2 acts at $t = \Delta t$, so that for $\Delta t < 0$ pulse-2 acts first and pulse-1 acts second, whereas for $\Delta t > 0$ pulse-1 acts first and pulse-2 second. Clearly, the time traces are asymmetric both in t and Δt. For all our measurements we have verified from the position of the TRS maximum that that the FWM-signal is emitted immediately after the second pulse. This behavior confirms that the exciton transition is predominantly homogeneously broadened at room temperature and therefore, that the FWM-signal corresponds to a free-induction decay. Furthermore, the self-consistency of the data was checked by numerically integrating for each Δt the TRS versus t, and comparing the result to the TIS versus Δt measured with a slow detector. Again, the agreement is excellent in our experiments.

In Fig. 8 we display the temporal profile of (a) the TRS versus the absolute time t at $\Delta t = 0$ and (b) the TIS versus Δt, for two total exciton densities, $N_x \approx 10^{11}$ cm^{-2} and $N_x \approx 4 \times 10^{11}$ cm^{-2}. For this set of experiments, the laser pulse duration was 98 ± 2 fs. Since the relevant information is contained in the lineshape, the two sets of curves have been normalized to unity and the unrelated time-axes have been shifted to bring the maxima into coincidence. The difference between the TRS and TIS profiles is evident: the former is clearly broader than the latter, with a slower rising edge and a significantly non-exponential trailing edge. This difference is density dependent and shows up specially on the trailing side of the profiles. Within a Δt series the total exciton density is constant and, therefore, all the TRS traces are

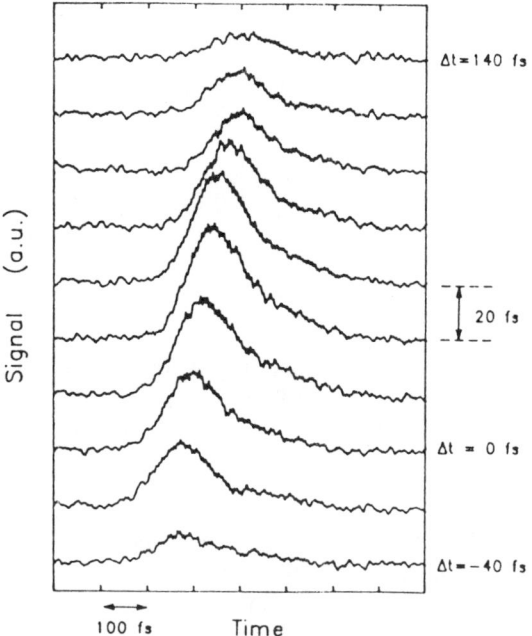

Figure 7. Time resolved four wave mixing signal versus absolute time t, for a series of Δt and a total exciton density $N_x \approx 3 \times 10^{11}$ cm^{-2}.

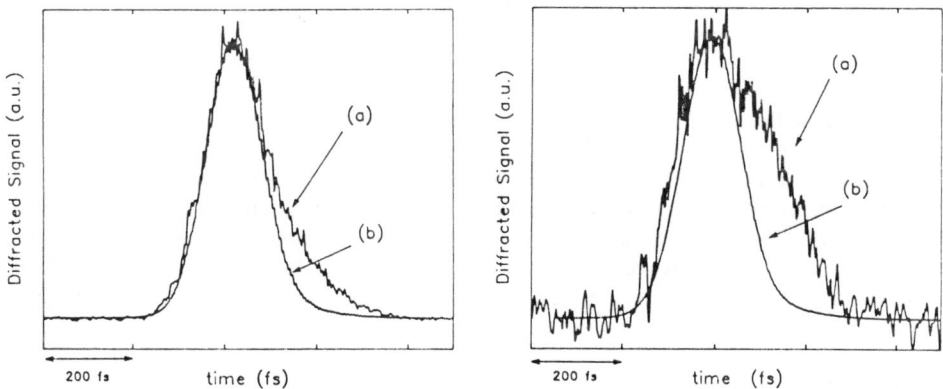

Figure 8. Comparison of the temporal profile of (a) the time resolved signal at Δt = 0 and (b) the time integrated signal, for two total exciton densities: $N_x \approx 4 \times 10^{11}$ cm^{-2} (left traces) and $N_x \approx 10^{11}$ cm^{-2} (right traces).

expected to have similar lineshapes (although their height depends on Δt) [19,20]. This is indeed what is observed at low densities, $N_x \approx 10^{11}$ cm^{-2}. At moderate density, $N_x \approx 2$–4×10^{11} cm^{-2}, noticeable changes in the temporal profile are seen within a Δt series as in Fig. 7. In

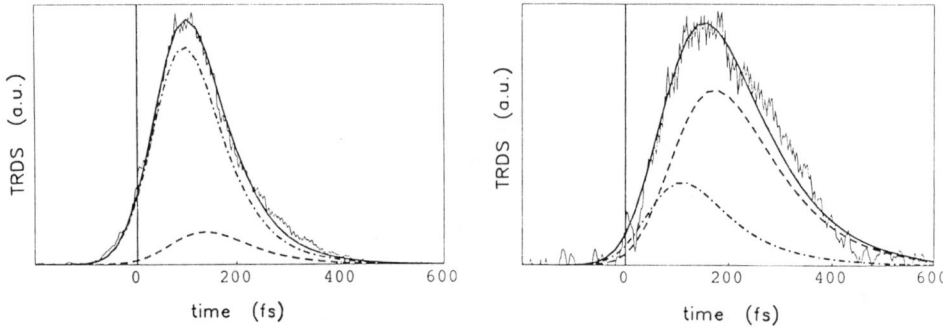

Figure 9. Fit of the temporal profiles shown in Fig. 8, using the model discussed in the text. The dashed-dotted lines give the (isolated-level-like) phase space filling contribution, the dashed lines give the exciton-exciton interaction contribution, and the smooth solid lines give their sum.

particular the sign of Δt is found to influence the temporal lineshape. As the exciton density increases further, the TRS and TIS profiles become more similar (compare Figs. 8(a) and 8(b)). Finally, at very high densities, $N_x \approx 10^{12}$–10^{13} cm^{-2}, where the band gap renormalization has completely washed out the exciton resonances, the TRS and the TIS become of the order of the laser pulse and are beyond our resolution. Then, the profile of all the traces of a Δt series become similar again. It is worth noting that in the experimental situation where the TRS is measured, the sample is in a steady state (fixed Δt) and the "time-resolved measurement" occurs after and is independent of the sample. This is in contrast with the conventional situation where the TIS is measured as Δt is varied, and thus the density, nature, and dynamics of the elementary excitations evolve during the measurement.

In order to be more quantitative, we have solved Eq. (4) numerically, for the nonlinear polarization, $P^{(3)}(t,k_s) = \mu^* \langle \psi^{(3)}(t,k_s) \rangle$ radiating in the direction k_s, using Gaussian laser pulses with a duration corresponding to that of our laser and accounting for the effect of upconversion in the time-resolved (TRS) measurement. Using this model (which involves only two fitting parameters: the exciton dephasing time, T_2, and the ratio of the two nonlinear terms, $v = V \psi_s^2$) we fit the TRS by imposing the severe constraint that all TRS curves in a Δt series must have the same origin of the absolute time t and a constant calibration. We find that it is impossible to fit the data with such a constraint if we retain only the PSF term in Eq. (4). An excellent fit is obtained, however, if both the XXI and PSF contributions are considered. This is shown in Fig. 9, where we fit the TRS temporal profiles of Fig. 8. The dashed-dotted lines give the PSF contribution, the dashed lines give the XXI contribution, and the smooth solid lines give their sum.

At low total exciton densities, $N_x \approx 10^{11}$ cm^{-2} (right traces in Figs. 8 and 9), the dephasing time, $T_2 = 190 \pm 5$ fs, and the ratio of non linearities, $v = 0.01 \pm 0.001$, remain essentially constant for all Δt. The contribution of the XXI to the total energy of the pulse emitted by the sample (which is proportional to the TRS area) dominates the emission. It is ~ 2.5 times larger than that due to PSF. As the total exciton density is increased, both T_2 and v become smaller. For example, for moderate density, as in the left traces of Figs. 8 and 9, $T_2 = 130 \pm 5$ fs, and the XXI contributes to only 14% of the emission. Interestingly, we find that the XXI contribution dominates as long as the exciton density does not exceed the exciton saturation density in the sample, $N_s \approx 3 \times 10^{11}$ cm^{-2} [1]. When this critical density is surpassed the XXI contribution decreases very rapidly and becomes negligible at high densities.

As mentioned above, at low density both T_2 and ν remained approximately constant for all the TRS fits within a single Δt series. At moderate densities, however, this was not found to be the case within a single Δt series (i.e., fixed N_x) both T_2 and ν had to be varied to fit the data. These observations can be understood in terms of the well established dynamics of excitons in quantum wells at room temperature [1]. Consider first the case of a total excitation density low enough that photo-generated excitons are in bound states (binding energy $\approx$ 10 meV for ~ 100 Å QW) and are spatially well separated. They interact effectively via the Coulomb potential and their dephasing time is determined by phonon collisions. Their environment is independent of the instantaneous density determined by the order in which the laser pulses arrive in the sample (i.e., Δt). At moderate densities, however, when the strongest pulse-2 arrives first in the sample, it generates a substantial number of excitons in the bound states. They are ionized by collisions with the energetic thermal phonons (phonon energy $\approx$ 36 meV for GaAs), in ~ 100–200 fs, generating e-h pairs in scattering states with a significant excess energy (~ 25 meV). The charged carriers, in turn, both shorten the relaxation time, T_2, owing to their larger effects on the neutral bound states [37], and screen the Coulomb potential. For the reverse time ordering, i.e., when the weaker pulse-1 arrives first, less e-h pairs are generated by the first pulse, the effects described above are less pronounced, and the sample remains closer to steady state during the FID emission. Finally, at very high densities, the band gap renormalizes so much during the laser pulses that the excitons are generated in scattering states, giving free e-h pairs immediately. They, of course, shorten the relaxation time below the experimental resolution. More importantly, however, they screen the Coulomb potential to the point that the XXI contribution to the emission is eliminated.

The fits also show that the maximum of the TRS is delayed with respect to the second pulse. We found that the delay is of the order T_2. This indicate that the time required to establish the coherent polarization wave within the sample (rise-time of the TRS) is directly related to the dephasing time T_2 which is usually associated only with the negative interferences that produce the signal decay.

4. Conclusion

We have explored the femtosecond dynamics of the nonlinear optical response of magneto-excitons, as the quasi-2D quantum well electronic states are further confined in quasi-0D by a strong magnetic field. We found that the magnetic field induces a restriction of the number of states available for transition which almost completely quenches the relaxation of nonthermal populations. This produces qualitatively different carrier dynamics that must be accounted for in any quasi-0D system or device. We also observed, in agreement with exact many-body theory, that at high field Coulomb correlation between magneto-excitons in the same state is quenched. It, however, persists between magneto-excitons in different states giving strong nonlinear responses. These results also show that a gas of magneto-excitons is a unique two-component many-body system that behaves very differently from the one-component systems such as an electron gas in the fractional quantum Hall effect or a Wigner crystal.

We have shown that, because of exciton-exciton interactions, semiconductors emit coherent light in a way fundamentally different from isolated atoms. We have found that the polarization wave scattering originating from this interaction is, in fact, dominant at low excitation densities.

At low and moderate densities it produces non-trivial effects that must be accounted for in a correct description of many phenomena occurring in these materials.

Acknowledgments

This work was performed in collaboration with S. Schmitt-Rink, J.B. Stark, W.H. Knox, S. Weiss, M.-A. Mycek, and J.-Y. Bigot. The work of DSC is supported by the Director, Office of Energy Research, Office of Basic Energy Sciences, Division of Materials Sciences of the US Department of Energy, under Contract No. DE-AC03-76SF00098.

References

1. S. Schmitt-Rink, D.S Chemla, and D.A.B. Miller, Adv. in Phys. **38**, 89 (1989).
2. O. Akimoto and H. Hasegawa, J. Phys. Soc. Jpn. **22**, 181 (1967).
3. M. Shinada and K. Tanaka, J. Phys. Soc. Jpn. **29**,1258 (1970).
4. S.R.E. Yang and L.J. Sham, Phys. Rev. Lett. **58**, 2598 (1987).
5. G.E.W. Bauer and T. Ando, Phys. Rev. B **38**, 6015 (1988).
6. H. Chu and Y.C. Chang, Phys. Rev. B **40**, 5497 (1989).
7. I.V. Lerner and Yu. E. Lozovik, Zh. Eksp. Teor. Fiz. **80**, 1488 (1981) [Sov. Phys. JEPT **53**, 763 (1981)].
8. D. Paquet, T.M. Rice, and K. Ueda, Phys. Rev. B **32**, 5208 (1985).
9. S. Schmitt-Rink and D.S. Chemla, Phys. Rev. Lett. **57**, 2752 (1986); S. Schmitt-Rink, D.S. Chemla, and H. Haug, Phys. Rev. B **37**, 941 (1988).
10. K. Leo, M. Wegener, J. Shah, D.S. Chemla, E.O. Göbel, T.C. Damen, S. Schmitt-Rink, and W. Schäfer, Phys. Rev. Lett. **65**, 1340 (1990).
11. M. Wegener, D.S. Chemla, S. Schmitt-Rink, and W. Schäfer, Phys. Rev. A **42**, 5675 (1990).
12. S. Schmitt-Rink, S. Mukamel, K. Leo, J. Shah, and D.S. Chemla, Phys. Rev. A **44**, 2124 (1991).
13. W. Schäfer, F. Jahnke, and S. Schmitt-Rink, in *Proceeding of the International Meeting on Optics of Excitons in Confined Systems*, Gardini Naxos, Italy 1991, Inst. Phys. Conf. Ser. No. 123, p. 273, and submitted to Phys. Rev. Lett.
14. A.V. Kuznetsov, Phys. Rev. B **44**, 8721 and 13381 (1991).
15. J.B. Stark, W.H. Knox, D.S. Chemla, W. Schäfer, S. Schmitt-Rink, and C. Stafford, Phys. Rev. Lett. **65**, 3033 (1990).
16. S. Schmitt-Rink, J.B. Stark, W.H. Knox, D.S. Chemla, and W. Schäfer, Appl. Phys. A **53**, 491 (1991).
17. J.B. Stark, W.H. Knox, and D.S. Chemla, Phys. Rev. Lett. **68**, 3080 (1992); Phys. Rev. B **46**, 7919 (1992).
18. C. Stafford, S. Schmitt-Rink, and W. Schäfer, Phys. Rev. B **41**, 10000 (1990).
19. M.-A. Mycek, S. Weiss, J.-Y. Bigot, S. Schmitt-Rink, and D.S. Chemla, Appl. Phys. Lett. **60**, 2666 (1992).
20. S. Weiss, M.-A. Mycek, J.-Y. Bigot, S. Schmitt-Rink, and D.S. Chemla, Phys. Lett. **69**, 2685 (1992).
21. W.H. Knox, C. Hirlimann, D.A.B. Miller, J. Shah, D.S. Chemla, and C.V. Shank, Phys.

Rev. Lett. **56**, 1191 (1986).
22. W.H. Knox, D.S. Chemla, G. Livescu, J.E. Cunningham, and J.E. Henry, Phys. Rev. Lett. **61**, 1290 (1988); W.H. Knox, in *Hot Carriers in Semiconductor Nanostructures*, edited by J. Shah (Academic Press, New York, 1992), p. 313.
23. C. Weisbuch, these proceedings.
24. A.B. Dzyubenko and Yu. E. Lozovik, J. Phys. A **24**, 415 (1991).
25. See also E.I. Rasbha and V. M. Apalkov, these proceedings.
26. S. Schmitt-Rink, D.S. Chemla, and D.A.B. Miller, Phys. Rev. B **32**, 6601 (1985).
27. N. Peyghambarian, H.M. Gibbs, J.L. Jewell, A. Antonetti, A. Migus, D. Hulin, and A. Mysyrowicz, Phys. Rev. Lett. **53**, 2433 (1984).
28. D. Hulin, A. Mysyrowicz, A. Antonetti, A. Migus, W. T. Masselink, H. Morkoc, H.M. Gibbs, and N. Peyghambarian, Phys. Rev. B **33**, 4389 (1986).
29. W. Schäfer, Adv. Solid State Phys. **28**, 63 (1988); W Schäfer, K.H. Schuldt, and R. Binder, Phys. Stat. Sol. B **150**, 407 (1988).
30. C. Ell, J.F. Muller, K. El Sayed, L. Banyai, and H. Haug, Phys. Stat. Sol. B **150**, 393 (1988); Phys. Rev. Lett. **62**, 304 (1989).
31. R. Zimmermann, Phys. Stat. Sol. B **146**, 545 (1988); R. Zimmermann and M. Hartmann Phys. Stat. Sol. B **150**, 365 (1989).
32. I. Balslev, R. Zimmermann, and A. Stahl, Phys. Rev B **40**, 4095 (1989).
33. R. Binder, S.W. Koch, M. Lindberg, N. Peyghambarian, and W. Schäfer, Phys. Rev. Lett. **65**, 899 (1990).
34. T. Yajima and Y. Taira, J. Phys. Soc. Jpn. **47**, 1620 (1979).
35. See, for example, L. Allen and J.H. Eberly, *Optical Resonances and Two Level Atoms* (Wiley, New York, 1975).
36. L. Schultheis, M.D. Sturge, and J. Hegarty, Appl. Phys. Lett. **47**, 995 (1985).
37. L. Schultheis, J. Kuhl, A. Honold, and C.W. Tu, Phys. Lett. **55**, 1635, (1986).
38. G. Noll, U. Siegner, S. Shevel, and E.O. Göbel, Phys. Rev. Lett. **64**, 792 (1990).
39. M.D. Webb, S.T. Cundiff, and D.G. Steel, Phys. Rev. Lett. **66**, 934 (1991).

FEMTOSECOND COHERENT SPECTROSCOPY OF SEMICONDUCTOR NANOSTRUCTURES

J. SHAH
AT&T Bell Laboratories
Holmdel, NJ 07733
USA

ABSTRACT. We discuss our recent results on femtosecond time-integrated and time-resolved four-wave-mixing (FWM) experiments in GaAs nanostructures. In the first part, we discuss femtosecond time evolution of FWM signals [1] from high quality intrinsic GaAs quantum wells. These results show unambiguously that the generally-neglected Coulomb interaction effects between excitons dominate the time-resolved and time-integrated FWM signals. The results also show evidence for further many-body interaction effects and interesting dependence on the polarizations of the beams. In the second part, we discuss investigation of Bloch oscillations in GaAs superlattices [2] using time-integrated FWM techniques. The results show oscillations with period linearly proportional to the applied electric field, in agreement with the expectations for Bloch oscillations.

1. Introduction

Following the excitation of a semiconductor by an ultrashort pulse, the photocreated excitations maintain, for some time after the photoexcitation, a definite phase relationship among themselves and the electromagnetic radiation creating them. Investigation of many interesting properties of this coherent regime using the techniques of femtosecond coherent spectroscopy is currently a very active field of research. We present in this paper a review of two of our recent experiments on semiconductor microstructures in the coherent regime. In the first, we address the question of how interaction between excitons affects the coherent response of GaAs quantum wells and in the second we present an investigation of the Bloch oscillations in a semiconductor superlattice. Since both these experiments involve femtosecond four-wave-mixing (FWM) techniques, we begin with a brief discussion of the previous results obtained by FWM experiments.

In the simplest of FWM experiments, two laser pulses propagating along $\mathbf{k}_1$ and $\mathbf{k}_2$ interfere in a sample to generate a grating which diffracts photons into the background-free direction $2\mathbf{k}_2-\mathbf{k}_1$. In most experiments performed to date, time-integrated signal (or total diffracted energy) is measured as the delay between the two beams is varied. Considerable information about atoms, molecules and solids has been obtained from such measurements [3–11]. Using this technique, the dephasing of excitons was first studied several years ago [12], and a number of studies on excitons have been reported during the last two years. These include the observations of quantum beats of excitons [13–16], coherent interactions of excitons resulting in unexpected FWM signals [17], and oscillatory FWM signals arising from coherent oscillations of an electronic wavepacket in a double well potential [18].

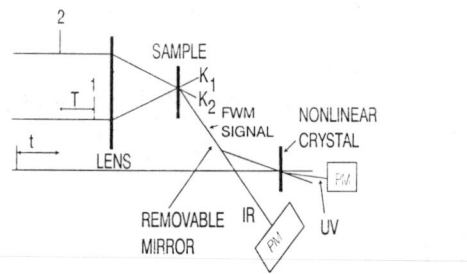

Figure 1. Schematic of the experimental configuration for time-integrated (TI-) and time-resolved (TR-) four-wave-mixing (FWM) experiments; M is a removable mirror and NL Crystal is a LiIO$_3$ crystal, and PM is a photomultiplier tube.

Large (> 300:1) resonance at the exciton peak was reported recently [19]. Additional information can be obtained by temporally resolving the diffracted signal, as shown by many experiments on spontaneous and stimulated photon echoes from localized states in solids [5,20,21]. Since the dephasing times of *intrinsic* states in semiconductors are in the femtosecond to picosecond range, time-resolved FWM (TR-FWM) from semiconductors has become possible only recently [1,22–24].

2. Experimental Technique

Figure 1 is a schematic diagram of the experimental arrangement for both time-integrated FWM (TI-FWM) and time-resolved (TR-FWM) experiments. The experiments were performed using either a tandem-pumped dye laser with tunable 300 fs puses or a 100 fs, tunable, passively mode-locked Ti-Sapphire laser. The sample is excited by two pulses propagating along k_1 and k_2 with linear or circular polarizations and the diffracted signal is measured in the phase-matched direction $k_d = 2k_2-k_1$. We measure the usual TI-FWM signal as a function of time delay T (positive when pulse 1 precedes pulse 2), and the TR-FWM signal as a function of time t following the pulse along k_2, by upconverting it with a part of the laser beam (beam 3) that did not go through the sample. Time t = 0 is determined with an accuracy of < 20 fs by upconverting the scattered light of the second beam into the direction of k_d.

3. Samples

We present here TR-FWM results on an intrinsic GaAs/Al$_{0.3}$Ga$_{0.7}$As multiple quantum well sample (170 Å well width and 10 periods) in which the exciton is nearly homogeneously broadened. The investigations of Bloch oscillations were performed on a GaAs/AlGaAs superlattice consisting of 91 periods of 95 Å wells and 15 Å barriers. The superlattice was in the i region of a p-i-n structure so that electric field can be applied by applying a reverse bias. Substrates were etched out in both samples to allow experiments in transmission geometry. The samples were mounted in a cryostat at 10 K.

4. Femtosecond Time-Resolved FWM in Intrinsic Quantum Wells

In order to put our results into proper perspective, we begin with a discussion of the independent two-level model [25] that has been used to explain the results not only in atomic systems but also in most semiconductors.

4.1. INDEPENDENT TWO-LEVEL MODEL

This model assumes that the results can be described in terms of an ensemble of two-level systems which do not interact with each other. We consider the case of (a) homogeneously and (b) inhomogeneously broadened non-interacting two-level system. For both cases, there is no TI-FWM signal at *negative* time delays much larger than the pulsewidths, and the TI-FWM signals peak at a delay comparable to the laser pulsewidths. For positive time delays, the TI-FWM signal decays with a time constant given by $T_2/2$ for case (a) and $T_2/4$ for the case (b). Therefore, apart from a difference in the slope, the two cases have similar behavior. For intermediate inhomogeneous broadening, the decay constant is related to T_2 by a factor between 1/2 and 1/4.

In contrast, the behavior of the TR-FWM is very different in the two cases. For case (a), the time-resolved signal is expected to peak within one or two pulsewidths of the arrival of the second pulse. This is the usual, prompt free-induction decay (FID) signal. For case (b) on the other hand, the TR-FWM peaks at a time equal to the time delay between the two pulses; this is the usual photon echo signal. For intermediate inhomogeneous broadening, it is possible that the peak occurs at time somewhat smaller than the time delay; in this case the temporal position of the peak depends on the time-delay. However, for weak inhomogeneous broadening, TR-FWM signal is expected within one or two pulsewidths of the arrival of the second pulse, as in case (a).

Recently, TI-FWM was observed at negative time delays larger than the pulsewidth [15], in contrast to the prediction of the independent two-level model discussed above. This negative time-delay signal was explained on the basis of interaction between excitons which has been treated rigorously using semiconductor Bloch equations [26–29]. To a first approximation, the interaction between excitons can be understood in terms of a local field model and can be shown to lead to the diffraction of polarization in addition to the diffraction of field, which in turn leads to a signal at negative time delays. Thus the breakdown of the independent two-level model is already known. However, these observations could not determine the relative strengths of the interaction-induced (II) signal and the normal (N) signal arising from the diffraction of the field. This is because the decay constant of the TI-FWM signal is the same for the II and the N signals.

We now present our results on femtosecond time-resolved experiments on intrinsic quantum wells that have allowed us to show unambiguously that the generally-neglected interaction induced signal in fact dominates by nearly two orders-of-magnitude.

4.2. EXPERIMENTAL RESULTS

Experimental results obtained by Kim et al. [1] are presented here. Figure 2 shows TR-FWM signals for linearly co-polarized beams at various time delays T, with the laser centered 4 meV below the heavy hole (HH) exciton absorption peak. Except for T = 0, the dominant peak occurs more than 1 ps after the second pulse. Thus the risetime of the signal is 10 times longer than the 100 fs pulsewidths. The position of the peak is constant in t at small T but begins to shift for T > 1 ps. Also, the temporal shape depends on T in a complicated manner. The decay constant of the TR-FWM emission is in good agreement with that of the TI-FWM.

Figure 3 shows the TR-FWM signal at T = 1 ps for three different temperatures. The peak moves to earlier t with increase in temperature showing that the position of the peak is not related to the pulsewidth or the time delay T but to the dephasing time T_2.

Figure 4 shows the TR-FWM signal at various time delays for linearly cross-polarized beams. The signal still peaks at long times, but the peak does not move with T and the additional peak near T is weak, in contrast to the data for co-polarized beams (Fig. 3). It should be noted that for negative time delays, the signal starts only after t = |T| as evident from the data at T = -1 ps in Figs. 2 and 4. This is simply because there is no diffraction until both beams arrive at the sample.

4.3. DISCUSSION: DOMINANCE OF INTERACTION EFFECTS

Our TR-FWM data provides a definitive answer concerning the relative strengths of the N and II signal. As we discussed above, the N signal peaks at one or two pulsewidths after the arrival of the second pulse. In contrast, when the dephasing time T_2 is longer than the pulsewidth, the signal arising from the diffraction of the polarization reaches a peak at approximately T_2. This is because the polarization persists for a time much longer than the pulse and decays with a time constant related to T_2. We note that the observed peak does indeed move with T_2 in our data (Fig. 3), supporting the model of interacting excitons. Furthermore, our data shows that the signal at the peak is 100 times stronger than the signal at t = t_p, the laser pulsewidth. Our data, therefore, lead to the conclusion that the II signal is about two orders of magnitude stronger than the N signal. Noting that the ratio of the II to N signal is given by $(VT_2/\hbar)^2$, and using T_2 = 1.3 ps from the data, we deduce that the interaction potential V is approximately 2–3 meV. We deduce from this that reducing T_2 to 0.1 ps will make the II and the N signals about equal.

An examination of our data in Figs. 2-4 shows that there are features in the data that cannot be explained by this simple local field model. In particular, there is an echo-like behavior that is evident at long time delays. An *intrinsic* echo-like behavior is predicted by the semiconductor Bloch equations [29,30] when the intensity of the first beam is such that effects such as bandgap renormalization become important. Under these conditions, the exciton binding energy decreases and the continuum states move toward the exciton resonance. The spectrum of the laser pulse then overlaps the continuum states, which give rise to an *intrinsic* photon echo. We have shown recently, using modulation-doped quantum wells, that the continuum states do indeed give rise to a photon echo [22]. This model provides a reasonable explanation of the observed echo-like signal. Also, the observed variation in the peak position of the II signal in TR-FWM signals is consistent with the calculations based on the semiconductor Bloch equations [26,27]. We note that both the echo and the dependence of the peak position on the time delay are weaker for the cross-polarized case than for the co-polarized case. This is due to a shorter effective T_2 for the cross-polarized case compared to the co-polarized case, as has been observed recently [31], and the fact that beam 2 does not see the changes induced by beam 1 due to the self-energy term, when the two beams are orthogonally polarized [30]. These

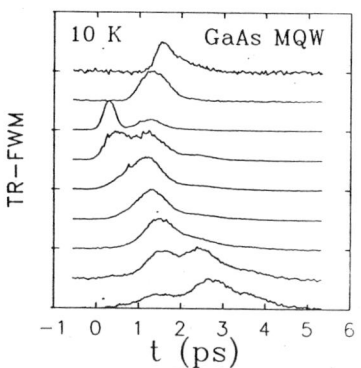

Figure 2. TR-FWM signals from a nearly homogeneously broadened GaAs MQW sample at 10 K for various time delays T using linearly co-polarized beams. Time delays vary from -1 ps for the top curve to 3 ps for the bottom curve in steps of 0.5 ps (From Kim et al. [1]).

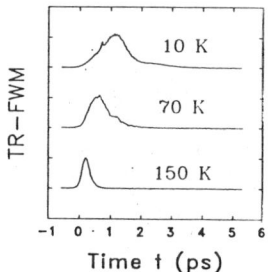

Figure 3. TR-FWM signals from a nearly homogeneously broadened GaAs MQW sample at three different temperatures for $T_2 = 1$ ps using linearly co-polarized beams (From Kim et al. [1]).

additional features in the data are under further investigation.

5. Investigation of Bloch Oscillations in GaAs Superlattices

Bloch oscillations in a solid subjected to a uniform electric field is a topic of considerable interest in solid state physics. We begin with a general introduction to Bloch oscillations and then present our results.

5.1. INTRODUCTION

In 1928, Bloch [32] predicted that, in a solid subjected to an electric field, an electron wavepacket initially centered around k_0 undergoes oscillatory motion in k-space, with a period of oscillation given by

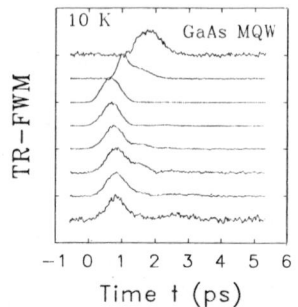

Figure 4. TR-FWM signals from a nearly homogeneously broadened GaAs MQW sample at 10 K for various time delays T using linearly cross-polarized beams. Time delays vary from -1 ps for the top curve to 3 ps for the bottom curve in steps of 0.5 ps (from Kim et al. [1]).

$$T_B = h/eFd, \qquad (1)$$

where F is the applied electric field and d is the periodicity of the solid. Since d is equal to the lattice constant (a few Å) in the bulk but can be of the order of a 100 Å in superlattices, it is clear that an oscillation period short compared to scattering times can be obtained at much smaller fields in superlattices than in bulk solids. Availability of high quality superlattices combined with this factor has led to a renewed interest in Bloch oscillations.

Assuming a one-dimensional solid for simplicity, the rate of increase of the electron quasi-momentum k is given by $\hbar \dot{k} = eF$. Since the rate of change of k is the same for all k making up the wavepacket, the wavepacket moves in k space with a constant velocity. When it reaches the boundary of the reduced Brillouin zone, it undergoes a Bragg reflection and resumes its motion. Therefore, it executes a periodic motion in the reduced-zone, the so-called Bloch Oscillation. While this is a semiclassical picture, it can be shown [33,34] that quantum mechanical treatment in terms of time evolution operator provides the same basic picture.

An alternative way of looking at Bloch oscillation is in real space. Assuming that the wavepacket is initially centered at some position x_0, its velocity is given by $\hbar v(k) = \partial E(k)/\partial k$. Assuming a one-dimensional band given by $E = E_0 - 2\Delta\cos(kd)$ so that 4Δ is the bandwidth, and also assuming an initial velocity $v_0 = 0$, it can be shown that v(t) and x(t) undergo a periodic oscillations such that $\dot{v} = -(2\Delta d/\hbar)\sin(2\pi t/T_B)$ and

$$x/d = (4\Delta/eFd)\sin^2(\pi t/T_B). \qquad (2)$$

Therefore, the position of the wavepacket also undergoes a periodic oscillation between 0 and x_{max}, where $x_{max} = 4\Delta/eFd$.

In order to obtain a quantum mechanical analog of this real space picture, one needs to know the eigenstates of the one dimensional solid in the presence of a uniform electric field. Momentum k is no longer a good quantum number and it was proposed by Wannier [35] that the eigenstates of a solid subjected to a constant electric field are discrete and evenly spaced with an energy separation eFd. The validity of this result, derived under the assumption of one-band model has been examined by several authors [36–43] and it is now believed that the results are valid provided one views these Wannier-Stark states as metastable states rather than true eigenstates and provided one accepts some other restrictions [41]. In fact, experimental observation of Wannier-Stark localization of superlattice miniband states has been reported recently [44,45].

Starting with Wannier-Stark states, it is relatively straightforward to show [41] in a tight-binding model that $P_i(t)$, the probability of finding an electron in the ith site at time t if it was centered at site 0 at t = 0, is given by $P_i(t) = J_i^2 [(4/f) \sin(\pi t/T_B)]$, where J_i is the Bessel function of order i and $f = eFd/\Delta$ and 4Δ is the miniband width for F = 0. Thus the probability $P_i(t)$ undergoes periodic oscillations with a period of the Bloch oscillations. In particular, $P_i(t)$ has a breathing behavior about the initial site.

Therefore, in this quantum mechanical picture, the wavepacket in real space may be viewed as a superposition of Wannier-Stark states. The initial wavepacket is determined by how it is created. Since each Wannier-Stark state has a different energy, the different Wannier-Stark states making up the wavepacket evolve at a different rate according to time-dependent Schrödinger equation, and lead to a temporal evolution of the wavepacket. Since the Wannier-Stark states making up the wavepacket are equally spaced, there will be small constructive interferences at multiples of the Bloch frequency (n/T_B, with n > 1), and a large constructive interference at the fundamental Bloch ($1/T_B$) frequency. This leads to a periodic motion of the wavepacket with a primary period given by the Bloch period and secondary periods given by sub-multiples of the Bloch period.

This description is very similar to the coherent oscillation of the electronic wavepacket in asymmetric double quantum well structure at a field such that the two lowest electron levels are brought into resonance [18]. In this case, optical excitation with an ultrashort laser pulse initially creates a non-stationary wavepacket, composed of a linear superposition of two eigenfunctions, in the wide well (WW) and this wavepacket undergoes a periodic oscillation because the phases of the two wavefunctions evolve at a different rate given by their eigenenergies. An analysis of the coherent oscillations in terms of a simple three-level model showed that the FWM signal from such a system should show a periodic oscillation as a function of time delay T, with the period equal to the period of the coherent oscillations [18]. Such coherent oscillations were indeed observed [18] by using time-resolved pump-and-probe as well as FWM experiments. Therefore, similar oscillations in FWM signal are expected for the case of Bloch oscillations. In fact, Zakharov and Manykin [39] have shown theoretically that FWM signal from a bulk solid under uniform electric field should exhibit a periodic oscillation as a function of the time delay between the two beams with an oscillation period given by T_B. Also, independent of our experiments, von Plessen and Thomas have proposed that such oscillation should be observable in semiconductor superlattices [34].

We review here results obtained by Feldmann et al. [2] using FWM experiments on a GaAs superlattice in a uniform electric field. The results clearly show periodic oscillations in theFWM signals, with period equal to the Bloch period T_B, and varying linearly with the applied electric field.

5.2. EXPERIMENTAL RESULTS

We present results for the sample described in Section 3. Figure 5(b) shows the peak energies of the S_{-1}, S_{-2} and S_{-3} transitions in the photocurrent spectra at various applied voltages. The notation S_i denotes interband transition involving hole in the 0th period with

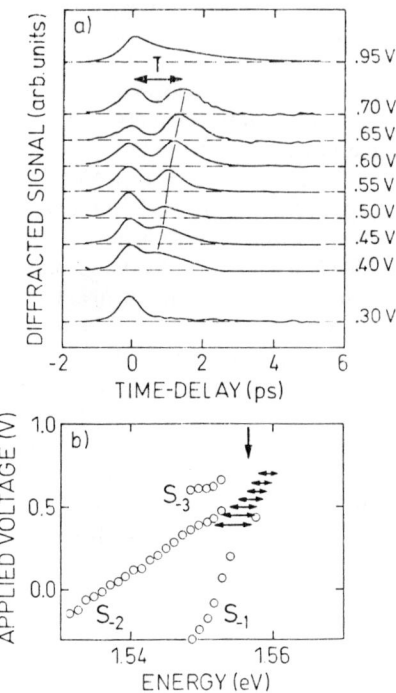

Figure 5. (a) TI-FWM signals from a GaAs superlattice at 10 K for various applied fields. (b) The energies of various peaks observed in the photocurrent spectra of the GaAs superlattice as a function of voltage (from Feldmann et al. [2]).

an electron in the ith period, where i is positive or negative and the electron Wannier-Stark states have energies given by $E_i = E_0 + ieFd$ in the simplest picture, neglecting excitonic effects. The measured interband transitions do indeed depend approximately linearly on the applied field F as expected from this simple model. This implies that excitonic effects [39] are not important.

When the laser is tuned so that its spectrum is centered at the energy shown by the downward arrow in Fig. 5(b), TI-FWM signal from the superlattice sample shows oscillations as a function of time delay T for several applied electric fields (Fig. 5(a)). The period of the oscillation increases as the forward applied voltage increases, i.e., as the netfield (sum of the applied field and the built-in field) decreases. The period of the oscillation at various applied voltages is plotted in Fig. 5(b) as horizontal double arrows located within the spectrum of the laser. We see that the magnitude of the period and its variation with voltage is in good agreement with the splitting of the Wannier-Stark states at various fields.

Figure 6 shows TI-FWM signal obtained from the same sample using a Ti:Sapphire laser with 110 fs pulsewidth [2] so that the laser spectrum encompasses more Wannier-Stark levels and the time resolution was better. A third peak is visible in the TI-FWM signal and the ratio of the intensities of the first to second peaks is larger.

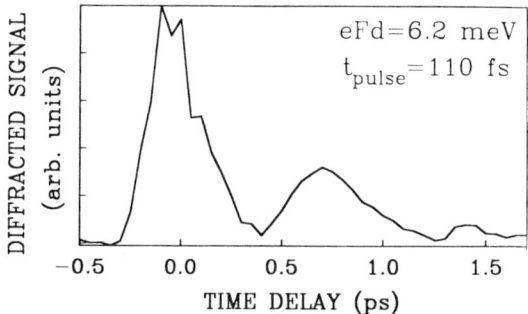

Figure 6. TI-FWM signals from a GaAs superlattice obtained by using 110 fs pulses from a Ti:Sapphire laser for an applied bias such that eFd = 6.2 meV (from Feldmann et al. [2]).

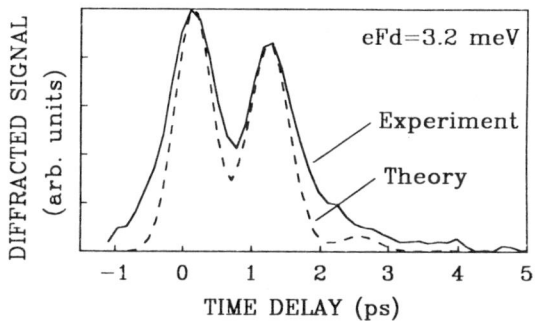

Figure 7. TI-FWM signals from a GaAs superlattice obtained by using 110 fs pulses from a Ti:Sapphire laser for an applied bias such that eFd = 3.2 meV. Also shown is a theoretical curve calculated by Thomas and collaborators (from Feldmann et al. [2]).

Figure 7 shows a comparison of the experimental results obtained with the Ti:Sapphire laser for a field such that eFd = 3.2 meV and a calculated results expected for the same case, assuming a dephasing time of 2 ps. The calculations were made by Thomas and coworkers for a finite superlattice excited by a 110 fs pulse. There is a reasonable agreement between the experiments and the calculations.

5.3. DISCUSSION: BLOCH OSCILLATIONS

It is clear that all the experimental observations are consistent with the expectations for Bloch oscillations. This leads to the reasonable conclusion that experimental results are manifestations of the much-sough-after Bloch oscillations. The question that needs to be answered is if the superlattice character is essential for understanding the experimental results or the results can be equally well explained by an identical structure with small number of periods N.

Our analysis indicates that the expected behavior of the TI-FWM signal as a function of the delay time T may not be too different for a double well structure (N = 2) and a superlattice. However, the magnitude of the oscillation period and its linear variation with the applied field show that the superlattice character is essential for understanding our results. A structure with small N and having identical well and barrier widths can be shown to have a minimum splitting between energy levels that is much larger than that deduced from the longest oscillation period. For example, the experimentally deduced minimum splitting is 2.6 meV whereas the calculated splitting between the levels for a double well with 95 Å well and 15 Å barrier is 10 meV. Furthermore, the splitting in a sample with $N = 2$ will vary quadratically with field near the resonance, in contrast to the observations. Similar arguments can be made for other structures with small N. The only way for a structure with a well-width of 95 Å and a barrier-width of 15 Å to have a splitting of 2.6 meV and a linear variation of the splitting with an applied electric field is to invoke superlattice effects.

A superlattice with N periods has a miniband made up of N states at zero applied field. The approximate condition for the validity of an equally spaced ladder of Wannier-Stark levels whose splitting varies linearly with the electric field is $eFdN > 4\Delta$, where 4Δ is the miniband width. Since $4\Delta = 21$ meV in our case, N must be greater than 8 to have the splitting vary linearly with the field for eFd as small as 2.6 meV; i.e., at least 8 periods of the superlattice must be involved to account for our observations. Using Eq. (5), we deduce that the spatial oscillations also cover approximately 8 periods at the smallest field investigated.

6. Summary

We have discussed two of our recent experiments in the coherent regime of semiconductor microstructures. The experiments were performed on an intrinsic quantum well sample and a superlattice sample using time-resolved and time-integrated four-wave-mixing techniques.

The results on the high quality intrinsic quantum well sample have provided new insights into the physics of coherent interactions in semiconductors. In particular we have shown that the independent two-level model used in analyzing most of the coherent experiments in semiconductors is inadequate, and that the generally-neglected Coulomb interaction between excitons dominates the coherent properties of semiconductors. This, however, does not invalidate all the conclusions drawn from the previous analyses because the variation of the FWM signal with the delay time is not affected by interaction effects except close to zero time-delay.

We have also presented FWM results on a GaAs superlattice in electric field which shows manifestations of Bloch oscillations. In particular, the FWM signal shows oscillations with period equal to the Bloch oscillation period, and the period varies linearly with the applied electric field. The results are in good agreement with calculations. The superlattice character of the sample is essential in explaining the results.

Acknowledgments

Work on time-resolved four-wave-mixing in intrinsic quantum wells [1] was in collaboration with DaiSik Kim, T.C. Damen, W. Schäfer, F. Jahnke, S. Schmitt-Rink, and K. Köhler, and work on investigation of Bloch oscillations in GaAs superlattices [2] was in collaboration with J. Feldmann, K. Leo, D.A.B. Miller, J.E. Cunningham, T. Meier, G. von

Plessen, A. Schulze, P. Thomas, and S. Schmitt-Rink.

References

1. D.S. Kim, J. Shah, T.C. Damen, W. Schäfer, F. Jahnke, S. Schmitt-Rink, and K. Köhler, Phys. Rev. Lett. **69**, 2725 (1992); D.S. Kim, Jagdeep Shah, W. Schäfer, and S. Schmitt-Rink, presented at NOEKS III, Bad Honef, Germany, May 17-20, 1992.
2. J. Feldmann, K. Leo, Jagdeep Shah, D.A.B. Miller, J.E. Cunningham, T. Meier, G. von Plessen, A. Schulze, P. Thomas, and S. Schmitt-Rink, Phys. Rev. B **46**, 7252 (1992).
3. Y.R. Shen, *The Principles of NonLinear Optics* (Wiley Interscience, New York, 1984).
4. M.D. Levinson, *Introduction to Nonlinear Laser Spectroscopy* (Academic Press, New York, 1988).
5. M.C. Webb, S.T. Cundiff, and D.G. Steel, Phys. Rev. Lett. **66**, 934 (1991); Phys. Rev. B **43**, 12658 (1991).
6. Erik T.J. Nibbering, Douwe A. Wiersma, and Koos Duppen, Phys. Rev. Lett. **66**, 2464 (1991).
7. P.C. Becker, H.L. Fragnito, J.-Y. Bigot, C.H. Brito-Cruz, R.L. Fork, and C.V. Shank, Phys. Rev. Lett. **63**, 505 (1989).
8. L. Schultheis, J. Kuhl, A. Honold, and C.W. Tu, Phys. Rev. Lett. **57**, 1635 (1986); Phys. Rev. Lett. **57**, 1797 (1986).
9. K. Leo, M. Wegener, J. Shah, D.S. Chemla, E.O. Göbel, T.C. Damen, S. Schmitt-Rink, and W. Schäfer, Phys. Rev. Lett. **65**, 1340 (1990); M. Wegener, D.S. Chemla, S. Schmitt-Rink, and W. Schäfer, Phys. Rev. A **42**, 5675 (1990); S. Schmitt-Rink, S. Mukamel, K. Leo, J. Shah, and D.S. Chemla, Phys. Rev. A **44**, 2124 (1991).
10. P.C. Becker, H.L. Fragnito, C.H. Brito-Cruz, R.L. Fork, J.E. Cunningham, J.E. Henry, and C.V. Shank, Phys. Rev. Lett. **61**, 1647 (1988).
11. J.-Y. Bigot, M.T. Portella, R.W. Shoenlein, J.E. Cunningham, J.E. Henry, and C.V. Shank, Phys. Rev. Lett. **67**, 636 (1991).
12. L. Schultheis, J. Kuhl, A. Honold, and C.W. Tu, Phys. Rev. Lett. **57**, 1635 (1986); Phys. Rev. Lett. **57**, 1797 (1986).
13. V. Langer, H. Stolz, and W. von der Osten, Phys. Rev. Lett. **64**, 854 (1990).
14. E.O. Göbel, K. Leo, T.C. Damen, J. Shah, S. Schmitt-Rink, W. Schäfer, J.F. Müller, and K. Köhler, Phys. Rev. Lett. **64**, 1801 (1990).
15. K. Leo, T.C. Damen, J. Shah, E.O. Göbel, and K. Köhler, Appl. Phys. Lett. **57**, 19 (1990).
16. B.F. Feuerbacher, J. Kuhl, R. Eccleston, and K. Ploog, Solid State Commun. **74**, 1279 (1990).
17. K. Leo, M. Wegener, J. Shah, D.S. Chemla, E.O. Göbel, T.C. Damen, S. Schmitt-Rink, and W. Schäfer, Phys. Rev. Lett. **65**, 1340 (1990); M. Wegener, D.S. Chemla, S. Schmitt-Rink, and W. Schäfer, Phys. Rev. A **42**, 5675 (1990); K. Leo, E. Göbel, T.C. Damen, J. Shah, S. Schmitt-Rink, W. Schäfer, J.F. Müller, K. Köhler, and P. Ganser, Phys. Rev. B **44**, 5726 (1991).
18. K. Leo, J. Shah, E.O. Göbel, T.C. Damen, S. Schmitt-Rink, W. Schäfer, and K. Köhler, Phys. Rev. Lett. **66**, 201 (1991); K. Leo, Jagdeep Shah, E.O. Göbel, T.C. Damen, S. Schmitt-Rink, and W. Schäfer, in *Proceedings of Topical Meeting on Picosecond Electronics and Optoelectronics*, edited by T.C.L.G. Sollner and Jagdeep Shah (Optical

Society of America, Washington D.C., 1991), p. 204.
19. D.S. Kim, J. Shah, J.E. Cunningham, T.C. Damen, S. Schmitt-Rink, and W. Schäfer, Phys. Rev. Lett. **68**, 1006 (1992).
20. G. Noll, U. Siegner, S. Shevel, and E.O. Göbel, Phys. Rev. Lett. **64**, 792 (1990).
21. L. Schultheis, M.D. Sturge, and J. Hegarty, Appl. Phys. Lett. **47**, 995 (1985).
22. D.S. Kim, J. Shah, J.E. Cunningham, T.C. Damen, W. Schäfer, M. Hartmann, and S. Schmitt-Rink, Phys. Rev. Lett. **68**, 2838 (1992).
23. M.-A. Mycek, S. Weiss, J.-Y. Bigot, S. Schmitt-Rink, D.S. Chemla, and W.S. Schäfer, Appl. Phys. Lett. **60**, 2666 (1992).
24. S. Weiss, M.-A. Mycek, J.-Y. Bigot, S. Schmitt-Rink, and D.S. Chemla, Phys. Rev. Lett. **69**, 2685 (1992); see also the paper by D.S. Chemla in these proceedings.
25. T. Yajima and Y. Taira, J. Phys. Soc. Jpn **47**, 1620 (1979).
26. S. Schmitt-Rink, D.S. Chemla, and H. Haug, Phys. Rev. B **37**, 941 (1988).
27. W. Schäfer, K.H. Schuldt, and R. Binder, Phys. Stat. Solid B **150**, 407 (1988); I. Balslev, R. Zimmermann, and A. Stahl, Phys. Rev. B **40**, 4095 (1989); M. Lindberg and S.W. Koch, Phys. Rev. B **38**, 3342 (1988).
28. C. Stafford, S. Schmitt-Rink, and W. Schäfer, Phys. Rev. B **41**, 10000 (1990).
29. M. Lindberg, R. Binder, and S.W. Koch, Phys. Rev. A **45**, 1865 (1992).
30. W. Schäfer, F. Jahnkne, and S. Schmitt-Rink, unpublished.
31. See, for example, K. Leo, J. Shah, S. Schmitt-Rink, and K. Köhler, in *Proceedings of the 7th International Symposium on Ultrafast Processes in Spectroscopy*, edited by A. Laubereau and A. Seilmeier (Inst. Physics, Bristol, 1992), p. 411.
32. F. Bloch, Z. Phys. **52**, 555 (1928).
33. A. Nenciu and G. Nenciu, Phys. Rev. **40**, 3622 (1989); G. Nenciu, Rev. Mod. Phys. **63**, 91 (1991).
34. G. von Plessen and P. Thomas, Phys. Rev. B **45**, 9185 (1992).
35. G.H. Wannier, Rev. Mod. Phys. **34**, 645 (1962).
36. J. Zak, in *Solid State Physics*, Vol. 27, edited by H. Ehrenreich and F. Seitz (Academic Press, New York, 1967), p. 1.
37. M. Saitoh, J. Phys. C **5**, 914 (1972).
38. M. Luban and J.H. Luscombe, Phys. Rev. B **34**, 3674 (1986).
39. D. Emin and C.F. Hart, Phys. Rev. B **36**, 7353 (1987).
40. F. Bentollosa, V. Grecchi, and F. Zironi, J. Phys. C. **115**, 7119 (1982).
41. G. Bastard and R. Ferreira, in *Spectroscopy of Semiconductor Microstructures*, edited by G. Fasol, A. Fasolina, and P. Lugli (Plenum Press, New York, 1989), p. 333.
42. J.B. Krieger and G.J. Iafrate, Phys. Rev. B **33**, 5494 (1986).
43. J.B. Krieger and G.J. Iafrate, Phys. Rev. B **35**, 9644 (1987).
44. E.E. Mendez, F. Agullo-Rueda, and J.M. Hong, Phys. Rev. Lett. **60**, 2426 (1988).
45. P. Voisin, J. Bleuse, C. Boouche, S. Gaillard, C. Alibert, and A. Regreny, Phys. Rev. Lett. **61**, 1639 (1988).
46. M.M. Dignam and J. Sipe, Phys. Rev. Lett. **64**, 1797 (1990).

FREE EXCITON RADIATIVE RECOMBINATION IN GaAs QUANTUM WELLS

B. DEVEAUD,[1] F. CLÉROT,[1] B. SERMAGE,[2] C. DUMAS,[2]
AND D.S. KATZER[3]
[1]*Centre National d'Études de Télécommunications, 22300 Lannion, France*
[2]*Centre National d'Études de Télécommunications, 92120 Bagneux, France*
[3]*Naval Research Laboratory,Washington, DC 20375, USA*

ABSTRACT. Radiative properties of free excitons in GaAs quantum wells are studied under resonant excitation. We first show that we do observe free excitons by their Lorentzian lineshape and their mobility in the plane of the well. Enhanced radiative recombination of the excitons, a consequence of the breakdown of the translational symmetry induced by the quantum well potential, is evidenced by the very short lifetime as well as by the strong intensity of the luminescence signal. Dephasing mechanisms, by transferring the excitons into non-radiative states, increase the observed lifetime. In the same way, the increase of the sample temperature, or of the exciton temperature by non-resonant excitation, increases the radiative decay time by reducing the exciton population close to k = 0. From our experiments, we deduce a radiative lifetime of 10 ± 4 ps in the absence of dephasing mechanisms.

1. Introduction

At low temperatures, linear and nonlinear optical properties of semiconductors are strongly affected by excitonic effects [1]. One of the interests of quantum wells comes from the persistence of strong excitonic resonances up to room temperature [2,3] allowing a number of nonlinear effects at this temperature [4]. Two-dimensional (2D) confinement of an exciton in a quantum well indeed leads to a shrinkage of the exciton Bohr radius accompanied by an increase of the oscillator strength and of the binding energy [5–7]. This, together with the reduction of the exciton phonon interaction, explains the longer dissociation time of excitons at room temperature.

The study of excitons is usually carried out by absorption studies, with the need, in the material system GaAs/AlGaAs, to remove by chemical etching the GaAs substrate which is absorbing at the wavelength of the exciton resonances. Luminescence, which does not need this technological processing of the sample, has therefore been quite widely used [8–11]. It is however not straightforward to compare directly the radiative behaviour of excitons in 2D and three-dimensional (3D) systems, as such systems should behave in a completely different way. We refer the reader to the book by Del Sole et al. [12] for an introduction to the modifications introduced by the finite size of the sample on the coupling of the exciton with the external photons.

2. Theoretical background

In a 3D material assumed to be infinite, the direct correspondence between one exciton state and one photon state precludes the application of the Fermi golden rule to describe the time behaviour of the exciton plus photon system. Hopfield [13] therefore proposed to describe the exciton-photon system as a whole and introduced the concept of excitonic-polariton.

Excitonic polaritons are stationary states of an infinite dielectric medium (and should not therefore exhibit any temporal evolution). Decay is either observed through phonon scattering [14] (as evidenced for example by Brillouin scattering [15]), or due to the existence of crystal surfaces where polaritons can transform into external photons [16]. The increased coupling, at the surfaces of a thin sample, may explain the "giant oscillator strength" of the exciton, as observed in GaAs [17]. The time resolved behaviour of excitonic polaritons has been extensively studied in quite a number of bulk materials: see Ref. [18] for a review.

The above description relies on the translational invariance of the crystal, and implies the conservation of momentum from the exciton to the photon. *For excitons confined in quantum wells, the coupling to photons is profoundly modified by the breakdown of translational symmetry in the growth direction* (hereafter labelled z). Excitons in quantum wells (or in surface layers of molecular crystals [19]) can thus couple to a whole one-dimensional (1D) distribution of photons. This coupling induces a finite radiative width for excitons with a wavevector $|k| < n\omega_0/c$ [20] (here ω_0 is the frequency of the photon at the exciton energy, n the material refractive index, and c the velocity of light). On the contrary, excitons with k above $n\omega_0/c$ should not recombine at all. The decay rate of near k = 0 free excitons in quantum wells has been predicted to be very fast: of the order of 10 ps [20–23].

If we go further down in dimensionality, when going from 3D to zero-dimensional (0D) systems, basically the volume over which the excitons integrates the elementary dipoles is reduced, and correlatively, the number of photon states available for radiative decay increases (see for example Hanamura [22]).

This large radiative decay rate is sometimes called *superradiant* [22,23], especially in molecular crystals, as it relies on the coherent nature of the exciton, which integrates the individual dipoles of the molecular sites over its coherence volume. We prefer not to use this qualification for Wannier excitons, as it might be interpreted as a coherence effect amongst a population of excitons, which is not the case (one exciton alone has a very large recombination rate). As a consequence of this need for coherence, and as the radiative region is very narrow (about 0.1 meV in GaAs quantum wells), the expected rapid radiative decay is only observable in the absence of perturbations and in particular if the phase coherence of the exciton is preserved long enough. Different authors [22–24] noted that the loss of phase coherence (for example due to the scattering by acoustical phonons if the temperature is too high) should prevent radiative recombination and thus increase the radiative lifetime from the expected value of about 10 ps to much longer times.

Apart from temperature effects, other perturbations may alter the decay of excitons: exciton-exciton or exciton-carrier scattering [25], but also any kind of imperfection in the quantum well and in particular interface roughness. Interface roughness will tend to localize the excitons and give rise to bound excitons with a lifetime of the order of 1 ns. As a result, and this was pointed out for example by Hanamura [22], the enhanced radiative decay of excitons should be very difficult to observe as it requires one single quantum well (QW) of the highest quality.

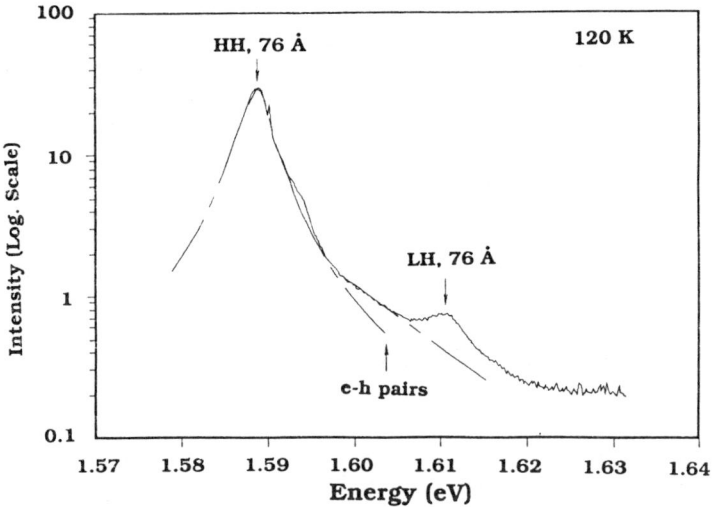

Figure 1. Continuous wave luminescence of one of the quantum wells used in this study (in log scale). The lineshape is mainly Lorentzian at all temperatures and densities (see the Lorentzian: dashed curve). At the same position, the low density, low temperature broadening is 1.5 meV and we show the spectrum at 120 K and rather high excitation density. The full width half maximum of the Lorentzian is 3.5 meV. We evidence on the high energy side the contribution of the light hole excitons, and a possible contribution of electron-hole pairs around 1.6 eV (see the fit to a Boltzmann tail with a temperature of 120 K).

3. Previous studies

It has been shown by Weisbuch et al. [26], using spin polarized luminescence and luminescence excitation, that the radiative recombination in quantum wells is dominated by free excitons. This is in sharp contrast with the behaviour of the highest quality available bulk materials where radiative recombinations are mainly governed by extrinsic channels, and where only small contribution of the polariton branches can be evidenced. The same team also proposed [27] that the linewidth of the excitons was dominated by the energy fluctuations due to the interface roughness. In the best samples, the interface roughness was estimated to extend over one monolayer only. Further improvements of the growth technique allowed the observation of splittings of the excitonic lines, due to an even better quality of the interface [28,29].

Time resolved studies by Göbel et al. [9], showed that the exciton lifetime (under non-resonant excitation conditions) is shortened when the well width is reduced. The observed lifetimes typically range from 300 ps up to 1 ns at low temperatures. It was also found that the lifetime is increasing almost linearly with temperature, as a result of the smaller exciton population in the region close to k = 0 where radiative recombination is possible [24]. These studies were followed by four-wave mixing experiments [30] where it has been shown that a relation existed between the radiative lifetime of excitons and their dephasing time.

In a recent letter [31], we gave evidence for the enhanced radiative decay of free excitons in a single quantum well. Not only did we monitor the decay of the luminescence intensity, but also the relative value of this intensity and the homogeneous linewidth. Under resonant excitation at 1.7 K, we observed a radiative decay time as short as 24 ps in a high quality GaAs/AlAs quantum well and showed that temperature (above 2 K), excitation density (above 3×10^9 cm^{-2}), as well as excitation detuning (by only 2 meV), by allowing phase breaking mechanisms, increase the luminescence lifetime and the homogeneous linewidth, and decrease the intensity. We will here both recall these early results and detail new experiments which confirm the concepts expounded above.

4. Sample characterization

We have studied several single GaAs/AlAs QWs grown by molecular beam epitaxy using growth interruption techniques [32]. These samples show a very high quality, which can be evidenced in different ways. We have given in Ref. [33] a description of the cw experiments used to show the quality of the samples as well as the technique used to obtain the time resolved data [34]. We only recall here the important results.

The luminescence of each well shows a very clear splitting of the excitonic transitions, with no Stokes shift between luminescence and luminescence excitation. We show in Fig. 1 the luminescence spectrum of a 76 Å quantum well, the position on the sample is chosen in order to minimize the contribution of the high energy line. The spectrum is taken at 120 K and rather high excitation density in order to evidence clearly the Lorentzian broadening of the transition. A small contribution from the electron-hole pair recombination is also observed. This evidences the fact that free excitons are indeed giving rise to the luminescence that we observe, and contrasts with the results on QWs with ordinary quality where the excitonic transitions show a Boltzmann tail at high energies [35].

The absence of localization has further been characterized by the time behaviour of the luminescence. To probe the ability of the excitons to move, at low temperatures, in the plane of the layers, we use the internal probes formed by the larger islands in the quantum well [36], where the exciton may localize and give a transition at a different energy. This movement can be evidenced by the time resolved luminescence behaviour shown on Fig. 2. In this experiment, we have excited a position of the sample where two lines are observed, corresponding roughly to 51 Å (labelled N-1) and 54 Å (labelled N). Exciting in the LH exciton transition of the N-1 region allows the production of excitons mainly in the N-1 zones of the sample. Subsequent observation of the intensity behaviour of the luminescence lines demonstrates the movement of the excitons from the N-1 ML wide zones to the N ML wide zones with a characteristic time of about 100 ps.

5. Time resolved studies

5.1. RESONANT EXCITATION

We show in Fig. 3 over three decades the time resolved behaviour of the heavy hole exciton luminescence resonantly excited at low density and low temperature. The decay behaviour is not perfectly exponential; the decay time constant is 25 ps at short times and 57 ps at long times. The homogeneous contribution, measured in the same conditions, is 0.2 meV. The excitation density is 20 µW, corresponding to a density of only 4×10^8 cm^{-2} photocreated

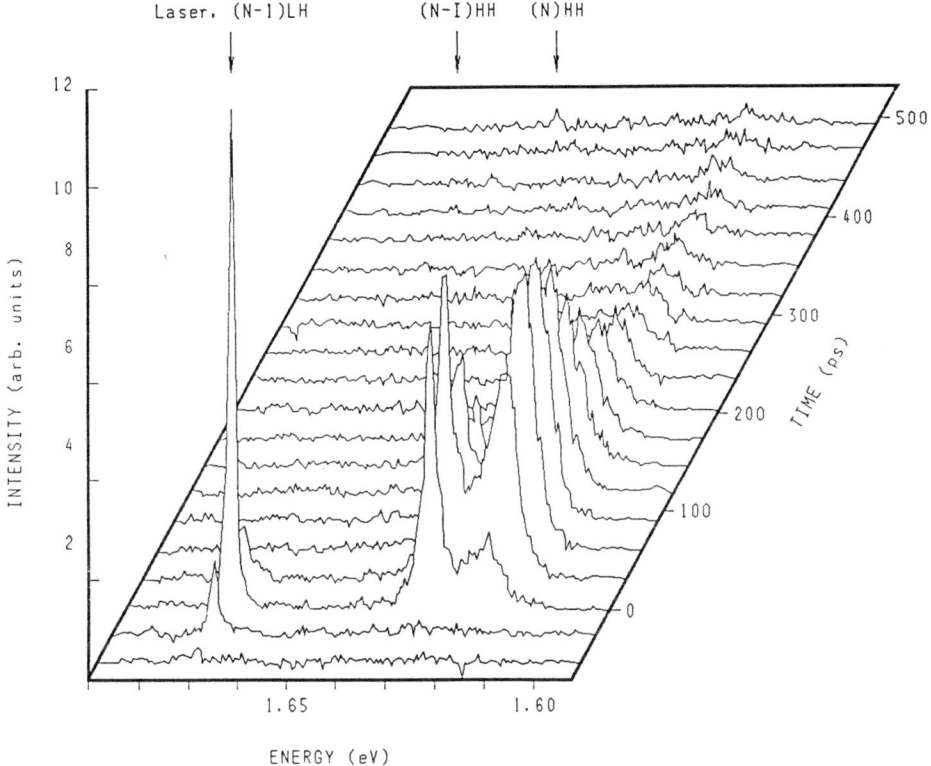

Figure 2. Time resolved behaviour of the N and N-1 exciton lines. The sample is excited resonantly in the LH transition corresponding to the N-1 zone of the sample. After emission of an optical phonon, all excitons transform into HH excitons in the N-1 zone. Transfer to the N zone occurs with a characteristic time of about 100 ps.

excitons. Note that not only the decay time is very short, but that the luminescence intensity is also very large. This allowed us to get time resolved results with excitation powers down to 2 μW only, although we only excite one quantum well. The increase of the decay time at long time may be due to the contribution of the sample temperature (10 K for this experiment) or to dephasing processes.

Diffracted light from the laser is much weaker than the luminescence under off resonance conditions. When the laser is brought to resonance, the diffracted intensity increases as a result of resonant Rayleigh scattering [37], but the exciton luminescence intensity increases even more due to the increased absorption and to the decrease of the radiative lifetime. The light scattered from the laser stays, therefore, weaker than the luminescence signal and cannot be responsible for the very short decays observed here.

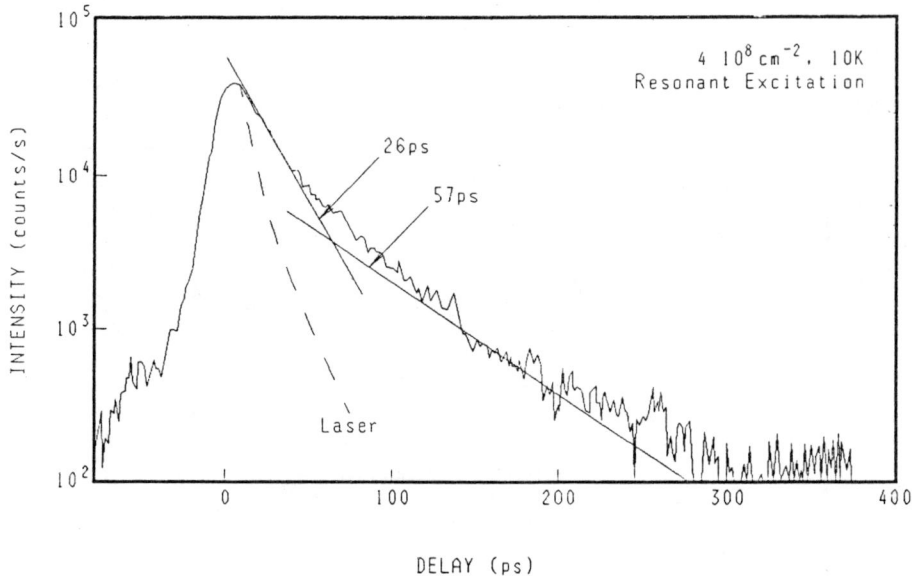

Figure 3. Time resolved luminescence of a resonantly excited HH exciton. The decay behaviour is not perfectly exponential; the decay time constant at short times is 25 ps and it is 57 ps at long times. The time resolution of the experiment is 8 ps (dashed line). The measured homogeneous contribution is 0.2 meV. The excitation intensity is 20 μW, corresponding to a density of about 4×10^8 cm^{-2}. The same decay time is observed over almost two orders of magnitude in excitation density around this value.

5.2. NON RESONANT EXCITATION

As soon as the excitation energy is detuned above the exciton energy, the decay time becomes longer, and a risetime develops. We show in Fig. 4 the time decay at 10 K of the excitonic luminescence for an excitation density of 2×10^9 cm^{-2}, as a function of the detuning between the pump and the exciton energies. All curves are recorded in the same conditions, and plotted in absolute number of counts. As is evidenced, a very small detuning is sufficient to induce a clear increase of the decay time constant. This effect is simply due to the fact that the temperature of the excitons is now increased (although the sample is still at low temperature). As a result, the proportion of excitons in the radiative zone (only 0.1 meV wide) gets smaller.

Simultaneously, when the detuning is increased above the exciton energy, a long risetime develops, which has already been observed by different authors [38,39] and attributed to exciton relaxation and cooling. We observe the same increase of the risetime, but with different features (see Fig. 5). The absolute value of the risetime is rather short in our case: we consider that this is a consequence of the high quality of the sample, preventing the heating of the exciton subsequent to the trapping into larger portions of the well [40]. The risetime is then only due to thermalization of initially hot excitons, and not to trapping of the excitons into lower energy regions of the well. At 42 meV, a strong resonance is observed that is due to

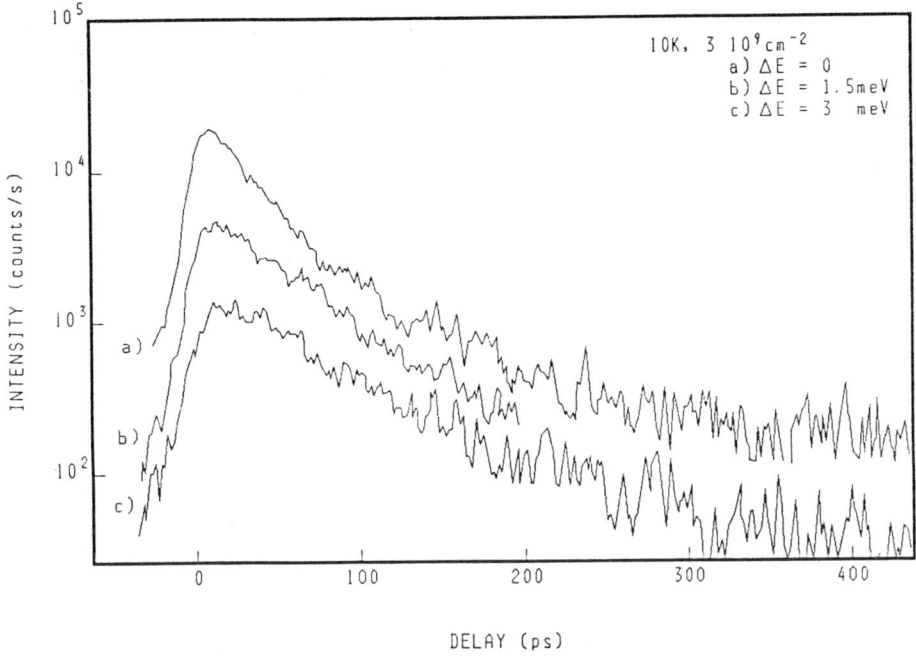

Figure 4. Exciton luminescence decays for different detunings between the laser and the exciton energy. The same scale is used for the different measurements: a decrease in intensity is observed simultaneously with the increase of lifetime for very small values of the detuning. Note the appearance of a delay for the position of the maximum of the luminescence signal.

resonant excitation of light hole excitons, which can transform into cold heavy hole excitons by emitting one optical phonon.

5.3. DENSITY EFFECTS

The decay time is also observed to increase under resonant excitation when the excitation density ϕ is increased above 3×10^9 cm^{-2}. In this case, due to the change in excitation power, we do not observe a decrease of luminescence intensity I_{lum}, but rather a decrease of the ratio $I_{lum}(t = 0)/\phi$. The time behaviour evidences two time constants, one at short times, which gets shorter when the excitation density is increased, and the other one at long times, which increases up to a limit close to 200 ps that is due to the structure of our samples (see below).

Lineshapes of the excitonic transition at 2.1 K, recorded 50 ps after resonant excitation at various densities, are reported in Fig. 6. The increase of the linewidth, and the Lorentzian contribution to the lineshape are clearly evidenced. At low temperature and low density, the linewidth is mainly Gaussian as a result of residual inhomogeneous broadening (for example potential fluctuations due to impurities in the barriers or at the interface). A small Lorentzian contribution can be resolved giving a homogeneous width of 0.34 meV. At higher densities, a good fit to the lineshape is obtained by keeping the Gaussian contribution constant and

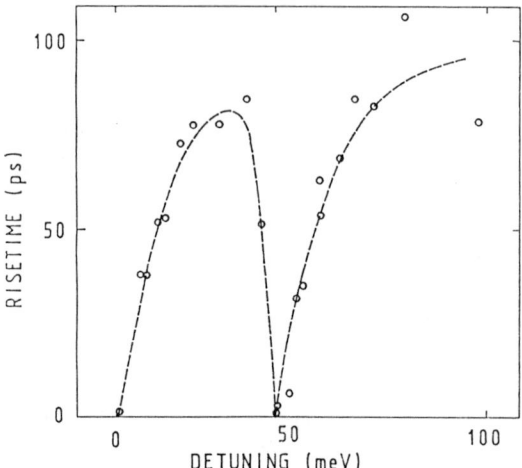

Figure 5. Risetime of the exciton luminescence as a function of the detuning between the laser energy and the exciton energy. The resonance effect at 42 meV corresponds to resonant excitation of light hole excitons, which can relax very quickly into heavy hole excitons.

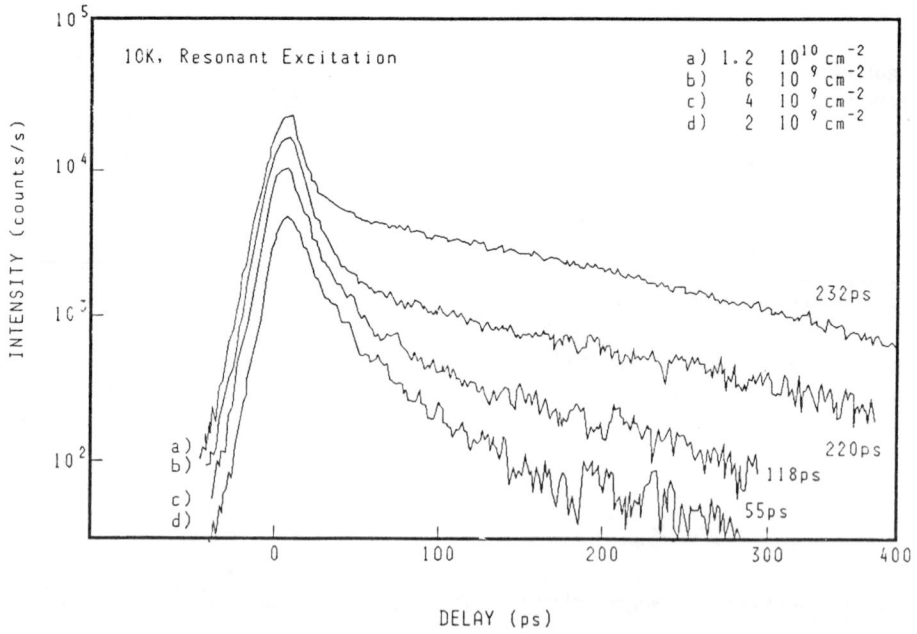

Figure 6. Exciton luminescence decays for different excitation densities. Note the increase of the long time decay time constant, and the decrease of the short time constant around the laser pulse.

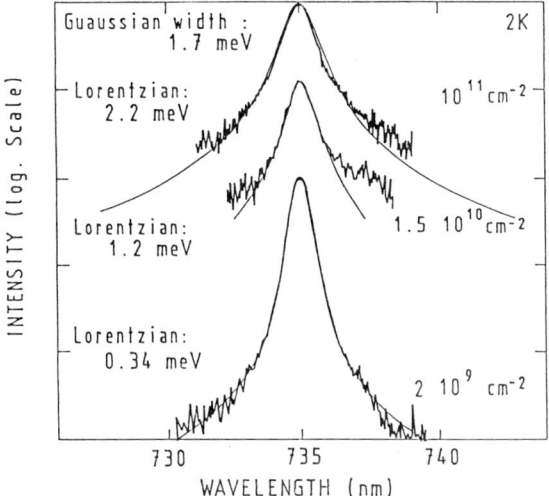

Figure 7. Excitonic lineshapes under resonant excitation at different densities. The fit is a convolution of a constant Gaussian linewidth (1.7 meV) and a density dependent Lorentzian linewidth.

increasing the Lorentzian linewidth (see the fits in Fig. 6). A moderate temperature increase, up to 50 K, also leads to a similar increase of the Lorentzian broadening.

5.4. TEMPERATURE DEPENDENT EFFECTS

Temperature effects have been studied, for resonant excitation conditions, using three different densities: very low, medium, and high density. At very low density (~ 4×10^8 cm^{-2} and below), a short radiative decay is observed at all temperatures between 10 and 60 K. The intensity of the signal does not change with temperature either. At high excitation density (above 5×10^9 cm^{-2}) a long decay time is already observed at low temperature and the increase of temperature does not change greatly the time behaviour in our samples where the maximum time (~ 200 ps) is limited by tunneling out of the double barrier structure.

Under medium excitation conditions (~ 2×10^9 cm^{-2}), a clear variation of the time behaviour with temperature is observed, as shown in Fig. 7. When the temperature is increased, the decay curve becomes shorter at short times and longer at long times (as in the case of increased excitation density). Correlatively, the luminescence intensity decreases at high temperatures.

5.5. DISCUSSION

5.5.1. *Density Effects*. We now discuss our results: at low temperature, under resonant excitation, upon increase of the excitation density ϕ, the dephasing time decreases because of exciton-exciton collisions [28]. The shorter dephasing time is clearly evidenced by the increased homogeneous linewidth and can be estimated by our fitting procedure. Even if we are at low temperature, the excitons can be scattered to different non radiative states: the triplet states, the longitudinal excitons, and the exciton states with k above $n\omega_0/c$. The respective importance of these states is difficult to assert. In Ref. [29], we have used a simple two level model to show that the scattering out of the radiative state with a time constant corresponding to the observed homogeneous width could indeed explain the observed variation of the decay times. Such a system has a short time solution decaying as $1/\tau = 1/\tau_0 + 1/\tau_{deph} + 1/\tau'$, which correspond to the decay at short times and cannot be fully resolved due to our limited time resolution, and a long time solution given by:

$$\tau_{lum} = \frac{1/\tau_0 + 1/\tau_{deph} + 1/\tau'}{1/\tau_0 \times 1/\tau'}, \qquad (1)$$

where τ_{deph} is the dephasing time, τ_0 the radiative lifetime in the absence of dephasing events, and τ' is a phenomenological parameter describing the scattering back to radiative states. From the fit of the observe decay times, we obtained for the radiative decay time, in the absence of scattering mechanism, a value of 10 ± 4 ps. This value is in quite good agreement with the theoretically predicted value of 8.4 ps: for our sample structure, the infinite well is a good approximation and we only correct Hanamura's value of 2.8 ps [25] by a factor of three to account for the refractive index of GaAs [26]. In our low density experiment, assuming a value of 10 ps for τ_0 and the same value for τ' and τ_{deph}, we get a short time solution of 3 ps, i.e., unobservable in our experimental conditions, and a long time solution of 30 ps, which approximately corresponds to our experimental result.

5.5.2. *Temperature Effects.* Temperature effects are more difficult to model. If we follow Ref. [22] or [23] and assume a thermalized exciton distribution at all times, the variation of the luminescence decay time should be linear in T, reflecting the proportion of excitons with wavevector $|k| < n\omega_0/c$. However, the phonon scattering time (for a single scattering event) is expected to be a few picoseconds at low temperatures [22] (1.5 ps at 9 K and 0.5 ps at 15 K). As a consequence, for resonant excitation, low temperature, and low density, we cannot safely assume a thermalized distribution.

Our experiments at very low excitation density indeed seem to show that the exciton gas never reaches a thermalized distribution, as no variation of the time behaviour is observed. As a matter of fact, for densities in the 10^8 cm^{-2} range, the average distance between excitons is of the order of 1 µm, and we are in the limit of a non-interacting gas, which does not reach a temperature. Then the scattering effects cannot be described as in a thermal distribution. For a temperature to be reached, a sufficient number of exciton-exciton collisions is needed in order to average the energies of the excitons.

We have modelled the scattering of a $k = 0$ exciton by acoustical phonons in the following way. We write the exciton-acoustical phonon interaction in the usual way:

$$H_{exc-ph} = \sum V^{\lambda\lambda'}(q) B^{\dagger}_{\lambda,k+q} B_{\lambda',k}(a_q + a^{\dagger}_{-q}), \qquad (2)$$

where the interaction potential $V^{\lambda\lambda'}$ is given by:

$$V^{\lambda\lambda'}(q) = \int d^2r \int dz_e \int dz_h \, \varphi^*_{\lambda}(r,z_e,z_h) \varphi_{\lambda}(r,z_e,z_h) \\ \times [U_{qe} \exp\{i(Q_e r + q_z z_h)\} - U_{qh} \exp\{i(Q_h r + q_z z_e)\}] \qquad (3)$$

and where $B^\dagger$ and $B(a_q, a^\dagger_{-q})$ are the exciton and photon creation and annihilation operators, $\varphi_\lambda(r,z_e,z_h)$ is the wavefunction of the exciton, Q_e and Q_h are the wave vectors of the electron and the hole making the exciton, and U_{qe} and U_{qh} the appropriate coupling coefficients, depending if we consider acoustical phonons or piezoelectrical phonons. As we use resonant excitation, we will only consider scattering of a k = 0 exciton to any possible state (we restrict ourselves to the 1s band of excitons). We compute the scattering times using the Fermi golden rule, and assume an infinite well for sake of simplicity.

For the wavevectors of interest to us, the scattering by piezoelectric phonons can be neglected as in 3D, and the main contribution comes from the acoustical phonons. We finally come to:

$$\tau_{scatt}(kT) = \frac{4h^3 \rho u^2 L}{3M(D_e - D_h)^2} \times \frac{1}{kT}, \qquad (4)$$

leading to a value of 50 ps for a temperature of 3 K, and of course 5 ps for a temperature of 30 K. The main conclusion is that the scattering time of k = 0 excitons is not appreciably shorter than the decay time. Furthermore, if we take an average value of 0.25 for the reduced mass of the exciton (see Andreani et al. [23]), scattering of the exciton will not bring the exciton out of the radiative zone. As a consequence, one scattering event is not enough to change the radiative behaviour of the exciton, and this explains why the radiative behaviour of the exciton does not change if we increase the temperature. Of course, such a picture is only true at very low excitation densities where exciton-exciton collision can be neglected.

Under medium excitation conditions, a thermalized distribution may more safely be assumed. However, the initial distribution is not a thermal distribution as the excitons are excited resonantly. Therefore, the initial decay corresponds to the time needed to reach a thermal distribution, and the subsequent decay corresponds to the recombination of an exciton gas thermalized at the temperature of the lattice. Figure 7 evidences that the long decay time indeed varies approximately linearly with temperature. The short time decay reduces with temperature as a result of the faster scattering out of the radiative states.

5.5.3. Possible Parasitic Effects: (i) Non-radiative Trapping. One might argue that our short lifetimes are simply due to localized excitons (on impurities or interface defects: as was assumed by Oberhauser et al. [41]). The short lifetime might have two explanations: short radiative lifetime of trap centres (giant oscillator strength of bound excitons [42]) or rapid capture by non-radiative centres. The lifetime increase with power or with temperature being simply due to the saturation of these localized states and to the release of free excitons. Indeed, the temperature and density where we observe the changes correspond to reasonable defect densities and localization energy. We discard this explanation for the following reasons:

1. First, it is only in the samples of the highest available quality that the very short lifetimes of excitons is observed. Reducing the quality of the sample, and thus increasing the density of possible trapped excitons, increases the radiative lifetime [39] in agreement with our model and contrary to what a model of localized excitons would predict. Furthermore, the radiative lifetime of localized excitons is expected to be of the order of 1 ns. It could be shorter only for highly localized traps then giving rise to large energy shifts. This is contrary to our observation of a negligible shift compared to the linewidth of only 2 meV. In the low density regime, we do not observe any change in lineshape nor

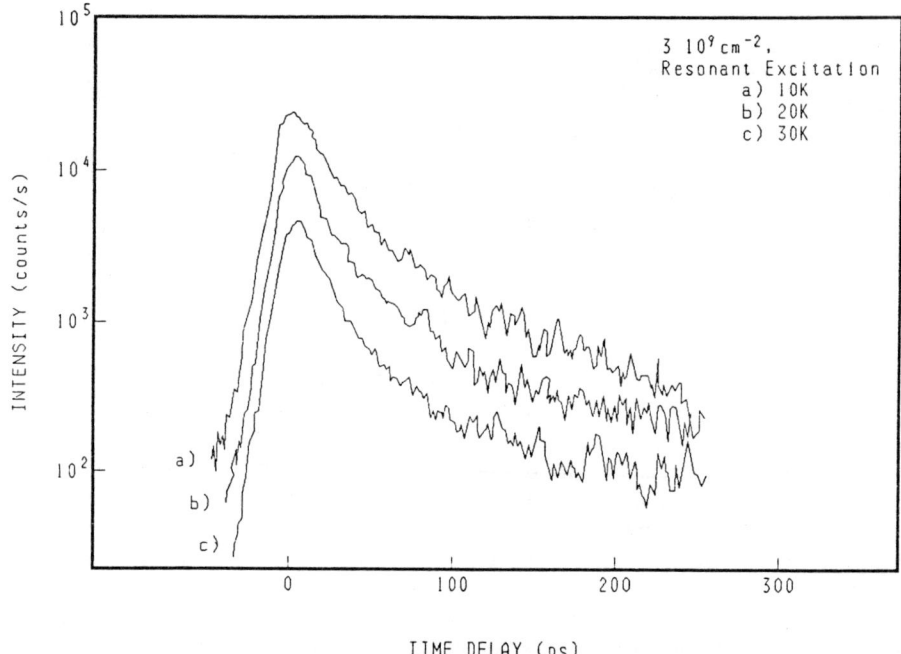

Figure 8. Exciton decay times as a function of the lattice temperature. As the temperature is raised, the excitons initially created at k = 0 leave the radiative zone. Therefore, a rapid decay is observed at short times when the exciton thermalize, and an increasingly long decay time at longer times. The excitation density is about 2×10^9 cm^{-2}. When the exciton density is 10 times larger, no changes in the decay times are observed up to 60 K.

    in peak position with time.
2. The assumption of rapid trapping on non radiative centres is contradicted by ourobservation of a diminution of the ratio $I_{lum}(0)/\phi$ upon density increase: in such a model, this ratio should stay approximately constant.

(ii) Rayleigh Scattering. It has also been argued [43] that the rapid transient at short times might be explained by Rayleigh scattering. Indeed, when the laser energy is brought to resonance with the exciton, a strong enhancement of the diffracted signal is observed, due to enhancement of the Rayleigh scattering [37]. At the same time, a strong enhancement of the luminescence intensity is also obtained as shown by the peak in the photoluminescence excitation spectrum, which is further increased in time resolved measurements by the decrease of the decay time. In our experimental conditions, we can show that the short times observed are luminescence lifetimes. We report time decays in conditions where the scattering from the laser (in off-resonance conditions) is much weaker than the luminescence signal. When the two signals are enhanced at resonance, the luminescence stays stronger. When we bring the laser closer and closer to the exciton resonance (see Fig. 6), the decay of the signal at the exciton energy becomes continuously shorter and shorter (correlatively, its intensity becomes stronger

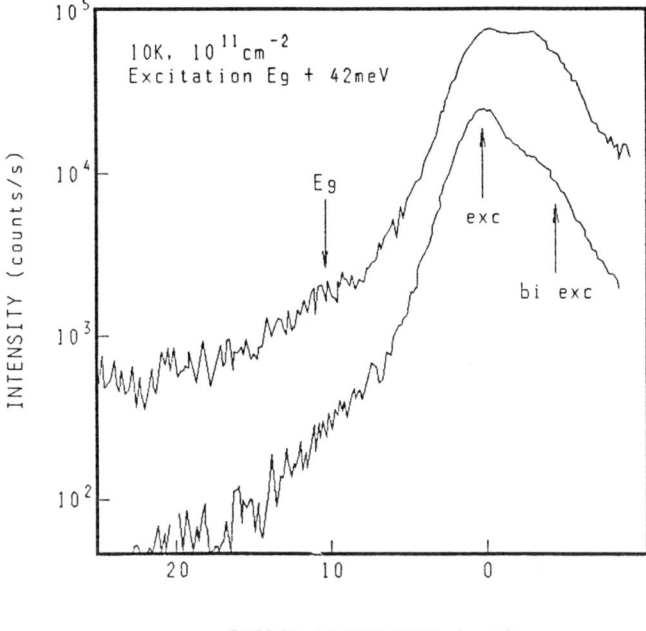

Figure 9. Luminescence spectra of a quantum well under high density excitation. The excitation energy is 42 meV above the band edge in order to allow fast creation of cold free electron-hole pairs. Upper curve, time window 0–20 ps; lower curve, time window 250–270 ps.

and stronger). If the signal was dominated by Rayleigh scattering, the variation should be opposite as the dephasing time gets shorter when the detuning is increased.

5.5.4. *Consequences of the Above Model: (i) Exciton Formation.* Formation of the exciton is a process which is more or less equivalent to exciton scattering. At low temperature, one electron and one hole will need a scattering event with an acoustical phonon to form an exciton. As a result of the very large transition probability of excitons, recombination of free electron-hole pairs will be difficult to observe. Basically, at a density of 10^9 cm^{-2} the radiative rate is four orders of magnitude larger for excitons than for free electron-hole pairs. On top of that must be added the population difference at low temperatures. The observation of free carrier recombination can nevertheless be achieved if we reduce the recombination probability of the excitons and increase that of free electron-hole pairs. Raising the temperature decreases the recombination probability not only of excitons but also of free electron-hole pairs. On the contrary, high excitation density favours the electron-hole pairs at the expense of excitons. A small contribution of electron-hole recombination can indeed be observed in Fig. 1 where very high excitation density was used.

We have thus performed time resolved measurements in order to evidence electron-hole pairs. Such transitions can very hardly be seen, but we report some evidence in Fig. 9. For this experiment, an excitation density of about 5×10^{10} cm^{-2} is used, with an energy corresponding to 42 meV above the band edge. By doing so, we create electrons that can relax very rapidly to the band edge by emitting one optical phonon. A small contribution of the electron-hole pairs can be evidenced on the spectrum at short times (see upper curve, time delay 0–20 ps). This contribution disappears on a time scale of the order of 100 ps (see lower curve, time delay 250–270 ps). If excitation energies not resonant with one optical phonon are used, i.e., if we create hot electrons, the signal is weaker and the estimated times are longer.

(ii) Directivity of the Signal. One of the consequences of the description in terms of free exciton recombination is the following: if only excitons very close to k = 0 do recombine, such excitons do not have any in-plane momentum. As a result, the emitted photons should have a k vector almost perpendicular to the plane of the quantum well. We have checked this effect on transitions where the exciton line is indeed a Lorentzian. Very preliminary experiments indeed show, especially by comparing the exciton luminescence to the GaAs buffer luminescence, that the exciton line can only be detected perpendicular to the substrate (within approximately 10 degrees, whereas this was 45 degrees for the buffer luminescence).

6. Conclusions

In summary, we have clearly evidenced the very fast radiative decay of excitons brought about by the breakdown in translational symmetry in quantum wells. The radiative lifetime deduced from our experiments is 10 ± 4 ps, in quite good agreement with theory. Our results easily explain the obvious difference between bulk GaAs samples where bound transitions dominate the spectrum even in the highest quality samples, and GaAs QWs where free exciton transitions dominate even in medium quality samples. Variations of the exciton radiative behaviour with temperature, excitation density, and detuning have been reported and explained.

7. Acknowledgments

The authors express their thanks to V. Agranovitch, I. Bar Joseph, G. Bastard, A. Chomette, B. Lambert, A. Regreny, and C. Weisbuch for useful discussions. D.S. Katzer is an O.N.T. Post Doctoral Fellow.

References

1. R.J. Elliott, Phys. Rev. **108**, 1384 (1957).
2. See, for example, D.S. Chemla, D.A.B. Miller, P.W. Smith, A.C. Gossard, and W. Wiegmann, IEEE J. Quantum Electron. **QE20**, 265 (1984).
3. W. Knox, R.L. Fork, M.C. Downer, D.A.B. Miller, D.S. Chemla, and C.V. Shank, Phys. Rev. Lett. **54**, 1306 (1985); D.S. Chemla and D.A.B. Miller, J. Opt. Soc. Am. B **2**, 1155 (1985).
4. For a review, see, for example, S. Schmitt Rink, D.S. Chemla, and D.A.B. Miller, Advances in Physics **38**, 89 (1989).
5. G. Bastard, E.E. Mendez, L.L. Chang, and L. Esaki, Phys. Rev. B **26**, 1974 (1982).
6. D.S. Chemla, Helv. Phys. Acta **56**, 607 (1983).
7. C. Weisbuch, in *Semiconductors and Semimetals*, edited by R. Dingle (Academic Press,

New York, 1987), p. 1.
8. C. Weisbuch, R.C. Miller, R. Dingle, A.C. Gossard, and W. Wiegmann, Solid State Commun. **37**, 29 (1981).
9. E.O. Göbel, H. Jung, J. Kuhl, and K. Ploog, Phys. Rev. Lett. **51**, 1588 (1983).
10. R.C. Miller and D.A. Kleinmann, J. Lumin. **30**, 512 (1985).
11. J. Lee, E.S. Koteles, and M.O. Vasell, Phys. Rev. B **33**, 512 (1986).
12. R. Del Sole, A. D'Andrea, and A. Lapiccirella, *Excitons in Confined Systems* (Springer-Verlag, Berlin, 1988).
13. J.J. Hopfield, Phys. Rev. **112**, 1555 (1958).
14. Y. Toyozawa, Suppl. Prog. Theor. Phys. **12**, 111 (1959).
15. C. Weisbuch and R.G. Ulbrich, Phys. Rev. Lett. **39**, 654 (1977).
16. See, for example, W.J. Rappel, L.F. Feiner, and M.F.H. Schuurmans in Ref. [12].
17. G.W. t'Hooft, W.A.J.A. van der Poel, L.W. Molenkamp, and C.T. Foxon, Phys. Rev. B **35**, 8281 (1987).
18. Ya. Aaviksoo, J. Lumin. **48 & 49**, 57 (1991).
19. Ya. Aaviksoo, Ya. Lippmaa, and T. Reinot, Opt. Spectrosc. (USSR) **62**, 419 (1987).
20. V.M. Agranovitch and O.A. Dubovskii, JETP Lett. **3**, 223 (1966).
21. M. Orrit, C. Aslangul, and P. Kottis, Phys. Rev. B **25**, 7263 (1982).
22. E. Hanamura, Phys. Rev. B **38**, 1228 (1988).
23. L.C. Andreani, F. Tassone, and F. Bassani, Solid State Commun. **77**, 641 (1990).
24. J. Feldmann, G. Peter, E.O. Göbel, P. Dawson, K. Moore, C.T. Foxon, and R.J. Elliott, Phys. Rev. Lett. **59**, 2337 (1987).
25. See, for example, A. Honold, L. Schultheis, J. Kuhl, and C.W. Tu, Phys. Rev. **40**, 6442 (1989).
26. C. Weisbuch, R.C. Miller, R. Dingle, A.C. Gossard, and W. Wiegmann, Solid State Commun. **37**, 219 (1981).
27. C. Weisbuch, R. Dingle, A.C. Gossard, and W. Wiegmann, Solid State Commun. **387**, 709 (1981).
28. B. Deveaud, J.Y. Emergy, A. Chomette, B. Lambert, and M. Baudet, Appl. Phys. Lett. **45**, 1078 (1984).
29. H. Sakaki, M. Tanaka, and J. Yoshino, Jpn. J. Appl. Phys. **24**, L417 (1985).
30. A. Honold, L. Schultheis, J. Kuhl, and C.W. Tu, Phys. Rev. B **40**, 6422 (1989).
31. B. Deveaud, F. Clérot, N. Roy, K. Satzke, B. Sermage, and D.S. Katzer, Phys. Rev. Lett. **67**, 2355 (1991).
32. D. Gammon, B.V. Shanabrook, and D.S. Katzer, Appl. Phys. Lett. **56**, 2710 (1990).
33. B. Sermage, B. Deveaud, F. Clérot, C. Dumas, and D.S. Katzer, J. Non-Linear Optics, to be published.
34. It is important to remember, as far as resonant experiments are concerned, that care is taken, during sample preparation and cooling, to ensure the surface cleanliness so that diffraction from the laser beam is reduced to a very low value. In the best conditions, this diffracted signal is much weaker than the luminescence signal. All spectra reported in this paper correspond to such a weak diffraction limit.
35. M. Colloci, M. Gurioli, A. Vinattieri, F. Fermi, C. Deparis, J. Massies, and G. Neu, Europhys. Lett. **12**, 417 (1990).
36. B. Deveaud, J. Shah, T.C. Damen, and C.W. Tu, Appl. Phys. Lett. **51**, 828 (1987).
37. J. Hegarty, M.D. Sturge, C. Weisbuch, A.C. Gossard, and W. Wiegmann, Phys. Rev. Lett. **49**, 930 (1982).

38. J.I. Kusano, Y. Segawa, Y. Aoyagi, S. Namba, and H. Okamoto, Phys. Rev. B **40**, 1685 (1989).
39. T.C. Damen, J. Shah, D.Y. Oberli, D.S. Chemla, J.E. Cunningham, and J.M. Kuo, Phys. Rev. B **42**, 7434 (1990).
40. M. Zachau, J.A. Kash, and W.T. Masselink, in *Quantum Optoelectronics*, Salt Lake City, March 1991 (Optical Society of America, Washington, 1991), Tech. Digest Series Vol. 7, p. 206.
41. D. Oberhauser, H. Kalt, W. Schlapp, H. Nickel, and C. Klingshirn, J. Lumin. **48 & 49**, 717 (1991).
42. E.I. Rashba and G.E. Gurgenishvili, Sov. Phys. Semicond. **4**, 759 (1962).
43. R. Eccleston, R. Strobel, W.W. Rühle, J. Kuhl, B.F. Feuerbach, and K. Ploog, Phys. Rev. B **44**, 1395 (1991).

TIME RESOLVED FOUR WAVE MIXING IN GaAs/AlAs QUANTUM WELL STRUCTURES

E.O. GÖBEL,[1] M. KOCH,[1] J. FELDMANN,[1] G. VON PLESSEN,[1]
T. MEIER,[1] A. SCHULZE,[1] P. THOMAS,[1] S. SCHMITT-RINK,[1†]
K. KÖHLER,[2] AND K. PLOOG[3]
[1]*Philipps-Universität Marburg, Fachbereich Physik and Wiss. Zentrum für Materialwissenschaften, Renthof 5, 3550 Marburg, Germany*
[2]*Fraunhofer Institut für Angewandte Festkörperforschung, 7800 Freiburg, Germany*
[3]*Paul-Drude-Institut für Festkörperelektronik, Hansvogteiplatz 5-7, O-1086 Berlin, Germany*

ABSTRACT. Femtosecond nonlinear laser spectroscopy is applied to study the dynamics of coherent optical excitations in semiconductors. In particular, beat phenomena of excitonic excitations in quantum wells and superlattice structures will be considered. We will demonstrate that four wave mixing experiments with time resolved detection of the nonlinear signal are able to distinguish between quantum beats and polarization interference. In addition, we report for a GaAs/AlAs superlattice the observation of quantum beats in the photon echo signal. The occurence of these beats is attributed to the simultaneous excitation of bound and localized excitons.

1. Introduction

Time resolved coherent laser spectroscopy has become a powerful tool for investigating the fundamental processes of light matter interaction. Whereas coherent laser spectroscopy by now is an almost traditional technique, as, for example, in atomic and molecular spectroscopy, it has only recently been established in the semiconductor area. This is basically due to the fact that phase relaxation of intrinsic optical excitations in semiconductors typically occurs on a picosecond or subpicosecond time scale. Application of coherent laser spectroscopy to semiconductors and semiconductor microstructures consequently requires temporal resolution in the femtosecond and picosecond regime.

A most powerful experimental technique for studying the dynamics of coherent optical excitations is transient four wave mixing (FWM) [1], as discussed in several contributions within these proceedings [2-4]. In these experiments the sample is excited by two or three subsequent pulses and the coherent nonlinear signal emitted into the phase matching direction is detected as a function of time delay between the incoming pulses. Although time integrated detection (i.e., employing a slow detector) already provides some basic information, for example, (within a factor of two) on the dephasing time of the transition under investigation, many fundamental questions can only be answered if the real time behavior of the nonlinear signal is known. The most evident example is the distinction between a free induction decay and a photon echo signal in the case of homogeneous and strong inhomogeneous broadening, respectively. Another example, which will be considered here, is the distinction between quantum beats and polarization interference.

Quantum beats occur in an electronic system with two (or more) optical transitions involving one (or more) common state, which, in most cases, is the system ground state. In contrast, polarization interference is observed in systems with two or more optical transitions with no state in common. Quantum beats thus reflect an intrinsic quantum mechanical interference, while polarization beats can be considered as a classical interference phenomenon occuring at the detector. Even though conceptionally this distinction is straightforward, in many real cases the distinction is ambiguous. The basic underlying physical question is whether there is an interaction between the levels involved in the different transitions or not.

In semiconductors, beat phenomena in the time integrated FWM signal have been observed for light and heavy hole exciton transitions in quantum wells [5,6], for free and bound exciton transitions in GaAs [7] and CdS [8,9], for excitons in AgBr split by an external magnetic field [10], and for excitons from the lower and upper polariton branch [11]. Oscillations in the polarization decay of a quantum well structure with two exciton transitions arising from spatially separated islands with monolayer thickness fluctuation have also been considered as quantum beats [12]; however, this already may serve as an example where an identification becomes questionable. If coupling of the spatially separated excitonic excitations exists, for example, due to dipole-dipole interaction, this indeed would correspond to quantum beats. In contrast, if there is no interaction between these two subsystems, only polarization interference can occur.

In the first part of this paper we demonstrate that time resolved detection of the FWM signal allows us to distinguish definitely between quantum beats and polarization interference. In the second part we will report the observation of quantum beats in the photon echo emitted from a GaAs/AlAs short period superlattice (SPS) sample, which allows us to resolve two distinct exciton transitions hidden within the inhomogeneously broadened emission and absorption spectra.

2. Experimental

The experimental set-up for the FWM experiments with time resolved detection of the nonlinear signal is schematically depicted in Fig.1 (a similar experimental set-up has also been described in Refs. [13–15]). We have applied the so-called two beam geometry, where two coherent laser pulses are focused onto the sample under directions corresponding to wavevectors k_1 and k_2, respectively. Pulse No. 2 can be delayed by a time τ with respect to pulse No. 1 by using a variable optical delay line. The nonlinear signal emitted coherently into the direction $2k_2 - k_1$ is up-converted in a 1 mm thick LBO crystal with parts of the pump pulse No. 1 split from the excitation pulse train and delayed by a time t (pulse No. 3). Thus, for any given delay time τ the real time response of the nonlinear signal can be obtained by scanning the delay time t. A mode-locked Ti:sapphire laser providing a 80 MHz pulse train of frequency tunable 100 fs pulses is used in the experiments, which are performed at a sample temperature of 5 K. We have employed various GaAs/AlAs quantum wells and short period superlattice structures which will be described in more detail below.

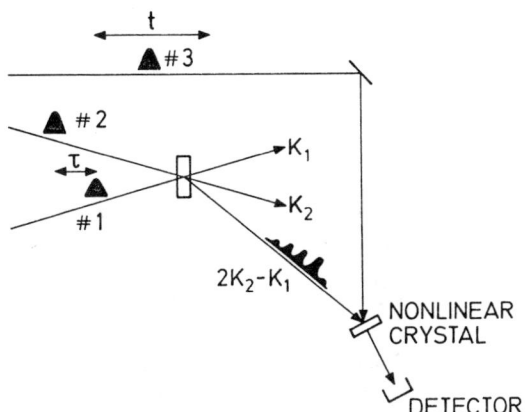

Figure 1. Schematic illustration of the transient FWM experiment with time resolved detection of the nonlinear signal.

3. Experimental results and discussion

3.1. QUANTUM BEATS VERSUS POLARIZATION INTERFERENCE

As pointed out in the introduction, simultaneous coherent excitation of two closely spaced optical transitions may lead to interference phenomena of the created polarization, which can be detected as a beating of the emitted FWM signal, provided the beat period is smaller than the dephasing time T_2 of the respective transitions. Yet, the case of quantum beats, which occurs if the two transitions have one electronic state in common, has to be distinguished from polarization beats, which is not a quantum mechanical interference phenomenon but rather related to classical optics. The fundamental difference between a three level system and two independent two-level systems with respect to the dynamics of the macroscopic polarization is illustrated in Fig. 2, where the time evolution of the complex polarization vector is illustrated for the two respective cases. It is assumed that in either case all electronic transitions are dipole allowed and excited simultaneously by a sufficiently short (i.e., spectrally broad) laser pulse. We are considering the dynamics of the macroscopic polarization for the special case corresponding to a time resolved FWM experiment, where the system is excited by two coherent pulses impinging onto the sample at times $t = 0$ and $t = \tau$. We then consider the subsequent dynamics of the nonlinear polarisation (in third order of the driving fields), which acts as a source for the detected signal. For the sake of clarity, we assume purely homogeneously broadened transitions and neglect dephasing as well as many body interactions. In the case of quantum beats (three level system), after the first pulse at $t = 0$, the macroscopic polarization of the given three level system with transition frequencies ω_1 and ω_2, respectively, can be represented by one single arrow moving in the complex plane according to [16]

$$P(t) \propto \cos\left(\frac{\Delta\omega \cdot t}{2}\right) \exp\left(i\left(\frac{\omega_1 + \omega_2}{2}\right)t\right). \tag{1}$$

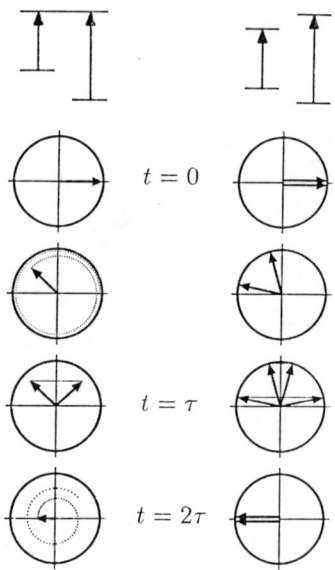

Figure 2. Illustration of the time evolution of the polarization vectors of a three level system (left) and two independent two level systems (right) in the FWM experiments described in the text. The time evolution is represented in the complex polarization plane with the horizontal and vertical axis corresponding to, respectively, the real and imaginary part of the polarization.

The quantum interference thus manifests itself in a periodic modulation of the magnitude of the polarization vector (beat frequency $\Delta\omega/2 = (\omega_1-\omega_2)/2$). The phase conjugation ($P \rightarrow -P^*$) induced by the second pulse at $t = \tau$ can be visualised as a mirror reflection of the polarization vector at the imaginary axis. From there on, the third order polarization develops further according to:

$$P^3(t,\tau) \propto -P^*(\tau)P(t-\tau)$$
$$= \cos\left(\frac{\Delta\omega \cdot \tau}{2}\right)\cos\left[\frac{\Delta\omega \cdot (t-\tau)}{2}\right]\exp\left[i\left(\frac{\omega_1+\omega_2}{2}\right)(t-2\tau)\right]. \qquad (2)$$

Consequently, the nonlinear signal, which is proportional to $|P^3|^2$, will, as a function of $t \geq \tau$, exhibit a maximum at $t = \tau$ and further maxima at $t = nT_b$, where T_b is the beat period $T_b = 2\pi/\Delta\omega$ and n is an integer.

The situation is significantly different or the case of polarization interference, as illustrated on the right hand side of Fig. 2. Here the polarization corresponding to the two independent transitions is characterized by two different arrows. Their independent development after the first pulse is described by:

$$P(t) = P_1(t) + P_2(t), \qquad (3)$$

where $P_j(t) = \exp(i\omega_j t)$. The action of the second pulse at $t = \tau$ again corresponds to a mirror reflection of both polarization vectors as shown in the figure. Depending on the instantaneous relative phase factor, the resulting polarization vector will be amplitude modulated with a period corresponding again to $\Delta\omega/2$. The resulting signal due to the nonlinear polarization will be determined by the superposition $|P^3|^2 = |P_1^3 + P_2^3|^2$. In any case, at $t = 2\tau$ the signal will exhibit a maximum, since at $t = 2\tau$ the relative phase factor will be zero again. This is actually the basic mechanism, which, in the case of an inhomogeneously broadened transition, leads to the occurence of a sharp photon echo at $t = 2\tau$. In fact, the two independent transition can be considered as the very first step towards inhomogeneous broadening. Since there will definitely be a maximum at $t = 2\tau$, further maxima will be observed at $t = 2\tau \pm nT_b$, with the restriction $t \geq \tau$. Plotting the position of the maxima of the nonlinear signal in a (t,τ)-plot thus allows unambigious distinction between quantum beats and polarization interference. It should finally be mentioned that inclusion of the so far neglected processes like dephasing, inhomogeneous broadening and many body effects basically set up a limited time window over which the beating is observable in real time.

In order to prove these ideas we have investigated two different quantum well structures, which have been designed in order to exhibit quantum beats or polarization interference, respectively [17]. For the case of quantum beats we have chosen the light and heavy hole exciton in a GaAs/AlGaAs quantum well structure. According to the spectral width of the Ti:sapphire laser pulses of about 20 meV, we have chosen a quantum well structure with quantum well thickness of 15 nm, where the corresponding splitting of the light and heavy hole exciton transition amounts to 5.4 meV with a heavy hole transition at about 1.533 eV. We have used a multiple quantum well structure with 40 periods. The GaAs substrate has been removed by selective etching in order to allow transmission experiments. The linear absorption spectrum of this sample (not shown here) clearly reveals the heavy and light hole exciton transitions. The spectral width of the heavy hole exciton luminescence line amounts to 0.8 meV, which is an indication of the high quality of the GaAs-AlAs interfaces.

The time behavior of the FWM signal is shown in Fig. 3 in a three dimensional plot, where the intensity of the FWM signal is plotted versus the time delays τ and t. For any given delay time τ, the FWM signal shows a free induction decay determined by the dephasing time T_2, which is about 1.3 ps for this particular sample (note that the signal decay yields a decay time of 0.65 ps). The observation of a free induction decay as opposed to a photon echo demonstrates the predominantly homogeneous nature of the optical transition. The free induction decay exhibits a pronounced beating with a beat period $T_b = 0.77$ ps, which corresponds to the light and heavy hole exciton splitting. The position in time, t, of the maxima increases with increasing τ, as is more clearly seen in a fan chart shown in the inset of Fig. 3. Here we have plotted the position of the various relative maxima of the FWM signal versus the time delays τ and t. This plot clearly shows the occurence of relative maxima at $t = \tau + nT_b$ as expected for a quantum beat signal (the missing of the first maximum at $t = \tau$ is attributed to Coulomb effects [14,18,19], as will be discussed in more detail in the contributions of Chemla [2] and Shah [3], respectively).

For the demonstration of polarization interference we have designed a quantum well structure containing two separate wells with different quantum well thicknesses. The particular GaAs/AlAs sample used in this experiment consists of 40 wells with thickness L_z of 9 nm separated by 10 nm thick AlAs barriers, followed by 40 wells with $L_z = 8$ nm which are again separated by 10 nm AlAs barriers. Due to the different quantum well

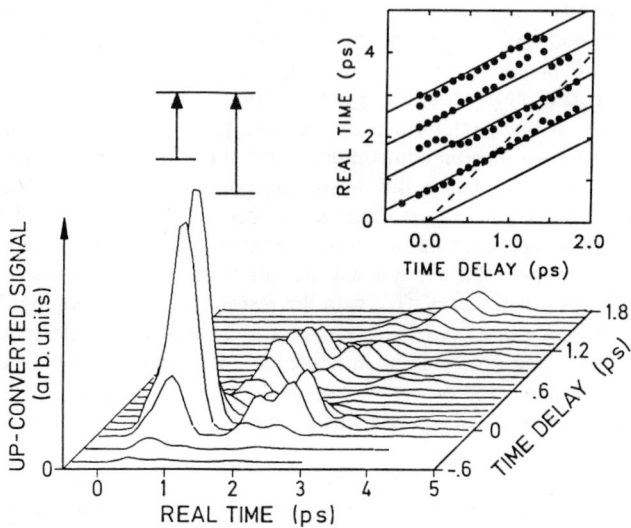

Figure 3. FWM signal versus time delay τ and real time t for the quantum well sample with $L_z = 15$ nm. The inset shows the position of the beat maxima (dots) as a function of delay time τ and real time t. The full lines represent the function $t = \tau + nT_b$ with n = 0,1,2,3,4 and $T_b = 0.77$ ps. For comparison, the dashed line corresponds to $t = 2\tau$.

thicknesses the lowest heavy-hole exciton transitions of the two subsets of multiple quantum wells are energetically separated by 12.9 meV. The light and heavy hole exciton splitting in this case is larger and amounts to 22 meV for the $L_z = 9$ nm wells. Thus selective excitation of the two heavy hole transitions of the two different quantum wells is possible without simultaneous excitation of the light and heavy hole transitions. The time behavior of the FWM signal for the case of simultaneous excitation of the two heavy hole exciton transitions corresponding to the quantum wells with $L_z = 8$ nm and $L_z = 9$ nm, respectively, is shown in Fig. 4. We again observe a beating in the intensity of the FWM signal with a period T_b, which now corresponds to the energetic separation of the two heavy hole exciton transitions. However, opposite to the results shown in Fig. 3, the FWM signal now always has a maximum at $t = 2\tau$. Again, this is seen more clearly in a fan chart, which is shown in the inset of Fig. 4. Besides the relative maximum at $t = 2\tau$, further maxima are observed at $t = 2\tau \pm T_b$. We thus conclude that this, in fact, corresponds to the case of polarization beats, implying that the heavy hole excitons in the two different multiple quantum well subsets do not interact with each other and can be regarded as two independent electronic systems. It should be pointed out that this does not yet imply that the beating of the FWM signal observed for quantum wells with island regions different in thickness by one monolayer cannot be attributed to quantum beats, since exciton coupling may exist under this condition. Yet, we do now know how to distinguish them and the appropriate experiment is under way.

It can also be seen quite nicely on the basis of Fig. 4 how a free induction decay develops into a photon echo signal for increasing inhomogeneous broadening. For the case of two distinct and independent transition energies the signal already peaks at $t = 2\tau$,

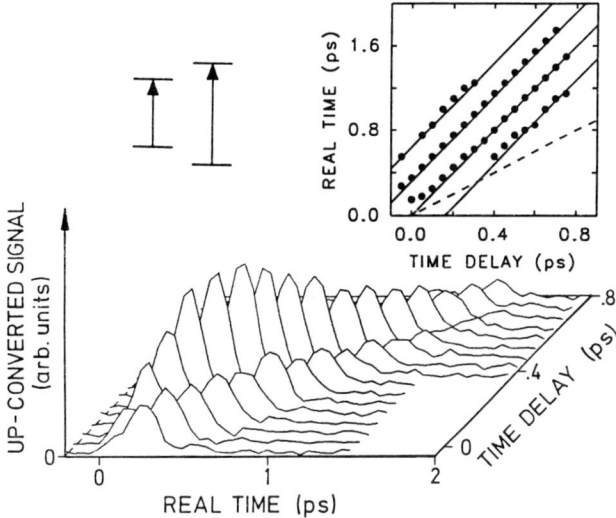

Figure 4. Same as Fig. 3 for the sample with separated quantum wells with thicknesses of 8 nm and 9 nm, respectively. However, the full lines in the inset now correspond to $t = 2\tau + nT_b$ with $n = -1,0,1,2$ and $T_b = 0.32$ ps, while, for comparison, the dashed line corresponds to $t = \tau$.

but at shorter and larger times, t, side maxima are still present. With increasing number of independent transitions, these side maxima will disappear due to destructive interference and only the maximum at $t = 2\tau$ will survive.

To summarize this part, we have demonstrated an experimental technique that allows a definite distinction between quantum beats and polarization interference. This distinction allows us to decide whether the different transitions involved do have states in common or not, which is not always possible from first principle considerations. We therefore believe that the technique of time resolved FWM with time resolved detection of the nonlinear signal might become a powerful method for studying coherent interactions of optical excitations in semiconductors.

3.2. QUANTUM BEATS IN A SHORT PERIOD SUPERLATTICE

One of the well-known features of quantum beat spectroscopy is the capability of resolving energy levels even in the case of strong inhomogeneous broadening where linear spectroscopy fails. In the case of strong inhomogeneous broadening, the FWM signal is expected to be emitted as a photon echo pulse with a temporal width determined either by the laser pulse width or the inverse spectral width of the inhomogeneous line, depending on which one is broader. For a strongly inhomogeneously broadened three-level system, quantum beats will be reflected in a periodic modulation of the photon echo intensity as a function of delay time τ with a modulation period again given by the inverse splitting of the center frequencies of the respective electronic transitions [20]. In contrast, polarization interference does not result in a modulation of the photon echo amplitude.

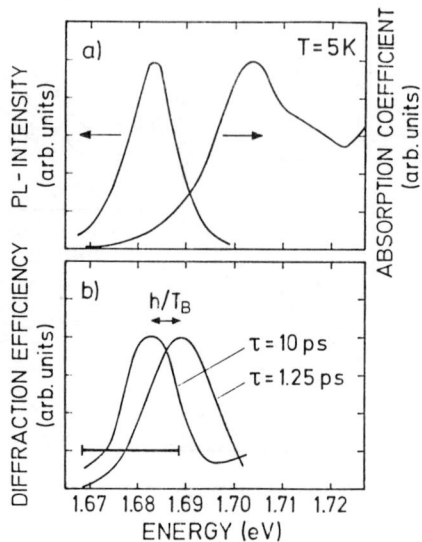

Figure 5. (a) Low temperature photoluminescence and absorption spectra of the GaAs/AlAs short period superlattice sample and (b) nonlinear response at two different delay times τ of 1.25 ps and 10 ps. The horizontal arrow in (b) shows the energy spacing h/T$_b$ obtained from the experimentally determined beat period T$_b$. The lower horizontal bar corresponds to the spectral width (FWHM) of the laser pulses.

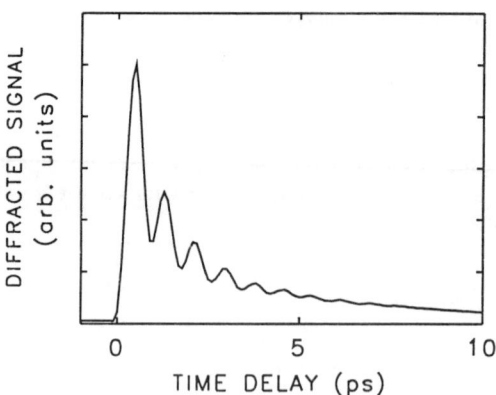

Figure 6. FWM signal of the GaAs/AlAs SPS sample measured time integrated as a function of time delay τ between the pump pulses.

We report here on FWM experiments performed on GaAs/AlAs short period superlattices. The particular sample investigated consists of 100 periods of 4.1 nm thick GaAs and 3 nm thick AlAs. The low temperature absorption and photoluminescence spectra are depicted in Fig. 5(a). The absorption peak at 1.705 eV corresponds to the n = 1

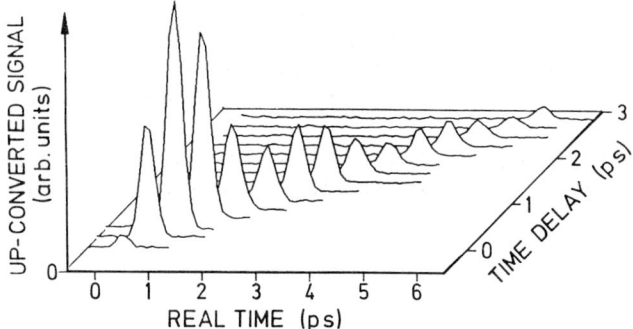

Figure 7. FWM signal of the GaAs/AlAs SPS-sample as a function of delay time τ and real time t.

heavy hole (hh) transition. The absorption shows a tail to lower energies indicating pronounced inhomogeneous broadening due to inter- and intra-well thickness fluctuations. The photoluminescence is spectrally located at the low energy side of the absorption peak and shows a broad unstructured band with about 12 meV full width at half maximum. The Stokes shift between the maximum of the hh-exciton in absorption and the luminescence maximum amounts to 22 meV.

The results of a FWM experiment with time integrated detection is shown in Fig. 6, where the intensity of the nonlinear signal is plotted versus time delay τ. The important features of these data are as follows. In the early time regime a pronounced beating with a period of $T_b = 0.8$ ps corresponding to an energy of 5.2 meV is observed. The decay of the beat amplitude is faster than the decay of the overall FWM signal. Finally, the maximum of the first beat structure does not occur at $\tau = 0$, but shows some delay of the order of 0.4 ps. From the experimental results we have to conclude that two different transitions separated by 5.2 meV are hidden within the inhomogeneously broadened absorption and emission bands. These two transitions must have different dephasing times. The faster one determines the decay of the beat amplitude while the slower one determines the overall decay of the FWM signal.

In order to decide whether the two transitions responsible for the beats are independent or do have (at least) one state in common, we have again time resolved the nonlinear signal. The experimental result is shown in Fig. 7, where again the nonlinear signal is plotted as a function of the time delays τ and t. In contrast to the examples shown above (Fig. 3 and Fig. 4), we now clearly observe the emission of a photon echo at $t = 2\tau$, demonstrating the dominant inhomogeneous broadening. The photon echo amplitude shows a periodic modulation that is responsible for the beat structure in the FWM experiment with time integrated detection (Fig. 6). This is clear evidence for quantum beats, i.e., the two transitions involved do have one state in common.

The nonlinear spectral response of the two transition involved can be determined by detecting the intensity of the diffracted signal as a function of laser photon energy at a delay time τ, where, respectively, the faster or slower component dominates. In Fig. 5(b) we show experimental results obtained for, respectively, $\tau = 1.25$ ps (corresponding to the first maximum in the time integrated FWM signal (cf. Fig. 6)) and $\tau = 10$ ps, where the beating is already damped out. Two distinct peaks with energy spacing corresponding exactly to the beat period can be clearly seen, yet the spectral width of these peaks is determined by the width of the laser spectrum. We attribute the higher and lower energy transition to localized and carbon acceptor bound exciton transitions, respectively [20]. The beat period in this case corresponds to the binding energy of the acceptor bound exciton. The value of 5.2 meV is consistent with reports in the literature [21]. It finally should be mentioned that the temporal width of the photon echo is slightly larger than the laser pulse width providing an independent measure of the inhomogeneous broadening, which amounts to 25 meV for the present sample [20].

4. Conclusion

In conclusion, we have demonstrated by two examples that time resolved detection of the nonlinear signal in a FWM experiment provides fundamental information on the detailed nature of optical excitations in semiconductors and their coherent interaction and dynamics. We have performed these experiments on different GaAs/AlAs quantum well structures in order to demonstrate the potential of this technique. In particular, we have demonstrated that it becomes possible to distinguish between quantum beats and polarization interference. Furthermore we have demonstrated that quantum beats in the case of strong inhomogeneous broadening are reflected in a periodic modulation of the photon echo amplitude.

Acknowledgment

We gratefully acknowledge expert technical assistance by M. Preis and the financial support through the Leibniz-Förderpreis.

References

†   Deceased.
1. See, for example, E.O. Göbel, in *Festkörperprobleme/Advances in Solid State Physics*, Vol. 30, edited by U. Rössler (Pergamon/Vieweg, Braunschweig, 1990), p. 269; K. Leo, E.O Göbel, T.C. Damen, J. Shah, S. Schmitt-Rink, W. Schäfer, J.F. Müller, K. Köhler, and P. Ganser, Phys. Rev. B **44**, 5726 (1991).
2. D. S. Chemla, these proceedings.
3. J. Shah, these proceedings.
4. D.G. Steel, S.T. Cundiff, H. Wang, M. Jiang, and V. Subramaniam, these proceedings.
5. K. Leo, T.C. Damen, J. Shah, E.O. Göbel, and K. Köhler, Appl. Phys. Lett. **57**, 19 (1990).
6. B.F. Feuerbacher, J. Kuhl, R. Eccleston, and K. Ploog, Solid State Commun. **74**, 1279 (1990).
7. K. Leo, T.C. Damen, J. Shah, and K. Köhler, Phys. Rev. B **42**, 11359 (1990).
8. H. Stolz, V. Langer, E. Schreiber, S. Permogorov, and W. von der Osten, Phys. Rev. Lett. **67**, 679 (1991).
9. K.H. Pantke, V.G. Lyssenko, B.S. Razbirin, H. Schwab, J. Erland, and J.M. Hvam, Phys.

Stat. Sol. (b) **173**, 69 (1992).
10. V. Langer, H. Stolz, and W. von der Osten, Phys. Rev. Lett. **64**, 854 (1990).
11. D. Fröhlich, A. Kulik, B. Uebbing, A. Mysyrowicz, V. Langer, H. Stolz, and W. von der Osten, Phys. Rev. Lett. **67**, 2343 (1991).
12. E.O. Göbel, K. Leo, T.C. Damen, J. Shah, S. Schmitt-Rink, W. Schäfer, J.F. Müller, and K. Köhler, Phys. Rev. Lett. **64**, 1801 (1990).
13. M.D. Webb, S.T. Cundiff, and D.G. Steel, Phys. Rev. Lett. **66**, 934 (1991).
14. S. Weiss, M.-A. Mycek, J.-Y. Bigot, S. Schmitt-Rink, and D.S. Chemla, Phys. Rev. Lett. **69**, 2685 (1992).
15. D.S. Kim, J. Shah, J.E. Cunningham, T.C. Damen, S. Schmitt-Rink, and W. Schäfer, Phys. Rev. Lett. **68**, 2838 (1992).
16. J. Feldmann, in *Festkörperprobleme/Advances in Solid State Physics*, Vol. 32, edited by U. Rössler (Pergamon/Vieweg, Braunschweig, 1992), p. 81.
17. M. Koch, J. Feldmann, G. von Plessen, E.O. Göbel, P. Thomas, and K. Köhler, Phys. Rev. Lett. **69**, 3631 (1992).
18. M. Lindberg, R. Binder, and S.W. Koch, Phys. Rev. A **45**, 1865 (1992).
19. W. Schäfer, F. Jahnke, and S. Schmitt-Rink, Phys. Rev. B **47**, 1217 (1993).
20. M. Koch, D. Weber, J. Feldmann, E.O. Göbel, T. Meier, A. Schulze, P. Thomas, S. Schmitt-Rink, and K. Ploog, Phys. Rev. B **47**, 1532 (1993).
21. See, for example, R.C. Miller, A.C. Gossard, W.T. Tsang, and O. Munteanu, Solid State Commun. **43**, 519 (1982).

SPIN DYNAMICS AND DIMENSIONALITY IN MAGNETIC SEMICONDUCTOR QUANTUM STRUCTURES

D.D. AWSCHALOM,[1] J.F. SMYTH,[1] AND N. SAMARTH[2]
[1]Department of Physics, University of California, Santa Barbara, CA 93106, USA
[2]Department of Physics, The Pennsylvania State University, University Park, PA 16802, USA

ABSTRACT. Complementary sets of low temperature magneto-optical experiments have been performed on a series of diluted magnetic semiconductor heterostructures in order to explore the effects of quantum confinement and reduced dimensionality on fundamental electronic and magnetic spin dynamics. In the first set, femtosecond time-resolved photoluminescence spectroscopy was used on a series of $Zn_{1-x}Mn_xSe/ZnSe$ spin superlattice structures to directly probe the spin-dependent dynamics as the carriers' confining potential was systematically controlled by a magnetic field. Exciton lifetimes and spin relaxation are found to be strongly dependent on both the energy and location of spin states in the heterostructures. In the second set of experiments, a variety of nonmagnetic $Zn_{1-y}Cd_ySe/ZnSe$ double quantum well heterostructures coupled by a thin diluted magnetic semiconductor barrier were studied by magneto-luminescence measurements. Spin splitting of the quantum confined ground state and the resulting large luminescence polarization were observed by tuning the barrier potential with a magnetic field. In the third set of experiments, femtosecond spectroscopic measurements of optically induced magnetization performed on $ZnTe/Cd_{1-x}Mn_xSe$ heterostructures reveal the formation and evolution of electron-based bound magnetic polarons. The magnetic spin dynamics on ultrashort time scales was found to vary dramatically with the degree of electron confinement. These experiments demonstrate the potential of spin-engineering to explore new carrier dynamics that are highly dependent on confinement and applied magnetic fields as well as the magnetic state of the host.

1. Introduction

The spin-dependent dynamics of carriers in semiconductors plays a fundamental role in the physics of heterostructures and is important in understanding both the equilibrium and nonequilibrium properties of such systems [1]. Recent investigations of carrier spin dynamics through the polarization analysis of interband optical transitions have provided considerable information about the time evolution of the spin states of electrons, holes, and excitons. For example, such studies have revealed new spin relaxation processes [2–10] and a variety of theories have been presented to explain these results [11–18]. In spite of this intense investigation, spin relaxation in quantum wells, and in particular the role of confinement, remains poorly understood [6]. Many quasi-two dimensional spin relaxation issues such as the effects of quantum well width and well depth on spin flip processes as well as many body effects like electron-hole exchange still need to be systematically investigated before a quantitative understanding of spin scattering in low dimensional systems can be formed.

A particularly attractive class of systems in which to address the connection between spin relaxation and quantum confinement is provided by quantum wells and superlattices containing diluted magnetic semiconductors (DMSs) [19]. These low dimensional DMS materials possess the convenient optical properties of III-V compounds, but also incorporates magnetic ions that can serve as spin scattering centers through carrier-ion interactions. In addition, the sp–d exchange, which is the exchange interaction between the p-like holes and the s-like electrons with the localized d-states of the magnetic ions, appears as a large effective g-factor (~ 1000) for the carriers at low temperatures. This large effective g-value acts as a magnetic field "amplifier" producing large Zeeman splittings between the carrier spin states in relatively modest applied magnetic fields. A wide variety of new DMS heterostructures have been fabricated using molecular beam epitaxy (MBE), with a great deal of freedom over band alignment, strain, and magnetic dilution, therein enabling one to separately vary the electronic and magnetic dimensionality [20]. Using bandgap engineering to tailor the electronic wavefunctions, one can independently control the interaction between the carrier spins and the localized magnetic moments, and the environment experienced by the magnetic ions themselves (i.e., paramagnetic, spin glass, and antiferromagnetic phases). These epitaxial materials preparations combined with low temperature experimental techniques such as femtosecond optical spectroscopy and integrated dc SQUID technology [21], which are discussed below, then enables one to obtain a real-time view of dynamical behavior in quantum magnetic systems.

In this paper we describe a set of experiments performed on three distinct DMS systems which are aimed at exploring the effects of quantum confinement and magnetic dimensionality on carrier spin dynamics in heterostructures. In the first section, results of femtosecond polarization spectroscopy from a series of "spin engineered" structures—the spin superlattice (SSL) [22–25]—are presented. In these systems, alternating layers of magnetic and nonmagnetic semiconductor are chosen such that there is essentially no quantum confinement of the carriers in zero magnetic field. The application of a magnetic field induces a spin dependent confining potential that eventually leads to a spatial and periodical separation of the opposite spin states of the electron and hole wavefunction. Analysis of the data shows that in low fields, excitons are confined to the magnetic quantum wells where spin flip scattering is directly observed on a time scale indicative of sp–d exchange interactions [15]. Then, with increasing field, the SSL forms and an extremely rapid spin flip is observed for carriers localized in the nonmagnetic wells.

The second set of experiments [26] are performed on a series of symmetric nonmagnetic double quantum wells (DQWs) coupled by a thin DMS barrier. The incorporation of magnetic spins into DQW heterostructures—in particular, into the separating barrier—allows one to probe the effects of spin dependent coupling, as well as to utilize the magnetic spins as "tracers" during tunneling events. Alternatively, the interaction of the carriers with the magnetic ions in the barrier can be used to optically explore the magnetic state of the barrier as a function of the barrier's thickness. Static magneto-optical photoluminescence experiments demonstrate an ability to "spin tune" the barrier potential with a magnetic field. Furthermore, measurements of the excitonic Zeeman energy splitting between spin states of the symmetric ground state as a function of the barrier's width indicates that the effective magnetic ion concentration depends upon the thickness.

The final set of experiments [5] described here presents the observation of magnetic polaron (MP) dynamics in magnetic quantum wells in new type-II $ZnTe/Cd_{1-x}Mn_xSe$ heterostructures. The type-II band alignment, with the holes spatially separated from the magnetic layers, creates a pure system in which electrons alone interact with the magnetic ions in a well defined geometry. In this case then, effects arising from electron-hole exchange can be neglected. In

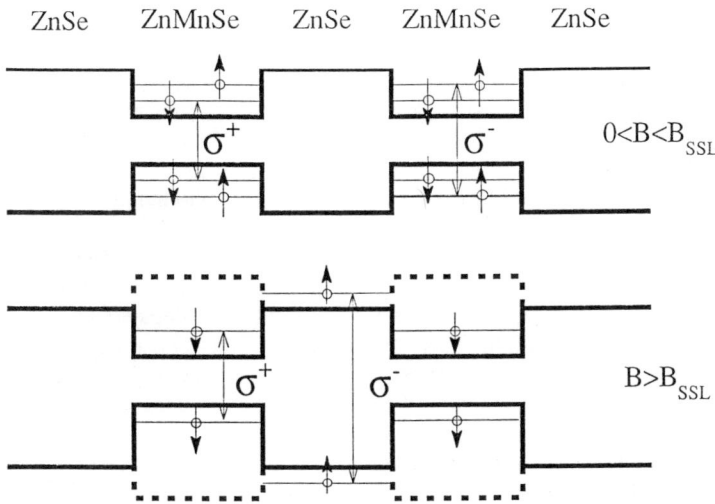

Figure 1. Schematic diagram of the band offsets and energy levels of the spin states for the two field-dependent regimes: $B < B_{SSL}$ where the carriers are confined to the magnetic quantum wells and $B > B_{SSL}$ where the spin states are spatially separated.

addition, the length scales are such that a two dimensional environment for the electronic degrees of freedom exists, while the magnetic ions continue to interact as in bulk. This enables the exploration of the dimensional dependence of donor-bound magnetic polaron behavior. Optically generated, spin-aligned carriers induce a magnetization, which was detected by a sensitive superconducting susceptometer. Strikingly, polarons were observed even for well widths as narrow as one-third the polaron diameter in the corresponding bulk material, with the in-plane diameter nearly unchanged. Moreover, an increase in the MP formation and decay times were observed with decreasing well dimension consistent with an effectively lower magnetic concentration seen by the electrons.

2. Spin Superlattices

The essential idea of a SSL is shown schematically in Fig. 1 [27]. Here a superlattice is composed of alternating layers of magnetic and nonmagnetic semiconductor such that the band offsets at zero applied field are small. Upon the application of a magnetic field, the large Zeeman splittings produce a large shift in the band edges of the magnetic semiconductor. By contrast, since there are no magnetic ions in the nonmagnetic layers, the splitting there is much smaller. When the Zeeman shift in the magnetic layers overcomes the zero field band offsets, the magnetic layers act as barriers for carriers in the spin up state and as quantum wells for carriers in the spin down state. Hence, electrons and holes with one spin state are confined in the magnetic layers while those with the opposite spin reside in the nonmagnetic layers. The carrier dynamics of such a novel structure may be expected to exhibit fundamental spin-dependent phenomena which are of contemporary interest in low dimensional systems.

The experiments use a series of $Zn_{1-x}Mn_xSe/ZnSe$ multilayer structures that were previously established as "spin superlattices" using magneto-optical absorption spectroscopy [23]. Each structure consists of a ten period sequence of {100 Å ZnSe/100 Å $Zn_{1-x}Mn_xSe$} grown on a (100) GaAs substrate after a 1 µm ZnSe buffer layer had been deposited. In addition, a 2000 Å ZnSe protective cap layer is grown to cover the structures. Sample growth is monitored in situ by 10 keV reflection high-energy electron diffraction and post growth characterization includes transmission electron microscopy, x-ray diffraction, and optical measurements. The Mn concentration in the series ranges from x = 0.05 to x = 0.09. Within this Mn composition range, the $Zn_{1-x}Mn_xSe$ bandgap "bows" [28] below that of the ZnSe gap and provides a minimal zero field offset (< 5 meV) [23] in both the conduction and valence bands. The $Zn_{1-x}Mn_xSe$ layers are coherently strained so that the in-plane lattice parameter matches that of the unstrained ZnSe layers, hence removing the heavy-hole–light-hole degeneracy, but only in the $Zn_{1-x}Mn_xSe$ layers. Note that the interband transitions in both the magnetic and nonmagnetic layers are type I. From the magneto-optical absorption studies the magnetic field for the formation of the SSL for these samples was determined to be $B_{SSL} \approx 0.5$ T and the details of the transition into a SSL have been reported elsewhere [23].

In the photoluminescence (PL) experiments, the samples were positioned in a variable temperature magneto-optical cryostat at 4.2 K using the Faraday configuration. Symmetry considerations show that the luminescence in the presence of the magnetic field is either left (σ^-) or right (σ^+) circularly polarized with the helicity determined by the spin state of the radiating exciton as shown in Fig. 1 [1]. To resolve the two circularly polarized components of the PL, and hence monitor the spin states directly, a $\lambda/4$ waveplate was used followed by a linear polarizer.

Time resolved PL spectroscopy is performed in these systems using the sum-frequency technique [29]. In this method, a laser pulse is used to excite the sample and a time delayed probe pulse interrogates a time slice of the collected luminescence as the two pass coincidentally through a thin nonlinear crystal, in this case β-barium borate (BBO). A small number of sum-frequency photons are then generated for those luminescence wavelengths that, along with the probe beam, satisfy the phase matching conditions of the crystal. In practice, "blue" femtosecond excitation pulses between 2.75 and 3.10 eV were generated by collinear, second harmonic generation in a thin BBO crystal using the transform limited output of a mode-locked titanium sapphire laser. Sum-frequency photons were efficiently generated using a mixing scheme in which an "unused" portion of the fundamental from the laser was mixed with collected luminescence in a second BBO crystal in order to probe the PL with a resolution of the order of the laser pulse width $\Delta t \approx 150$ fs. The signal photons were detected by a cooled photomultiplier tube equipped with a monochromator to provide additional energy resolution. Furthermore, a streak camera with $\Delta t \approx 10$ ps resolution was used in experiments that required relatively less time resolution than that obtained using the sum-frequency method.

The field dependent time-integrated PL from a SSL with x = 0.09 Mn is shown in Fig 2(a). Here, and in the rest of the data for the SSLs, the sample was excited by a linearly polarized beam with photon energy E = 2.945 eV, which creates equal numbers of spin up and spin down carriers. This polarization resolved σ^+ component of the heavy hole PL from the $Zn_{1-x}Mn_xSe$ layers results from the recombination of a spin J = +3/2 hole with a spin S = +1/2 conduction electron in the magnetic quantum well. With increasing field the confining potential for this excitonic spin state increases and produces a decrease in the ground state energy, which is seen as a shift to lower energy in the PL peak intensity (Fig. 2(b)). The energy shift is accompanied by a growth in the PL intensity due to larger electron-hole correlations with stronger localization. As mention previously, the tunability of the confining potential results from the

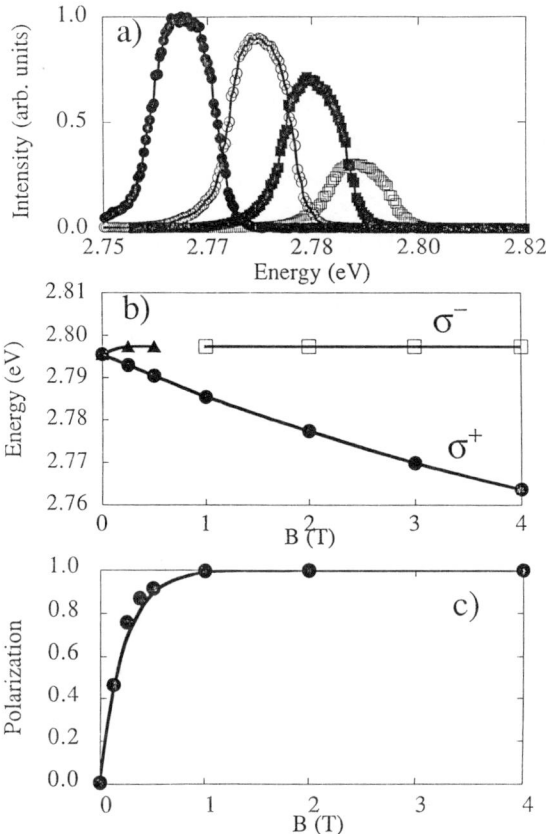

Figure 2. (a) Time-integrated σ^+ component of the PL from a $Zn_{0.91}Mn_{0.09}Se/ZnSe$ SSL (B = 1/4 (□), 1 (■), 2 (○), 4 (●) T). (b) Energy of the σ^+ and σ^- luminescence peak with applied magnetic field. Open squares represent PL from the ZnSE buffer layer. (c) Field dependence of the time-integrated polarization.

large Zeeman splitting of the band edges via the exchange interaction between the charge carriers and the Mn^{2+} ions. While the σ^+ PL peak shifts to lower energy, the σ^- PL peak energy moves to higher energy with increasing field, since it originates from carriers whose spin is opposed to the applied magnetic field. Thus for B < 0.5 T, the quantum well potential depth for this spin state decreases with increasing field resulting in a larger excitonic Bohr radius, and a decrease in the observed σ^- PL intensity has been observed. For B ≈ 0.5 T, where the SSL is formed, the σ^- component has dramatically decreased and is now indistinguishable from the ZnSe buffer layer PL [27]. Unlike PL in SSLs with strain in both magnetic and nonmagnetic layers [24,25], no σ^- component from the SSL was observed at higher fields, which, as can be inferred from the net polarization $P = (\sigma^+ - \sigma^-)/(\sigma^+ + \sigma^-)$ shown in Fig. 2(c), has implications for spin scattering in the nonmagnetic layers. Note that measurements on other SSLs in the series showed qualitatively similar behavior [27], so from hereon we will only

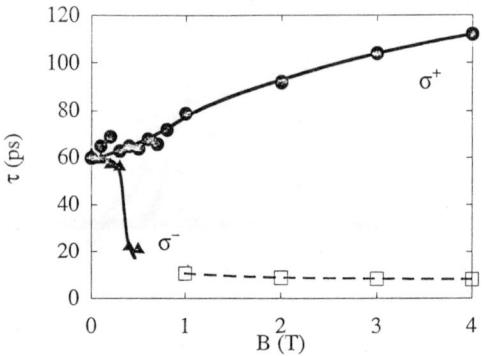

Figure 3. σ^+ and σ^- PL lifetimes with applied magnetic field. Open squares represent PL lifetime from the ZnSe buffer layer.

focus on the sample with x = 0.09.

In order to directly probe the physical processes that underlie the steady-state data, a series of field-dependent time-resolved PL measurements were conducted. Figure 3 shows the radiative life times of the polarization resolved σ^+ and σ^- components of the PL as a function of applied magnetic field. This particular data is obtained utilizing the streak camera with the monochromator set at the peak of the luminescence intensity of Fig. 2. The lifetime values were calculated through single exponential fits to the streak camera spectra. We find that both τ^+, the lifetime of the σ^+ PL, and τ^-, the σ^- PL lifetime, are field dependent. In low fields, B < 0.5 T, where both spin states are confined to the magnetic layers, the field dependence is relatively weak. With increasing field, an increase in both τ^+ and the σ^+ intensity is observed. This unexpected increase in lifetime may be attributed to a decrease in scattering of the excitons at the interfaces [30] as the spin down carriers become more localized in the field induced $Zn_{1-x}Mn_xSe$ wells. This decrease in scattering may also contribute to the increase in PL intensity shown in Fig. 2. In sharp contrast, a dramatic decrease in τ^- was observed in the vicinity of B ≈ 0.5 T where the SSL formation occurs. The decrease coincides with the spatial separation of the carriers in the SSL and therefore most likely results from a fast spin flip of the carriers in the ZnSe "wells" which then requires a change in the carrier localization from the nonmagnetic layers to the magnetic wells before recombination via the spin-up (σ^-) state. This fast spin relaxation then results in the polarization shown in Fig. 2.

The data of Fig. 3 are accumulated under low excitation density (< 1 mW) where the radiative lifetimes were nearly independent of excitation power and energy. At carrier densities greater than 10^{10} cm^{-2}, the exciton lifetime was found to increase due to thermal effects [30]. The field and power dependent lifetimes are in stark contrast with the static PL linewidths which are independent of field and power over the range investigated. Since many dynamical phenomena are often estimated from static measurements, this demonstrates the significance of time resolved data in determining the underlying physical processes that are responsible for the time averaged observations in these systems. In this regard, the present work indicates that the traditional static polarization data (Fig. 2) result from changes in *both* the intensities and lifetimes of σ^+ and σ^-—a characteristic that is not typically considered in interpretations of PL polarization [17, 25].

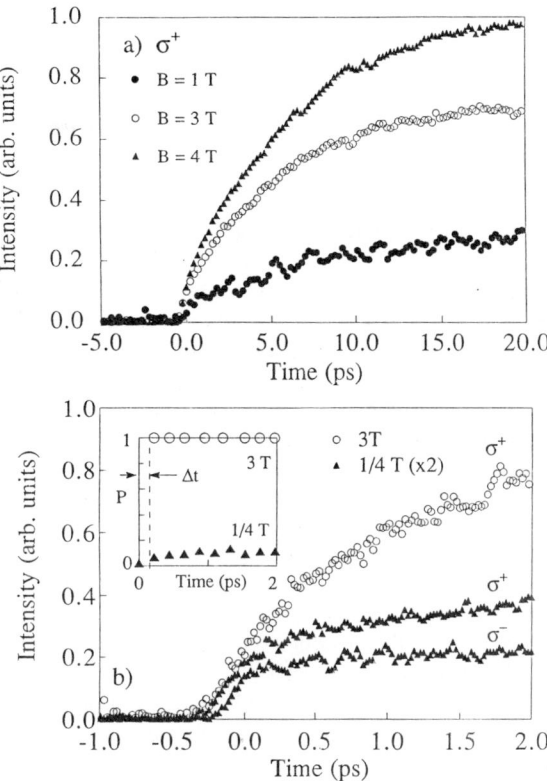

Figure 4. (a) Time- and polarization-resolved PL with the spin states spatially separated. (b) The initial two picoseconds of the luminescence for the two different field (i.e., $B < B_{SSL}$ and $B > B_{SSL}$) regimes and the net polarization (inset).

The remarkable change in exciton dynamics observed upon formation of the SSL is more clearly resolved by femtosecond upconversion spectroscopy. Shown in Fig. 4 is the onset of the σ^+ PL for $B > B_{SSL}$ with the detection energy set to the peak of the luminescence intensity. These data depict the nonequilibrium relaxation of excitons into the ground state as the quantum well depth within the SSL is varied in a single sample. The curves may be normalized by their respective fields to show that the carrier cooling is independent of confining potential. For fields $B > 0.5$ T where the SSL has formed and the spin states for the σ^- are localized within the ZnSe layers, no σ^- intensity from the SSL was detected within the resolution of the experiments. Therefore, the resulting polarization is instantaneous (inset to Fig. 4(b)). Since the system is excited with linearly polarized light, this indicates an extremely rapid spin flip relaxation of carriers from the ZnSe layers to the magnetic quantum wells on a time scale much shorter than the zero-field exciton lifetimes. Recall that the heavy and light holes are degenerate in the unstrained ZnSe layers, and thus mixing of the hole states may play an important role in the spin relaxation process, which could explain the behavior seen here [17]. This is in contrast to the dynamics in low fields shown in Fig. 4(b) ($B = 0.25$ T) where both spin states are localized to the $Zn_{1-x}Mn_xSe$ wells and both σ^+ and σ^+ PL is observed. The luminescence gives rise to the time resolved polarization with B = 0.25 T shown in the inset to Fig. 4(b), which is indicative of spin relaxation times from spin scattering due to sp–d exchange interactions [15]. It should be noted that the ultrafast temporal behavior of the PL in Fig. 4 is qualitatively similar to that seen in other DMSs [3] as well as high quality GaAs/AlGaAs heterostructures [31].

3. Magnetically Coupled Double Quantum Wells

A variety of experiments on III-V semiconductor coupled double quantum wells have shown these structures to be ideal systems in which to explore fundamental dynamical phenomena such as tunneling [32–34]. As mentioned in the introduction, the development of MBE for II-VI DMS provides a unique opportunity to combine the physics of semiconductors with magnetism in such quantum structures. Not only does the incorporation of magnetic spins into the thin barrier between two wells allow one to explore the effects of spin dependent coupling but also the giant Zeeman splittings of the barrier's band edges enables the continuous tuning of the barrier potential for a specific spin state with an applied magnetic field. This yields a systematic investigation of variable coupling between two wells in a single sample under flat band conditions. Alternatively, the interaction of the confined carriers with the magnetic ions in the barrier can be used to optically explore the magnetic state of the barrier (i.e., paramagnetic, spin-glass, and antiferromagnetic).

In this section we report on the results of static magneto-optical photoluminescence experiments with II-VI double quantum wells coupled by a thin DMS layer and demonstrate an ability to spin tune the barrier potential [26]. A schematic of the new DQW heterostrucure is shown in Fig. 5. The quantum wells and the outer barriers are composed of nonmagnetic $Zn_{1-y}Cd_ySe$ and ZnSe, respectively, while the barrier separating the wells is magnetic $Zn_{1-x}Mn_xSe$. By adjusting the magnetic dilution, the coupling barrier potential can be less than, equal to, or greater than that of the outer ZnSe layers. This results from the band gap bowing of the $Zn_{1-x}Mn_xSe$ with Mn composition (as mentioned previously). The energy levels of the symmetric ground state of this DQW are spin degenerate in zero applied field (Fig. 5(a)). This degeneracy is lifted in the presence of a magnetic field (Fig. 5(b)) since the Zeeman splitting of the DMS layer band edges results in a spin-dependent barrier potential. As a result, with increasing field, spin-up carriers experience an increase in the barrier potential, thus expelling their wavefunctions out from the barrier into the wells. The stronger confinement of the carrier wavefunction increases the energy level for these spin states. Conversely, spin-down carriers experience a decrease in barrier potential, and their wavefunctions spread as the coupling between the wells increases, thus decreasing the energy level for this spin state.

The samples in this study were grown by MBE on (100) GaAs and their compositions and layer thicknesses were determined using the procedures describe above. A typical series of samples consists of single pairs of 40 Å nonmagnetic $Zn_{1-y}Cd_ySe$ quantum wells [y ≈ 0.24 and energy gap E_g ≈ 2.45 eV] separated by a thin (12 Å, 32 Å) $Zn_{1-x}Mn_xSe$ magnetic barrier [x ≈ 0.25 and band gap E_g ≈ 2.87 eV]. The DQW is coherently strained between a micron thick ZnSe buffer layer and a 2000 Å ZnSe cladding layer. The PL measurements were performed using the Faraday configuration as described earlier. Frequency doubled linearly polarized light from the titanium sapphire with energy E_e ≈ 2.55 eV was used to photoexcite the samples. The incident power of less than 10 μW was focused to a 50 μm spot on the sample and

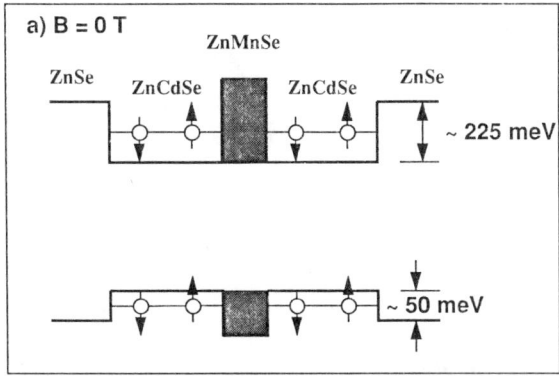
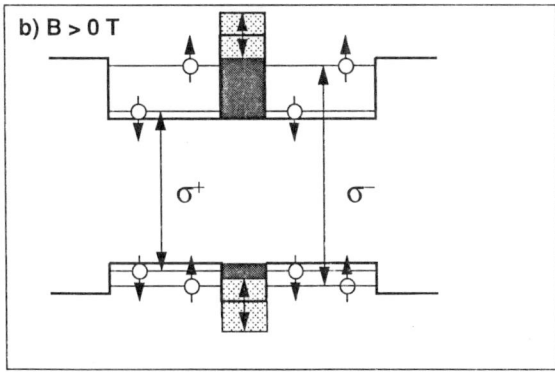

Figure 5. Schematic diagram of the band alignment of a ZnSe/Zn$_{1-y}$Cd$_y$Se/Zn$_{1-x}$Mn$_x$Se DMS coupled double quantum well. Note that the valence-band potential profile shows only the heavy-hole states.

amplitude modulated for phase sensitive detection. Polarization resolved PL spectra were collected with a lock-in amplifier and a 0.64 m monochromator equipped with a photomultiplier tube.

The interplay of electronic and magnetic couplings is seen in Fig. 6, which shows the polarization resolved photoluminescence from a sample with a barrier width L_B = 12 Å in applied magnetic fields of 0 and 4 T at low temperature. In zero field, the PL spectra for the ground state excitons for the two polarizations have equal intensity and identical energy profiles. With increasing field, a significant difference in the net polarization develops from Zeeman splitting the two-fold-degenerate heavy-hole and conduction-band states, resulting in two allowed transitions (one for each polarization—see Fig. 5). At B = 4 T (Fig. 6(b)) the spectral peaks for the two polarizations are separated in energy by ~ 10 meV and the σ^- peak intensity has become substantially smaller that of the σ^+ peak. In this case, the net polarization $P = (\sigma^+ - \sigma^-)/(\sigma^+ + \sigma^-) = 0.76$, demonstrating a remarkably large carrier polarization in a structure with only a few monolayers (12 Å) of magnetic material. Surprisingly, these results are highly dependent upon excitation energy and power, and further experiments are under way to determine the cause of these dependencies.

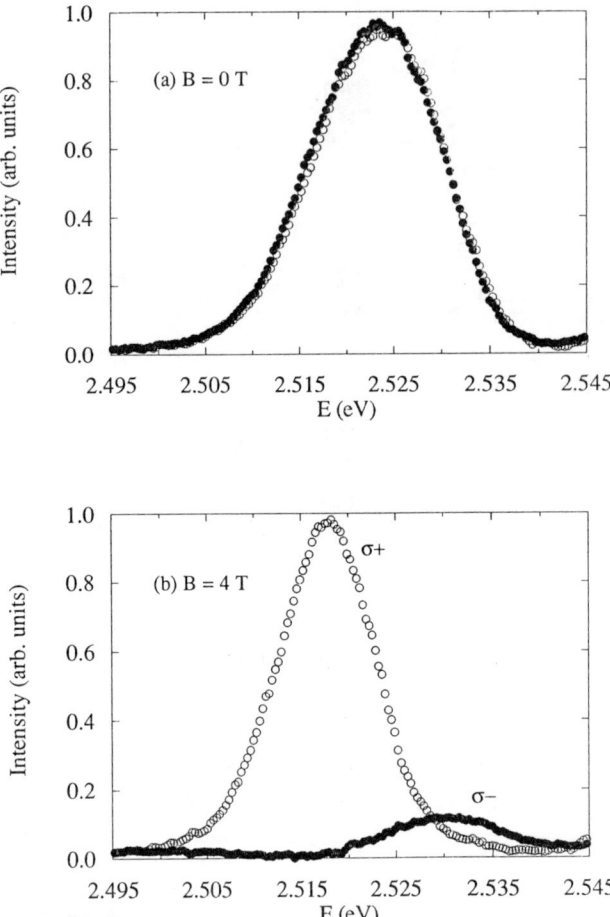

Figure 6. Polarization-resolved time-integrated PL spectra of a symmetric coupled DQW at T = 4.5 K in (a) zero applied field and (b) B = 4 T. Open (solid) symbols represent the σ^+ (σ^-) component of the PL.

The interaction between the photoexcited carriers and the Mn^{2+} ions in the DMS barrier produce large energy splittings for modest applied magnetic fields, which are in sharp contrast to the case of coupled DQWs with a nonmagnetic barrier. Figure 7 shows the peak energy in the polarization resolved PL as a function of applied magnetic field for two samples with a DMS barrier and for a third sample with a nonmagnetic ZnSe barrier. The samples with the DMS barrier display strong energy splittings, with the largest separation for the narrowest barrier, whereas no splitting is observed (within experimental uncertainty) in the energy levels of the sample with the nonmagnetic barrier. A qualitative analysis [26] of the relative size of the energy shifts indicate that the effects of magnetic dimensionality and/or exitonic properties are more prominent in the $L_B = 12$ Å sample. Note that as a result of the larger band gap of the

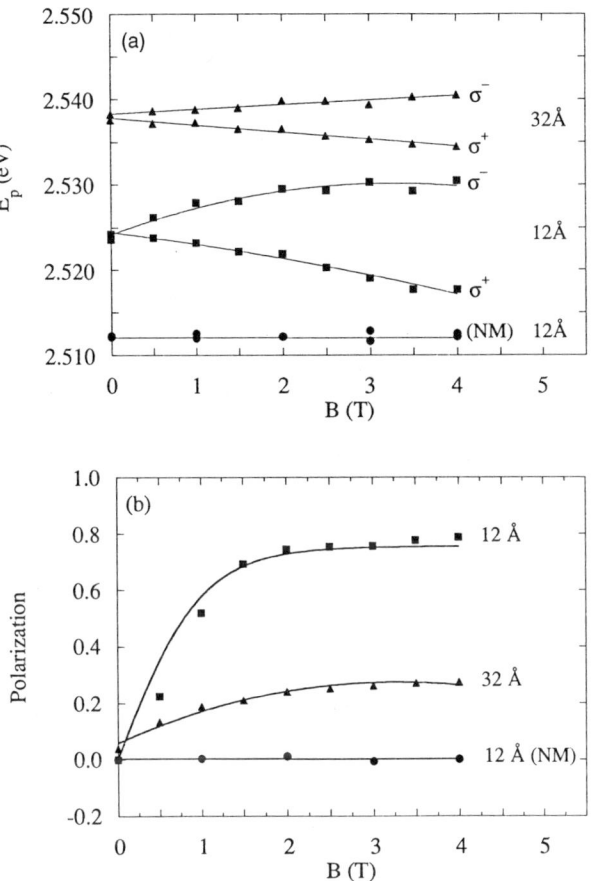

Figure 7. Peak energy of the polarization resolved components of the PL as a function of applied magnetic field for a variety of coupled DQWs. Note that for clarity the L_B = 32 Å result has been shifted up by 15 meV and the nonmagnetic L_B = 12 Å result shifted down by 15 meV. (b) The polarization of the luminescence as a function of applied field.

DMS barrier, the zero field ground state energy for the L_B = 12 Å DMS coupled sample is greater than its nonmagnetic counterpart, since the higher potential barrier decreases the coupling between the wells. In addition, the zero-field energy for the 32 Å barrier is greater than the 12 Å DMS barrier, again since the symmetric ground-state energy level decreases with increasing coupling. In this case, however, the decrease of interwell coupling is due to spatial separation instead of potential height.

The spin splitting of the energy levels is also evident in the field dependent intensities of the polarization resolved PL and their resulting polarizations as shown in Fig. 7(b). In each DMS sample, a decrease in the intensity of the σ^- PL component is observed with a slight increase in the σ^+ PL intensity. The resulting polarization (Fig. 7(b)) is qualitatively consistent with the spin splitting of the energy levels of Fig. 7(a). In marked contrast, no net polarization was

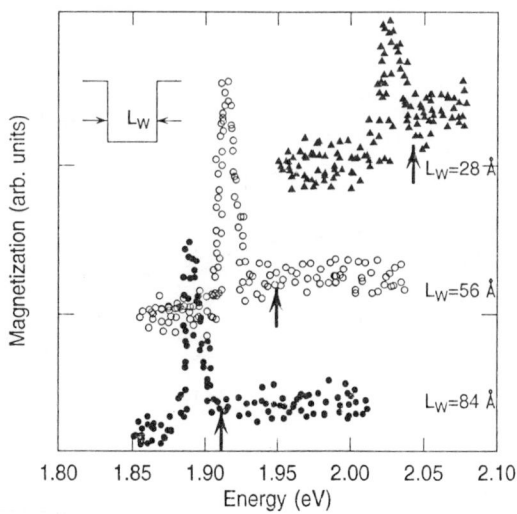

Figure 8. Magnetic excitation spectra of three multiple-quantum-well heterostructures having well widths of 28, 56, and 84 Å, and ZnTe barrier thickness of 160 Å. The y-axis tick marks represent the relative zeros of the magnetization. T = 1.5 K; power = 100 µW; number of periods = 30, 15, and 10, respectively.

observed in the nonmagnetically coupled double quantum well.

4. Polaron Dynamics

Multiple-quantum-well samples of $Cd_{1-x}Mn_xSe$ confined by ZnTe barriers were grown by MBE upon 2-µm-thick ZnTe buffer layers on (100) GaAs substrates. The sample growth and characterization were similar to those discussed previously above. Optical absorption and reflection spectra and the absence of excitonic luminescence indicated a type-II band alignment in $ZnTe/Cd_{1-x}Mn_xSe$, with the electrons confined to the $Cd_{1-x}Mn_xSe$ layers and the holes to the ZnTe layers. Samples constituting a series of well widths (28, 56, and 84 Å) confined by fixed width (160 Å) barriers were studied for two different magnetic ion concentrations (x = 0.13, 0.23).

An integrated dc SQUID microsusceptometer [21] is used for direct measurements of the optically induced magnetization. For dynamical experiments, the time dependence of this magnetization was recorded by a pump-time-delayed-probe technique, which takes advantage of the nonlinear magneto-optical susceptibility of the Mn^{2+} ion sublattice [35]. The samples were excited by circularly polarized 150 fs laser pulses, thereby instantaneously creating spin polarized carriers. For the measurements presented here, the sample and SQUID were immersed in superfluid ^{4}He at 1.5 K. The coupled energy sensitivity of the SQUID system corresponds to an ability to detect ~ 10^4 aligned ionic spins. Other details of the experimental set-up have been published elsewhere [5].

Polaron formation in these structures is demonstrated in Fig. 8. The time-averaged

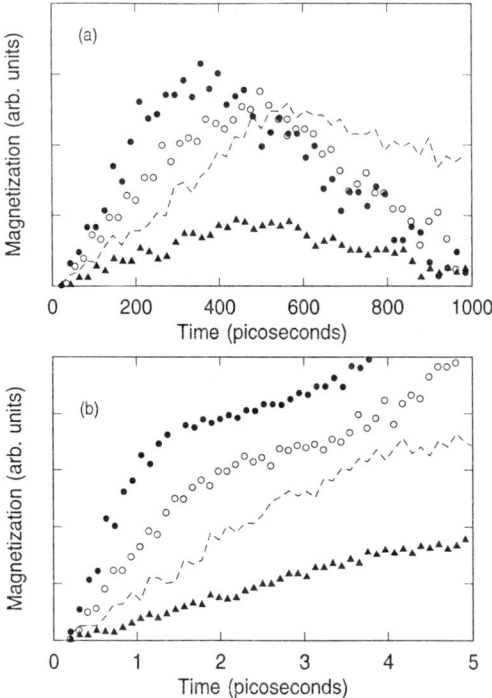

Figure 9. Time dependence of the optically induced magnetization as determined by time resolved magnetic spectroscopy for above band-gap excitation. (a) Spin-lattice relaxation of the magnetization of the samples from Fig. 8. The dashed line represents an 84 Å 13%-Mn well structure. (b) Initial rise of the magnetization in response to 150-fs excitation pulses.

magnetization induced by circularly polarized light is shown for the x = 0.23 series of structures as a function of excitation energy in the vicinity of the quantum well ground state level, yielding a magnetic excitation spectrum analogous to the more familiar PL excitation spectra. A well-defined peak due to magnetic polarons was observed in each case, including the wells that are narrow in comparison to the estimated polaron diameter (90 Å) in bulk material. The energy dependence of the peaks, as well as the intensities, are well described by consideration of the energy dependent spin dynamics and the initial random distribution of Mn^{2+} spin orientation before illumination [5]. The binding energy of the polarons shifts the magnetic spectrum peak several millielectronvolts below the lowest excitonic energy level (n = 1) in the well as determined by photoreflectance (and indicated by the arrows in the figure).

Figure 9 illustrates the time dependence of the magnetization induced by subpicosecond optical pulses as a function of the quantum well width and magnetic concentration, for excitation energies above the polaron peaks in Fig. 8. Figure 9(a) shows the spin-lattice relaxation of the magnetization. These signals evolve long after the electronic spins have depolarized. We find a significant dependence of the relaxation rate on well thickness. This effect can be described by a decrease in the effective magnetic ion concentration with decreasing thickness. A comparison between wide well samples with different Mn concentration (the solid circles and the dashed line in Fig. 9(a)) suggests that the actual dependence of the relaxation rate on concentration is approximately linear in this regime. This concentration dependence then indicates that some form of cross relaxation involving pairs or larger clusters of ions is taking place leading to the observations in Fig. 9(a).

In Fig. 9(b) we show the rate at which the magnetization grows to its saturation value, which is determined by spin-flip scattering with the optically excited carriers. The magnetization onset follows the same trend as a function of well dimension as the decay times of Fig. 9(a). The rate at which this magnetization grows to its saturation value is determined in part by the strength of the spin-flip scattering with the optically excited carriers. Although this growth can be limited by the spin-lattice relaxation of carriers via other mechanisms, in the present situation the electron-ion mechanism should dominate, because the Mn concentration is high and because exchange effects on the electrons are minimized by the real-space separation of electrons and holes. The pisosecond time scales are consistent with spin-flip times calculated [15] and measured for other semimagnetic II-VI systems [36]. The significance of these time dependencies and the role of magnetic polaron formation in the dynamics have been described previously in Ref. [5].

5. Conclusions

Results from a series of experiments on a variety of DMS heterostuctures have been presented. Striking spin dependent dynamics of photo-excited carriers have been observed for carriers located in magnetic wells confined by nonmagnetic barriers, for carriers localized in nonmagnetic wells with magnetic barriers, and when the carriers where separated into both magnetic and nonmagnetic regions. These results should stimulate further experimental and theoretical investigations into the large parameter space of spin dynamics in magnetic quantum structures that should provide a foundation to spin dynamics in 2D systems.

Acknowledgments

We would like to thank D.A. Tulchinsky, H. Luo, and J.K. Furdyna for their help. This work was supported in part by NSF Grant DMR 92-07567.

References

1. F. Meier and B.P. Zachachrenya, *Optical Orientation* (North-Holland, Amsterdam, 1984); S. Schmitt-Rink, D.S. Chemla, and D.A.B. Miller, Adv. Phys. **38**, 89 (1989).
2. R.C. Miller and D.A. Kleinman, J. Lumin. **30**, 520 (1985).
3. M.R. Freeman, D.D. Awschalom, J.M. Hong, and L.L. Cheng, Phys. Rev. Lett. **64**, 2430 (1990).
4. M. Kohl, M.R. Freeman, D.D. Awschalom, and J.M. Hong, Phys. Rev. B **44**, 5923 (1991).
5. D.D. Awschalom, M.R. Freeman, N. Samarth, H. Luo, and J.K. Furdyna, Phys. Rev. Lett. **66**, 1212 (1991).
6. T.C. Damen, L. Viña, J.E. Cunningham, and J. Shah, Phys. Rev. Lett. **67**, 3432 (1991).
7. Ph. Roussignol, P. Rolland, R. Ferreira, C. Delalande, G. Bastard, A. Vinattieri, L.

Carraresi, M. Colocci, and B. Etienne, Surf. Sci. **267**, 360 (1992).
8. J.B. Stark, W.H. Knox, and D.S. Chemla, Phys. Rev. B **46**, 7919 (1992).
9. S. Bar-Ad and I. Bar-Joseph, Phys. Rev. Lett. **68**, 349 (1992).
10. I. Brener, W.H. Knox, K.W. Goossen, and J.E. Cunningham, Phys. Rev. Lett. **70**, 319 (1993).
11. A.E. Ruckenstein, S. Schmitt-Rink, and R.C. Miller, Phys. Rev. Lett. **56**, 504 (1986).
12. M.I. D'yakonov and V. Yu. Kachorovskii, Fiz. Teckh. Poluprovodn. **20**, 178, (1986) [Sov. Phys. Semicond. **20**, 110 (1986)].
13. L.J. Sham, J. Phys. (Paris), Colloq. **48**, C5-381 (1987).
14. A. Twardowski and C. Hermann, Phys. Rev. B **35**, 8144 (1987).
15. G. Bastard and L.L. Chang, Phys. Rev. B **41** 7899 (1990).
16. T. Uenoyama and L.J. Sham, Phys. Rev. Lett. **64**, 3070 (1990).
17. T. Uenoyama and L.J. Sham, Phys. Rev. B **42**, 7114 (1990).
18. R. Ferreria and G. Bastard, Phys. Rev. B **43**, 9687 (1991).
19. J.K. Furdyna and J. Kossut, *Diluted Magnetic Semiconductors*, Semiconductors and Semimetals, Vol. 25 (Academic Press, San Diego, 1988).
20. J.M. Hong, D.D. Awschalom, F. Agullo-Rueda, and L.L. Chang, J. Cryst. Growth **111**, 1016 (1991).
21. D.D. Awschalom, J.R. Rosen, M.B. Ketchen, W.J. Gallagher, A.W. Kleinsasser, R.L. Sandstrom, and B. Bumble, Appl. Phys. Lett. **53**, 2108 (1988).
22. M. von Ortenberg, Phys. Rev. Lett. **49**, 1041 (1982).
23. N. Dai, H. Luo, F.C. Zhang, N. Samarth, M. Dobrowolska, and J.K. Furdyna, Phys. Rev. Lett. **67**, 3824 (1991).
24. W.C. Cho, A. Petrou, J. Warnock, and B.T. Jonker, Phys. Rev. Lett. **67**, 3820 (1991).
25. W.C. Cho, A. Petrou, J. Warnock, and B.T. Jonker, Phys. Rev. B **46**, 1041 (1992).
26. J.F. Smyth, D.D. Awschalom, N. Samarth, H. Luo, and J.K. Furdyna, Phys. Rev. B **46**, 4340 (1992).
27. J.F. Smyth , D.A. Tulchinsky, D.D. Awschalom, N. Samarth, H. Luo, and J.K. Furdyna, submitted to Phys. Rev. Lett.
28. R.B. Bylsma, W.M. Becker, and J. Kossut, Phys. Rev. B **33**, 8207 (1986).
29. H. Mahr and M.D. Hirsch, Opt. Comm. **13**, 96 (1975); J. Shah, IEEE J. Quantum Electron. **24**, 276 (1988).
30. G. Bastard, *Wave Mechanics Applied to Semiconductor Heterostructures* (Les Editions de Physique, France, 1988).
31. T.C. Damen, J. Shah, D.Y. Oberli, D.S. Chemla, J.E. Cunningham, and J.M. Kuo, Phys. Rev B **42**, 7434 (1990).
32. M.M. Dignam and J.E. Sipe, Phys. Rev. B **43**, 4084 (1991), and references therein.
33. P. Bonnel, P. Lefebvre, B. Gil, H. Mathieu, C. Deparis, J. Massies, G. Neu, and Y. Chen, Phys. Rev. B **42**, 3435 (1990).
34. B. Deveaud, A. Chomette, F. Clerot, P. Auvray, A. Regreny, R. Ferreira, and G. Bastard, Phys. Rev. B **42**, 7021 (1990), and references therein.
35. J. Warnock and D.D. Awschalom, Jpn. J. Appl. Phys. **26**, 819 (1987).
36. M.R. Freeman, D.D. Awschalom, and J.M. Hong, J. Appl. Phys. **67**, 5102 (1990).

COHERENT OPTICAL PHENOMENA OF QUANTUM WELL EXCITONS IN A MAGNETIC FIELD

I. BAR-JOSEPH, S. BAR-AD, O. CARMEL, AND Y. LEVINSON
Department of Physics
The Weizmann Institute of Science
Rehovot, Israel

ABSTRACT. We present results of differential absorption and four wave mixing experiments in GaAs quantum wells in a magnetic field. Quantum beats between the exciton Zeeman levels and a rotation of the photon echo polarization are observed and analyzed. We show that exciton spin orientation is maintained for several tens of picoseconds. Fast oscillations of the differential absorption and the four wave mixing signal are observed and attributed to bi-excitonic nonlinearity.

1. Introduction

The nonlinear optical properties of intrinsic semiconductors quantum wells (QWs) at energies close to the fundamental gap are dominated by the interaction of light with excitons. This interaction was intensively investigated during the last decade, yielding valuable information on the exciton state and its dynamics in quasi two-dimensional (2-D) structures. In recent years there is a growing interest in extending the use of nonlinear optical spectroscopy into studying coherent phenomena in these structures [1,2]. Using phase sensitive techniques, and taking advantage of the very fine time resolution which is available with short pulse lasers, the dynamical manifestations of phase coherence and dephasing processes of the exciton state can be studied. An interesting dimension of these studies is the ability to create coherent superpositions of states and to probe their time evolution. This will result in the observation of quantum interference effects, and the detection of small energy splitting. Our purpose in this paper is to demonstrate how these techniques are implemented into the study of coherent exciton states in a magnetic field.

The application of a magnetic field normal to the layers of a QW structure strongly affects the excitonic interaction and the optical properties associated with it [3]. The excitonic enhancement and the constant density of states of the 2-D continuum change into discrete magneto-exciton states associated with the different Landau levels, and the absorption spectrum is modified accordingly. Furthermore, the excitons are confined into a small volume giving rise to a stronger oscillator strength on one hand and a weaker dipole moment on the other. The degeneracy of the electronic spin states is removed, similarly to the Zeeman splitting in atoms, and two non-degenerate exciton states, which are coupled to light, σ^- and σ^+, are formed.

It is commonly believed that the coherence of exciton states can be understood in terms of this simple picture, and the more complex interaction of electrons and holes, which form a hydrogen molecule-like state, known as the bi-exciton, is negligible in bulk GaAs and QWs. The reason is that the binding energy of the GaAs bi-exciton is very small, of the order of a meV, such that this state is not stable even at low temperatures. Indeed, photoluminescence studies show a small peak at energies lower than that of the exciton [4]. Transient nonlinear optical interactions, however, are not as sensitive to the bi-excitonic lifetime, and can be affected by the possibility to create this metastable state. As we shall show in this paper the formation of bi-excitons can give rise to interesting interference effects.

In the following we shall present extensive experimental investigations of coherent nonlinear optical phenomena in a magnetic field [5–7]. We begin by a discussion of excitonic effects which are primarily due to the removal of the spin degeneracy. In addition to the various interference effects which are observed, such as quantum beats and photon echo rotation, valuable information on spin dynamics could be derived. In particular we show that spin orientation of excitons could be maintained for several tens of picoseconds, and that there are evidences for two sequential spin flip processes, a faster one which we associate with that of the holes and a slower one associated with the electrons, before complete spin relaxation is obtained. We then proceed to discussing bi-excitonic effects. We shall present results of both time resolved differential absorption, commonly termed "pump-probe" experiments, and four wave mixing (FWM) in which oscillatory phenomena are observed. A comprehensive model explains these oscillations as reflecting quantum interference between excitonic and bi-excitonic terms in the time dependent third-order susceptibilities.

The experiments are conducted in a magneto-optical cryostat, equipped with a split coil magnet of 7 T, at 4 K. The laser system consists of a mode locked Nd:YAG laser and a synchronously pumped two-jet dye laser. We concentrate on two nonlinear techniques, differential absorption and two-beam FWM, with a time resolution of about 1 ps. The experimental setup of these two experiments is very similar, with two pulses hitting the sample with a variable delay. The difference is that while in the differential absorption experiment the absorption change of the second delayed pulse is detected, in the FWM we measure the time integrated intensity of the signal emitted in the direction $2\mathbf{k}_2-\mathbf{k}_1$, where $\mathbf{k}_1$ and $\mathbf{k}_2$ are the propagation directions of the first and second pulse, respectively. A special emphasis is given to the polarizations of the two incident light pulses and to that of the detected signal. Due to the absorption selection rules, circularly polarized light interacts with only one of the excitons, while linearly polarized light interacts with a superposition of the σ^- and σ^+ excitons.

We have studied three different GaAs/AlGaAs samples: a square multiple quantum well (MQW) sample, a stepped MQW sample, and a superlattice. The square MQW sample has 20 periods of 80 Å GaAs wells separated by 300 Å AlGaAs barriers (x = 0.3), with a strong heavy hole excitonic absorption line at 1.570 eV. The stepped MQW sample has 100 stepped wells, with adjacent layers of 30 Å GaAs and 100 Å AlGaAs (x = 0.1), separated by 100 Å AlGaAs barriers (x = 0.3). The structure is embedded in the intrinsic region of a p-i-n structure. The wave functions associated with the first energy levels of the electron and holes are confined mainly in the 30 Å GaAs layer. The first electron energy level in this structure coincides with the potential step of the well, which makes the structure prone to large potential disorder due to well width and alloy fluctuations. Indeed, the absorption spectrum of that sample shows a wide (~ 6 meV), inhomogeneously broadened, heavy-hole excitonic peak at 1.617 eV. The superlattice sample consists of 100 periods of alternating

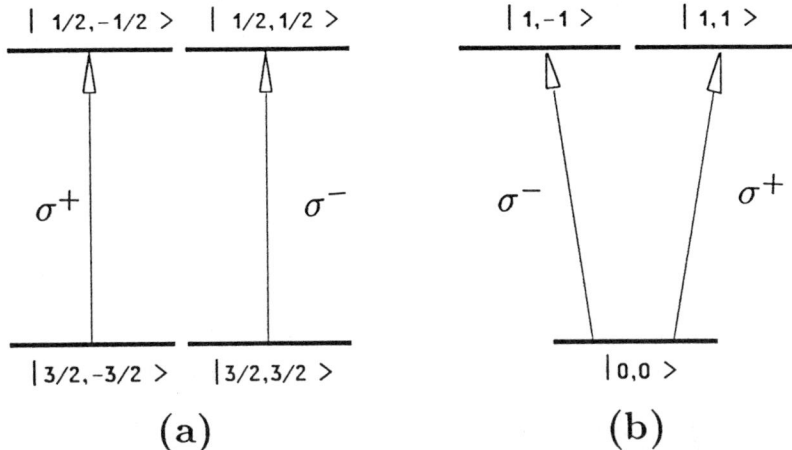

Figure 1. (a) Electronic and (b) excitonic representations of the energy levels associated with optical transitions from the heavy hole valence band to the conduction band of a GaAs QW.

30 Å layers of GaAs and AlGaAs (x = 0.3), with the heavy hole excitonic absorption line at 1.683 eV. This structure is also embedded in the intrinsic region of a p-i-n junction. The substrates of all samples are removed to allow transmission measurement through the sample.

2. Exciton Effects

At zero magnetic field the excitonic system can be considered as a three level system (Fig. 1), with two degenerate excited states, corresponding to the σ^- and σ^+ excitons, and a ground state which is the vacuum state. The σ^+ exciton is made of a $|1/2,-1/2\rangle$ electron and a $|3/2,-3/2\rangle$ heavy hole, and the σ^- is similarly composed of $|1/2,-1/2\rangle$ electron and $|3/2,-3/2\rangle$ heavy hole. One can therefore represent the σ^- and σ^+ exciton states by a total angular momentum of $J = 1$, with $J_z = 1$ for the σ^+ exciton and $J_z = -1$ for the σ^- exciton. The vacuum ground state total angular momentum in this representation is $J = 0$. As a magnetic field is applied the two exciton states become non-degenerate, with a splitting which is a few tenths of a meV for moderate magnetic fields (0–10 T). Under experimental conditions where the bandwidth of the laser pulse (typically 1 meV in our experiment) is large compared to this splitting, and when the laser light is linearly polarized, these two exciton states are simultaneously driven and a coherent superposition is prepared. A second linearly polarized pulse probes the evolution in time of this superposition, and the measured physical quantity can be either the change of absorption or the emitted FWM signal. In this part we shall present experimental results of both differential absorption and time resolved FWM, where the coherence between the exciton states is manifested, and describe a quantitative analysis of these results. We shall conclude this part with some conclusions on exciton spin dynamics.

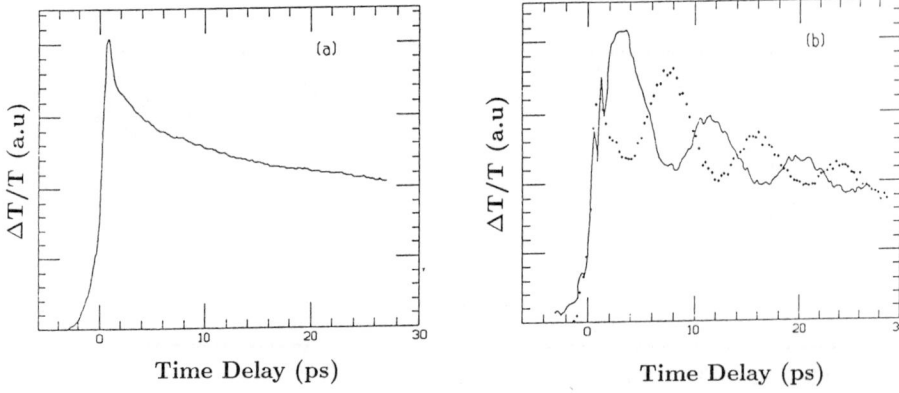

Figure 2. Temporal evolution of $\Delta T/T$ at the heavy hole exciton of the stepped MQW sample, in magnetic fields of (a) B = 0 and (b) B = 5 T. The solid curves were measured with cross linear pump and probe polarizations (CP), while the dotted curves are for the case of parallel linear polarizations (PP).

2.1. QUANTUM BEATS OF ZEEMAN SPLIT EXCITONS

Figure 2(a) shows the relative change in transmission, $\Delta T/T$, at B = 0 for the stepped MQW sample, pumped and probed with roughly equal intensities of approximately 5 W/cm^2, slightly below the heavy hole exciton absorption peak (1.619 eV). Since $\Delta T \ll 1$ ($\approx$5%) this signal is directly proportional to the exciton absorption change $\Delta\alpha$. The decay of the signal reflects the spectral diffusion of the low energy, localized excitons. This behavior is strongly modified as a magnetic field normal to the layers is applied. The solid curve in Fig. 2(b), taken at B = 5 T with linear cross polarizations (CP) of the pump and the probe, exhibits a slow rise (relative to the pulse width), followed by very deep, damped oscillations [5]. These oscillations have a period of 8 ps, with the first minimum occurring a full period after t = 0. Up to six oscillations were observed at a field of 5 T. The damping time constant, ~ 20 ps, is approximately the same as the decay time for the B = 0 case. The signal then levels off, and decays with a very slow time constant of hundreds of picoseconds, consistent with a recombination process. A similar temporal behavior was observed for the parallel polarizations (PP) case. The pump-probe signal for PP, also at B = 5 T, is plotted as a dotted curve in Fig. 2(b). It can clearly be seen that there is a π phase shift with respect to the CP case. Less pronounced oscillations were observed with the superlattice sample: the phase of the oscillations is similar to that measured in the stepped MQW sample and the oscillations period is slightly longer. No oscillations were detected in the square MQW sample.

The experimental results change dramatically when circular polarizations are used. Whenever the pump or the probe is circularly polarized the oscillations disappear. We also repeated the experiment with the cryostat rotated by 90 degrees, so that the magnetic field was in the plane of the layers and the laser beams were normal to them. No oscillations were observed in any of the polarizations, in any of the samples.

Oscillations with the same period are observed at the FWM signal. Figure 3 shows the

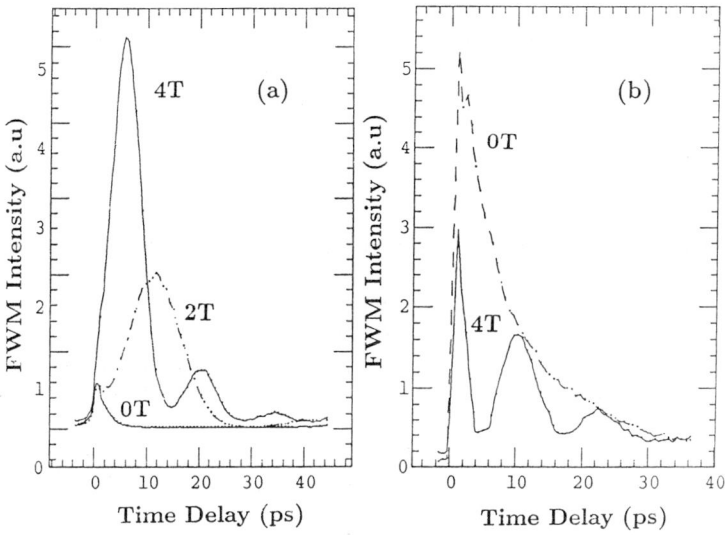

Figure 3. Temporal evolution of the FWM intensity at the heavy hole exciton for (a) CP and (b) PP experimental configurations.

time integrated FWM intensity as a function of the delay time between the two pulses. As has been observed by several groups in the past [8], the signal decays at zero magnetic field much faster when the exciting pulses are CP (Fig. 3(a)), as compared to the case where the exciting pulses are PP (Fig. 3(b)). This difference is especially manifested at the low energy side of the exciton. As the magnetic field is turned on the time evolution of both the PP and the CP signals becomes oscillatory. The period of the oscillations is inversely proportional to the magnetic field strength, and is identical to that observed in the differential absorption measurement. The decay time of the envelope of these oscillations for PP pulses is approximately the same as in zero magnetic field. This is not the case for the CP signal. The decay time of the oscillations is much longer than at zero magnetic field, and in fact becomes identical to the PP decay time. As can be seen, here also there is a π phase shift between the oscillations in the CP and the PP cases.

These oscillations are clearly quantum beats between the non-degenerate σ^+ and σ^- exciton states, which are coherently driven by the laser field, and the frequency of the oscillations is related to the Zeeman splitting between these levels. Similar oscillations were observed in the resonance fluorescence of bound excitons in CdS [9]. The dramatic change in the decay time of the FWM signal in the CP case demonstrates that the vanishing of the echo in zero magnetic field is due to a cancellation of the σ^+ and the σ^- contributions to the signal. At a finite magnetic field, when the degeneracy of the two states is removed, this cancellation changes into a beating behavior.

To calculate the absorption change or the intensity of the emitted signal it is necessary to evaluate the induced polarizability by the two pulses <P>. The total absorbed intensity in the sample can then be written as $\mathbf{E} \cdot d\mathbf{P}/dt$, where $\mathbf{E}$ is the total field amplitude, and the emitted FWM intensity as $|P|^2$. The calculation of <P> is done using a density matrix approach, which is usually applied to atomic systems [10,11]. The relation between <P>

and the density operator $\rho(t)$ is given by $<P> = \text{Tr}\{\mathbf{p}\,\rho(t)\}$, where $\mathbf{p}$ is the electric dipole moment operator. In the same manner that it is done for atomic systems one can calculate the density operator for the excitonic system, after a sequence of two pulses of light that are separated by a time interval τ. Following this procedure one obtains a general expression for $<P>$

$$<P> = 1/2 \, \Sigma \, \exp[-i\{(t-\tau)\Delta\omega_{1,2} - \tau\Delta\omega_{3,4}\}]\, <j_1 m_1 |\mathbf{p}| j_2 m_2> \qquad (1)$$

$$\times <j_2 m_2 | \sin A_2 \mathbf{p} | j_3 m_3><j_3 m_3 | \sin 2A_1 \mathbf{p} | j_4 m_4><j_4 m_4 | \sin A_2 \mathbf{p} | j_1 m_1> + \text{c.c.},$$

where $\Delta\omega_{1,2} = -\omega_1 + \omega_2 - \omega$, $\Delta\omega_{3,4} = -\omega_3 + \omega_4 - \omega$, $A_1 = e_1 \int F_1(t)\,dt$, $A_2 = e_2 \int F_2(t)\,dt$, $F_1(t)$ and $F_2(t)$ are the envelopes of the first and second pulse, respectively, and e_1 and e_2 are the corresponding polarizations. j and m are the quantum numbers for the total angular momentum and its projection along the z axis, respectively. The indices m_1, j_1 and m_3, j_3 are summed over all allowed ground state values and m_2, j_2 and m_4, j_4 are summed over all allowed excited state values. In the particular case of the excitonic level diagram, the levels 1 and 3 are degenerate and correspond to the ground state, and the levels 2 and 4 are the σ^- and σ^+ excitons, respectively.

For inhomogeneously broadened QWs, the photon echo intensity is proportional to $|P(2\tau)|^2$ and its polarization is parallel to the direction of $P(2\tau)$. When the polarizations of the two pulses are parallel to each other (PP), say x polarization, we get, by summing over all the combinations of m_2 and m_4, the polarizability in the x direction:

$$P_x(2\tau) \sim <p> \sin^2(A_2 <p>) \sin(2A_1 <p>)[1+\cos(\omega_1 - \omega_2)\tau], \qquad (2)$$

where $<p>$ is the matrix element $<0,0|p_z|1, \pm 1>$, $\omega_1 - \omega_2 = 2g_{exc}\mu_B B/\hbar$, g_{exc} is the exciton Lande factor, B is the magnetic field strength, and μ_B is Bohr magneton. At zero magnetic field the echo intensity decays as the time interval between the two pulses increases, with a time constant proportional to the dephasing time. At finite magnetic field the quantum beats of the polarizability gives rise to damped oscillations in the echo intensity.

When the polarizations of the two pulses are orthogonal to each other, say y polarization for the first pulse and x polarization for the second pulse, and we measure the FWM single at the polarization of the first pulse, we find

$$P_y(2\tau) \sim <p> \sin^2(A_2 <p>) \sin(2A_1 <p>)[1-\cos(\omega_1-\omega_2)\tau]. \qquad (3)$$

This means that in the CP case, when the magnetic field is off, the induced polarizability vanishes and there is *no echo signal*. When the magnetic field is turned on the degeneracy is lifted, and the echo is recovered! The echo intensity is modulated in time with a decaying envelope given by the same dephasing process as in the PP case.

2.2. THE ECHO ROTATION

An interesting effect is observed when analyzing the polarization of the emitted FWM signal in the PP case. To demonstrate this we fixed the delay time between the exciting pulses and ramped the magnetic field between 0 to 7 T. Figure 4 shows the intensity of the FWM signal as a function of the magnetic field at a polarization parallel or perpendicular to the

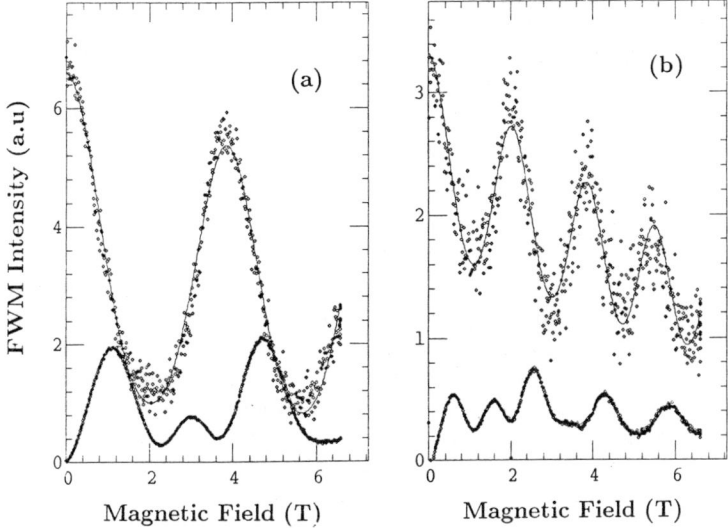

Figure 4. The FWM intensity as a function of magnetic field for fixed delays of (a) 11 ps and (b) 24 ps in PP configuration. In each frame two traces are shown: the upper trace corresponds to the case where the signal is analyzed at the same polarization as the exciting pulses, while the lower trace corresponds to the case in which the signal is analyzed at the perpendicular polarization. The upper curves are fitted to Eq. (2) with a magnetic field dependent g_{exc}.

exciting pulses, for several delays between the exciting pulses. Clearly, a signal appears at a polarization, that is normal to that of the exciting pulses. This polarization rotation is not a manifestation of a Faraday effect, but rather a result of a precession of the induced dipole moment in the magnetic field.

The rotation of the echo polarization is readily obtained by calculating P_y for the case of two identical polarizations x, using Eq. (1), with the appropriate choice of polarizations. Evaluating this expression for the exciton non-degenerate energy levels scheme yields

$$P_y(2\tau) \sim <p> \sin^2(A_2<p>) \sin(2A_1<p>) \sin(\omega_1 - \omega_2)\tau. \qquad (4)$$

Equations (2) and (4) describe the main experimental features shown in Fig. 4. The dependence of the echo polarization on the magnetic field and the delay comes from the argument of the oscillatory functions, $\omega_1 - \omega_2 = 2g_{exc}\mu_B B/\hbar$, such that a precession in the x-y plane is obtained. It is interesting to note that as a consequence, at certain values of B the echo signal vanishes. It also follows from Eqs. (3) and (4), by taking $|P|^2$, that there is a factor of two between the period of oscillations in the x and the y polarizations. This behavior is indeed seen in the experiment, but some discrepancies appear. As can be seen, the rotated signal is indeed doubled in frequency at low fields, but oscillates at the same frequency as the non-rotated signal at higher fields. It seems that there is an additional term in P_y, which depends linearly on B, that is responsible for the observed pattern of oscillations. The origin of this linear term is not clear. It should be noted that the rotation of the echo was also observed in an atomic system, and explained along similar lines [11].

Finally, from analysis of the temporal and the B dependent measurements it is seen that the oscillation frequency increases with the magnetic field, indicating that g_{exc} is B dependent. A fit to the experimental data can be seen in Fig. 4, where we took g_{exc} as constant fit parameter plus a linear dependence in B. The best fit value for g_{exc} is 0.76 + 0.016B, where the energy splitting between the σ^+ and the σ^- excitons was assumed to be $2g_{exc}\mu_B B$.

2.3. EXCITON SPIN DYNAMICS

An implicit assumption in the discussion above, of the quantum beats and the other coherent effects, is that spin orientation of σ^- the σ^+ excitons is maintained for a period which is larger or equal to the dephasing time. A spin flip process would have washed out the coherence between the two exciton states and diminish the observed coherent effects. One can therefore conclude that in the samples in which the above coherent phenomena were observed, the excitons are spin oriented for a few tens of picoseconds. To confirm this observation in a direct way, we have conducted a pump probe experiment with circularly polarized pulses. Unlike the linear polarizations experiments, where a coherent superposition of the two exciton states was prepared, the circular polarizations allow exciting and probing the population in each of the spin states separately.

Figure 5 shows the temporal evolution of the signal when the stepped MQW sample is degenerately pumped and probed with circularly polarized pulses, just below the heavy hole excitonic absorption peak. In Fig. 5(a) we show the change of absorption of a σ^+ probe and that of a σ^- probe, following excitation by a σ^+ pump. In Fig. 5(b) we plot the σ^- signal with better definition. The data shown in the figure was taken at a zero magnetic field with pump intensity of approximately 30 W/cm^2 (corresponding to an exciton density of 1.5×10^{10} cm^{-2}). The first few tens of picoseconds after excitation are characterized by a sharp decrease of the σ^+ signal and a sharp rise of the σ^- signal. The two curves then converge gradually and coincide approximately 350 ps after excitation, indicating that complete spin relaxation has been reached by then. This behavior suggests the existence of two sequential processes. In Fig. 5(c) we plot the log of the difference between the two signals of Fig. 5(a), as a measure of the net relaxation rate of the difference between the σ^+ and the σ^- exciton populations. Two time constants, approximately 50 ps and 120 ps, are clearly observed. We associate the short one with the spin relaxation of the heavy hole, and the long one, after which the two spin populations are equalized, with the electron spin relaxation. The same general behavior, of two time constants and a decay of the σ^- signal at long delays, was observed when the pump intensity was lowered over two decades. The superlattice sample also shows a clear double-exponent behavior, with somewhat different time constants (35 ps and 250 ps for the holes and electrons, respectively). The square MQW sample, on the other hand, behaves differently: the rise of the σ^- signal is instantaneous and the hole spin relaxation cannot be resolved. The σ^- and σ^+ signals approach the same level after 400 ps. The difference between the square MQW sample and the two other samples shows up in another form: the pump-probe signal in that sample does not show any Stokes shift relative to the absorption spectrum, contrary to the case of the two other samples. The existence of such a shift between the linear and the nonlinear responses is an indication of the existence of localized excitons in the sample.

Optical spectroscopy has been extensively implemented for the research of spin

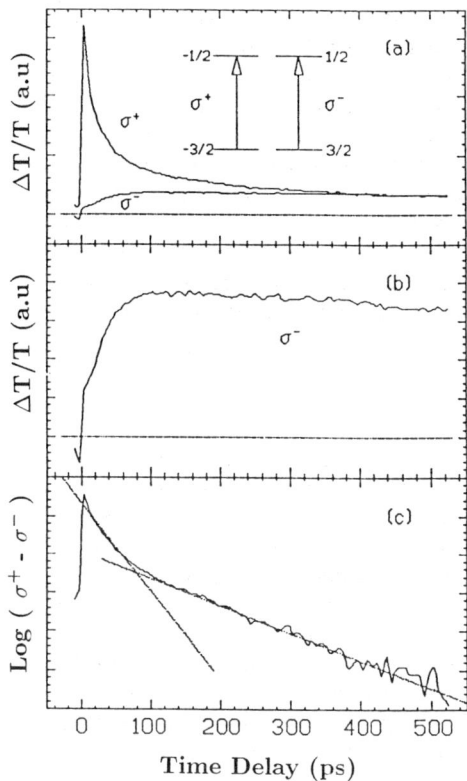

Figure 5. The differential absorption signal slightly below the heavy-hole exciton of the stepped MQW sample, with σ^+ pump polarization. (a) The signals for σ^+ and σ^- probe at B = 0. (b) The σ^- signal with better definition. (c) The logarithm of the difference between the σ^+ and σ^- signals, as a measure of spin orientation.

relaxation processes in GaAs QWs, mainly by way of measuring the degree of polarization in the photoluminescence (PL) emitted following cw excitation by polarized light [4,12]. The electron spin relaxation time, deduced with the assumption of instantaneous hole spin relaxation during thermalization [13], was found to be 200 ps at 4 K. This is not very different from the spin relaxation times reported for bulk GaAs and AlGaAs [14]. Time resolved optical spectroscopy allows separation between the initial depolarization, resulting from the thermalization process, and that of low energy carriers at the bottom of the band. Correlating time resolved and cw PL spectroscopy, Freeman et al. have shown that spin lifetimes cannot reliably be extracted from cw spectra [15]. They have also found that the depolarization rate is sample dependent. Damen et al. have used modulation doped samples to separate between electron and hole spin relaxation [16]. They found a 150 ps relaxation time for the electron spin in p-type modulation-doped QWs and 4 ps relaxation time for the holes in n-type QWs.

Theoretical studies predict a significant suppression of the spin flip rate for free heavy hole states near the zone center due to the removal of the degeneracy between heavy and light hole bands [17–19]. The hole spin relaxation time which is inferred from our measurement is indeed much longer than the relaxation time in bulk GaAs, which is practically instantaneous. However, present theories for hole spin relaxation do not address the excitonic case in general, and the effect of localization in particular. It is thus difficult to perform a direct comparison between our results and the theory. It is also important to clarify the importance of electron-hole exchange in spin flip processes. Assuming a weak exchange interaction between electrons and holes we can apply the results which were calculated for free holes in Ref. [19] to the case of excitons. Performing a weighted average of the calculated spin relaxation rates of the holes over all k states which contribute to the 2-D excitonic wavefunction (for the appropriate well width) we get a relaxation rate which is in a reasonable agreement with the experimental results.

3. Bi-exciton Effects

Focusing on the instantaneous rise of the signal in the oppositely handed circularly-polarized pump-probe experiment in the stepped MQW sample, a very striking phenomenon is observed. Figure 6 is a close-up of the first few picoseconds of σ^- signal, with 3 W/cm^2 pump and probe intensities. The sudden change of absorption is accompanied by deep damped oscillations. Similar oscillations were measured with the superlattice sample, but none were observed with the square MQW sample. Three main features are apparent in Fig. 6: a negative dip, damped oscillations, and a final positive signal (decreased absorption). All three features exist throughout the low energy side of the excitonic absorption line. The period of the oscillations is independent of both the intensity and the excitation energy. The oscillations period was measured to be 1.6 ps for the stepped QW sample and 2 ps for the superlattice sample. There is, however, a weak dependence of the oscillation frequency on the magnetic field strength and direction. We have measured a slight increase of the frequency when the field was oriented in one direction normal to the sample and a similar decrease when the field was changed to the opposite direction (by reversing the direction of the current in the magnet). This difference in frequency is in a very good agreement with the Zeeman splitting measured by differential absorption or FWM.

Oscillations with the same period were observed in a FWM experiment with CP pulses. Figure 7 shows measurements of this signal on a short time scale, for various values of magnetic field. Oscillations with 1.6 ps period are clearly observed, and they become significantly more visible as the magnetic field is increased. Similar observations, at zero magnetic field, have been recently reported by Lovering et al. [20] and Pantke et al. [21].

We examined several possible mechanisms to explain the observed oscillations, which are indicative of the existence of a fundamental energy gap of about 2.5 meV. Removal of spin degeneracy due to spin-orbit coupling can exist in principle in asymmetric structures but the energies of the σ^+ and the σ^- transitions remain degenerate. Moreover, the observation of the oscillations in the symmetric superlattice, under flat band conditions, contradicts this explanation. In addition, the effect is too small (by orders of magnitude) to explain such a large energy gap. The insensitivity of the oscillation period to the energy and intensity of the excitation tends to reject any explanation which is based on optical nutation.

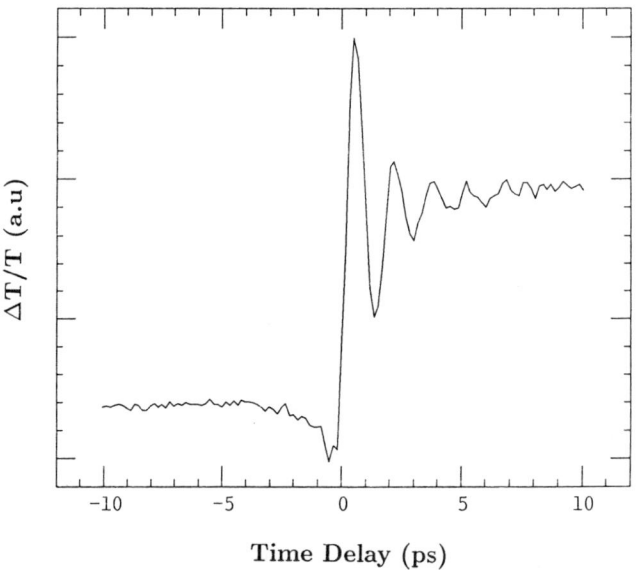

Figure 6. The transient signal measured in a pump-probe experiment with pump and probe of oppositely handed circular polarizations. We associate the observed oscillations and initial dip with the formation of singlet bi-excitons.

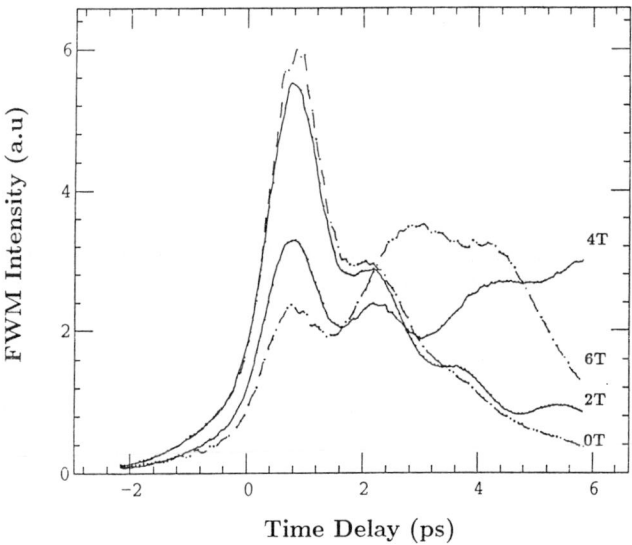

Figure 7. Oscillations of bi-excitonic origin in the FWM intensity of CP pulses. The four traces were measured in magnetic fields of 0, 2, 4, and 6 T.

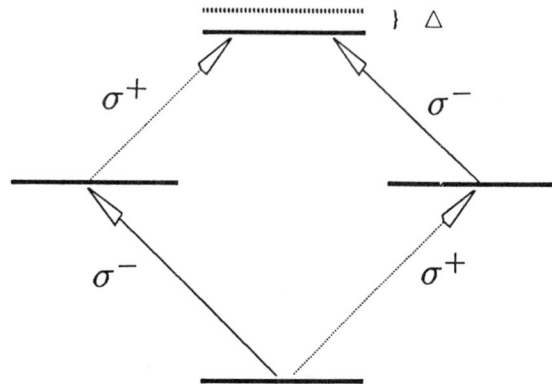

Figure 8. A four level system used in the interpretation of our experimental results. The two excitonic states, corresponding to oppositely-handed circular polarization, are coupled through both the ground state (vacuum) and the singlet bi-exciton upper state, with binding energy delta. The allowed optical transitions are also indicated.

The fact that the oscillations are observed in the pump probe experiment when the pump and probe are of opposite circular polarizations raises the possibility that the oscillations are of bi-excitonic origin. Excitons created by a σ^- probe at a location where σ^+ excitons were created by the pump could bind with these pump excitons to create a bi-exciton molecule. The relevant energy level diagram should therefore be updated to the one described in Fig. 8, where the bi-exciton energy is $2\varepsilon - \Delta$, ε is the exciton energy, and Δ is the bi-exciton binding energy. Like the case of a hydrogen molecule, it is assumed here that the two excitons which can bind have opposite J_z, i.e., the bi-exciton is formed from a σ^+ and a σ^- exciton.

To understand how the creation of a bi-excitonic molecule can give rise to the observed signal, let us distinguish between two cases: the first is when the two pulses do not overlap in time and the second is when they do. In the first case, a renormalization by the pump of the available states for absorption of the probe occurs: absorption can occur not only from the ground state to the excitonic level but also from the populated excitonic level to the bi-excitonic one. Therefore, as available states for σ^- absorption are eliminated at the exciton energy ε, other states are created at a lower energy $\varepsilon - \Delta$. The net result will therefore be reduced absorption at the exciton energy, as indeed observed in the experiment. To understand the origin of the oscillations it is necessary to consider the second case, of overlapping pulses, and to find the time dependent suceptibilities. The calculated third order susceptibility includes terms in which transitions to and from the bi-excitonic level are involved and terms in which only excitonic transitions are involved, and the interference of these terms is responsible for the observed oscillations. In that sense the observed oscillations are not ordinary quantum beats, of two excited states with a common ground state, but rather a manifestation of a quantum interference of different paths. It is important to note that this interference occurs only while the pulses overlap, and its decay does not reflect a dephasing process but rather the diminishing of the pulses overlap.

We have conducted a detailed analytical calculation of this effect, taking into account the pulse shape and the inhomogeneous broadening. We have found that the oscillations

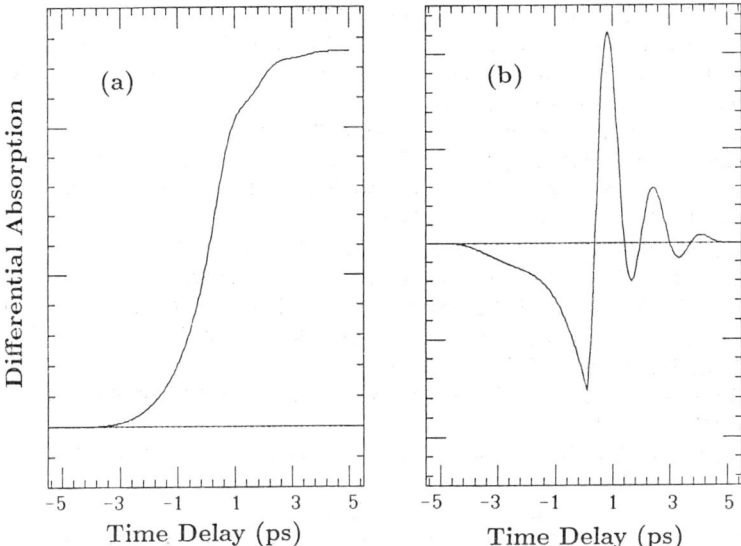

Figure 9. Results of a numerical simulation of the transient pump-probe signal with pump and probe of oppositely-handed circular polarizations, using the four level system shown in Fig. 8, and a 1.2 ps autocorrelation single-sided exponential pulse with an abrupt trailing edge. Frame (a) shows the sum of all the calculated terms, while frame (b) shows only the oscillatory term.

amplitude depends on the assumed pulse shape, and increases for an asymmetric envelope. The inhomogeneous broadening was found to eliminate the dependence of the oscillation frequency on the detuning of the laser frequency from the exciton line, leaving only the binding energy Δ as the oscillation frequency. Figure 9 shows the calculated change of absorption for a single-sided exponential pulse shape, where we have separately plotted the oscillating and the non-oscillating parts. Assuming equal dipole moments for the excitonic and the bi-excitonic transitions, the ratio between the two terms is strongly in favor of the non-oscillating part, giving a signal with only weak oscillations (Fig. 9(a)). If we assume a stronger dipole moment for the bi-excitonic transition, the weight of the oscillatory part increases and there is a better agreement with the experimental results. Our results imply that the bi-exciton binding energy in these samples is 2–2.7 meV. This value is larger than the upper limit of 2 meV for narrow QWs, which is predicted by calculations. It is, however in agreement with recent PL measurements in narrow QWs, which give values of 1.5–2.8 meV for the bi-exciton binding energy.

References

1. I. Schultheis, J. Kuhl, A. Honold, and C.W. Tu, Phys. Rev. Lett. **57**, 1635 (1986); **57**, 1797 (1986).
2. E.O. Gobel, K. Leo, T.C. Damen, J. Shah, S. Schmitt-Rink, W. Schafer, J.F. Muller,

and K. Kohler, Phys. Rev. Lett. **64**, 881 (1990).
3. J.C. Maan, in *Physics and Applications of Quantum Wells and Superlattices*, edited by E. Mendez and K. von Klitzing, NATO Advanced Study Institute, Ser. B, Vol. 170 (Plenum, New York, 1987) p. 347.
4. R.C. Miller and D.A. Kleinman, J. Lumin. **30**, 520 (1985).
5. S. Bar-Ad and I. Bar-Joseph, Phys. Rev. Lett. **66**, 2491 (1991). After publication we have discovered an error in the calibration of the magnetic field strength. The pubished value of B = 4 T is acutally 5 T.
6. S. Bar-Ad and I. Bar-Joseph, Phys. Rev. Lett. **68**, 349 (1992).
7. O. Carmel and I. Bar-Joseph, Phys. Rev. B, to be published.
8. H.H. Yaffe, Y. Prior, J.P. Harbison, and L.T. Florez, in *Quantum Electronics Laser Science*, OSA Technical Digest Series **11**, 196 (1991); S.T. Cundiff, H. Wang, and D.G. Steel, Phys. Rev. B **46**, 7248 (1992).
9. H. Stolz, V. Langer, E. Schreiber, S. Permogorov, and W. von der Osten, Phys. Rev. Lett. **67**, 679 (1991).
10. J.G. Gordon, C.H. Wang, C.K.N. Patel, R.E. Slusher, and W.J. Tomlinson, Phys. Rev. **179**, 294 (1969).
11. T. Baer and I.D. Abella, Phys. Rev. A **16**, 2093 (1977).
12. C. Weisbuch, R.C. Miller, R. Dingle, A.C. Gossard, and W. Wiegmann, Solid State Commn. **37**, 219 (1981).
13. R.C. Miller, D.A. Kleinman, W.A. Nordland, and A.C. Gossard, Phys. Rev. B **22**, 863 (1980).
14. A.H. Clark, R.D. Burnham, D.J. Chadi, and R.M. White, Phys. Rev. B **12**, 5758 (1975) and Solid State Commun. **20**, 385 (1976); G. Fishman and G. Lampel, Phys. Rev. B **16**, 820 (1977); K. Zerrouati, F. Fabre, G. Bacquet, J. Bandet, J. Frandon, G. Lampel, and G. Paget, Phys. Rev. B **37**, 1334 (1988).
15. M.R. Freeman, D.D. Awschalom, J.M. Hong, and L.L. Chang, in *Proceedings of 20th International Conference on Physics of Semiconductors*, edited by E.M. Anastassakis and J.D. Joannopoulos (World Scientific, Singapore, 1990), p. 1129.
16. T.C. Damen, L. Viña, J.E. Cunningham, J. Shah, and L.J. Sham, Phys. Rev. Lett. **67**, 3432 (1991).
17. A. Twardowski and C. Hermann, Phys. Rev. B **35**, 8144 (1987).
18. T. Uenoyama and L.J. Sham, Phys. Rev. Lett. **64**, 3070 (1990).
19. R. Ferreira and G. Bastard, Phys. Rev. B **43**, 9687 (1991).
20. D.J. Lovering, R.T. Philips, G.J. Denton, and G.W. Smith, Phys. Rev. Lett. **68**, 1880 (1992).
21. K.-H. Pantke, D. Oberhauser, V.G. Lyssenko, G. Weimann, C. Klingshirn, and J.M. Hvam, in *Quantum Electronics Laser Science*, OSA Technical Digest Series **13**, 14 (1992).

COHERENT NONLINEAR LASER SPECTROSCOPY OF EXCITONS IN QUANTUM WELLS

D.G. STEEL, S.T. CUNDIFF, AND H. WANG
The Harrison M. Randall Laboratory of Physics
The University of Michigan
Ann Arbor, MI 48109
USA

ABSTRACT. Coherent nonlinear laser spectroscopy of excitons in quantum wells based on frequency domain four wave mixing and picosecond photon echoes shows the presence of long dephasing times due to disorder. Measurements demonstrate that these excitons relax primarily by phonon assisted migration between localization sites. However, measurements of the polarization dependence of the nonlinear response shows the simple picture of excitons is inadequate for describing the interaction. Furthermore, the measurements indicate that there are two distinct resonant contributions to the nonlinear response which are characterized by different relaxation, saturation and transport properties.

1. Introduction

The strong excitonic resonance observed in optical spectroscopy of semiconductors just below the bandedge is the result of the Coulomb interaction between electrons and holes. Quantum confinement of the carriers in semiconductor heterostructures results in an increase in the transition energy as well as an increase in the electron-hole correlation leading to an increase in the binding energy and electronic oscillator strength. The persistence of a strong resonance even at room temperature has resulted in considerable interest in these materials for potential optoelectronic applications based on the strong nonlinear optical susceptibility associated with a resonant response.

However, complete understanding of the nonlinear optical response has been difficult. In contrast to simple atomic systems, the excitations are described by extended states subject to the Pauli exclusion principle and characterized by strong Coulomb interactions. In simple atomic systems, optical interactions have been described by optical Bloch equations and more recently by modified optical Bloch equations [1]. In an effort to parallel this successful formalism for semiconductors, a set of effective optical Bloch equations has been developed which reveals that although many similar coherent nonlinear optical effects are present in semiconductors, their physical origin and interpretation is considerably more complex (for a review see [2,3]). In an ideal semiconductor, the origin of the nonlinear response is dominated by effects of Coulomb screening and the Pauli exclusion principle including phase space filling and exciton-exciton interactions. Coulomb screening and effects of the Pauli exclusion principle lead to manybody contributions to the nonlinear response causing such effects as optically induced changes in the oscillator strength as well as shifts in the transition energy and the bandedge. In the presence of quantum confinement, Coulomb screening effects are reduced due to the reduced dimensionality [2]. The large optical density of these systems is also not common in simple low density atomic systems and leads to coherent coupling of excitons through exciton-exciton interactions which are similar to local field effects. The qualitative change in behavior can be seen by extending the simple two level model [4] which can lead to so-called "negative" delay signals and nonexponential decays. This effect is closely related to the Lorentz-Lorenz law in gases [5] and was first noted in coherent transient effects in atomic systems in accumulated photon echo experiments [6] and recently reported in semiconductor systems [7-9]. Of more fundamental interest is the presence of strong Coulomb coupling in semiconductors. This can lead to dramatic modification of the nonlinear optical response for sufficiently strong excitation, as demonstrated by recent numerical calculations [10,11].

In semiconductor heterostructures, the description of nonlinear optical behavior has been greatly complicated by the presence of disorder which is also an issue of fundamental interest [12]. In quantum wells fluctuations in the confinement potential due to interface roughness lead to localization of the exciton where the exciton envelope function decays exponentially in space. The optical response in these systems becomes inhomogeneously broadened, resulting in an increase in the linear absorption linewidth due to small fluctuations in the exciton transition energy. In addition exciton localization also results in reduced phonon scattering leading to a decrease in the homogeneous linewidth (i.e., decreased dephasing rate) compared to an ideal quantum well at low temperature where the linewidth and energy relaxation are determined by phonon scattering along the two-dimensional dispersion curve (~ 1 ps [13]). For localized excitons, energy relaxation proceeds by phonon assisted migration between localization sites and thermal activation to delocalized states [14,15].

In this paper we describe a series of nonlinear optical measurements based on frequency domain cw four-wave mixing and picosecond transient four-wave mixing. The measurements show a rich diversity in the coherent nonlinear optical phenomena which can be observed in these systems but also demonstrate the complexity induced by disorder. The results show that existing theoretical understanding in these systems is inadequate for describing the observations.

2. High Resolution Nonlinear Spectroscopy of Localized Excitons

Early insight into the effects of disorder in semiconductor heterostructures was obtained in studies of exciton luminescence which confirmed the emission was inhomogeneously broadened due to the corresponding potential fluctuations at the interface [16]. These studies were followed by results that indicated that this interface roughness is characterized by a scale length large compared to the Bohr radius of the exciton (see, for example [17,18]); however, more recent studies by chemical lattice imaging have shown the presence of monolayer flat islands with a smaller spatial extent of 50-100 Å [19]. While the results are presently somewhat controversial [20,21], they have led to the proposal that at least two scale lengths for interface roughness are required in order to account for the observations [22]. Further evidence that the island size distribution is bimodal has been recently presented in luminescence and Raman scattering measurements [23]. Details of the interface roughness also depend on specific growth processes, such as interrupted or non-interrupted growth [18], or whether GaAs is grown on $Al_xGa_{1-x}As$ or $Al_xGa_{1-x}As$ is grown on GaAs [24].

Localized and extended excitons have qualitatively different relaxation properties [15]. Even at low temperature (< 10 K), localized excitons do not remain truly localized, instead they can migrate among localization sites by emitting or absorbing acoustic phonons. Such phonon assisted migration was first proposed to explain the slow and non-exponential energy relaxation observed in time resolved luminescence measurements in a GaAs quantum well (QW) structure [25] and directly observed in InGaAs/InP QW where all excitons are localized by alloy disorder [26] and in GaAs QW where excitons are localized due to interface disorder [27,28]. At higher temperatures, excitons can absorb phonons with sufficient energy to become activated to delocalized states at higher energies. The activation process has been observed in a number of measurements such as spectral hole burning [29], resonant Raleigh scattering [30], and resonant Raman scattering [31]. Estimates of the activation energy have suggested that the onset for the delocalized exciton in GaAs QW structures is near the absorption line center [32]. In contrast, decay of the delocalized exciton is determined by the exciton-phonon scattering along the energy-momentum dispersion curve and the exciton recombination. Furthermore, delocalized excitons also experience elastic scattering from potential fluctuations, which introduces additional dephasing due to the decay of the polarization. Indeed as we show below in typical quantum wells, the dephasing time is > 45 ps whereas in near ideal quantum well, the dephasing time is an order of magnitude shorter [13].

To study the nonlinear optical response in these systems, experiments were performed on various GaAs quantum well structures. Typical samples were comprised of 96 Å GaAs wells and 98 Å $Al_{0.3}Ga_{0.7}As$ barriers. The substrate was removed for transmission and four wave mixing measurements. The number of periods was either 10 or 60. In 60 period wells, the typical linear absorption linewidth of hh1 was 2 meV with a 1 meV Stokes shift in the luminescence. In the 10 period sample, the absorption linewidth was 0.9 meV and there was a negligible Stokes shift in the luminescence (< 0.2 meV).

In the first set of measurements, we use a new method of saturation spectroscopy based on cw frequency domain four-wave mixing to demonstrate spectral hole burning and phonon assisted migration of excitons in these systems. Since the theoretical basis for these spectroscopic measurements is discussed in detail elsewhere [33], we summarize the basic approach for completeness. The experimental configuration is based on the backward four-wave mixing geometry where two beams $E_1(\omega_1, k_1)$ and $E_2(\omega_2, k_2)$ separated by an angle Θ interact in the sample with a third beam $E_3(\omega_3, k_3)$ ($E_1 \| E_2 \perp E_3$) through the resonant third order susceptibility to generate a signal beam $E_s(\omega_s, k_s)$ proportional to $\chi^{(3)}(\omega_s = \omega_1 - \omega_2 + \omega_3):E_1E_2^*E_3$. In these experiments, k_3 is counter-propagating with respect to k_1 and the signal counter-propagates with respect to k_2. In a simple physical picture for the response of this system (though not necessarily in general as we see in the time domain measurements below), $E_1 \cdot E_2^*$ results in a spatial and temporal modulation of the exciton population, which modifies the optical response of the sample and leads to a traveling wave grating or modulation of the absorption and dispersion. The coherent nonlinear signal arises from scattering of the third beam, E_3, from the grating. Spectroscopic information can be obtained by studying the dependence of the nonlinear optical response on the relative frequency detuning, absolute frequency detuning, the electric field polarization, the input beam intensities and the relative angle of the different input beams, especially the angle between k_1 and k_2. The observed line shapes and corresponding physical information that is obtained depends on which frequency is tuned.

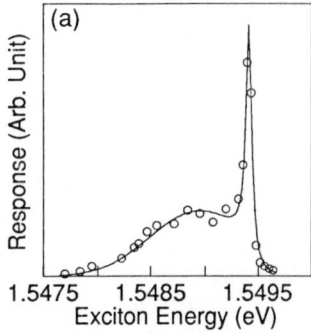 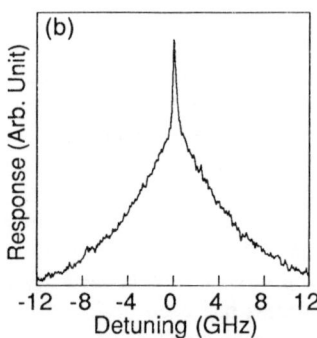

Figure 1. (a) The four-wave mixing response as a function of the third beam frequency at 5 K showing the spectral hole with a width determined by the homogeneous width. The four-wave mixing response also shows a broad pedestal at lower energy due to the quasi-equilibrium distribution of excitons produced by exciton migration. This migration may also contribute to the width of the narrow feature. (b) The four-wave mixing response at 10 K as a function of the detuning between the first and second beams. Unlike Fig. 1(a), this lineshape reflects the energy relaxation rate. The broad feature measures the rate of decay of the narrow spectral hole in Fig. 1(a) while the narrow feature measures the rate of decay of the quasi-equilibrium distribution giving rise to the pedestal.

In general, tuning ω_1 or ω_2 probes population and grating relaxation, while tuning ω_3 provides a measure of the homogeneous line shape similar to spectral hole burning of the spectral redistribution of the excitation.

In Fig. 1(a), we show the four wave mixing response obtained by scanning the frequency of the third beam (ω_3) [27, 34]. This enables a direct detection of the spectral hole produced by excitons excited by $\mathbf{E}_1 \cdot \mathbf{E}_2^*$ at energy E (1.5 meV below the hh1 absorption line center) as well as detecting any excitons scattered from energy E to E' due to spectral diffusion. The nonlinear response is corrected for sample absorption. The narrow resonance in the response corresponds to exciton spectral hole burning. The width of the hole gives an exciton homogeneous line width $2\gamma_{ph} \sim 0.03$ meV, corresponding to a dephasing time of 45 ps, assuming no contribution due to spectral diffusion. (In the absence of spectral diffusion, the observed width is $4\gamma_{ph}$.) The broad Stokes shifted feature is due to the quasi-equilibrium distribution of the exciton population produced by the migration of excitons from energy E to E'. The response is a function of the steady state exciton population assuming all excitons in the spectral region concerned give rise to the same cw nonlinear response. The line shape in Fig. 1(a) can be fit to a simple model which neglects migration to states above the excitation energy. The calculation is based on the nonlinear optical response of a simple two level system and assumes a Gaussian distribution for the quasi-equilibrium population of excitons that have migrated to states below the excitation energy. The spectral profile of the hole-burning resonance is assumed to be Lorentzian. The result is plotted as the solid line in the figure. A small but finite detuning of order 100 kHz is set between fields 1 and 2 to reduce contributions to the data from slower components (discussed elsewhere [35]).

In the simple picture of spectral hole burning in the presence of spectral diffusion, the lifetime of the excitation created by $\mathbf{E}_1 \cdot \mathbf{E}_2^*$ is characterized by two time scales. The initial excitation decays on a short time scale due to spectral diffusion. However, assuming equilibrium is established on a time scale short compared to the recombination time, a second time scale is observed due to decay of the equilibrium population; i.e., the recombination time. The presence of two time scales is clearly seen in Fig. 1(b) (T = 10 K) where we show the four-wave mixing lineshape obtained by tuning ω_1. Recall that tuning ω_1 varies the frequency difference $\omega_1-\omega_2$ which determines the oscillation time of the grating. The signal decreases when the frequency difference exceeds the relaxation rate (similar to the Fourier transform picture of a time-resolved pump-probe measurement). The narrow resonance corresponds to a lifetime of 1 ns and represents the decay time of the quasi-equilibrium exciton distribution (the broad feature in Fig. 1(a)). The wide feature corresponds to a lifetime of 45 ps and is the inverse of the spectral diffusion rate. Indeed, setting the detuning of $\omega_1-\omega_2$ large compared to the width of the narrow feature greatly reduces the contribution of the broad feature in Fig. 1(a) and in fact further narrows the sharp resonance, confirming that the broad feature in Fig. 1(a) corresponds to spectral diffusion.

The theoretical model for phonon assisted exciton migration was developed recently by Takagahara [14]. In the model, excitons resonantly excited are in a non-equilibrium state and can migrate to other sites by emitting or absorbing acoustic phonons. Migration is possible due to the overlap of the exciton wave function in different sites when the inter-site distance is small. When the inter-site distance is much greater than the localization length the process occurs by inter-site dipole-dipole coupling. The typical magnitude of participating phonon wave vectors is within a few times of the inverse of the localization length corresponding to phonon energies of order 0.01 to 0.1 meV. The theory predicts a distinctive temperature dependence for the migration rate. At low temperatures, the dependence is described by $\exp(\beta T^\alpha)$. In this expression, β is positive and independent of temperature but is expected to increase with the exciton energy and depends on details of interface roughness; α is estimated to be between 1.6 and 1.7. The predicted temperature dependence is quite different from that of variable range hopping used by Mott to interpret electronic conduction in the localized regime [36]. The difference has been attributed to the long-rang nature of the inter-site interaction and the phonon emission process involved in the migration of the localized exciton [15]. This temperature dependence has been observed in transient hole burning experiments in an InGaAs/InP QW where all excitons are localized by alloy disorder [26]. Figure 2 shows the temperature dependence of the exciton migration rate measured at 0.6 meV and 1.5 meV below the absorption line center. The data are in good agreement with the phonon assisted migration theory discussed above with a = 1.6. The measurement indicates that the dominant contribution to relaxation of the localized exciton is phonon assisted migration up to a temperature of 15 K. The slow scattering rates and observed temperature dependence are quite different from the faster rates and linear temperature dependence reported in material with reduced disorder [13].

3. Picosecond Coherent Optical Spectroscopy

The strong inhomogeneous broadening of the hh1 linear absorption spectrum compared to the intrinsic homogeneous width suggests that such a system is ideal for observing a classical photon echo. In the classical picture of the three pulse transient resonant four wave mixing response, a homogeneously broadened system gives rise to a free polarization

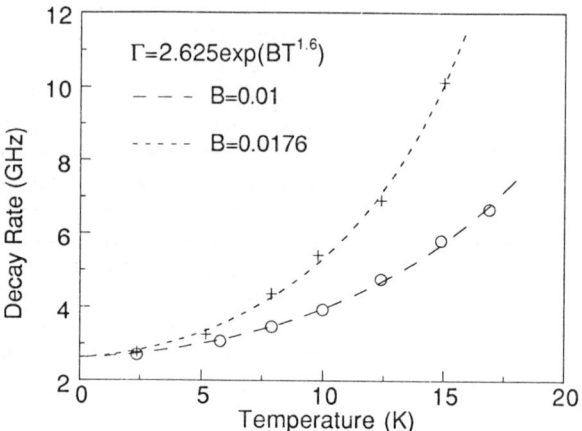

Figure 2. The temperature dependence of the exciton decay rate. Circles and crosses are data obtained at 1.5 meV and at 0.6 meV below line center respectively. Dashed lines are fit to the theory of phonon assisted migration.

decay signal which is coincident in time with the third excitation pulse. However, in the presence of strong inhomogeneous broadening, the fields from the different resonances initially interfere, destroying the free polarization decay. At a time after the third pulse given by the time between the first two pulses, a rephasing occurs giving rise to a constructive interference and coherent emission designated as an echo. Recent theoretical work has predicted that in an ideal semiconductor, exciton-exciton interactions and coupling to the continuum lead to complex temporal structure in the emitted radiation in a coherent transient four-wave mixing experiment producing time delays in the emission as seen in a photon echo experiment [11]. However, in a classical two pulse (spontaneous) or three pulse (stimulated) photon echo, an inhomogeneously broadened resonant system produces a time delay in the emitted signal with respect to the last pulse that is given by the time between the first two pulses. In addition, the emission is a single peak with a width determined by the inhomogeneous width. Early work on GaAs multiple QW demonstrated the existence of a spontaneous photon echo [37] while more recently, transient four-wave mixing in the mixed crystal CdSSe demonstrated a stimulated photon echo [38].

Using the identical geometry as in the above cw measurements, we have examined the time resolved emission generated in a three pulse transient four-wave mixing experiment in the limit where the Rabi flopping frequency, $\mu E/\hbar$ is much less than the exciton binding energy [39]. A picosecond laser system (pulsewidth adjustable between 200 fs and 8 ps) was used to excite the system. A portion of the laser beam was mixed with the coherent signal to produce a frequency up-converted signal for time resolution. Figure 3(a) shows the time resolved emission where the inset confirms that the time delay in the emission is given by the time delay between the first two pulses. In these experiments, the normal time ordering is E_2 followed by E_1 followed by E_3. (Note that the contributions due to polarization scattering which give rise to so-called negative time signals [4,7,8,11] for E_1 followed by E_2 are not observed in these experiments on time scales of T_2 since local field effects decay on the time scale of T_2^*, the inverse of the inhomogeneous width). The solid

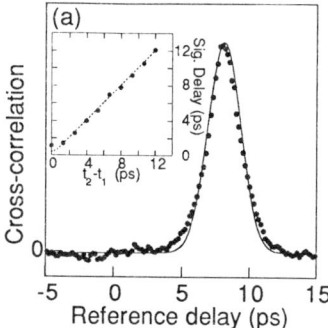 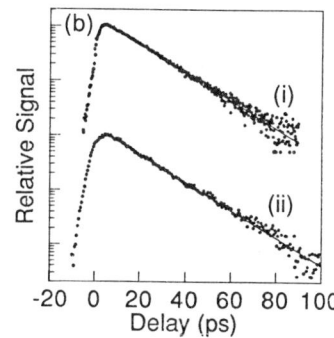

Figure 3. (a) The time resolved transient four-wave mixing response demonstrating delay of the emission. The inset shows a linear dependence on incident pulse interval characteristic of a classic photon echo. (b) The transient four-wave mixing signal as a function of delay between first two pulses (curve (i)), and second and third pulses (curve (ii)). Straight lines are fit to exponential behavior, yielding a dephasing time of 69 ps for curve (i) and an excitation relation time of 35 ps for curve (ii).

line in Fig. 3(a) represents a numerical integration of the optical Bloch equations including finite pulsewidth effects where the only fitting parameters are the amplitude and the inhomogeneous width (in the limit of delta-function pulses, the temporal width of the echo is given by $\Delta t = 4\hbar \sqrt{2} \ln 2/\Delta E$ where Δt is the echo pulse width and ΔE is the full width half maximum of the Gaussian inhomogeneous distribution in energy). The fit gives ΔE = 2.25±0.25 eV, in excellent agreement with the 2.2 eV linear absorption width of this particular sample. Figure 3(b) shows the excellent single exponential decays associated with the energy relaxation rate and the dephasing rate corresponding to 35 ps and 69 ps, respectively. Based on this approach, we have reported measurements of the energy and temperature dependence of the relaxation rates further confirming the details of phonon assisted migration in disordered systems [28].

Through the above discussion, we have provided a self-consistent picture of the nonlinear optical response and associated dynamics. It is clear that disorder plays a major role in these systems since the dynamical processes which we have been investigating are controlled by the nature of disorder. In the above experiments all fields were linearly co-polarized, however, several studies have recently reported that the dephasing rate changes dramatically in two pulse transient four-wave mixing experiments in a multiple QW when $E_2 \perp E_1$ [40,41]. In Fig. 4 we demonstrate a fundamental change under these conditions in the time resolved emission by comparing the time resolved emission in a co-polarized experiment (Fig. 4(a)) to that for $E_2 \perp E_1 \parallel E_3$, $E_s \parallel E_2$ (Fig. 4(b)). The data show that in the former case the emission time depends on the interval between E_1 and E_2 while in the second case it does not. [The weak echo signal evident in Fig. 4(b) arises from a resonance which is 2.5 meV Stokes shifted from the absorption peak. On resonance the strength of the signal is comparable to the free polarization decay, however, the contribution is observable in Fig. 4(b) despite the large detuning because the dephasing time of this signal, 70 ps, is much longer than the dephasing time of the primary peak and hence decays far more slowly with increasing time delay between the first and second pulse.] Both the free polarization decay Fig. 4(b) and co-polarized stimulated photon echo Fig. 4(a) are resonant

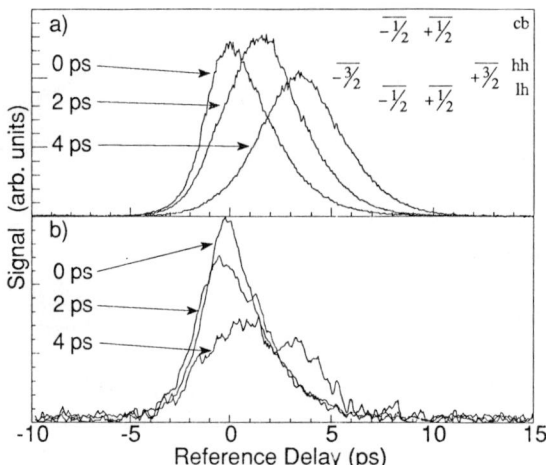

Figure 4. The time resolved transient four-wave mixing signal for (a) co-polarized fields and (b) $E_1 \perp E_2 \| E_3, E_s \| E_1$. The response for a series of delays between E_1 and E_2 are given in each panel. The inset in the upper panel is the energy level structure for the magnetic substates in a GaAs multiple QW.

with hh1 and show very similar spectral dependence; after correction for absorption, they are nearly coincident with the absorption maximum. The free polarization decay is approximately two orders of magnitude weaker than the co-polarized stimulated photon echo at low intensity at the peak of the response. Based on the above classical picture, it is clear that the signal for $E_1 \| E_2$ is due to an inhomogeneously broadened resonance while the signal for $E_1 \perp E_2$ is due to a homogeneously broadened resonance.

This behavior is completely unexpected. Based on the energy level structure (see Fig. 4(a) inset) no difference in the nature of the signal is expected other than the polarization. In GaAs, quantum confinement lifts the valence band degeneracy at k = 0 resulting in heavy-hole–light-hole splitting, which for these experiments is large compared to the laser bandwidth and the Rabi flopping frequency. When the axis of quantization is taken to be perpendicular to the barriers, the heavy-hole valence band has only two degenerate magnetic substates given by m = ±3/2 levels while the conduction band substates are given by m = ±1/2. The selection rules for optical excitation are $\Delta m = \pm 1$. Since linearly polarized fields propagating parallel to the axis of quantization can be taken as a superposition of two circularly polarized fields, we can designate the exciton created by a σ_+ polarized field as |+1> and the exciton created by a σ_- polarized field as |-1>. The signal is produced by the scattering of E_3 off the grating produced by $E_1^* E_2$. We see then that for linear co-polarized E_1 and E_2, two spatial gratings consisting of |+1> and |-1> excitons are created and are spatially coincident. For $E_1 \perp E_2$, the two gratings are exactly spatially out of phase. However, E_3 must also be decomposed into circularly polarized components, each of which independently scatters from one of the gratings, resulting in a linearly polarized signal field. Hence the change in the nonlinear response in Fig. 4 is clearly unexpected. (This picture is for a homogeneously broadened system but still provides insight for an inhomogeneously broadened system. Although there is no net

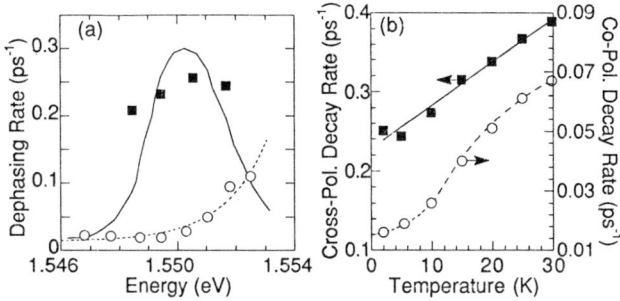

Figure 5. Dependence of dephasing rates on (a) energy and (b) temperature. Solid squares are for $E_1 \perp E_2$ and open circles for $E_1 \| E_2$. In (b) the dashed line is the fit to theoretical predictions for excitons migrating between localization sites, and the solid line is the fit for single phonon scattering.

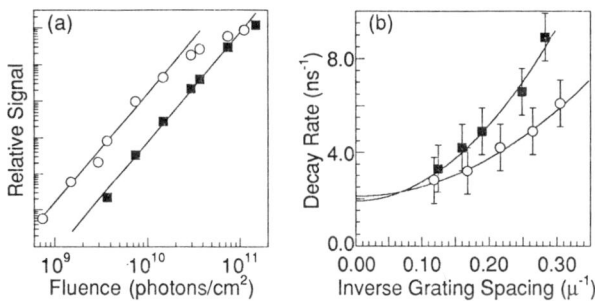

Figure 6. (a) Signal strength as a function of incident intensity; straight lines are for cubic behavior. (b) Decay of grating as a function of grating spacing; lines are fit to the quadratic dependence expected for spatial diffusion. In both, solid squares are for $E_1 \perp E_2$ and open circles for $E_1 \| E_2$.

grating, there is a grating for each frequency group within the inhomogeneous distribution. Each group results in a component of the scattered signal, and it is the phase relationship between these components which results in the signal delay.)

In Fig. 5(a) we compare the energy dependence of the dephasing rates for these two components of the response. The dephasing rate for the prompt signal ($E_1 \perp E_2$) is much greater than the delayed signal, 0.20-0.25 ps^{-1} and essentially energy independent. The dephasing rate for the delayed signal ($E_1 \| E_2$) varies from 0.016±0.001 ps^{-1} on the low energy side to 0.1±0.02 ps^{-1} on the high energy side. The increase in the dephasing rate with increasing photon energy has been reported earlier [27,28,42] and was suggested to be due to the increase in scattering rate for localized excitons near the mobility edge. The temperature dependence of these dephasing rates also differs considerably. In Fig. 5(b) we see that the temperature dependence of the dephasing rate of the delayed component follows the behavior for phonon assisted migration and thermal activation of localized excitons as described above. However, the dephasing rate for the prompt signal shows a linear temperature dependence of the form $\gamma_{ph} = \gamma_0 + \gamma^* T$, where $\gamma^* = 4$ μeV/K. We note that the magnitude of γ_0, the linear dependence on temperature and the magnitude of γ^* are similar to that observed by Schultheis in a single quantum well which was believed to be of very high quality [13]. The sample in that work was homogeneously broadened with no Stokes shift in the luminescence. Based on the quality of the sample, the homogeneous broadening of the hh1 resonance even at low temperature and the linear temperature dependence of the dephasing rate, Schultheis et al. proposed that the scattering mechanism was single phonon scattering [43] along the 2-D dispersion curve of an exciton described by an extended (delocalized) wave function.

It is, of course, difficult to determine experimentally if a wave function is truly extended. However, there are several additional experimental observations in the current experiments which are consistent with the proposal that the resonance involved in the emission in Fig. 4(b) arises from excitons with extended wave functions. In Fig. 6(a) we show that for fixed pulse delay and energy, the magnitude of both signals vary cubically as expected with the incident beam energies (the solid line corresponds to a slope of n=3) for a true four-wave mixing process. However, at densities above 2×10^8 excitons-cm^{-2}-layer^{-1} (corresponding to a fluence of 2×10^{10} photons-cm^{-2}, where the density was calculated based on the absorbed energy) the stimulated photon echo emission deviates from cubic behavior and begins to saturate while there is no evidence of saturation in the prompt signal. Such saturation of the delayed signal which arises due to localized excitons would be expected since there are a finite number of localization sites. The saturation intensity of the nonlinear response in an ideal quantum well where all states are extended is considerably higher than that seen for localized states. At sufficiently high excitation intensity, the co-polarized signal is comprised of both a prompt signal and an echo [39] but as the intensity increases, the delayed signal completely saturates, leaving only the free polarization decay.

Transport measurements also provide an indication of the origin of the differences in these two signals. To examine this aspect of the behavior, we studied the angle dependence of the excitation relaxation rate to probe the spatial "washout" time due to motion of the excitation. This is the typical approach in transient grating experiments to measure spatial diffusion [44]. In these experiments measurement of the signal as a function of the delay between the second and third fields measures the decay rate of the excitation induced by the first two fields. The total signal decay is given by $\Gamma_{pop} = \gamma_{rec} + \Gamma_{SD} + \Gamma_D$ where γ_{rec} is the recombination rate, Γ_{SD} is the spectral diffusion rate and Γ_D is the spatial diffusion rate given by $4\pi^2 D/\Lambda^2$. Λ is the grating spacing: $\Lambda = \lambda/n2\sin(\theta/2)$. It is expected that delocalized states should have a larger D than localized states. Figure 6(b) shows $2\Gamma_{pop}$ as a function of inverse grating spacing for the free polarization decay and the stimulated photon echo, fitting to a quadratic dependence yields D = 5.2 for the stimulated photon echo and D = 10.1 for the free polarization decay. This data is taken at an exciton energy 0.5 meV below line center and at a temperature of 15 K. The measured difference is larger in other samples. This difference is clear evidence of the greater mobility of the excitons which are responsible for the free polarization decay signal. The diffusion coefficient for localized excitons is dependent on the exciton energy [42] and this measurement is taken at a point where it is beginning to undergo a transition from its low energy (small D) to high energy (large D) values. The ratio of the two diffusion coefficients is sample dependent as expected where in a 10 period sample, the ratio is as large as 3.5.

The difference in the diffusion coefficients supports the earlier data suggesting that the emission shown in Fig. 4 arises from two different resonant effects and the larger dephasing rate and increased mobility observed for the resonance studied with orthogonal excitation indicates this exciton may not be as strongly localized. Indeed, the conclusion that the prompt signal arises from a homogeneously broadened resonance is not completely accurate in view of recent theoretical work [11]. These results show that for low intensity resonant excitonic excitation in the transient four-wave mixing geometry, the emission time does not depend on the time between the first two pulses. At higher excitation density, additional temporal structure is observed at later times which does depend on the time between the first two pulses. The intensity dependence shown in this model for this delayed structure is opposite that reported in the above discussion. However, the intensities for the observation in these experiments are all well below that predicted in the theoretical work for observing the additional temporal structure. In fact, we believe that the emission in Fig. 4(b) is indeed the low intensity response predicted in reference [11].

We note that the proposal that these resonances are distinguished by different degrees of localization is certainly unexpected in the usual picture of localized and delocalized states [45]. However, an alternative explanation may be that these resonances coincide in energy but not in space, a possibility consistent with the recent proposals that interface roughness is characterized by a bimodal distribution in scale lengths [22,23]. However, even in material characterized as high quality, we have shown that these effects can dominate optical studies.

We also note that the disappearance of the echo for orthogonal polarization remains unexplained. Experimentally, the absence of the signal demonstrates the absence of a grating. Recalling that there are two induced gratings associated with |+1> and |-1>. excitons, respectively, we note that complete spatial mixing of the two gratings on a time scale of the measurement would indeed destroy the signal. One such mechanism is spin relaxation. However, measurements show the spin relaxation time in this system is on the order of 40 ps, in agreement with other measurements [46,47], far too slow to account for the observed behavior.

Additional insight into these results is seen in the experimental study of the unexpected intensity dependence and depolarization of the four-wave mixing response. In these experiments, the first two fields E_1 and E_2 are circularly co-polarized σ_+. Hence, in the simple picture where m is a good quantum number, only |+1> excitons are created. Using a *linearly* polarized field for E_3, we would expect that only the σ_+ component of the E_3 polarization would be scattered, giving rise to a complete circularly polarized signal. In Fig. 7(a,) we plot the polarization given by the intensity difference between left and right circularly polarized light normalized to the sum. At low intensity, we see the surprising result that the polarization of the system is lost; i.e., the scattered signal is linearly polarized. As the excitation intensity increases, the polarization of the system is recovered and the emitted radiation is circularly polarized. While the physical basis for this response is not understood, the disappearance of the echo with orthogonally linearly polarized fields is now clear. Since linearly polarized light couples equally well to the |+1> and |-1> exciton gratings, there is no effective modulation of the optical response and hence no scattering of E_3. Alternatively, the scattering of radiation from the |+1> exciton grating is exactly out of phase with the radiation scattered from the |-1> exciton grating, resulting in complete destructive interference.

The origin of this behavior is not fully understood, but it is unlikely that dynamical effects such as relaxation can explain the above behavior. Incoherent transient spectroscopy pump-probe measurements using circularly polarized radiation show that the

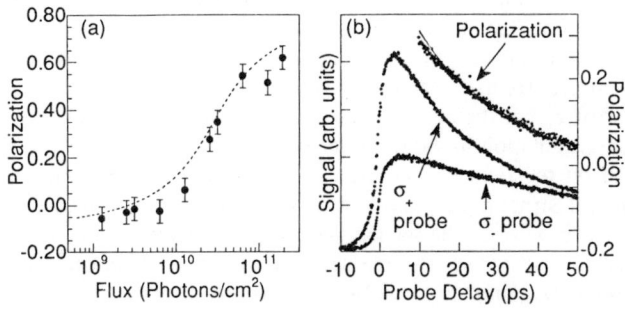

Figure 7. (a) Polarization $(I_+ - I_-)/(I_+ + I_-)$ of signal field as a function of incident intensity; dashed line is to guide the eye. (b) Transient absorption response for a circularly polarized (σ_+) pump. Upper curve is the decay of polarization (right axis, offset for clarity).

coupling appears to be instantaneous, as seen in Fig. 7(b). Similar results are obtained with 200 fs pulses, though the sign of the response depends on pulsewidth and detuning. It is unlikely that any relaxation effects would occur on such a short time scale. Alternatively, the above discussion has been based on the assumption that m is a good quantum number. Recently, it has been shown in numerical studies in an ideal GaAs quantum well, that band mixing is not strong enough to account for the observed coupling in the nonlinear response although mixing of the magnetic substates may occur due to interface roughness [48]. An alternative explanation is based on disorder enhanced exciton-exciton interactions. Nondegenerate pump-probe measurements using orthogonally circularly polarized light show a pump induced increased absorption due to a Stokes shifted resonance when the pump is tuned just below the hh1 absorption line center. The Stokes shift is of order 1–1.5 meV and is constant with respect to the pump frequency rather than the hh1 resonance. These measurements show that the net coupling for co-rotating and oppositely rotating circularly polarized fields is comparable and may explain the behavior observed in the photon echo experiments. Recent evidence for biexciton effects in the nonlinear response have been reported in both quantum beat experiments [47] and in transient four-wave mixing [49].

4. Summary

It is clear from the energy level structure of excitons in semiconductors that there exists a strong similarity between excitons and simple atomic systems. However, because of the strong Coulomb interaction in these systems and the strong Fermionic nature of the carriers, the interaction of coherent light with these resonances gives rise to complexities that can not be simply described in terms of traditional theory based on optical Bloch type equations. The development of a more complete theoretical picture as discussed elsewhere in this book has provided considerable insight into this problem and has enabled the application of traditional spectroscopy methods such as four-wave mixing for the study of these systems. In spite of this progress, it is equally clear that the optical interaction in these systems is greatly complicated not only by the effects of disorder but possibly by higher order coupling between excitons, which may lead to biexciton formation. These effects are clearly seen by the deviations between theory and experiment in the low excitation density limit where the experiments in this paper have been performed.

Acknowledgment

This work was supported by AFOSR, ARO, and NSF.

References

1. P.R. Berman and R.G. Brewer, Phys. Rev. A **32**, 2784 (1985).
2. S. Schmitt-Rink, Phys. Rev. B **32**, 6601 (1985).
3. H. Haug and S.W. Koch, *Quantum Theory of the Optical and Electronic Properties of Semiconductors* (World Scientific, Singapore, 1990).
4. M. Wegener, D.S. Chemla, S. Schmitt-Rink, and W. Schäfer, Phys. Rev. A **42**, 5675 (1990).
5. R.P. Srivastava and H.R. Saidi, J. Quant. Spectrosc. Radiat. Transfer. **16**, 301 (1976).
6. S. Saikan, H. Miyamoto, Y. Tosaki, and A. Fujiwara, Phys. Rev. B **36**, 5074 (1987).
7. B. Fluegel, N. Peyghambarian, G. Olbright, M. Lindberg, S.W. Koch, M. Joffre, D. Hulin, A. Migus, and A. Antonetti, Phys. Rev. Lett. **59**, 2588 (1987).
8. K. Leo, M. Wegener, J. Shah, D.S. Chemla, E.O. Göbel, T.C. Damen, S. Schmitt-Rink, and W. Schäfer, Phys. Rev. Lett. **65**, 1340 (1990).
9. K. Leo, E.O. Göbel, T.C. Damen, J. Shah, S. Schmitt-Rink, W. Schaefer, J.W. Müller, K. Köhler, and P. Ganser, Phys. Rev. B **44**, 5726 (1991).
10. M. Lindberg and S.W. Koch, Phys. Rev. B **38**, 3342 (1988).
11. M. Lindberg, R. Binder, and S.W. Koch, Phys. Rev. A **45**, 1865 (1992).
12. P. Lee and T.V. Ramakrishnan, Rev. Mod. Phys. **57**, 287 (1985).
13. L. Schultheis, A. Honold, J. Kuhl, K. Kohler, and C.W. Tu, Phys. Rev. B **34**, 9027 (1986).
14. T. Takagahara, Phys. Rev. B **32**, 7013 (1985).
15. T. Takagahara, J. Lumin. **44**, 347 (1989).
16. C. Weisbuch, R. Dingle, A.C. Gossard, and W. Wiegmann, Solid State Comm. **38**, 709 (1981).
17. R.C. Miller, C.W. Tu, S.K. Sputz, and R.F. Kopf, Appl. Phys. Lett. **49**, 1245 (1986).
18. C.W. Tu, R.C. Miller, B.A. Wilson, P.M. Petroff, T.D. Harris, R.F. Kopf, S.K. Sputz, and M.G. Lamon, J. Crys. Growth **81**, 159 (1987).
19. A. Ourmazd, D.W. Taylor, J. Cunningham, and C.W. Tu, Phys. Rev. Lett. **62**, 933 (1989).
20. B. Deveaud, B. Guenais, A. Poudoulec, and A. Regreny, Phys. Rev. Lett. **65**, 2317 (1990).
21. A. Ourmazd and J. Cunningham, Phys. Rev. Lett. **65**, 2318 (1990).
22. C.A. Warwick, W.Y. Jan, A. Ourmazd, and T.D. Harris, Appl. Phys. Lett. **56**, 2666 (1990).
23. D. Gammon, B.V. Shanabrook, and D.S. Katzer, Phys. Rev. Lett. **67**, 1547 (1991).
24. T. Tanaka and H. Sakaki, J. Cryst. Growth **81**, 153 (1987).

25. Y. Masumoto, S. Shionoya, and H. Kawaguchi, Phys. Rev. B **29**, 2324 (1984).
26. J. Hegarty, K. Tai, and W.T. Tsang, Phys. Rev. B **38**, 7843 (1988).
27. H. Wang, M. Jiang, and D.G. Steel, Phys. Rev. Lett. **65**, 1255 (1990).
28. M.D. Webb, S.T. Cundiff, and D.G. Steel, Phys. Rev. B **43**, 12658 (1991).
29. J. Hegarty and M.D. Sturge, J. Lumin. **31&32**, 494 (1984).
30. J. Hegarty, M.D. Sturge, C. Weisbuch, A.C. Gossard, and W. Wiegmann, Phys. Rev. Lett. **49**, 930 (1982).
31. J.E. Zucker, A. Pinczuk, D.S. Chemla, and A.C. Gossard, Phys. Rev. B **35**, 2892 (1987).
32. J. Hegarty, L. Goldner, and M.D. Sturge, Phys. Rev. B **30**, 7346 (1984).
33. H. Wang and D.G. Steel, Phys. Rev. A **43**, 3823 (1991).
34. H. Wang and D.G. Steel, Appl. Phys. A **33**, 514 (1991).
35. M. Jiang, H. Wang, and D.G. Steel, Appl. Phys. Lett. **61**, 1301 (1992).
36. N.F. Mott and E.A. Davis, *Electronic Processes in Non-Crystalline Materials* (Clarendon Press, Oxford, 1979), 2nd ed.
37. L. Schultheis, M.D. Sturge, and J. Hegarty, Appl. Phys. Lett. **47**, 995 (1985).
38. G. Noll, U. Siegner, S.G. Shevel, and E.O. Göbel, Phys. Rev. Lett. **64**, 792 (1990).
39. M.D. Webb, S.T. Cundiff, and D.G. Steel, Phys. Rev. Lett. **66**, 934 (1991).
40. H.H. Yaffe, Y. Prior, J.P. Harbison, and L.T. Florez, *Quantum Electronics Laser Science*, 1991 Technical Digest Series (Optical Society of America, Washington, 1991).
41. K. Leo, J. Shah, S. Schmitt-Rink, and K. Köhler, *VIIth International Symposium on Ultrafast Processes in Spectroscopy*, Bayreuth, 1991 (Institute of Physics, Bristol, 1992).
42. J. Hegarty and M.D. Sturge, J. Opt. Soc. Am. B **2**, 1143 (1985).
43. T. Miyata. Jr., Phys. Soc. Japan **31**, 529 (1971).
44. H.J. Eichler, P. Günter, and D.W. Pohl, *Laser-Induced Dynamic Gratings* (Springer-Verlag, Berlin, 1986), p. 256.
45. L. Fleishman and P.W. Anderson, Phys. Rev. B **21**, 2366 (1980).
46. T.C. Damen, L. Viña, J.E. Cunningham, J. Shah, and L. J. Sham, Phys. Rev. Lett. **67**, 3432 (1991).
47. S. Bar-Ad and I. Bar-Joseph, Phys. Rev. Lett. **68**, 349 (1992).
48. S. Koch, Private communication (1992).
49. B.F. Feuerbacher, J. Kuhl, and K. Ploog, Phys. Rev. B **43**, 2439 (1991).

THEORY OF SPIN DYNAMICS OF EXCITONS AND FREE CARRIERS IN QUANTUM WELLS

L.J. SHAM
Department of Physics
University of California, San Diego
La Jolla, CA 92093-0319, USA

ABSTRACT. A survey of the theoretical understanding of the spin relaxation processes in quantum wells is given in relation to the recent time-resolved polarization measurements. A new theory is described of the spin dynamics and relaxation of the exciton through the exchange interaction, momentum scattering, and valence band mixing. Spin relaxation processes for the conduction electron and for the valence hole are analyzed for their importance in quantum wells.

1. Introduction

In non-linear optical processes, in which one could include the conventional luminescence as well as the pump and probe spectroscopy, the observed optical intensity depends critically on the excited carrier energy, momentum, and spin relaxation. In particular, the intensities of polarized light depend on the spin relaxation. Recent rapid progress in the time-resolved measurements of polarization in luminescence [1-5] and in absorption of optically excited states [6-9] in semiconductor heterostructures has stimulated an examination of the theory of spin dynamics of the excited carriers, particularly since an understanding of the spin relaxation of the carriers is essential to the interpretation of the observed polarization. This workshop provides an ideal forum for the discussion of the relevant spin relaxation mechanisms. I propose to survey the theories here to promote such discussions.

The basic mechanisms for spin relaxation have been studied in connection with the optical orientation in bulk semiconductors [10-14] but in quantum wells the dependence of each mechanism on system properties differs from the bulk and, therefore, the relative importance of the various mechanisms in given ranges of temperature and carrier density changes. The distinction of the spin dynamics in quantum wells from the bulk arises from three factors: the quantum confinement leading to subband electronic structures, and to the enhancement of the excitonic interaction, and the high mobility samples. An example of the confinement effect is the slow-down of the valence hole spin relaxation in the quantum well compared with the bulk, leading to a different and more satisfactory explanation of the cw luminescence polarization in quantum wells [15]. The enforced close proximity of the electron-hole pair leading to a strong exciton binding in a quantum well is also expected to lead to a strong spin relaxation due to exciton exchange [4,16]. Mechanisms which depend on motional narrowing [17], such as the D'yakonov-Perel (DP) mechanism [18], are sensitive to carrier momentum relaxation and should, therefore, play quite a different role for the high-mobility quantum well than for the dirty bulk semiconductor.

In the following, we shall examine in turn the physics of the spin relaxation mechanisms of the exciton, of the conduction electron, and of the valence hole, and by relating the theory to the polarization experiments assess the degree of understanding in this early stage of the spin dynamics in quantum wells.

2. Preliminaries on Carrier Spins

Optical transitions across the fundamental gap depend not only on the energy difference of the conduction and valence subband states but also on the nature of their wave functions. The transition matrix element between a conduction subband state (nc,s), where n denotes the band index and s its spin component, and a valence subband state (jv,m_J), where j denotes the band index and m_J its spin component, is given by

$$M(ncs; jvm_J) = \langle cs | \mathbf{p} \cdot \mathbf{A} | vm_J \rangle \langle f_{ncs} | f_{jvm_J} \rangle, \qquad (1)$$

where the first factor on the right is the matrix element of the dot product of the momentum vector and vector potential of the light between the Bloch states at the band edges of the bulk material and the second factor is the overlap of the envelope wave functions. In a III-V direct gap semiconductor, the conduction band edge is Γ_6 and the index s represents the z-component of the angular momentum of spin 1/2, where the z-axis has been chosen along the growth axis. The valence band edge is Γ_7 and the index m_J is isomorphous to the z-component of the angular momentum of spin 3/2.

We classify the light with circular polarization σ- and σ+ respectively with polarization vectors

$$A_\pm = A_x \pm i A_y. \qquad (2)$$

We assign photon angular momentum m_p values 1 for the σ+ polarization, -1 for the σ- polarization, and 0 for the π-polarization with vector A_z. Then a straightforward evaluation of the matrix element of $\mathbf{p} \cdot \mathbf{A}$ yields the selection rule which expresses the angular momentum conservation along the z-axis:

$$s - m_J = m_p. \qquad (3)$$

Since an exciton is composed of a conduction electron and a valence hole, it is more convenient to represent the hole state by the basis of time-reversed states of the valence band edge Bloch states with the index

$$m_h = - m_J. \qquad (4)$$

Photoluminescence may be regarded as a three-step process [Fig. 1(a)]: (1) excitation in which a photon is absorbed raising an electron from a valence subband to a conduction subband; (2) relaxation in which the electron in the conduction subband and the hole in the valence subband relax to quasi-equilibrium states near the respective band edges; (3) recombination in which the electron-hole pair annihilates emitting a photon.

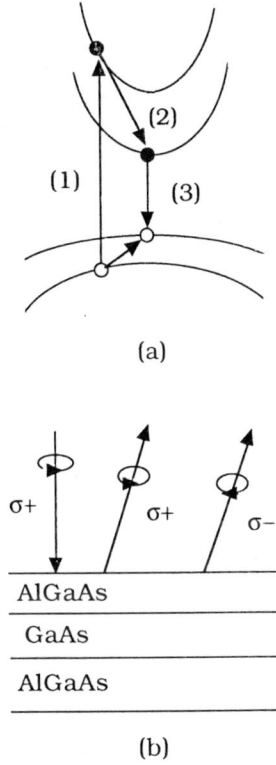

Figure 1. (a) Three steps in the optical processes of luminescence across the band gap. (b) An arrangement for luminescence polarization measurement.

The schematic arrangement shown in Fig. 1(b) shows a typical way to measure the luminescence polarization. A circularly polarized light, say σ+, is normally incident on the quantum well. The emitted light at a small angle to the normal is passed through a polarizer and the intensity of the light for each polarization σ+ and σ- is measured. The fractional difference of the intensities of the two polarizations is defined as the polarization:

$$P = \frac{I_+ - I_-}{I_+ + I_-}. \tag{5}$$

The luminescence intensities of the two polarizations depend on the initial spin populations [15] of conduction electrons and valence band holes created in step (1) of Fig. 1(a) and spin relaxation of these carriers in two stages: during the energy and momentum relaxation of step (2) and before recombination in step (3). We shall consider first the spin relaxation in the second stage. Such spin relaxation of electron-hole pairs near the band gap is most relevant for resonant excitation of excitons [3], for low-energy excitation luminescence, and for absorption probe at the exciton energy after pulse pumping [7].

3. Exciton Spin Relaxation

In an undoped quantum well, the electron-hole pair near the gap forms an exciton. Since the heavy-hole band usually has the highest energy in an unstrained well layer, the total spin of the heavy-hole exciton, from Eqs. (3) and (4), can assume the values $m_p = \pm 1, \pm 2$. The spin ± 1 exciton states are optically active and the ± 2 states are inactive. The exciton spin can relax via the spin relaxation of either constituent, electron or hole, or via simultaneous spin flip of both constituents. The likely mechanism for the electron, DP, (more about it in Sec. 4) depends on momentum relaxation, which becomes less likely for a free exciton. The valence hole can relax its spin through energy relaxation because of the heavy and light hole mixing, which is again restricted by phase space near the top of the heavy-hole band. The most likely mechanism for flipping simultaneously the electron and hole spins is via the exchange interaction.

In an optically excited exciton, spin relaxation of either constituent, electron or hole, would render the exciton optically inactive. Thus, the diminishing of one excited polarization, $\sigma+$, would not increase the other polarization, $\sigma-$. In the pump and probe experiment of Tackeuchi et al. [7], the decrease of the parallel polarization $\sigma+$ is accompanied by a compensating increase of the opposite polarization $\sigma-$ on the tens of picoseconds time scale, much shorter than the recombination time. Roussignol et al. [5] also claimed to observe exciton polarization relaxation and not free electron or hole relaxation. Both of these experiments confirm the theoretical consideration above, which argues for simultaneous spin flip of the electron and hole constituents of the exciton. Interpretation of the polarization relaxation in terms of an instantaneous hole spin relaxation and an electron spin relaxation time equal to the polarization time is, then, incorrect.

Bar-Ad and Bar-Joseph [9] have interpreted their observation of two relaxation times of the polarization as a fast relaxation of the hole and slow relaxation of the electron. Their data do show the unremarked fact that the $\sigma+$ component drops faster than the rise of the $\sigma-$ component, which may indicate the spin relaxation of hole, but the analysis of the spin relaxation times must be related to the decline of the intensities of the two polarizations separately and not to the polarization given by Eq. (5) alone. This can also be tested by looking at the time dependence of absorption of a linearly polarized light [19]. However, the fact that the two relaxation times were observed only in a superlattice and in a stepped well but not in a clean quantum well indicates defect scattering effects rather than the intrinsic behavior of the exciton.

The theory of relaxation of exciton spin through the exchange interaction is not as straightforward as one might think at first blush. I give here a preliminary report of the theory by Maialle, de Andrada e Silva, and Sham [20]. I discuss the essential physics and refer the reader to the paper for details. First, we need the exchange interaction for the Γ_6/Γ_7 exciton in a quantum well, which has been worked out [21–24]. For the optically active heavy-hole exciton, primarily we need the coupling between the $(s, m_h) = (-1/2, +3/2)$ state and the $(+1/2, -3/2)$ state.

The exchange interaction may be separated into a long-range part and a short-range part, illustrated by the Feynman diagrams in Fig. 2 [25]. Figure 2(a) shows a typical example of the attractive interaction process which binds a conduction electron and a valence hole. Figure 2(b) shows that the exchange interaction which contains the long-range Coulomb interaction is the dipole interaction of the electron-hole pair and involves the annihilation of the pair at one place inducing the creation of another pair somewhere else. Figure 2(c) gives an example of higher order processes which contributes only to the short range part of the exchange. The umklapp part of the Coulomb interaction in Fig. 2(b) also contributes to the short-range exchange. The short-range part of the exchange interaction is of the form

$$V_{sr} = \Delta [- (2/3) \, \boldsymbol{\sigma} \cdot \mathbf{J} + (1/2)], \tag{6}$$

where $\boldsymbol{\sigma}$ and $\mathbf{J}$ are the electron and hole spin vectors. The exchange constant Δ is enhanced over the bulk value by the square of the ratio of the exciton wave functions at zero relative distance between the electron and hole times the square of the overlap subband wave functions along the z direction [24]. The long-range part is of the form [22,23]

$$\mu \, (2\pi e^2/K) \, K_\alpha K_\beta, \tag{7}$$

where K_α is a component of the total momentum of the exciton. The constant μ also contains the exciton wave function at zero distance and the overlap integral in the z direction. The matrix elements of the long-range exchange between the different spin states (s, m_h) are non-zero except when one of the states $s + m_h = \pm 2$ is involved. Each matrix element conserves angular momentum: for example, the one between (-1/2, 3/2) and (1/2, -3/2) is proportional to $(K_x - iK_y)^2$. The increase in the spin angular momentum of +2 is compensated by the decrease of the center of mass orbital angular momentum of equal magnitude about the z-axis.

The coupling between two optically active states is analogous to a magnetic field in the xy-plane, which will flip the exciton spin. If the center of mass momentum of the exciton is changed by scattering by defects or by phonons, the resulting fluctuating effective magnetic field causes spin relaxation. This is analogous to the motional narrowing effect in the nuclear spin relaxation in a metal [17]. Thus, the inverse spin relaxation time of the exciton is proportional to the time correlation of the long-range exchange matrix element or to the autocorrelation of the square of the momentum, $(K_x - iK_y)^2$, or equivalently to the momentum relaxation time. The faster the momentum changes, the long the spin relaxation is. The condition for the validity of this treatment is the momentum scattering rate has to be larger than the spin precession rate. The spin relaxation by the long-range exchange requires finite center of mass momentum. The final wave vector acquired from the excitation light gives a negligible contribution. However, in resonant excitation of excitons, thermalization of the exciton population [3] gives the exciton sufficient momentum for the long-range exchange to be effective. The inverse proportionality of the spin relaxation time to the momentum relaxation could also explain the sample dependence of the polarization relaxation [1].

The spin angular momentum conservation of the short-range exchange, Eq. (6), means that there is no direct matrix element which flips the exciton spin from (-1/2, 3/2) to (1/2, -3/2) or vice versa. Only a combination of the short-range exchange with the heavy and light hole mixing can flip the exciton spin. For example, as depicted in Fig. 2(d) one can connect the heavy-hole exciton (-1/2, 3/2) to the light-hole exciton (1/2, 1/2) and then by the Luttinger Hamiltonian for the valence bands from (1/2, 1/2) to the heavy-hole exciton state of the opposite spin, (1/2, -3/2). The latter step involves non-zero hole momentum. Thus, the momentum scattering again leads to a motional narrowing type of spin relaxation.

Our estimate [20] for the GaAs/AlGaAs quantum well in Ref. [3] shows the short-range exchange contribution to be an order of magnitude less than the long-range part. With momentum scattering estimated from the width of the exciton line and the exciton temperature deduced in [3], we obtain a spin relaxation time about half that of the measured time of 50 ps. Since the exciton exchange increases in strength with decreasing well width, the spin relaxation rate, proportional to the square of the exchange interaction, is expected to increase with

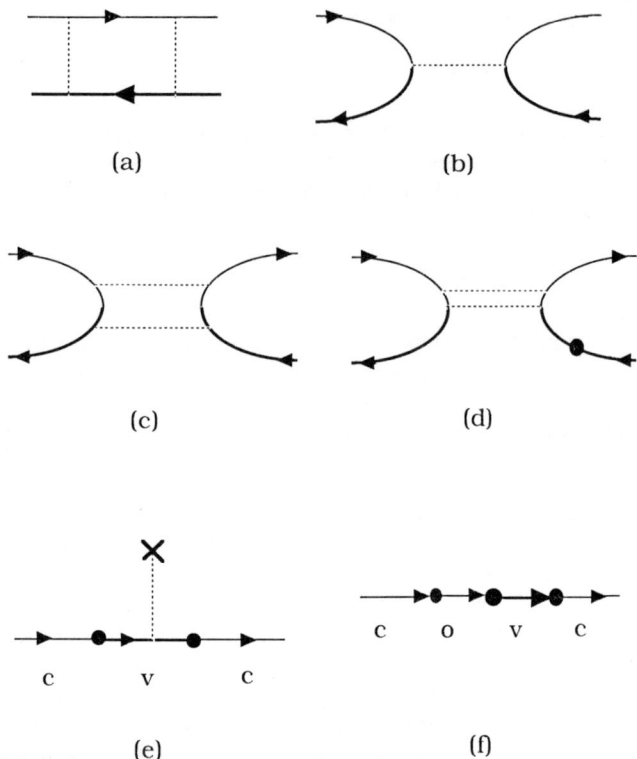

Figure 2. Diagrammatic representation of exciton and electron spin relaxation processes. A thin line with an arrow represents a conduction electron propagation, a thin line a valence hole, a dotted line Coulomb interaction, a dotted line ending in a cross scattering by an impurity, and a dot band mixing via the **k**·**p** Hamiltonian. (a) An example of electron-hole attraction (second order in this case) which provides exciton binding. (b) The lowest order exciton exchange which includes both the long-range interaction and the umklapp contribution to the short-range part. (c) An example of a higher-order exchange process which contributes to the short-range part only. (d) Short-range exchange followed by band mixing in the valence band (see text). (e) An Elliott-Yafet process for electron spin relaxation. The letters c and v stand for conduction and valence bands. (f) A D'yakonov-Perel process. The letter o stands for a band other than the Γ_6 and Γ_7 bands bordering the gap.

decreasing well width. However, the dependence is weakened somewhat by the decrease of the momentum scattering rate. Our calculated well-width dependence for the series of wells measured in Ref. [5] is in good agreement with the measured well width dependence of the polarization rate.

A strong magnetic field normal to the well can be used to affect the polarization [26]. Since the magnetic field will also change the strength of the exchange interaction, the study of the magnetic field dependence of the exciton spin relaxation due to exchange should be very interesting.

4. Conduction Electron Spin Relaxation

In step (2) of the luminescence, as the electron and the hole relax toward quasi-equilibrium, their spins may also flip, resulting in partial depolarization. The spin relaxation of the electron during this stage contributes to the determination of polarization in an undoped quantum well and is also essential in determining the polarization in a p-doped quantum well in which the hole Fermi sea with compensated spin populations dominates over the low density of photo-excited holes.

Commonly, three mechanisms for the electron spin relaxation are considered [11]. Two of them involve electron momentum scattering by impurity or acoustic phonon plus spin-orbit interaction. In the Elliott-Yafet (EY) mechanism [27], a conduction electron has, through the **k·p** term, a component of the valence band wave function. Impurity or acoustic phonon scattering may connect different valence band states, which connect to conduction band states of opposite spins. This process is illustrated in Fig. 2(e). In bulk wide-gap semiconductors, which include GaAs but not InSb, the EY process is weak because of the band gap denominator in the **k·p** term and is considered not competitive in comparison with the DP process. Since the latter also involves the band gap factor, the difference in strength of the two processes in a bulk semiconductor must be quantitative. The EY mechanism should be reconsidered for the quantum well.

In the DP mechanism [18] already mentioned above, the asymmetry of the constituent atoms in the III-V compound with the zinc blende structure leads to the possibility of a term cubic in k which connects the conduction electron states of opposite spins [28]. This is illustrated in Fig. 2(f): three **k·p** terms connect a spin-up, say, conduction band state to bands other than the Γ_6 conduction bands or the Γ_7 valence bands, in turn to the Γ_7 valence bands which couple to the spin-down conduction state. This provides an effective magnetic field Ω, which flips the electron spin. The **k·p** terms provide the effective magnetic field with cubic k dependence, e.g.,

$$\Omega_x \propto k_x (k_y^2 - k_z^2). \tag{8}$$

The electron spin along the growth axis, the z direction, is flipped by the inplane field, $\Omega_x \pm i\Omega_y$. In a quantum well, the k_z^2 term yields the energy of the conduction subband measured from the bottom of the well and is expected to give [29] an enhancement factor of the zero-point energy in the well over the typical kinetic energy in the bulk. Because of the k-dependence, the spin-flip effective magnetic field fluctuates as the electron momentum is changed by scattering. This causes spin relaxation in the same manner as the motional narrowing spin relaxation mechanism [17]. In particular, the spin relaxation rate is proportional to the momentum relaxation time. The DP mechanism is analogous to the exciton exchange mechanism discussed in Sec. 3 in the motional narrowing aspect, but differs in the physical origin of the spin-flip.

In the time-resolved luminescence polarization study of the electron spin relaxation by Damen et al. [4] of a GaAs/AlGaAs quantum well of width 60 Å and p-doped with a density of 4×10^{11} cm^{-2}, a spin relaxation time of 150 ps at 10 K is deduced, which is about a third to a quarter of that in a p-doped bulk GaAs [30] with comparable hole density of 6×10^{17} cm^{-3}. If we take the D'yakonov-Kachorovskii enhancement factor for the spin relaxation rate to be the square of the ratio of the zero-point energy in the quantum well to the electron energy at the Fermi wave-vector of the three dimensional system, the ratio is less than unity for the systems under comparison [4]. The factor of three to four increase in going from 3D to 2D then has to be accounted for by the decrease in momentum scattering, which is possible but requires separate confirmation.

A careful evaluation by Bastard and Ferreira [31] of the DP term including the momentum scattering by 10^{10} cm^{-2} impurities yields a spin relaxation of about 200 ps at 10 K for a well width of 60 Å and for an electron density of 10^8 cm^{-2}. Scaling the electron energy up to that at the Fermi wave-vector of the p-doped well in Ref. [4] would reduce the electron spin relaxation time by an order of magnitude. The first term on the right of Eq. (8), which was neglected in Ref. [31] would have to be reinstated. To reach agreement with experiment, the momentum relaxation time would have to be adjusted. We do not know without further characterization if such momentum scattering is reasonable or not in the quantum well under study.

A third mechanism for the electron spin relaxation is the exchange scattering with a hole suggested by Bir, Aronov, and Pikus (BAP) [32]. The difference with the exciton exchange is the replacement of the exciton wave function at zero separation by the electron-hole scattering cross-section. In the limit of strong hole momentum scattering by impurities or phonons, i.e., when the hole momentum relaxation rate is larger than the precession rate of the equivalent magnetic field which flips the electron spin via the exchange scattering, the electron spin relaxation is again of the motional narrowing class, with the electron spin relaxation time inversely proportional to the hole momentum relaxation time. In the opposite limit of weak hole scattering, the electron spin relaxation rate is given by the Fermi golden rule of the exchange scattering with holes averaged over the hole population.

To estimate the BAP rate for the electron in a p-doped quantum well [4], we compare the 2D rate with the corresponding 3D system. In 3D, the spin relaxation rate is inversely proportional to the exchange scattering cross-section squared. In 2D, it is proportional to the square of the inverse cross-section times the overlap integral of the square of the conduction and valence subband wave functions in the z direction. A crude estimate using the Born approximation for the exchange scattering via a Thomas-Fermi screened potential and the subband wave functions in an infinite well yields the ratio of spin relaxation rates:

$$\tau_{3D} / \tau_{2D} = 3\pi / (2k_f L), \tag{9}$$

where k_f is the Fermi wave vector in the 3D system and L is the quantum well. The enhancement factor for the 2D relaxation rate is about three for the p-doped quantum well in Ref. [4] described above, which led us to favor the BAP mechanism as a possible way to explain the electron spin relaxation in the quantum well.

5. Valence Hole Spin Relaxation

Because of the valence band complex, the hole spin relaxation is very different from the electron one. The strong band mixing together with the spin-orbit interaction means that as a hole changes its momentum state it will change the admixture of the spin 3/2 components. Thus, in three dimensions, holes are assumed to depolarize instantly [10,11]. However, removal of the heavy and light hole degeneracy by uniaxial stress prolongs the hole spin relaxation rate [33]. The finite hole spin relaxation time plays an important role in the luminescence polarization spectra of quantum wells, particularly in explaining the polarization reversal at the second heavy-hole exciton in going from p-doped to n-doped [15]. A relaxation time of 5 ps has been measured in an n-doped well [4].

In a recent theory of hole relaxation [15], it is noted that the doubly degenerate valence subband states at the same inplane wave vector are related to each other with the z dependent part of the envelope functions of the form

$$(f_{3/2}, f_{1/2}, f_{-1/2}, f_{-3/2}) \quad \text{and} \quad (f^*_{-3/2}, -f^*_{-1/2}, f^*_{1/2}, -f^*_{3/2}), \tag{10}$$

which are related by time-reversal, as can be seen from the structure of the Luttinger Hamiltonian. Note that the x and y dependent parts of the two states are the same and not time-reversed states of each other. In a symmetric quantum well, the parity with respect to the mirror operation about the center of the well gives the parity of the components in the form

$$(+, -, +, -) \quad \text{and} \quad (-, +, -, +) . \tag{11}$$

We name the first state even parity and the second odd parity. In a scattering event that changes the energy and wave vector of the valence state, there is a finite probability of scattering into one or the other of the two degenerate states. The two channel scatterings have to be followed. The spin populations of the holes can then be calculated from the spin components of the states.

A number of hole scattering mechanisms have been examined. Scattering by the optical phonon is unimportant because in recent experiments the energy range of interest for the holes is less than the optical phonon energy (of the order 30 meV). Scattering by shake-up of the electron Fermi sea is estimated [34] to contribute about 10% of the total spin scattering rate in an n-doped well of density of about 10^{12} cm^{-2}. The most important source of scattering is by acoustic phonons, both by deformation and piezo-electric coupling [15]. Evaluation for the n-doped quantum well studied in Ref. [4] has yet to be carried out. Calculated spin relaxation time due to impurity scattering [27] gives 20 ps for the quantum well width of 50 Å and n-density of 3×10^{11} cm^{-2}, somewhat larger than the measured value of 4 ps [4], but lack of knowledge of the impurity content in the quantum well prevents a more quantitative comparison.

6. Conclusion

We have examined a number of spin relaxation mechanisms for excitons, free electrons, and holes. The principal mechanism for the exciton is exchange interaction, which flips the exciton spin and which depends on the exciton momentum, whose fluctuation gives rise to spin relaxation. For electrons, all three mechanisms depend on interaction with holes, relying on the holes' large p-orbital spin-orbit interaction to change the electron spins. The mechanisms mostly belong to the motional narrowing class. If the momentum relaxation is due to scattering by phonons, then temperature dependent studies can elucidate the physical causes. If the momentum relaxation is due to scattering by impurities, then the sample dependence can be explained, but quantitative comparison between calculated and measured spin relaxation become difficult. The valence hole relaxes its spin primarily through the spin-orbit coupling to the momentum.

At this early stage of study in this field of spin relaxation in the heterostructures, while the theory gives a very good account of some qualitative aspects of the polarization, quantitative comparison between theory and experiment has yet to be made. Study of changes induced by varying the system conditions such as using magnetic field and stress will be helpful.

Acknowledgments

My part of the research described above is supported by the National Science Foundation, Grant No. DMR 91-17298. I wish also to acknowledge helpful discussions with my collaborators, especially, Marcelo Maialle, Erasmo de Andrada e Silva, Jag Shah, and Ted Damen.

References

1. M.R. Freeman, D.D. Awschalom, J.M. Hong, and L.L. Chang, Phys. Rev. Lett. **64**, 2430 (1990); *Proc. 20th Intl. Conf. on Physics of Semiconductors*, edited by E.M. Anastassakis and J.D. Joannopoulos (World Scientific, Singapore, 1990), p. 1129.
2. T.C. Damen, J. Shah, D.Y. Oberli, D.S. Chemla, J.E. Cunningham, and J.M. Kuo, Phys. Rev. B **42**, 7434 (1990).
3. T.C. Damen, K. Leo, J. Shah, and J.E. Cunningham, Appl. Phys. Lett. **58**, 1902 (1991).
4. T.C. Damen, L. Viña, J.E. Cunningham, J. Shah, and L.J. Sham, Phys. Rev. Lett. **67**, 3432 (1991).
5. Ph. Roussignol, P. Rolland, R. Ferreira, C. Delalande, G. Bastard, A. Vinattieri, L. Carraresi, M. Colocci, and B. Etienne, Surface Science **267**, 360 (1992).
6. J.B. Stark, W.H. Knox, and D.S. Chemla, Phys. Rev. Lett. **65**, 3033 (1990).
7. A. Tackeuchi, S. Muto, T. Inata, and T. Fujii, Appl. Phys. Lett. **56**, 2213 (1990).
8. M.D. Webb, S.T. Cundiff, and D.G. Steel, Phys. Rev. Lett. **66**, 934 (1991).
9. S. Bar-Ad and I. Bar-Joseph, Phys. Rev. Lett. **68**, 349 (1992).
10. M.I. D'yakonov and V.I. Perel, in *Optical Orientation*, edited by F. Meier and B.P. Zakharchenya (North-Holland, Amsterdam, 1984), Chap. 2.
11. G.E. Pikus and A.N. Titkov, in *Optical Orientation*, edited by F. Meier and B.P. Zakharchenya (North-Holland, Amsterdam, 1984), Chap. 3.
12. R. Planel and C. Benoit à la Guillaume, in *Optical Orientation*, edited by F. Meier and B.P. Zakharchenya (North-Holland, Amsterdam, 1984), Chap. 8.
13. C. Hermann and C. Weisbuch, in *Optical Orientation*, edited by F. Meier and B.P. Zakharchenya (North-Holland, Amsterdam, 1984), Chap. 11.
14. G.E. Pikus and E.L. Ivchenko, in *Excitons*, edited by E.I. Rashba and M.D. Sturge (North-Holland, Amsterdam, 1982), Chap. 6.
15. T. Uenoyama and L.J. Sham, Phys. Rev. Lett. **64**, 3070 (1990); Phys. Rev. B **42**, 7114 (1990).
16. M.J. Snelling, E. Blackwood, R.T. Harley, P. Dawson, and C.T.B. Foxon, to be published.
17. C.P. Slichter, *Principles of Magnetic Resonance* (Harper & Row, New York 1963), p. 154.
18. M.I. D'yakonov and V.I. Perel, Sov. Phys. JETP **33**, 1053 (1971).
19. H. Stolz, D. Schwarze, W. von der Osten, and G. Weimann, Superlattices and Microstructures, **6**, 271 (1989).
20. M. Maialle, E.A. de Andrada e Silva, and L.J. Sham, to be published.
21. Y. Chen, B. Gil, P. Lefebvre, and H. Mathieu, Phys. Rev. B **37**, 6429 (1988).

22. B. R. Salmassi and G.E.W. Bauer, Phys. Rev. B **39**, 1970 (1989).
23. L.C. Andreani and F. Bassani, Phys. Rev. B **41**, 7536 (1990).
24. U. Rössler, S. Jorda, and D. Broido, Solid State Commun. **73**, 209 (1990).
25. L.J. Sham and T.M. Rice, Phys. Rev. **144**, 708 (1966).
26. J.B. Stark, W.H. Knox, and D.S. Chemla, Phys. Rev. B **46**, 7919 (1992).
27. R.J. Elliott, Phys. Rev. **96**, 266 (1954); Y. Yafet, Solid State Phys. **14**, 1 (1963).
28. G. Dresselhaus, Phys. Rev. **100**, 580 (1955).
29. M.I. D'yakonov and V. Yu. Kachorovskii, Sov. Phys. Semicond. **20**, 110 (1986).
30. K. Zerrouati, F. Fabre, G. Bacquet, J. Bandet, J. Frandon, G. Lampel, and D. Paget, Phys. Rev. B **37**, 1334 (1988).
31. G. Bastard and R. Ferreira, Surface Science **267**, 335 (1992).
32. G.L. Bir, A.G. Aronov, and G.E. Pikus, Sov. Phys. JETP **42**, 705 (1975).
33. G. Lampel, A.N. Titkov, and V.I. Safarov, in *Proceedings of the 14th International Conference on the Physics of Semiconductors*, edited by B.L.H. Wilson (Institute of Physics, London, 1978), p. 1031.
34. L.J. Sham, J. de Physique **48**, C5-381 (1987).

ALL-OPTICAL BISTABILITY IN ASYMMETRIC QUANTUM WELLS

S. SCANDOLO[1,2] AND F. TASSONE[1]
[1]Scuola Normale Superiore, Piazza dei Cavalieri 7, I-56126 Pisa, Italy
[2]Institut Romand de Recherche Numerique en Physique des Materiaux (IRRMA)
PHB-Ecublens, CH-1015 Lausanne, Switzerland

ABSTRACT. A new mechanism for all-optical bistability in type-II asymmetric quantum wells is presented. It is based on the redshift of the interband transition energy induced by the intrinsic photogenerated electric field. Increasing-absorption bistability is shown to take place in the absence or presence of excitonic effects. Realistic examples in III-V and II-VI compounds are also presented and switching intensities of the order of tens of kW/cm^2 for response times in the nanosecond range are estimated.

1. Introduction

Semiconductor structures of reduced dimensions have proven in recent years to be suitable for many optoelectronic-device applications, such as tunable laser sources and photodetectors [1], their flexibility being due to the large freedom in the structural parameters (composition, dimensions, doping, etc.). Among these applications, optical bistability has been extensively studied both theoretically and experimentally in quantum-well (QW) structures, with very promising results [2].

Bistable operation may be obtained in different device configurations. Refractive bistability may result for sufficient shift of the modes of an external cavity upon variation of the index of refraction in the active medium [2]. Operation in the high-refraction and low-absorption region and external-cavity tuning may become delicate points in device design. Absorptive bistability arises from the increase of the absorption coefficient with the carrier density. In the self-electro-optic effect device (SEED) this increase is provided by an external feedback mechanism which connects the photogenerated carrier density to the applied electric field [3]. Many-body effects such as exciton bleaching and band-gap renormalization may also produce this absorption increase without the use of any external feedback; nevertheless, high excitation densities are usually required to reach the bistability regime [4]. Recently proposed intrinsic mechanisms based on space-charge separation may overcome these limitations [5–8]. In these systems a charge distribution is created by optical excitation due to the asymmetry of the structure. Large intrinsic electric fields are therefore generated. Elementary electrostatics considerations tell us that the effect of space-charge separation associated with a single optical transition is to increase the transition energy with the density of excitations, so that a blueshift of the transition occurs, as observed in Refs. [7] and [9]. Therefore, increasing absorption may be obtained either by intersubband absorption [5] or by excitonic absorption [7]: in both cases the photon energy has to be tuned above the resonant peak, where the slope of the absorption is negative. In the first case, for standard III-V compounds, intersubband transitions occur in the far-infrared spectrum, well below the region of interest for optical fibre communications. In the second case, bistability can be achieved, but with extremely long recovery times [10].

We propose a different type-II asymmetric-QW structure working on interband transitions and on the following mechanism: the electric field initially created by optical excitation is reversed by the relaxation of the electrons and the holes to lower levels, so that the intrinsic electric field *decreases* the transition energy of the higher levels. The redshift overcomes the limitations of negative-slope absorption, and bistability can be achieved both in the presence or absence of excitonic features in the absorption spectrum.

In Sec. 2 we model excitation, relaxation and recombination of the carriers in the QW system by a simple set of rate equations with a few phenomenological parameters. We consider both the case of an exciton resonance and of a step-like spectrum (no excitons). In Sec. 3 we provide realistic examples of the two situations in small-gap III-V compounds and in large-gap II-VI compounds. In Sec. 4 we discuss the results.

2. Model Structure and Optical Bistability

We consider the type-II QW structure shown in Fig. 1. The $h2 \to e2$ transition is optically excited by an external beam of frequency ω and intensity I. Electrons and holes subsequently relax to the sublevels e1 and h1 respectively; the relaxation time τ is taken for simplicity to be equal for electrons and holes. Finally electrons and holes recombine radiatively with a characteristic time T supposed to be equal for all the recombination paths shown in Fig. 1.

We first consider the case where excitonic effects may be neglected. A step-like homogeneously broadened spectrum is then assumed for the optical transitions. The following rate equations for the electron and hole populations in the two levels are obtained:

$$\dot{n}_2 = \frac{NI}{\pi T I_2^{(s)}} \left[\frac{\pi}{2} + tg^{-1}\left(\frac{E - \delta E(n_1)}{\Gamma}\right) \right] - \frac{n_2}{T} - \frac{n_2}{\tau},$$

$$\dot{n}_1 = \frac{N}{T} \frac{I}{I_1^{(s)}} - \frac{n_1}{T} + \frac{n_2}{\tau},$$

(1)

where n_1, n_2 are the electron (and hole) populations in levels 1 and 2, N is the sheet density of atoms, Γ is the characteristic phenomenological broadening, T is the recombination time, τ the relaxation time, and $E = \hbar\omega - \left(E_2^e - E_2^h\right)$ is the detuning. $I_{1(2)}^{(s)}$ are the saturation intensities given by

$$I_{1(2)}^{(s)} = \frac{c\hbar^3 nN}{8\pi\mu_{cv}^2 \mu |\langle h1(2)|e1(2)\rangle|^2 T},$$

(2)

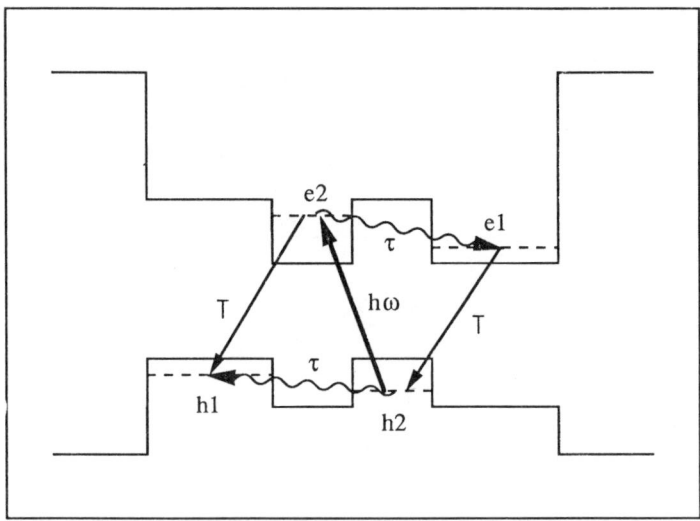

Figure 1. Quantum well model structure and energies of electron and hole states. Relaxation and recombination paths are shown with their characteristic times.

where n is the index of refraction of the medium, μ_{cv} is the dipole matrix element betweenvalence and conduction Bloch functions, and μ is the reduced mass of electrons and holes. Finally, δE is the density dependent Stark shift, which in lowest order is given by

$$\delta E = \frac{4\pi e^2}{\varepsilon}[\langle e2|z|e2\rangle - \langle h2|z|h2\rangle]n_1 \equiv \Delta E \frac{n_1}{N}. \qquad (3)$$

A more accurate evaluation of the Stark shift, which requires the integration of the Poisson equation in the structure, yields negligible correction to Eq. (3), because of the small overlap of positive and negative charge in the structure.

A further assumption is fundamental for our model, i.e., $\tau \ll T$. As a consequence of this assumption the competition of relaxation and recombination leads to a large population in the ground state n_1, favouring the formation of a strong intrinsic electric field, which shifts the considered transition as given in Eq. (3). The same hypothesis enables us to disregard the recombination term for the population n_2.

We search for optical bistability in stationary conditions, $\dot{n}_1 = \dot{n}_2 = 0$. We then obtain from Eq. (1) the following implicit equation:

$$\frac{I_2^{(s)} n_2 T}{IN\tau} = \frac{1}{2} + \frac{1}{\pi} tg^{-1}\left[\frac{E}{\Gamma} + \frac{\Delta E}{\Gamma}\left(\frac{I}{I_1^{(s)}} + \frac{n_2 T}{N\tau}\right)\right], \qquad (4)$$

Equation (4) gives n_2 as a function of I, for a given value of the detuning E. Bistability is achieved if a range of intensities exists where multiple values of n_2 are found. This range defines switch-on and switch-off intensities, I_{on} and I_{off}, respectively. A straightforward derivation of Eq. (4) shows that $I_{on}, I_{off} > I_2^{(s)} \pi \Gamma / \Delta E$ [11]. We call $I_{lim}^{(s)}$ this lower bound for the switching intensities. We also find a threshold value of the detuning $\overline{E} \cong [(1/2) + I_2^{(s)}/I_1^{(s)}] \pi \Gamma$, below which the bistable behavior is absent.

In the case when the excitonic effects can not be neglected, different rate equations for n_1 and n_2 hold, where the Lorentzian line-shape replaces the step-like one in Eq. (1). These equations are:

$$\dot{n}_2 = \frac{N}{T} \frac{I}{I_2^{(e)}} \frac{\Gamma^2}{(E - \delta E(n_1))^2 + \Gamma^2} - \frac{n_2}{T} - \frac{n_2}{\tau},$$

$$\dot{n}_1 = \frac{N}{T} \frac{I}{I_1^{(e)}} - \frac{n_1}{T} + \frac{n_2}{\tau}. \tag{5}$$

Here the density dependent Stark shift is again given in Eq. (3), and Γ is the phenomenological homogeneous broadening of the excitonic peak. The saturation intensity at the exciton peak, $I_2^{(e)}$, turns out to be

$$I_2^{(e)} = I_2^{(s)} \frac{\pi}{8} \frac{\Gamma}{Ry^*} \left(\frac{\rho}{a_B^*}\right)^2, \tag{6}$$

where Ry* is the effective exciton Rydberg, a_B^* is the effective Bohr radius, and ρ is the exciton radius in the layer plane. The saturation intensity of the e1–h1 transition is calculated at the exciton continuum and given by

$$I_1^{(e)} = I_1^{(s)} \frac{e^{\pi \alpha}}{\cosh(\pi \alpha)}. \tag{7}$$

The second factor of the right hand side of Eq. (7) is the two-dimensional Sommerfeld factor, where $\alpha = (k_{\parallel} a_B^*)^{-1}$.

In stationary conditions, I and n_2 are related by:

$$\frac{I}{I_2^{(e)}} \frac{N\tau}{n_2 T} \Gamma^2 = \left[E - \Delta E \left(\frac{n_2 T}{N\tau} + \frac{I}{I_1^{(e)}} \right) \right]^2 + \Gamma^2. \tag{8}$$

Here again we find a lower bound to the switching intensities, $I_{lim}^{(e)}$. In the case of $I_2^{(e)} \ll I_1^{(e)}$, an approximation justified in the extreme excitonic limit when $\Gamma \ll Ry^*$, we derive this lower bound to be $I_{lim}^{(e)} = 2.7 I_2^{(e)} \Gamma / \Delta E$; in the same approximation a threshold value $\overline{E} = \sqrt{3} \Gamma$ for the detuning is obtained.

3. Results for Realistic Structures

Realistic structures corresponding to the previously considered cases are investigated. The energy levels and wavefunctions are calculated within the effective mass approximation by

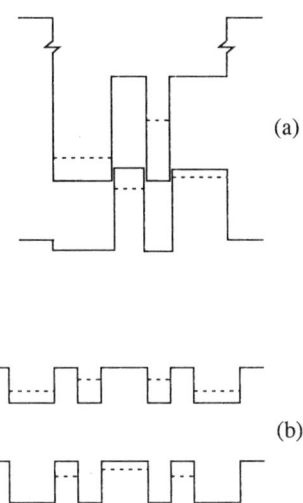

Figure 2. Examples of realistic structures: (a) AlSb-InAs-GaSb-AlSb quantum well; (b) ZnSe-ZnTe superlattice. Band profiles and energies of the electron and hole states are shown.

solving a one-dimensional Schrödinger equation with the transfer-matrix method [12].

The first structure is shown in Fig. 2(a) and is composed by 46 Å of GaSb, 18 Å of $InAs_{0.9}Sb_{0.1}$, 18 Å of GaSb, 36 Å of $InAs_{0.9}Sb_{0.1}$, confined by barriers of AlSb; the stoichometric composition is chosen to match lattice constants to those of GaSb. In these small-gap ternary compounds the effective Rydberg is smaller than the broadening Γ, so that excitonic features are negligible in the absorption spectrum. The step-like case therefore applies. Non-parabolicity effects have been included in the calculations since they are not negligible in these materials [12]. We find the h2–e2 transition energy at 0.5 eV, and 10 Å for the dipole moment in Eq. (3). Relaxation times are expected to be in the picosecond range [13], while the recombination time T can be expected to be around 10 ns at room temperature [14]; a ratio $T/\tau = 10^2$ and a phenomenological broadening of 10 meV have been assumed. Using these values of the parameters we find $I_2^{(s)} = 500$ kW/cm^2 and. $I_{lim}^{(s)} = 3.5$ kW/cm^2. In Fig. 3 we plot a set of hysteresis loops obtained for different detunings E.

The second structure is the ZnSe-ZnTe superlattice (SL) shown in Fig. 2(b). Thin (13 Å) and large (18 Å) layers of both ZnSe and ZnTe are arranged in such a way that the basis period is composed of two QWs similar to those described above, but with opposite asymmetry. Exciton effects dominate the optical spectrum of these large-gap materials, and the exciton model of Sec. 2 applies. Exciton wavefunctions and energies are calculated by minimization of the exciton hamiltonian with respect to their radius ρ in the layer plane. For the considered SL we obtain a radius $\rho = 1.1\ a_B^*$ and a transition energy h2 → e2 of 2.1 eV. In the absence of experimental investigations, a recombination time of 2 ns is obtained by an appropriate scaling of the III-V radiative lifetimes. A phenomenological broadening of 5 meV is assumed. Correspondingly, we obtain $I_2^{(e)} = 5$ MW/cm^2 and $I_{lim}^{(e)} = 3$ kW/cm^2. Hysteresis loops for two values of the detuning are shown in Fig. 4.

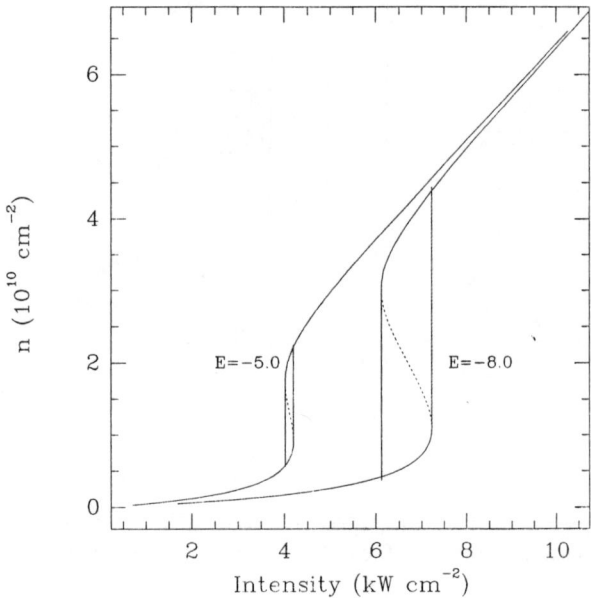

Figure 3. Hysteresis loops obtained for two different values of the detuning E (in units of Γ), for the III-V structure described in the text.

Both in the step-like model and in the exciton model we find increasing switch-on intensities together with an enlargement of the hysteresis loop with increasing detunings. We notice however that when the device is switched on we have $\delta E \approx E$, and the linear approximation Eq. (3) probably becomes inappropriate for too large δE. The density n_2 in correspondence to the bistable region is of the order of 10^{10} cm^{-2}. This shows that other many-body effects such as band-edge renormalization and line broadenings may be safely disregarded in the calculation.

4. Discussion and Conclusions

For practical applications, a multiple structure of type II quantum wells may be used in order to obtain an easily detectable bistability in the transmitted signal. A study of this structure may be carried out with the use of a set of equations all similar to Eq. (1), each corresponding to a given well. Coupling between these equations comes from the transmitted light intensities entering in each of them. A detailed numerical analysis of the system is therefore needed to find switching intensities, but their order of magnitude should be the same as the single quantum well case.

A dynamical analysis of the switching was not carried out; nevertheless it may be reasonably assumed that a recovery time of the order of T will be found. If this is the case, a remarkably good device performance may be expected. A more extensive research on different materials to realize this type of structure and optimization may also substantially

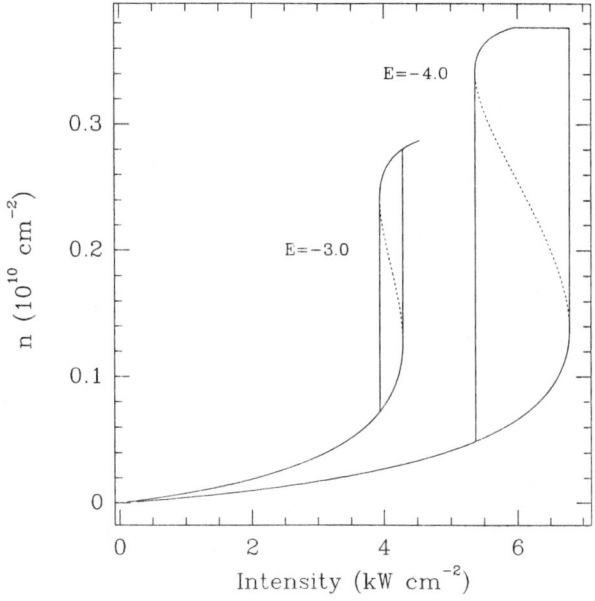

Figure 4. Hysteresis loops obtained for two different values of the detuning E (in units of Γ), for the II-VI structure described in the text.

improve this initial result.

Another important characteristic of real QW structures, which needs to be included into a more realistic model of bistability, is the inhomogeneity of the well thicknesses. It results in a position-dependent gap and exciton energy, and of the detuning E. As a consequence, different regions of the sample should exhibit different thresholds and hysteresis loops. If the diffusion of the carriers in regions of different thicknesses also becomes important, different regimes of operation are to be expected. In fact, these processes produce an evolution of the carrier density distribution over the sample into a non-uniform one for switching times comparable to the characteristic diffusion time, and a more complicated set of rate equations is needed to describe the system. Nevertheless we are confident that the nonlinear mechanism presented here will always result in bistable behaviour of the device for appropriate choice of the operative parameters.

As a concluding remark we stress that a novel mechanism to achieve intrinsic bistability has been proposed, based on the energy shifts produced by the strong intrinsic electric field associated with the relaxed photogenerated charges in type II quantum well structures.

Acknowledgment

We acknowledge useful suggestions by M. Nägele and J. Gutowski, and we thank F. Bassani for his continuous encouragement.

References

1. M. Jaros, *Physics and Applications of Semiconductor Microstructures* (Oxford University Press, New York, 1989).
2. *Optical Nonlinearities and Instabilities in Semiconductors*, edited by H. Haug (Academic, London, 1988).
3. D.A.B. Miller, D.S. Chemla, T.C. Damen, T.H. Wood, C.A. Burrus, A.C. Gossard, and W. Wiegmann, IEEE J. Quantum Electronics **QE-21**, 1462 (1985).
4. H.M. Gibbs, *Optical Bistability: Controlling Light with Light* (Academic, New York, 1985).
5. J. Khurgin, Appl. Phys. Lett. **54**, 2589 (1989).
6. M. Seto and M. Helm, Appl. Phys. Lett. **60**, 859 (1992).
7. P. Dawson, I. Galbraith, A.I. Kucharska, and C.T. Foxon, Appl. Phys. Lett. **58**, 2889 (1991).
8. L.M. Weegels, E.-J. Vonken, J.E.M. Haverkort, M.R. Leys, and J.H. Wolter, in *Proceedings of the International Meeting on Optics of Excitons in Confined Systems*, edited by A. D'Andrea, R. Del Sole, R. Girlanda, and A. Quattropani, IOP Conf. Series, Vol. 123, p. 297 (1991).
9. L.M. Weegels, J.E.M. Haverkort, M.R. Leys, and J.H. Wolter, Superlatt. and Microstruct. **10**, 143 (1991).
10. I. Galbraith, P. Dawson, and C.T. Foxon, Phys. Rev. B **45**, 13499 (1992).
11. S. Scandolo and F. Tassone, in *Proceedings of the International Workshop on Nonlinear Optics and Excitation Kinetics in Semiconductors*, Phys. Stat. Solidi (b) **173**, 453 (1992).
12. G. Bastard, *Wave Mechanics Applied to Semiconductor Heterostructures* (Les Éditions de Physique, Les Ulis, 1988).
13. S.A. Gurwitz, I. Bar-Joseph, and B. Deveaud, Phys. Rev. B **43**, 14703 (1991).
14. M. Colocci, M. Gurioli, A. Vinattieri, F. Fermi, C. Deparis, J. Massies, and G. Neu, Europhysics Lett. **12**, 417 (1990); M. Gurioli, A. Vinattieri, M. Colocci , C. Deparis, J. Massies, G. Neu, A. Bosacchi, and S. Franchi, Phys. Rev. B **44**, 3115 (1991).

MAGNETO-OPTICS OF SHALLOW IMPURITIES IN SUPERLATTICES

F.M. PEETERS, J.M. SHI, AND J.T. DEVREESE*
Universiteit Antwerpen (UIA)
Department of Physics
Universiteitsplein 1
B-2610 Antwerpen, Belgium

ABSTRACT. The ground and excited states of shallow impurities in quantum wells and superlattices in the presence of a magnetic field along the growth axis are studied. We discuss the neutral D^0-center and the charged D^--center. The importance of electron-phonon interaction and band non-parabolicity in realistic systems is pointed out. The theoretical results are compared with recent experimental data on the $GaAs/Al_xGa_{1-x}As$ system.

1. Introduction

Thanks to recent developments in crystal-growth techniques [1] such as molecular-beam epitaxy (MBE) and metal-organic chemical vapor deposition (MOCVD) it has become possible to grow semiconductor heterostructures, quantum wells, and superlattices that can be selectively doped on the scale of a lattice constant. The optical and transport properties are strongly influenced by these dopant impurities and therefore knowledge of the effect of the confining potential barriers on the electronic states of the impurities is important [2,3].

At low doping concentration the dopants are far apart and the problem is reduced to a one-electron problem, which is equivalent to the study of the states of a three-dimensional (3D) hydrogen-like atom in the presence of a one-dimensional (1D) potential. The energy of the impurity levels will depend on the position of the impurity relative to the 1D potential, the height of the potential barriers, the width of the wells, and the material parameters. Applying an external magnetic field allows one to tune the separation between the energy levels of the donor. If this energy separation equals the energy of a longitudinal-optical (LO) phonon, resonant interaction occurs whose strength is a very sensitive function of: (a) the LO-phonon frequency (ω_{LO}) and (b) the electron LO-phonon interaction strength (α). Thus infrared spectroscopy on such donor states in high magnetic fields gives us information on the electron-phonon interaction. In real device applications, which usually operate at high temperature or under large driving electric fields, the electron mobility is limited by this LO-phonon interaction process and therefore an accurate knowledge of it is very important.

Shallow impurities (also called D^0-centers) in semiconductors are also a laboratory for the study of atomic-like structures under extreme conditions. This can be understood by looking at the natural units of the problem: (a) the length the effective Bohr radius; $a^*_B = \hbar^2\varepsilon/m^*e^2$, with ε the dielectric constant of the semiconductor, m^* the electron band mass and e the electronic charge, (b) the energy the effective Rydberg; $R^* = e^2/2\varepsilon a^*_B$, and (c) for the magnetic field;

TABLE 1. Scales in the problem of a H-atom and a D^0-center in GaAs. B_0 is the magnetic field at which $\gamma = 1$ or $R^* = \hbar\omega_c/2$.

	H-atom	D^0-center
ε	1.0	12.5
m^*/m_e	1.0	0.067
R^* (meV)	1.4×10^4	5.8
a^*_B (Å)	0.53	98.7
B_0 (T)	2.3×10^5	6.7

$\gamma = e\hbar B/2m^*cR^*$, which is the magnetic energy ($\hbar\omega_c/2$) divided by the Coulomb energy (R^*). Here, B is the magnetic field strength, $\omega_c = eB/m^*c$ the cyclotron energy, and c the velocity of light. In Table 1 we compare the different scales for a hydrogen atom in vacuum with those of a donor in GaAs. Notice that because of the small band mass and the large dielectric constant in GaAs the energy scale is reduced by a factor of 2.4×10^3 and the length scale is enhanced by a factor of 200. As a consequence, the effect of a magnetic field is much stronger on donor states than on a H-atom. For example, studying the D^0-center in GaAs at a magnetic field of B = 6.7 T gives us information on the states of a hydrogen atom under a B-field of 2.3×10^5 T, which is unattainable in terrestrial laboratories.

The aim of the present work is to study the electronic energy levels of neutral and charged donors in superlattices and quantum wells in the presence of a magnetic field directed along the growth axis. Because our ultimate aim is to compare our theoretical results with experimental data, we will follow an approach such that the different effects present in a semiconductor (e.g., electron band mass discontinuity at the interfaces, finite barrier height, band non-parabolicity, electron-phonon interaction,...) are included as well as possible. A variational form for the electronic wave functions is taken, which is chosen such that it describes the exact wave function as closely as possible but which is not too complex so that it can still be used as input for other calculations, like the polaron effect.

2. Energy Levels of the D^0-center

Most of the theoretical calculations [4-9] of the binding energy of donor states address the single-quantum well problem. This is a good approximation for the case of a weakly coupled superlattice, i.e., wide and/or high barriers, and when the donor is near the center of the well. Chaudhari [10] has generalized this calculation to a well-center donor in interaction with two adjacent wells. Lane and Greene [11] generalized this calculation to a superlattice with arbitrary position of the donor, but with an uniform effective electron mass throughout the superlattice. Recently Helm et al. [12] extended these calculations to higher excited states (i.e., 1s, 2s, $2p^\pm$, $2p_z$), where the spatial dependence of the electron mass was included.

Here we consider the general case of a superlattice in an external magnetic field directed along the growth axis, where the donor is allowed to be in an arbitrary position in the superlattice. The wave function of the donor states will be determined by a variational calculation. In the absence of an electron-phonon interaction the problem is, in the effective mass approximation, described by the Hamiltonian

$$H = H_z + H_{x,y} + H_c, \quad (1)$$

where

$$H_z = -\frac{\hbar^2}{2}\frac{\partial}{\partial z}\frac{1}{m^*(z)}\frac{\partial}{\partial z} + V(z), \quad (2a)$$

$$H_{x,y} = -\frac{1}{2m^*(z)}\left(\hbar\frac{\partial}{\partial x} - i\frac{eB}{2c}y\right)^2 - \frac{1}{2m^*(z)}\left(\hbar\frac{\partial}{\partial y} - i\frac{eB}{2c}x\right)^2, \quad (2b)$$

$$H_c = -\frac{e^2}{\varepsilon(z)}\frac{1}{\sqrt{x^2+y^2+(z-z_I)^2}}, \quad (2c)$$

which describes a hydrogen atom at $z = z_I$ in the presence of a magnetic field and a superlattice potential $V(z)$ in the z direction. The magnetic field is taken along the z axis. Notice that the azimuth angular quantum number m, which is connected to the angular operator $L_z = -i\hbar\partial/\partial\phi$, commutes with the Hamiltonian, so m is still a good quantum number in this system for all field strengths. For impurities located at the centre of the quantum well, the z-type parity $\pi_z = (-)^p$ is also a good quantum number. But the other 3D equivalent quantities n and l are no longer good quantum numbers.

A variational calculation will be made for the energy levels of this hydrogenic-type problem. The variational wave function is taken to be the product of two functions

$$\Psi(x,y,z) = F_n(z)G_\mu(\rho, z-z_I, \phi), \quad (3)$$

where $F_n(z)$ is a solution of the Schrödinger equation $H_z F_n(z) = E_{z,n} F_n(z)$ with zero average momentum, and which can be obtained in an analytic form. The second part describes the localized donor state around the donor position $(0,0,z_I)$ with $(\rho = (x^2+y^2)^{1/2}, \phi)$ the cylinder coordinates in the x–y plane. In the present paper we describe the localized part of the wave function by a Gaussian form [13]

$$G_{m,j}(\rho, z, \phi) = \rho^{|m|} e^{im\phi} e^{-\alpha_j \rho^2 - \beta_j z^2}, \quad (4)$$

where (α_j, β_j) are variational parameters that are determined by minimizing the energy: $E_\mu = \langle\psi_\mu|H|\psi_\mu\rangle/\langle\psi_\mu|\psi_\mu\rangle$. Notice that Eq. (4): (a) still has a z-dependence, which is often neglected for quasi-two dimensional donor states, (b) this term $\exp(-\beta_j z^2)$ corrects for taking only the $k_z = 0$ state in $F_n(z)$, and (c) allows for a magnetic field dependent asymmetry in the width of the wave function along and perpendicular to the magnetic field. For low magnetic field it is expected that the present Gaussian form is inadequate because we know that the Coulomb potential leads to an exponential decay of the wave function. We have used an improved functional form in which the exponentials in Eq. (4) were replaced by $\exp(-\alpha_j\rho^2 - \beta_j\sqrt{\rho^2+z^2})$. This function indeed gave a slightly lower energy (typically a few percent) for small magnetic fields. For large magnetic fields, which is of particular interest to us, the magnetic field is

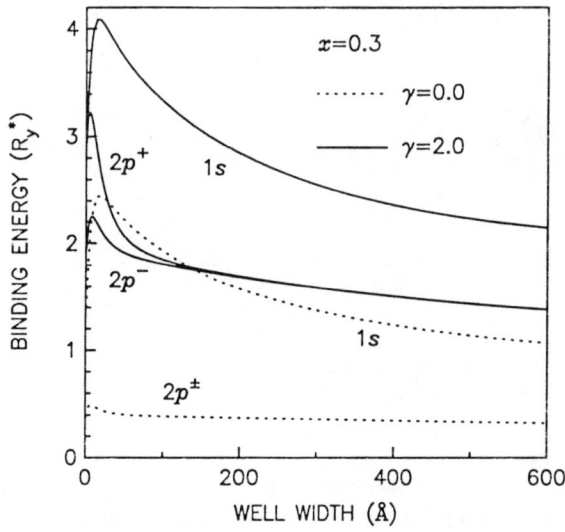

Figure 1. The binding energy of the 1s and $2p^{\pm}$ states of a donor in the middle of a quantum well as function of the well width for two values of the magnetic field.

dominant and the wave function should be Gaussian as was found. In previous work one has taken a more sophisticated expression for Eq. (4) that is expected to give a result which is closer to the exact expression. For example Greene et al. [7] considered a linear combination of Gaussians, and Larsen [14] took an exponential form with the correct (x,y)–z asymmetry.

In the present work we consider GaAs/Al$_x$Ga$_{1-x}$As superlattices with electron band mass $m^*/m_e = 0.067 + 0.083x$ and barriers of height $V_0(eV) = 0.693x + 0.22x^2$. The well width is denoted by w and b is the barrier width. The dielectric constant $\varepsilon = 12.5$ is taken to be the same over the whole superlattice and we take $x = 0.3$ unless otherwise stated.

Confining the electron wave function increases the binding energy, because the electron is forced closer to the donor, in case the donor is in the center of the quantum well. For example, in the limit of two-dimensions (2D) the binding energy increases by a factor of four and the width of the wave function reduces by a factor of two. Because of the finite height of the Al$_x$Ga$_{1-x}$As barriers, an increase by a factor of four can never be realized, as with decreasing quantum well width the electron wave function starts to spill over into the barriers. For $x = 0.3$ the maximum binding energy is about 2.4 times the 3D binding energy, which occurs for w ≈ 30 Å. This is illustrated in Fig. 1, where we show the binding energy of the 1s and $2p^{\pm}$ states as function of the well width of a quantum well for two different magnetic field strengths. Note that the binding energy is defined as $E_\nu^{binding} = E_\nu - E_z - \hbar\omega_c(n+1/2)$ where $n = 0$ for $\nu = $ 1s, 2p⁻ and $n = 1$ for $\nu = 2p^+$, and E_z is the subband energy of the lowest quantum well state. Notice also that the binding energy increases with increasing magnetic field strength, which is due to the magnetic squeezing of the electron wavefunction in the xy-direction.

Figure 2 shows a contour plot of the electron density in the (x,z)-plane for the 1s and the $2p^{\pm}$ states of a donor in the center of the well for zero magnetic field and for $\gamma = 2$ for the extreme case of a quantum well (w = 100 Å) and of a strongly coupled superlattice of 100/11 Å. Notice that the magnetic field compresses the wave function considerably in the quantum well plane,

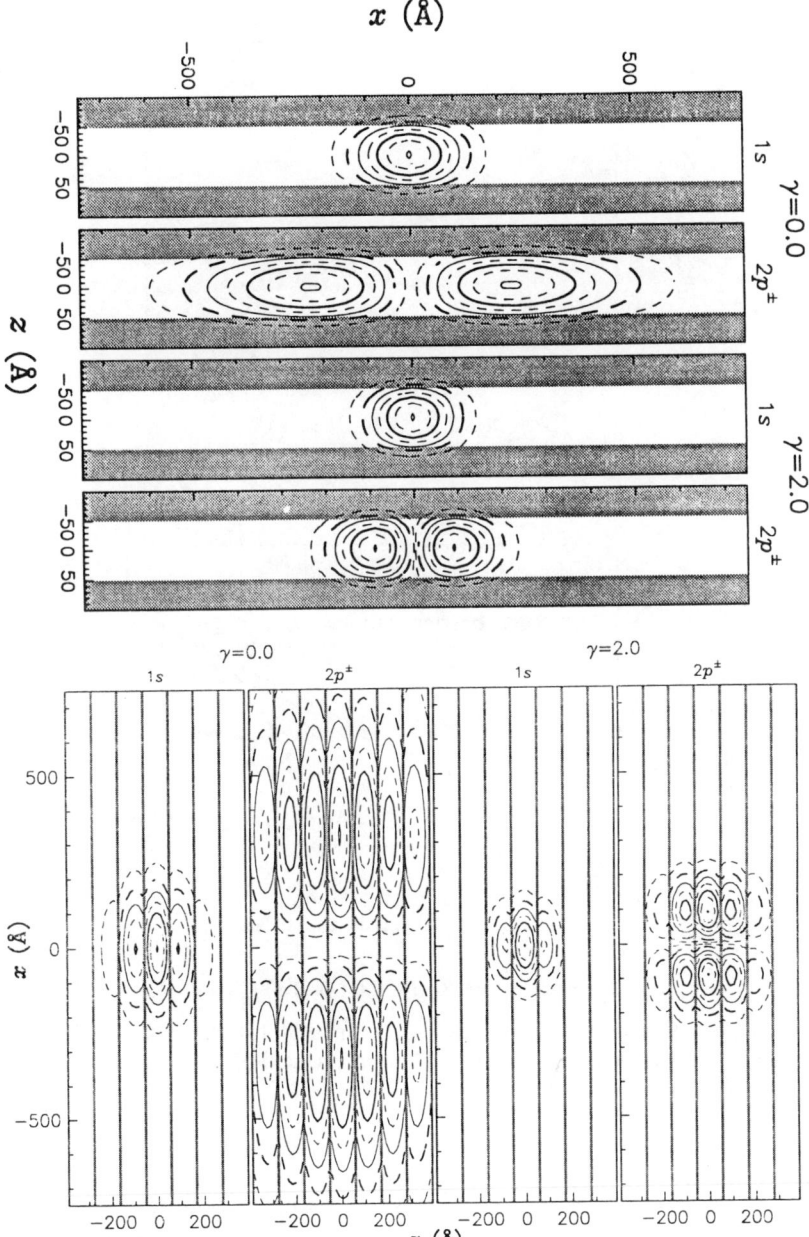

Figure 2. Contour plot of the electron density of the well-center donor states 1s and 2p$^\pm$ for two magnetic field values $\gamma = 0$ and $\gamma = 2$ and for a quantum well of w = 100 Å and a strongly coupled superlattice with w = 100 Å and b = 11 Å.

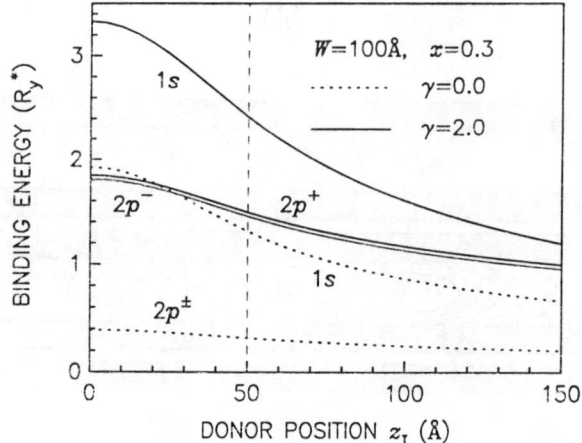

Figure 3. The binding energy of the 1s and $2p^{\pm}$ states for two values of the magnetic field as function of the donor position z_I measured from the centre of the 100 Å wide quantum well.

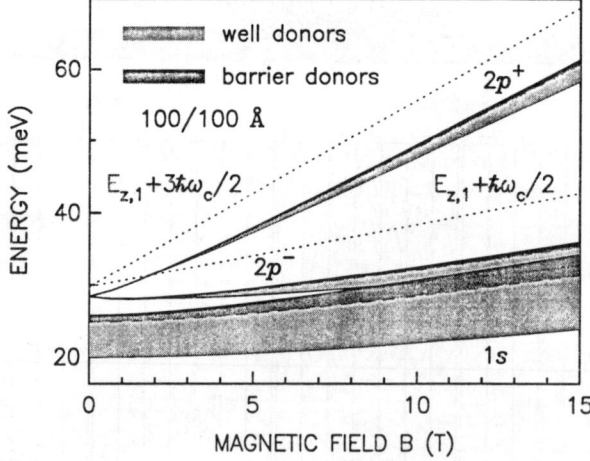

Figure 4. Energy bands corresponding to the 1s, $2p^-$ and $2p^+$ states of a uniformly doped GaAs/Al$_{0.3}$Ga$_{0.7}$As superlattice as function of the magnetic field. The dotted lines correspond to the first two Landau levels of free electrons.

but at the same time the electron wave function is also compressed in the z-direction. The reason is that the magnetic field confines the electron motion in the plane perpendicular to the magnetic field; the width of the electron wave function, which first was of order a_B^*, becomes of order $l_B = \sqrt{e\hbar/cB}$ in the limit of high magnetic fields. Thus on the average the electron is closer to the donor center and consequently the Coulomb binding energy increases, which in turn leads to a reduction of the extent of the wave function along the magnetic field.

The energy levels of the donor states depend on the distance of the donor with respect to the barriers in the superlattice. The levels have minimum energy (or largest binding energy) when the donor is situated near the center of the well and maximum energy (or smallest binding energy) for a donor deep in the barrier in the case of a quantum well or near the center of the barrier in the case of a superlattice. This is illustrated in Fig. 3 for a quantum well of width 100 Å and for two different values of the magnetic field. If experimentally the superlattice is uniformly doped, the different donor energy levels are broadened into bands. We have illustrated this by the shaded areas in Fig. 4 for a 100/100 Å superlattice, where we made a distinction between donors in the quantum well and donors in the barrier. For comparison we have also plotted the two lowest Landau levels for free electrons. At the edge of these bands the density of states is singular, and in a cyclotron resonance experiment one will see absorption peaks corresponding to transitions between the edges of those bands, i.e., from well-center donors and barrier-center donors.

3. Polaron Correction

The superlattices under investigation are weakly polar and as a consequence the electron energy is influenced by the polarization of the medium around the electron. Two main effects are important [15]: (a) a static polarization of the polar lattice around the average position of the electron that will slightly increase the binding energy, which is of order α, the electron-phonon coupling constant (this effect increases with increasing localization of the electron wave function, i.e., with increasing magnetic field), and (b) a resonance effect that occurs when the energy separation between two donor levels equals the LO-phonon energy. This leads to a splitting of the energy levels [16], which in 2D is of order $\alpha^{1/2}$ and in 3D of order $\alpha^{2/3}$. Both effects are to leading order in the electron-phonon coupling correctly described by the improved Wigner Brillouin perturbation theory (IWBPT) [16,17]

$$\Delta E_i = -\sum_j \sum_{\vec{q}} \frac{|<j;\vec{q}|H_I|i;\bar{o}>|^2}{\hbar\omega_{\vec{q}} + E_j - E_i - \Delta_i}, \quad (5)$$

where H_I is the Hamiltonian describing the electron-phonon interaction, $\Delta_i = \Delta E_i - \Delta E_{1s}$, and $|i;\vec{q}>$ describes a state composed of an electron with unperturbed energy E_i and a LO-phonon with momentum $\hbar\vec{q}$ and energy $\hbar\omega_{\vec{q}}$. In principle, we have to include all donor states in the sum $\sum_j$, which is a formidable task. We have limited ourselves to the lowest three donors states 1s, 2p$^-$, and 2p$^+$, which are most relevant for our purposes.

Previous work [18-23] on the polaron correction to donor states in confined structures was limited to the case of a single quantum well. In Refs. [18-20] the polaron correction to the 1s state of a donor in the center of a quantum well was calculated. Osório et al. [21] used IWBPT to calculate the polaron correction to the 1s $\rightarrow$ 2p$^+$ transition and investigated the splitting below the LO-phonon frequency for donors in the center and at the interface of the quantum well. Their wave function is identical to Eq. (4) but with $\beta_j = 0$. Recently we showed [13] that for donors at the interface (and in the barrier) the quantum well approximation to the superlattice system is inadequate. In Ref. [22] band non-parabolicity was neglected, and in Ref. [23] only the polaron correction to the 2p$^\pm$ state was considered within WBPT, which is known to give the wrong pinning behaviour [16,17].

In Fig. 5 we depict the energy levels of the three lowest states together with the same levels shifted over an LO-phonon energy (dashed curves) and the levels corrected for the polaron

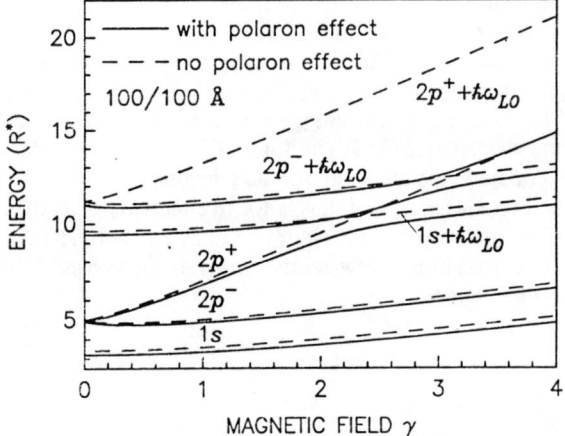

Figure 5. The energy levels of a well-center donor in a 100/100 Å superlattice as a function of the magnetic field, with (solid curves) and without (dashed curves) electron-phonon interaction.

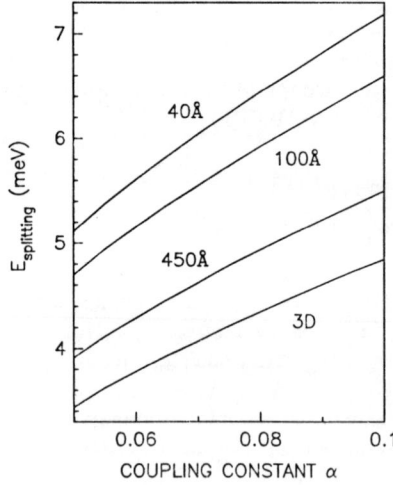

Figure 6. The polaron resonant splitting at $E_{2p^+} - E_{1s} = \hbar\omega_{LO}$ as function of the electron-phonon coupling constant for different values of the well width and for the 3D case.

effect (solid curves) for a donor in the center of a quantum well of a superlattice with w = b = 100 Å. We see clearly: (a) the lowering of the energy of the donor states, and (b) the lifting of the degeneracy at the crossing of two energy levels, which leads to a splitting of those energy levels.

The magnitude of the splitting of the cyclotron resonance peak at $E_{2p^+} - E_{1s} = \hbar\omega_{LO}$ is plotted in Fig. 6 as function of the electron-phonon coupling constant for a donor in the center

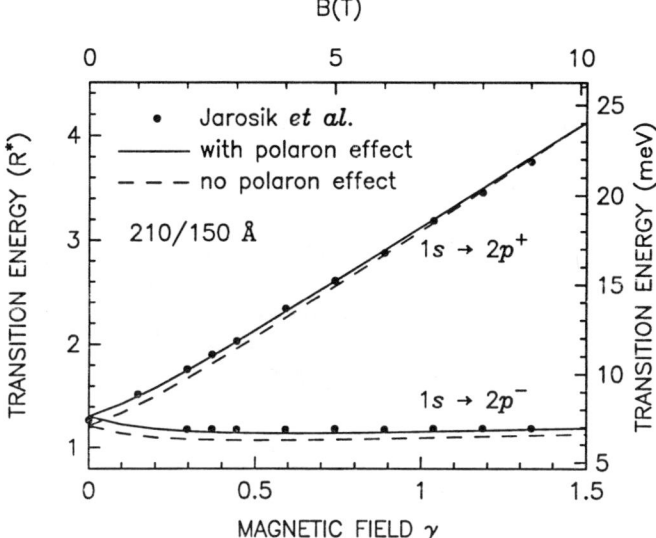

Figure 7. The 1s → 2p$^\pm$ transition energy with (solid curves) and without (dashed curves) polaron correction for well-center donors in a 210/150 Å superlattice. The solid dots are the experimental results of Jarosik et al. [25].

of the quantum well. Notice that the splitting and consequently the effective electron-phonon interaction increases with increasing confinement. We know from previous work [38] that the non-resonant one polaron correction at zero magnetic field increases with decreasing well width. From Fig. 6 we learn that also the resonant interaction has a similar behavior. It is also interesting to notice that even for a quantum well width of 450 Å the splitting is still appreciably larger than for the bulk case.

At present there exist several experimental results [24–34] on the magnetic field dependence of the donor transition energy in GaAs/Al$_x$Ga$_{1-x}$As superlattices with different w and b. In Fig. 7 we compare our theoretical results with the experimental results of Jarosik et al. [25] for the 1s → 2p$^\pm$ transitions in a 210/150 Å superlattice for magnetic fields below resonance. The agreement is excellent and notice that the polaron correction is small. Recently Cheng et al. [32] measured the transition energies near resonance of a well-center donor in a superlattice of 125/125 Å. The results are depicted in Fig. 8. The splitting of the energy levels due to the electron-phonon resonance is apparent in the region 14 T < B < 24 T. The splitting occurs due to degeneracy of the 2p$^+$ level with first the 1s + $\hbar\omega_{LO}$ and, for higher fields, with the 2p$^-$ + $\hbar\omega_{LO}$. For B > 10 T, band non-parabolicity becomes important as is apparent from the difference between the solid (nonparabolic band) and dashed (parabolic band) curves shown in Fig. 8. We included this effect by assuming an energy dependent mass: $m_w(E)/m_e = 0.0665 + 0.0436E + 0.236E^2 - 0.147E^3$, where E is the single-electron energy in eV.

Up to now we have assumed that the phonons are those of bulk GaAs. The assumption was that the superlattice does not influence the phonons. It is well-known [35–37] that in a quantum well the phonon modes are altered because of: (a) the finite extent of the quantum well material, which leads to a selection of phonons with discrete wave vectors $q_z = n\pi/w$, n = 1,2,..., and (b) the presence of the barrier material, which leads to interface modes. These different

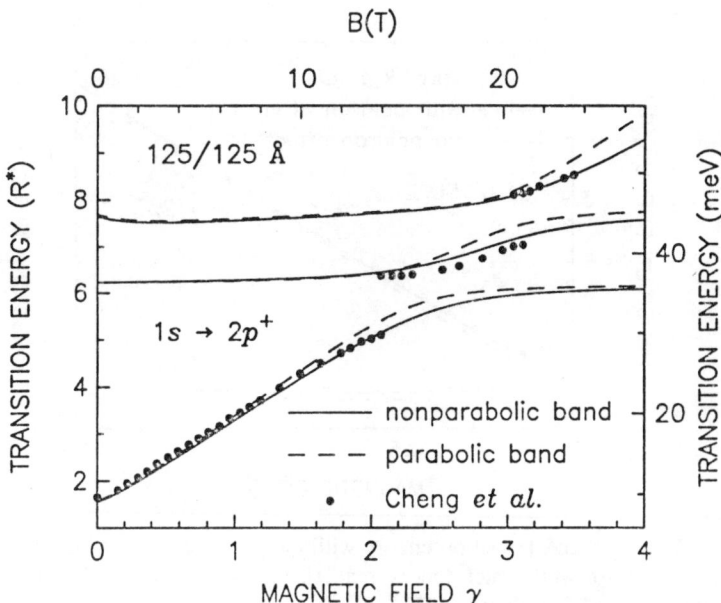

Figure 8. The energy of the 1s → 2p$^+$ transition for center-well donors in a 125/125 Å superlattice as function of the magnetic field with (solid curves) and without (dashed curves) inclusion of band non-parabolicity. The solid dots are the experimental results of Cheng et al. [32].

phonon modes have been observed with Raman scattering. In an early work [38] we found that at zero magnetic field and for free electrons the difference between the polaron correction as calculated from bulk phonons is very close to a calculation where slab and interface phonons were included. Only for w < 50 Å is there a small difference, which is still less than 10%. In the presence of a magnetic field it was suggested [23] that near resonance, interface phonons influence the position of the resonance energy, because of the finite dispersion of these modes. For example, the resonance energy should be pushed downward. And as a consequence, the cyclotron resonance near the LO-phonon resonance is expected to be a sensitive probe of the modification of the phonon modes in quantum wells and superlattices. This was recently clearly demonstrated theoretically for electrons in GaAs/AlAs quantum wells [49].

Analysis [39] of free electron cyclotron resonance data in GaAs heterostructures near resonance could be explained by interaction with bulk phonons. No interface phonons had to be invoked. In those experiments the many-particle effects influence strongly the size of the polaron correction and because no experimental data were available above resonance, no definite conclusions could be made about the relevance of interface phonons in the polaron correction in heterostructures. In the present situation of D-centers we have essentially a one electron problem and the complication of screening and the occupation effect are not present. Furthermore because one has the added flexibility of growing different samples where the donor position is nearer or further away from the interface, one would expect that the importance of the electron-interface phonon interaction can be varied.

4. The D⁻ Center

In previous sections we have studied the ground and excited states of a neutral donor in a magnetic field. One can ask if it is possible to add an extra electron to this system and create a D⁻ ion. This problem is analogous to the H⁻ ion which has been extensively studied in astrophysics [40]. Classically, H⁻ should not exist, but quantum mechanically, due to exchange and correlation, it turns out that H⁻ has still one and only one bound state for B = 0. When B > 0, H⁻ possess bound excited states. The dissociation energy of H⁻ is 0.75 eV and thus a factor of 18 smaller than for a neutral H ion. The negative hydrogen ion is of great importance for the opacity of the atmosphere of the sun, because the dissociation energy 0.75 eV ≈ 8700 K is near the temperature of the solar atmosphere. In the past, different calculations have been performed to obtain the binding energy. For our purpose we mention the variational calculation of Chandrasekhar [41] who proposed the wave function

$$\Psi(\vec{r}_1, \vec{r}_2) \sim (e^{-ar_1 - br_2} + e^{-br_1 - ar_2})(1 + c|\vec{r}_1 - \vec{r}_2|), \tag{6}$$

and found a = 1.075, b = 0.478, c = 0.312, and E_b = 0.0518R^* = 0.704 eV. Notice that this wave function almost represents an unscreened hydrogen atom plus a very loosely bound outer electron.

The electronic properties of an isolated D⁻ center in bulk semiconductors [42–44] has been studied extensively in the presence of a magnetic field. Polaron corrections to the bulk D⁻ ground state was studied by Adamowski [45]. Here we are interested in the properties of the D⁻ center placed in a quantum well in the presence of a magnetic field. Pang and Louie [46] used a diffusion quantum Monte Carlo technique to calculate the binding energy of a D⁻ center in a quantum well in the presence of a magnetic field along the growth axis. Xia and Quinn [47] employed the local spin density functional method to calculate the D⁻ ground state energy and solved the Schrödinger equation numerically to find the D° states. The latter results compare well with the experimental data of Jarosik et al. [25] for a quantum well. In both calculations the electron-phonon interaction was not included.

Here we will give a variational calculation in order to obtain the binding energy and the transition energies of a D⁻ center in a magnetic field. The correction due to electron-phonon interaction will be incorporated within second-order perturbation theory. Our wave function has the form

$$\Psi(\vec{r}_1, \vec{r}_2) = \Phi_i(\vec{r}_1)\Phi_0(\vec{r}_2) \pm \Phi_0(\vec{r}_1)\Phi_i(\vec{r}_2), \tag{7}$$

where the inner orbital Φ_i always has the same form as that of the donor ground state, and the outer orbitals Φ_0 vary depending on the states: for the symmetric s-state it has the same functional form as Φ_i, and for the anti-symmetric $p^t_\pm$ states it is similar to the donor $2p^\pm$ states. For the ground state we took the plus sign in Eq. (7) and the minus sign for the $p^t_\pm$ states (parallel spins for the two electrons). We found that the excited states $p^t_\pm$ with anti-parallel spins are not bound and therefore will not be considered in the following. Recently Larsen and McCann [50] have argued that these states maybe important in cyclotron resonance experiments as quasi-resonant states. Their calculation was for a strict 2D system.

We took $\Phi(x,y,z) \sim F_0(z)\rho^{|m|}\exp(im\phi - \xi\rho^2 - \eta r)$ with $F_0(z)$ the wavefunction of the lowest subband in the quantum well. For the moment we limit ourselves to the case of a

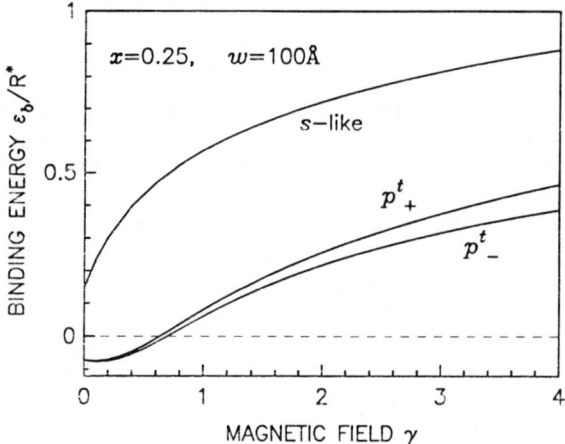

Figure 9. The binding energy of the three lowest states of a D⁻ center as function of the magnetic field. The donor is in the center of the quantum well with width w = 100 Å.

quantum well. Thus our wavefunction contains four variational parameters. The results of the present variational calculation are shown in Fig. 9. Notice that for $\gamma < 0.7$ T there are no bound excited states within the present variational approach. The binding energy is defined through

$$E_b = E_v^{binding} = E_{1s}^0(D^0) + E_z + \hbar\omega_c(n+1/2) - E^0(D_v^-), \quad (8)$$

where n = 1 for the p_+^t-state and n = 0 for the others.

For the ground state, we give in Fig. 10 an idea of the extent of the wavefunction of each of the electrons in the D⁻ system. We plot $\sqrt{1/\xi}$ for the inner and outer electron and compare it with a similar quantity for the D⁰-center (dashed curve) and with the cyclotron radius (dotted curve). Notice that the presence of the second electron in the D⁻ system pushes the inner electron closer to the donor center as compared to the electron in the D⁰ system. Because the two electrons in the D⁻ system are indistinguishable, the present distinction between inner and outer electrons is artificial. Indeed a calculation of the charge density from the full wavefunction (Eq. (7)) would give a uniform s-like distribution.

The polaron correction to the ground state of the D⁻ state was obtained through second-order perturbation theory and is given by

$$\Delta E_s = \sum_f \sum_{\vec{q}} |V_q|^2 \frac{|<f|e^{-\vec{q}\cdot\vec{r}_1} + e^{-i\vec{q}\cdot\vec{r}_2}|s>|^2}{E_s^0 - E_f^0 - 2\hbar\omega_{\vec{q}}}, \quad (9)$$

where V_q is the electron-phonon matrix element, s indicates the ground state of the D⁻ center with energy E_s^0 in the absence of electron-phonon interaction, and f indicates all the possible final states of the two-electron problem. In our calculation we only included the following states: the ground state of the D⁻ state and the ground state of the D⁰ state together with a free

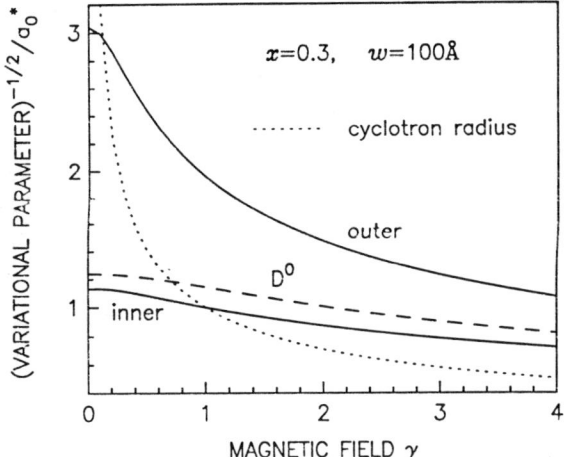

Figure 10. The extent of the inner and outer electron in the ground state of a D⁻ center. The results are compared with the extent of an electron in a D⁰-center and with the cyclotron radius.

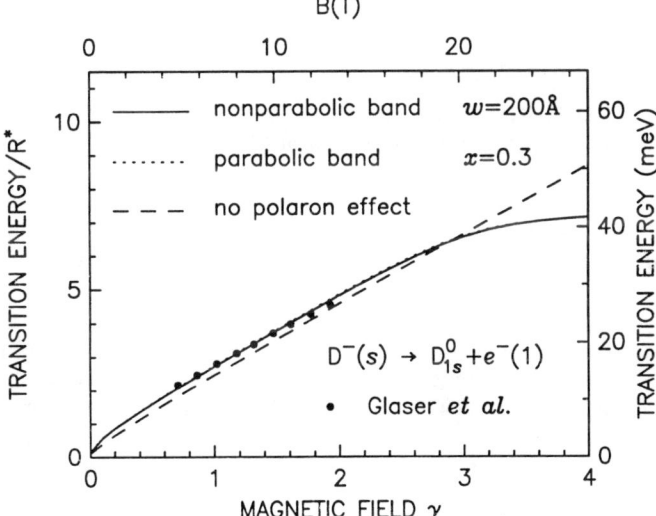

Figure 11. The cyclotron transition energy of a D⁻-center in a quantum well of width w = 200 Å as function of the magnetic field. The dots are the experimental results of Glaser et al. [24].

electron in the lowest two Landau levels. The excited $p^t_\pm$ states of the D⁻ center do not contribute, because the spin matrix element with the ground state of the D⁰ state is zero.

In a far-infrared magnetoabsorption experiment no spin reversal is possible and the $|s> = |1s, 1s> \rightarrow |p^t_+> = |1s, 2p^+>$ transition is forbidden. It is believed that one observes the transition from the D⁻ ground state to the neutral D⁰ state plus a free electron, which, due to the angular momentum selection rule, is in the first excited Landau level. It is believed [48] that

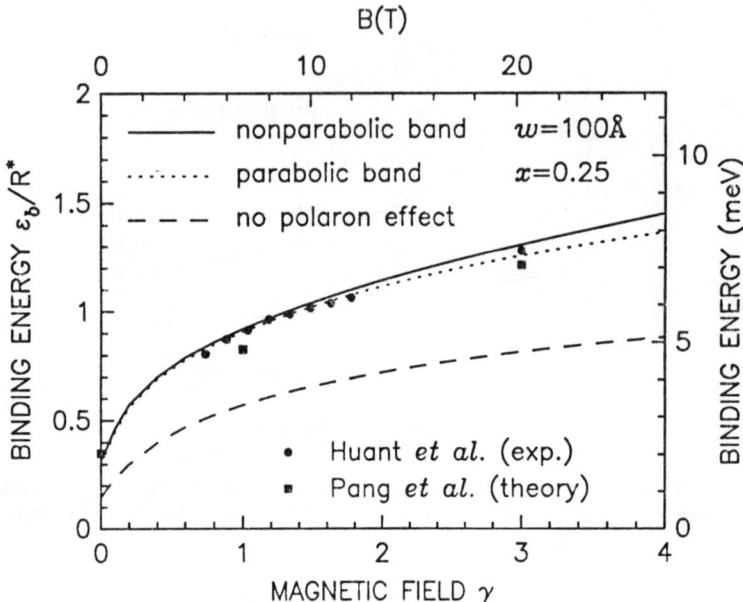

Figure 12. The binding energy of a D⁻-center as function of the magnetic field in a quantum well of width w = 100 Å. The solid circles are the experimental results of Huant et al. [48]. The solid squares are the theoretical results of Pang and Louie [46].

such transition energies were observed by Glaser et al. [24] and we compare them in Fig. 11 with the present theoretical results. Notice that the polaron effect on the transition energies are not significant in the magnetic field range B < 20 T for the case of a quantum well with width w = 200 Å.

Recently Huant et al. [48] performed magneto-photoconductivity measurements on multi-quantum-well structures and was able to observe directly the binding energy of the D⁻ center. The results are shown in Fig. 12 together with our theoretical results. Notice that the polaron correction is significant and has to be included in order to explain the experimental results. The reason is that the binding energy of the D⁻ center is small and any small correction to it looks like a relatively large contribution. Furthermore the D⁻ state is like a *bipolaron* and we know that in such a system the electron-phonon contribution is proportional to -4α instead of -2α as for two separate electrons. The reason is that in a bound system the two electrons feel not only their own lattice-polarization cloud but also the polarization cloud of the other electron. The numerical results of Pang et al. [46] are also included for reference. The results of Pang et al. [46] do not include the electron-phonon correction and thus if one includes this polaron correction they will be significantly above the experimental binding energies. We expect that our theoretical result significantly underestimates the binding energy at zero magnetic field. But with increasing magnetic field our variational wave function is expected to be reliable, as is confirmed by the experimental observation and by our earlier work. In a previous calculation [51] we took the wave function $\Phi(x,y,z) \sim F_0(z)\rho^{|m|}\exp(im\phi - \xi\rho^2 - \eta z^2)$ and found that the difference with the present result is small for $\gamma < 2$.

5. Conclusions

We have shown that shallow impurities in semiconductors are an interesting laboratory for atomic systems, which one can study in a parameter space that is not accessible otherwise. For example, for D-centers in GaAs the effective magnetic field is about a factor 10^5 larger as compared to the case of a H atom.

The study of electronic states of such impurities at high magnetic fields gives us information on the electron-phonon interaction in the host material. We are able to obtain information on the strength of this interaction and the relevant phonon frequencies responsible for it. Furthermore, knowledge is obtained of the band non-parabolicity near the bottom of the conduction band. These physical parameters are important input parameters for the simulation of electronic transport through, for example, GaAs based devices like HEMTs.

In the present study we have relied on a variational calculation of the donor states. Effects due to band mass discontinuity along the barriers, the finite height of the barrier, the influence of adjacent quantum wells, and the electron-phonon interaction are included. We expect that those different effects have been included with an accuracy on a 5–10% level.

Acknowledgments

Collaboration with M. Helm on the work on the strongly coupled superlattice is gratefully acknowledged. During the course of this work we have benefited from stimulating discussions with S. Haunt, C. Langerak, B. McCombe, J. Singleton, and S. Louie. One of us (FMP) is supported by the National Science Foundation of Belgium. This work is supported by F.K.F.O. (Fonds voor Kollektief Fundamenteel Onderzoek, Belgium), project No. 2.0093.91 and "Diensten voor de Programmatie van het Wetenschapsbeleid" (Belgium) under contract No. IT/SC/24.

References

[*] Also at University of Antwerp (RUCA), B-2020 Antwerp and University of Technology, NL-5600 MB Eindhoven.
1. See, for example, E.E. Mendez and K. von Klitzing, *Physics and Applications of Quantum Wells and Superlattices* (Plenum Press, New York, 1987).
2. W.T. Masselink, Y.C. Chang, H. Morkoç, D.C. Reynolds, C.W. Litton, K.K. Bajaj, and P.W. Yu, Solid State Electron. **29**, 205 (1986).
3. B.D. McCombe, N.C. Jarosik, and J.M. Mercy, in *Two-dimensional Systems: Physics and New Devices*, edited by G. Bauer, G. Kuchar, and H. Heinrich (Springer-Verlag, Berlin, 1986), p. 156.
4. B. Bastard, Phys. Rev. B **24**, 4714 (1981).
5. C. Mailhiot, Y.C. Chang, and T.C. McGill, Phys. Rev. B **26**, 4449 (1982).
6. R.L. Greene and K.K. Bajaj, Solid State Commun. **45**, 825 (1983).
7. R.L. Greene and K.K. Bajaj, Phys. Rev. B **31**, 913 (1985); **37**, 4604 (1988).
8. K. Jayakumar, S. Balasubramanian, and M. Tomak, Phys. Rev. B **34**, 8794 (1986).
9. C. Priester, G. Bastard, G. Allan, and M. Lannoe, Phys. Rev. B **30**, 6029 (1984).
10. S. Chaudhari, Phys. Rev. B **28**, 4480 (1983).

11. P. Lane and R.L. Greene, Phys. Rev. B **33**, 5871 (1986).
12. M. Helm, F.M. Peeters, F. DeRosa, E. Colas, J.P. Harbison, and L.T. Florez, Phys. Rev. B **43**, 13983 (1991); Surf. Sci. **263**, 518 (1992).
13. J.M. Shi, F.M. Peeters, G.Q. Hai, and J.T. Devreese, Phys. Rev. B **44**, 5692 (1991).
14. D.M. Larsen, Phys. Rev. B **44**, 5629 (1991).
15. F.M. Peeters and J.T. Devreese, Physica Scripta T **13**, 282 (1986).
16. F.M. Peeters and J.T. Devreese, Phys. Rev. B **31**, 3689 (1985).
17. G. Lindemann, R. Lassnig, W. Seidenbuch, and E. Gornik, Phys. Rev. B **28**, 4693 (1983).
18. A. Erçelebi and M. Tomak, Solid State Commun. **54**, 883 (1985).
19. M.H. Degani and O. Hipólito, Phys. Rev. B **33**, 4090 (1986).
20. B.A. Mason and S. Das Sarma, Phys. Rev. B **33**, 8379 (1986).
21. F.A.P. Osório, M.Z. Maialle, and O. Hipólito, in *Proc. of the 20th International Conference on the Physics of Semiconductors*, edited by E.M. Anastassakis and J.D. Joannaopoulos (World Scientific, Singapore, 1990), p. 1017.
22. C.D. Hu and Y.H. Chang, Phys. Rev. B **40**, 3878 (1989).
23. D.L. Lin, R. Chen, and T.F. George, Phys. Rev. B **43**, 9328 (1991).
24. E. Glaser, B.V. Shanabrook, R.L. Hawkins, W. Beard, J.M. Mercy, B.D. McCombe, and D. Musser, Phys. Rev. B **36**, 8185 (1987).
25. N.C. Jarosik, B.D. McCombe, B.V. Shanabrook, J. Comas, J. Ralsto, and G. Wicks, Phys. Rev. Lett. **54**, 1283 (1985).
26. Y.H. Chang, B.D. McCombe, J.M. Mercy, A.A. Reeder, J. Ralston, and G.A. Wicks, Phys. Rev. Lett. **61**, 1408 (1988).
27. S. Huant, W. Knap, G. Martinez, and B. Etienne, Europhys. Lett. **7**, 159 (1988).
28. B.D. McCombe, A.A. Reeder, J.M. Mercy, and G. Brozak, in *High Magnetic Fields in Semiconductor Physics II*, edited by G. Landwehr (Springer-Verlag, Berlin, 1989), p. 258.
29. S. Huant, R. Stepniewski, G. Martinez, V. Thierry-Mieg, and B. Etienne, Superlattices and Microstructures **5**, 331 (1989).
30. G. Brozak, B.D. McCombe, and D.M. Larsen, Phys. Rev. B **40**, 1265 (1989).
31. S. Huant, W. Knap, R. Stepniewski, G. Martinez, V. Thierry-Mieg, and B. Etienne, in *High Magnetic Fields in Semiconductor Physics II*, edited by G. Landwehr (Springer-Verlag, Berlin, 1989), p. 293; S. Huant, S.P. Najda, W. Knap, G. Martinez, B. Etienne, C.J.G.M. Langerak, S. Singleton, R.A.J. Thomeer, G. Hai, F.M. Peteers, and J.T. Devreese, in *Proc. of the 20th Int. Conf. on the Physics of Semiconductors*, edited by E.M. Anastassakis and J.D. Joannopoulos (World Scientific, Singapore, 1990), p. 1138.
32. J.P. Cheng, B.D. McCombe, G. Brozak, J. Ralston, and G. Wicks, in *Quantum Well and Superlattice Physics III*, SPIE Proceedings **1283**, 281 (1990).
33. J.P. Cheng and B.D. McCombe, Phys. Rev. B **42**, 7626 (1990).
34. J.P. Cheng, B.D. McCombe, and G. Brozak, Phys. Rev. B **43**, 9324 (1991).
35. J.J. Licari and R. Evrard, Phys. Rev. B **15**, 2254 (1977).
36. L. Wendler, Phys. Status Solidi B **129**, 513 (1985).
37. N. Mori and T. Ando, Phys. Rev. B **40**, 6175 (1989).
38. G.Q. Hai, F.M. Peeters, and J.T. Devreese, Phys. Rev. B **42**, 11063 (1990).
39. F.M. Peeters, X.G. Wu, and J.T. Devreese, Solid State Commun. **65**, 1505 (1985); F.M. Peeters, X.G. Wu, J.T. Devreese, C.J.G.M. Langerak, J. Singleton, D.J. Barnes, and R.J. Nicholas, Phys. Rev. B **45**, 4296 (1992).
40. H.A. Bethe and E.E. Salpeter, in *Quantum Mechanics of One- and Two-Electron Atoms*, (Springer-Verlag, Berlin, 1957), p. 154.

41. S. Chandrasekhar, Astrophys. J. **100**, 176 (1944).
42. D.M. Larsen, Phys. Rev. B **20**, 5217 (1979).
43. D.M. Larsen, Phys. Rev. Lett. **42**, 742 (1979).
44. A. Natori and H. Kamimura, J. Phys. Soc. Jpn. **47**, 1550 (1979); **44**, 1216 (1978).
45. J. Adamowski, Phys. Rev. B **39**, 13061 (1989).
46. T. Pang and S.G. Louie, Phys. Rev. Lett. **65**, 1635 (1990); S.G. Louie and T. Pang, in *New Horizons in Low Dimensional Electron Systems* (a Festschrift in honour of H. Kamimura), edited by H. Aoki, M. Tsukada, and M. Schlüter (Kluwer, Dordrecht, 1992).
47. X. Xia, X. Zhu, and J.J. Quinn, Phys. Rev. B **45**, 1341 (1992) ; X. Xia and J.J. Quinn, Phys. Rev. B **46**, 12530 (1992).
48. S. Huant, S.P. Najda, and B. Etienne, Phys. Rev. Lett. **65**, 1486 (1990); A. Mandray, S. Huant, and B. Etienne, Europhys. Lett. **20**, 181 (1992).
49. G.Q. Hai, F.M. Peeters, and J.T. Devreese, in *Proc. of the Conf. on the Application of High Magnetic Fields in Semiconductor Physics* (Chiba, Japan, 3–7 August 1992), Physica B **183** (1993)).
50. D.M. Larsen and S.Y. McCann, Phys. Rev. B **45**, 3485 (1992).
51. F.M. Peeters, J.M. Shi, and J.T. Devreese, in *Proc. of the CAP-NSERC Workshop on Excitations in Superlattices and Multi-Quantum Wells* (London, Ontario, Canada, 28 July – 3 August 1991).

ENHANCED LINEAR ELECTRO-OPTIC ANISOTROPY IN GaAs QUANTUM WELLS

S.H. KWOK,[1] H.T. GRAHN,[2] K. PLOOG,[2]* AND R. MERLIN[1]
[1]The Harrison M. Randall Laboratory of Physics, The University of Michigan, Ann Arbor, Michigan 48109-1120, USA
[2]Max-Planck-Institut für Festkörperforschung, Heisenbergstrasse 1, W-7000 Stuttgart 80, Federal Republic of Germany

ABSTRACT. Photoluminescence due to nominally forbidden excitons in GaAs quantum-well structures exhibits large differences between [110] and [1$\bar{1}$0] polarizations in the presence of electric fields parallel to the growth axis [001]. Allowed transitions show no noticeable anisotropy. Our observations relate to the linear electro-optic (Pockels) effect. The anisotropy decreases with increasing field. Results can be qualitatively accounted for by perturbation theory.

1. Introduction

The optical properties of semiconductor quantum well (QW) structures exhibit unique features reflecting confined geometries. In particular, electric fields perpendicular to the layers [1–10] lead to many interesting effects such as the quantum-confined Stark effect [1–4] and Wannier-Stark ladders [6–9]. In this work, we report on a novel phenomenon relying on quantum confinement which relates to Pockels linear electro-refraction. The new effect manifests itself in a giant field-induced biaxial anisotropy and associated odd dependence on the field. Enhanced Pockels-like anisotropy is observed in photoluminescence (PL), primarily from excitons such as e_2h_1 showing weak coupling to photons at zero field [11] (e_ih_j denotes states with one electron and one hole in the ith and jth subbands respectively). Consistent with recent near-gap measurements of linear electro-optic coefficients in GaAs-Al$_x$Ga$_{1-x}$As QW structures [12], we find no enhancement at the dominant e_1h_1 PL.

2. Experimental Details and Sample Parameters

The two QW structures used in the experiments were grown by molecular-beam epitaxy on a (001) n$^+$-GaAs substrate. They consist of 40 periods of 90-Å GaAs/40-Å AlAs and 100 periods of 131-Å GaAs/79-Å Al$_{0.35}$Ga$_{0.65}$As, respectively. The superlattice with AlAs (AlGaAs) barriers was sandwiched between heavily doped GaAs (Al$_{0.5}$Ga$_{0.5}$As) layers in a p-i-n configuration. PL measurements were performed at $T = 2$ K using the $\lambda = 676.4$ nm line of a Kr$^+$ laser. At high excitation density P, our samples break into electric field domains characterized by the alignment of levels in neighboring wells leading to sequential resonant tunneling [13]. In the interest of clarity, the discussion will focus on data at low P

(uniform field) for the GaAs-AlAs QW and at high P (domain pattern) for the GaAs-$Al_{0.35}Ga_{0.65}As$ sample.

3. Theoretical Considerations

The Pockels effect refers to *linear* electric field induced optical birefringence shown by non-centrosymmetric crystals. In zinc-blende semiconductors, fields along [001] lead to biaxial behavior [14] where the anisotropy in the plane perpendicular to the field distinguishes [110] from [1$\bar{1}$0]. Although it is important for some applications, the effect remains relatively small up to the breakdown field ($\sim 10^6$ Vcm^{-1}). To facilitate our discussion on enhanced biaxial anisotropy, we focus first on the symmetry arguments underlying the problem. Let F be the z-component of the electric field and consider photons which are polarized perpendicular to the field in the direction of the unit vector ($\sin\theta\sin\alpha$, $\sin\theta\cos\alpha$, $\cos\theta$). Consistent with the D_{2d} point group symmetry of the QW, the intensity I emitted (or absorbed) can be written as

$$I(\theta,\phi,F) = \sin^2(\theta)I_\parallel^+(F) + \sin^2(\theta)\cos(2\phi)I_\parallel^-(F) + \cos^2(\theta)I_\perp^+(F), \qquad (1)$$

where I^+ and I^- ($|I_\parallel^-| \leq I_\parallel^+$ and $0 \leq I_\perp^+$) are, respectively, the even and odd parts of I associated with the directions perpendicular ($\perp$) and parallel ($\parallel$) to the layers; $\phi = \alpha - \pi/4$ is the angle between the electric vector and [110]. At $F = 0$, the QW structures are already uniaxial. Pockels-like effects require $I_\parallel^- \neq 0$ and follow from the absence of inversion symmetry. To measure the degree of in-plane anisotropy, we define the polarization ratio

$$\rho = \frac{I_{[110]} - I_{[1\bar{1}0]}}{I_{[110]} + I_{[1\bar{1}0]}} = \frac{I_\parallel^-}{I_\parallel^+}. \qquad (2)$$

Notice that the corresponding ratio for [100] and [010] is equal to zero.

A qualitative understanding of what determines the magnitude of ρ can be gained from first-order perturbation theory. The analysis discriminates between excitons e_ih_j with $i = j$ (electric-dipole allowed in the envelope-function approximation) and $i \neq j$ (forbidden). As we shall see, only the latter may exhibit enhanced anisotropy. The two main contributions to $I_\parallel^-$ [15] are: (i) an electronic coupling between E_0 states and higher-lying $\Gamma_{7,8}^c$ states of the conduction band [16] and (ii) a lattice term due to field-induced sublattice displacement leading to heavy-light hole mixing. The corresponding hybridization parameters are (i) eFa_0/Δ and (ii) $Dd_{14}F/\delta$. Constants $\Delta \approx 3-4.5$ eV [16], $D \approx 6$ eV, and $d_{14} = -2.7 \times 10^{-10}$ V^{-1}cm [17] represent, respectively, the separation between the $\Gamma_{7,8}^c$ states and the lowest conduction (Γ_6^c) and the top valence ($\Gamma_{7,8}^v$) band states at Γ, the [111] deformation potential, and the piezoelectric coefficient of GaAs; a_0 is the (hydrogen) Bohr radius and δ is a measure of the heavy-light hole splitting. Clearly, $eFa_0/\Delta \sim Dd_{14}F/\delta \ll 1$ for typical fields.

For allowed excitons, the oscillator strength at $F = 0$ is comparable to the value for bulk GaAs [11]. Given that the electric-dipole matrix elements that matter are all of the same order, one can show that $(d\rho/dF)_{F=0} \sim 2ea_0/\Delta \sim 2Dd_{14}/\delta$, i.e., $\rho \ll 1$ in the range of interest. Nominally forbidden excitons behave quite differently, for their oscillator strength

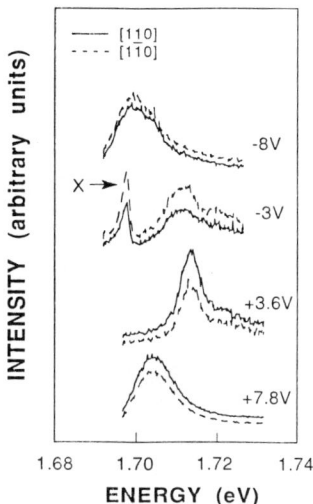

Figure 1. Polarized PL spectra of the GaAs-AlAs sample showing e_2h_1 recombination at various voltages. Negative values correspond to reverse biases. Peak labeled X is due to transitions involving electron AlAs (X) states. The data show large differences between [110] and [1$\bar{1}$0] intensities at low biases. Measurements were performed at $T = 2$ K. The excitation wavelength is $\lambda = 676.4$ nm and the power density is $P = 4$ Wcm^{-2}.

originates in **q**-dependent mixing of heavy and light holes [11] (**q** is the wave vector component parallel to the layers). The fact that mixing parameters vary widely among the various branches [11] provides the path for enhancement. An illuminating example relevant to our experiments is the case of e_2h_1. Because the h_1-branch does not couple much to l_2 or other branches (l_s indicates light-hole sth-subband states), the oscillator strength of e_2h_1 at $F = 0$ is negligible. On the other hand, strong l_1-h_2 hybridization at **q** ≠ 0 results in comparable strengths for the (**q** = 0 forbidden) e_2l_1 and (**q** = 0 allowed) e_2h_2 transitions. The field couples h_1 and l_1 through the sublattice displacement. Thus, if $\gamma\{l_1|h_2\}$ and $\nu\{l_2|h_1\}$ measure the relevant **q**-induced admixture of the corresponding subbands, $(d\rho/dF)_{F=0} \sim (\gamma/\nu) 2Dd_{14}/\delta$ and, therefore, $\rho(e_2h_1) \gg \rho(e_kh_k)$ at low fields because $\gamma \gg \nu$.

4. Results and Discussion

Photoluminescence spectra of the GaAs-AlAs structure showing pronounced anisotropy are reproduced in Fig. 1. The main broad feature corresponds to the e_2h_1 exciton. This assignment is based primarily on calculations of QW energies as a function of electric field. The identification is further supported by a combination of photocurrent and PL experiments revealing maxima for the current and the PL intensity at the particular bias for which e_1 and e_2 become aligned [18]. Other than e_2h_1, X is due to recombination of AlAs(X) electrons with holes at GaAs(Γ). This peak exhibits a blue shift that is nearly linear on the field and with a slope that is consistent with our assignment [19]. X couples to and

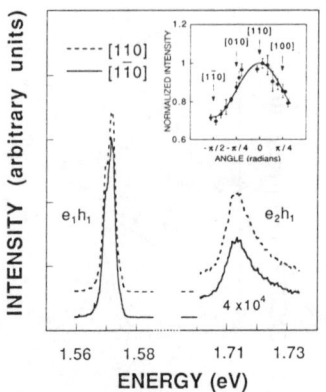

Figure 2. Polarized PL spectra at $V = 4$ V. Inset: dependence of the e_2h_1 recombination intensity on the angle ϕ (see text); solid line is $\cos(2\phi)$.

its polarization properties are similar to those of e_2h_1 (see below). The situation resembles that of the well-documented problem of X-e_1h_1 hybridization [19].

A comparison between the polarization behavior of e_1h_1 and e_2h_1 is shown in Fig. 2. As already mentioned, the dominant e_1h_1 emission at the gap, being 10^3–10^4 times more intense than e_2h_1, is isotropic within experimental error. The field-induced biaxial behavior follows closely the group-theory predictions contained in Eq. (1); see inset of Fig. 2. Specifically, we find that [110] and [1$\bar{1}$0] are the pair of mutually orthogonal axes exhibiting the largest effect and that these directions exchange their roles after a field reversal. In addition, the comparison between [100]- and [010]-polarized spectra reveals no appreciable differences.

Both the strength and the field-dependence of the anisotropy support our treatment of the phenomenon. These characteristics are displayed in the measurements of the polarization ratio which are plotted in Fig. 3. Clearly, the large magnitude of the effect and the unusual decrease in anisotropy with increasing F are beyond interpretations based solely on bulk properties. However, they can be accounted for by QW effects. At higher fields, the response is no longer linear. Here, we find it appropriate to use

$$\rho = \frac{F/F_1}{1 + (F/F_2)^2} \qquad (3)$$

providing a convenient parametrization [20]. According to our earlier discussion, $F_1 \sim [\nu\{l_2|h_1\}/\gamma\{l_1|h_2\}]\delta/(2Dd_{14})$ for e_2h_1. Moreover, since the F-dependence of $I_\parallel^+$ relies mainly on the behavior of the envelope function—like in the quantum-confined Stark effect [1,2]—one obtains the estimate $F_2 \approx h^2/8\pi^2 MeL^3$ where L is the well width and M is the exciton mass. We note that Eq. (3) applies to nominally forbidden transitions. The sign in the denominator reflects the fact that the associated overlap between the electron and the hole envelope functions increases with increasing field. The curve in Fig. 3 is a best fit using Eq. (3) at $F_1 = 2.46 \times 10^4$ Vcm^{-1} and $F_2 = 1.43 \times 10^4$ Vcm^{-1}. According to the above arguments, the latter values are consistent with the parameters of our sample. The linear term translates into $\nu\{l_2|h_1\}/\gamma\{l_1|h_2\} \approx 4 \times 10^{-3}$, which is not unreasonable. Unfortunately, ρ could not be determined for $|F| \leq 3 \times 10^4$ Vcm^{-1} as the oscillator strength

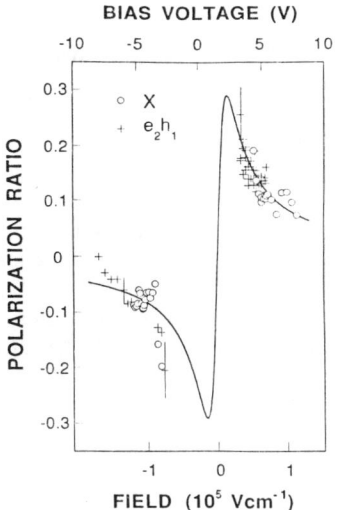

Figure 3. Polarization ratio [Eq. (2)] for X and e_2h_1 PL as a function of the z-component of the electric field F and bias V. Here, $F = (V-V_0)/l$ where $V_0 \approx 1.5$ V is the built-in voltage and $l \approx 0.57$ μm is the width of the intrinsic region of the device. The solid line is a least-squares fit to the data (see text). Experimental parameters are given in Fig. 1. Results are for the GaAs-AlAs heterostructure.

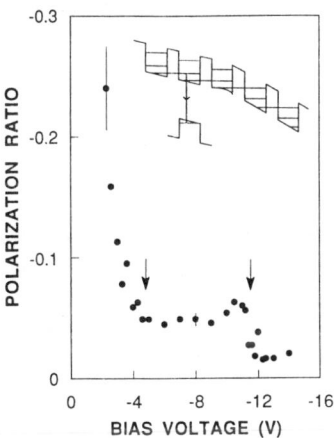

Figure 4. PL polarization ratio for e_2h_1 in the GaAs-Al$_{0.35}$Ga$_{0.65}$As sample at $P = 160$ Wcm^{-2} ($T = 2$K and $\lambda = 676.4$ nm). Results correspond to the high-P domain regime. Arrows denote biases for which the field is uniform (single domain). The field pattern in the region delimited by the arrows is shown in the inset; PL originates mainly from the low-field domain.

of e_2h_1 becomes negligible. It is in this range that the linear contribution is expected to dominate.

To conclude, we consider as an application high-P results for the GaAs-Al$_{0.35}$Ga$_{0.65}$As sample. Anisotropy data for e_2h_1 PL are depicted in Fig. 4. At arbitrary voltage, this structure has been shown to break into electric-field domains characterized by alignment of specific conduction subbands in neighboring wells [13,21]. The inset shows the potential profile for the range inside the arrows. Here, the low-field domain e_1-e_2 coexists with the high-field domain resulting from e_1-e_3 alignment. Unlike e_1h_1, the e_2h_1 spectra show not a doublet but a single line. This reflects the much larger e_2 population of the low-field domain arising from e_1-e_2 resonant tunneling. Our measurements complement and are consistent with data obtained using other techniques [13,21]. In particular, ρ is nearly constant in the coexistence region and, as the domain pattern changes, it varies rapidly to signal the disappearance of the domain boundary. The latter phenomenon is not well understood. Because ρ increases at low fields, our method holds promise for studies of electric fields in a range where standard e_1h_1-PL measurements are not very sensitive [13].

Acknowledgments

This work was supported by the U.S. Army Research Office (Contract No. DAAL-03-92-G-0233), the University Research Initiative Program (Contract No. AFOSR-90-0214), and the Bundenministerium für Forschung und Technologie of the Federal Republic of Germany.

References

* Present address: Paul-Drude-Institut fur Festkörperelektronik, Hausvogteiplatz 5-7, 0-1086 Berlin, Federal Republic of Germany.
1. D.A.B. Miller, D.S. Chemla, T.C. Damen, A.C. Gossard, W. Wiegmann, T.H. Wood, and C.A. Burrus, Phys. Rev. B **32**, 1043 (1985), and references therein.
2. L. Viña, R.T. Collins, E.E. Mendez, and W.I. Wang, Phys. Rev. Lett. **58**, 832 (1987), and references therein.
3. D.A.B. Miller, D.S. Chemla, T.C. Damen, A.C. Gossard, W. Wiegmann, T.H. Wood, and C.A. Burrus, Appl. Phys. Lett. **45**, 13 (1984).
4. T.H. Wood, C.A. Burrus, D.A.B. Miller, D.S. Chemla, T.C. Damen, A.C. Gossard, and W. Wiegmann, IEEE J. Quantum Electron. **21**, 117 (1985).
5. D. Fröhlich, R. Wille, W. Schlapp, and G. Weimann, Phys. Rev. Lett. **59**, 1748 (1987).
6. J. Bleuse, G. Bastard, and P. Voisin, Phys. Rev. Lett. **60**, 220 (1988).
7. E.E. Mendez, F. Agulló-Rueda, and J.M. Hong, Phys. Rev. Lett. **60**, 2426 (1988).
8. P. Voisin, J. Bleuse, C. Bouche, S. Gallard, C. Alibert, and A. Regreny, Phys. Rev. Lett. **61**, 1639 (1988).
9. K.W. Goossen, J.E. Cunningham, and W.Y. Jan, Appl. Phys. Lett. **59**, 3622 (1991).
10. See, for example, J.B. Krieger and G.I. Iafrate, Phys. Rev. B **33**, 5494 (1986).
11. See, for example, W.T. Masselink, P.J. Pearah, J. Klem, C.K. Plug, H. Morkoç, G.D. Sanders, and Y.-C. Chang, Phys. Rev. B **32**, 8027 (1985).
12. A. Jennings, C.D.W. Wilkinson, and J.S. Roberts, Semicond. Sci. Technol. **7**, 60 (1992).

13. H.T. Grahn, H. Schneider, and K. von Klitzing, Phys Rev. B **42**, 2890 (1990), and references therein.
14. See, for example, J.F. Nye, *Physical Properties of Crystals* (Clarendon Press, Oxford, 1985), Chap. 13.
15. S. Zekeng, B. Prevot, and C. Schwab, Phys. Stat. Sol. (b) **150**, 65 (1988); I.P. Kaminow and W.D. Johnston, Phys. Rev. **160**, 519 (1967).
16. See, for example, M. Cardona, N.E. Christensen, and G. Fasol, Phys. Rev. B **38**, 1806 (1988).
17. See, for example, S. Adachi, J. Appl. Phys. **58**, R1 (1985).
18. H.T. Grahn, H. Schneider, W.W. Rühle, K. von Klitzing, and K. Ploog, Phys. Rev. Lett. **64**, 2426 (1990).
19. M.-H. Meynadier, R.E. Nahory, J.M. Worlock, M.C. Tamargo, J.L. de Miguel, and M.D. Sturge, Phys. Rev. Lett. **60**, 1338 (1988).
20. S.H. Kwok, H.T. Grahn, K. Ploog, and R. Merlin, Phys. Rev. Lett. **69**, 973 (1992).
21. S.H. Kwok, E. Liarokapis, R. Merlin, and K. Ploog, in *Light Scattering in Semiconductor Structures and Superlattices*, edited by D.J. Lockwood and J.F. Young (Plenum, New York, 1991), p. 491.

SPECTROSCOPY ON LATERAL SUPERLATTICES

J.P. KOTTHAUS
Sektion Physik
Ludwig-Maximilians-Universität
Geschwister-Scholl-Platz 1
D 8000 München 22
Germany

ABSTRACT. Field-effect devices on GaAs and Si with laterally periodic gates serve to impose widely tuneable one- and two-dimensional superlattice potentials on two-dimensional electron systems and make it possible to investigate all regimes between weakly perturbing and strongly confining lateral superlattices. Studies of the intraband electronic excitations at far-infrared frequencies provide information on collective modes in lateral superlattices, on strength and form of the lateral confining potentials, as well as on the mechanisms of electronic interactions. A magnetic field applied perpendicularly to the plane of the electron system is used to study the transition from lateral confinement to quasi-two-dimensional behavior. Here recent results are discussed that show the effects of anisotropy and non-parabolicity of the confining potential on the intraband modes. With these experiments one spectroscopically gains information on electronic interactions both within a wire or dot as well as across the superlattice period. Most experimental results are discussed in terms of semiclassical models based on ballistic electron motion in the lateral superlattices.

1. Introduction

In recent years the experimental study of the effects of lateral superlattice potentials on the electronic properties of two-dimensional electron systems (2DES) has been an increasingly active area of research [1]. The fabrication of high quality lateral superlattices with periods in the range of few hundred nanometers has become possible by combining high resolution lithography using electron beams or ultraviolet (UV) lasers with subsequent deposition and etching techniques. Here we will focus on voltage-tuneable lateral superlattices in which suitably structured gate electrodes serve to tune the relative strength of the lateral superlattice potential with respect to the Fermi energy E_F. The concept of such devices is illustrated in Fig. 1 for the case of a square lateral superlattice [2]. Ideally, the device should be tuneable from the weakly modulated case (Fig. 1(a)) via the situation in which the superlattice amplitude just exceeds the Fermi energy (Fig. 1(b)) resulting in arrays of antidots to the strongly modulated case (Fig. 1(c)), in which electron dots are formed in the pockets of the superlattice potential. Note that the sketch in Fig. 1 implies that the electron density can adjust to screen large portions of the external superlattice potential. In all cases discussed in the following this will be true since the period of the superlattice potential is larger than the screening length in the modulated electron system.

One way to realize such tuneable superlattice devices is sketched in Fig. 2 [3,4]. A suitably patterned insulator which may be a photoresist layer or a patterned undoped GaAs

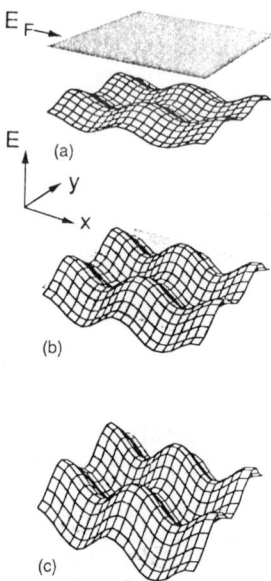

Figure 1. Sketch of a two-dimensional electron gas with Fermi energy E_F subjected to a two-dimensional sinusoidal superlattice potential for the characteristic cases of (a) weak modulation, (b) an array of antidots, and (c) an array of dots. (From Ref. [2].)

spacer separates a metallic gate electrode from a two-dimensional electron system induced via modulation doping at an AlGaAs-GaAs interface. Figure 2 also displays a low-temperature capacitance voltage trace of such structure. At zero gate voltage the electron system is essentially two-dimensional with a weak potential modulation imposed by the modulated distance between the Schottky gate and the hetero-interface. With increasingly negative gate voltage the relative amplitude of the superlattice potential and hence the periodic density modulation of the electron system increase as sketched in Fig. 1(a) but the differential capacitance stays essentially constant as long as there remain electrons below all parts of the gate. The sudden decrease of the capacitance at lower gate voltage then reflects the formation of antidots (see also Fig. 1(b)). Further lowering of the gate potential finally depletes electrons except in the pockets of the superlattice potential and creates arrays of quantum dots (Fig. 1(c)) which are formed around the final step-like decrease of the gate capacitance. Here it is worth noting that homogeneous dot arrays will only form if laterally uniform electron depletion is made possible via some remaining electron conduction in the direction perpendicular to the electron plane [3]. This has to be assured by suitable layer design. Similarly lateral superlattices can also be defined in MOSFET-like devices on Si and InSb in which a patterned gate electrode is forward biased to induce electron superlattices [1].

In the following we will discuss how far-infrared spectroscopy can be used to study the dynamic response of such laterally modulated or confined electron systems. We will start with a review of the intraband infrared excitations in parabolically confined electron systems that are well understood. We then will discuss some recent experiments which

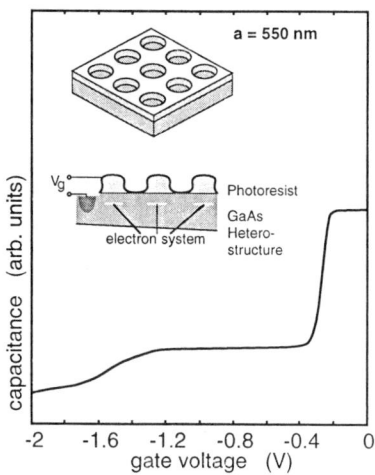

Figure 2. Capacitance-voltage trace at T = 4.2 K for a field-effect device on GaAs designed to investigate square antidot arrays of period a = 550 nm as sketched in the inset. (From Ref. [4].)

employ anisotropy of the confined electron systems and anharmonicity of the confining potential to achieve a more detailed understanding of the dynamic response of individual laterally confined electron systems as well as of strongly interacting periodic electron systems, i.e., true lateral superlattices.

2. Parabolically Confined Electron Systems

In strongly confining lateral superlattices such as arrays of quantum wires [5] and quantum dots [6] the dominant infrared excitation is now well understood as reflecting the collective center of mass motion of all electrons in the bare confining potential. Fig. 3(a) displays typical far-infrared transmission spectra of a square array of quantum dots each containing about 50 electrons as a function of magnetic field B applied perpendicularly to the plane of the superlattice [3]. The magnetic field lifts the two-fold degeneracy of the infrared mode at B = 0 and gives rise to a bulk-like high frequency mode and an edge-like low frequency mode. As illustrated in Fig. 3(b) the magnetic field dispersion of these modes is well described by:

$$\omega_\pm = \sqrt{\omega_0^2 + (\omega_c/2)^2} \pm \omega_c/2, \qquad (1)$$

where ω_0 denotes the B = 0 resonance frequency and $\omega_c = eB/m^*$ the cyclotron frequency. Note that for parabolically confined electrons this dispersion relation can be derived within a single particle model either classically [7] or quantum mechanically [8] or can be obtained from electrodynamics as the collective mode of a laterally confined plasma [9]. Surprisingly, first experimental studies of quantum dot arrays on InSb with very few electrons showed that the characteristic infrared frequency ω_0 is essentially independent of

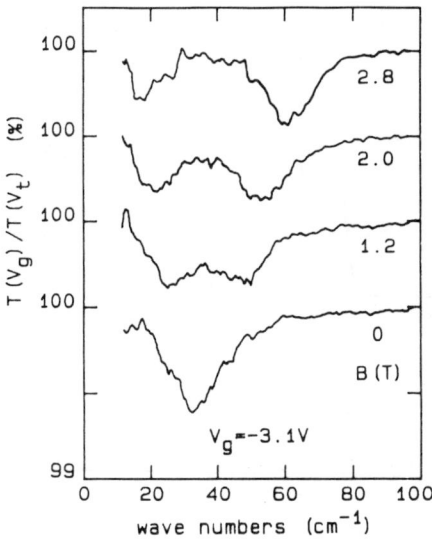 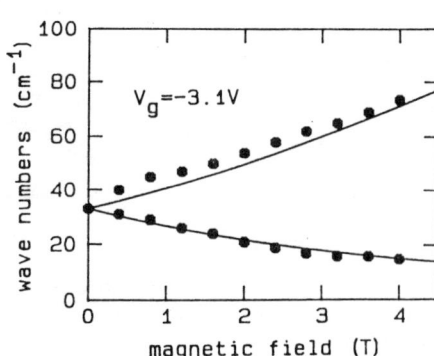

Figure 3. Magnetic field dependence of the far-infrared response of an array of quantum dots on GaAs (period a = 450 nm, electron number per dot $n_0 \approx 50$): (a) normalized transmission spectra at T = 2 K and (b) measured (dots) and calculated (solid line, Eq. (1)) resonance positions. (From Ref. [3].)

the number of electrons in the dot as illustrated in the inset of Fig. 4 [10]. This is now well understood as reflecting the generalized Kohn theorem initially introduced by Brey et al. [11] to explain the infrared modes of electrons confined in a quasi-three-dimensional parabolic quantum well. This theorem states that electrons that are confined in a bare parabolic confinement potential and driven by a spatially uniform high frequency field respond only at a resonance frequency that reflects the center of mass motion of all electrons and thus at B = 0 the curvature of the bare confining potential. For a fixed bare potential it therefore is independent of electron-electron interactions. This is strictly true if the bare confining potential is parabolic, independent of the dimensionality of the electron system [12]. In heterostructure devices in which mobile electrons are laterally confined by electrostatic potentials the bare potential is induced by external charges on the gate, on exposed surfaces, in the modulation doped donor layer, and in depletion regions. Since these external charges are mostly at distances larger than or comparable to the typical width of the confined quantum wire or dot the bare potential is rather smooth and can be well approximated by a parabola [13]. Since the positive and negative charges in these devices often outnumber the confined electrons it is also no longer surprising that the curvature of the bare confining potential and hence the frequency of the center of mass motion of the electrons is not strongly affected by the number of confined electrons as observed in quantum dots on InSb [6], Si, and GaAs [3,14].

Though the infrared excitations in wires and dots correspond to a collective motion of all carriers the center of mass motion can be well described in a single particle model if the

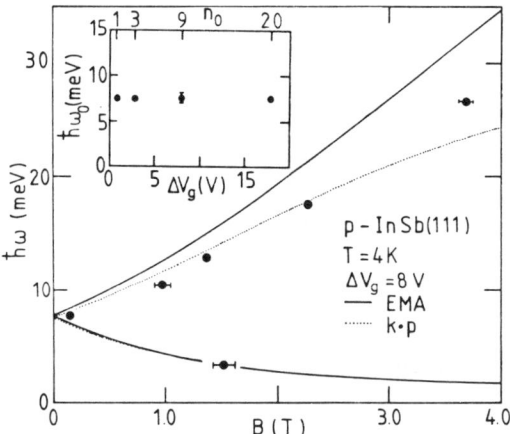

Figure 4. Magnetic field dependence of the infrared resonance energies in a quantum dot array on InSb (period a = 250 nm, dot diameter about 100 nm). The experimental data (dots) are compared with a calculated dispersion assuming a parabolic conduction band (solid line, Eq. (1)) and one including non-parabolicity (dotted line). The inset gives the dependence of the measured resonance energy at B = 0 on gate voltage above threshold and corresponding average electron number n_0 per quantum dot. (From Ref. [10].)

strength of the potential is adjusted to reflect the corresponding bare potential. A particularly fruitful approach to describe the infrared response within a single particle model has been suggested by Lorke [2,15]. He calculates the infrared response within a billiard-type model of ballistic motion in a suitably chosen potential. In this model initial positions and momentum directions of the electrons are chosen at random and a sufficiently large number of electron trajectories at constant total energy corresponding to the Fermi energy are calculated from the equations of motions for a time interval that roughly corresponds to the elastic scattering time expected from the mobility of the unperturbed 2DES. A Fourier analysis of the velocity components of the ballistic electronic motion then yields the infrared response. Figure 5 shows a spectrum calculated for a parabolic confinement potential with curvature chosen to match the experimentally observed B = 0 resonance position in Fig 3. This model not only yields resonance positions in excellent agreement with Eq. (1) but also describes the lineshape asymmetry observed in corresponding experimental spectra (see Fig. 3) between the upper and the lower mode better than previous classical [7] and quantum mechanical models [10]. The particular beauty of the billiard model is that it can be easily extended to non-parabolic potentials and yields qualitative description of experimental results for the fundamental modes that essentially reflect the center of mass motion of the electrons better than most other models as will be discussed below. As a single particle model it naturally cannot account for differences between the bare and the self-consistent potential that reflects the screening properties of the confined electrons.

Though the information on the curvature of the bare confining potential that is provided from infrared transmission experiments is a valuable result that cannot be easily extracted from magnetotransport studies it is highly desirable to also be able to extract information in

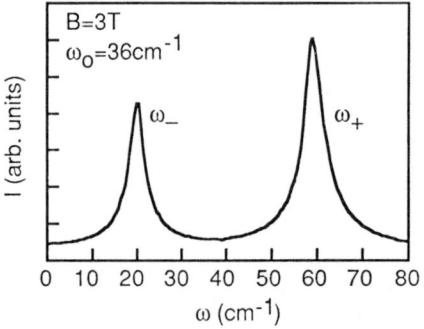

Figure 5. Calculated power spectrum of the infrared absorption for parabolically confined electron dots on GaAs in a magnetic field B. (From Ref. [15].)

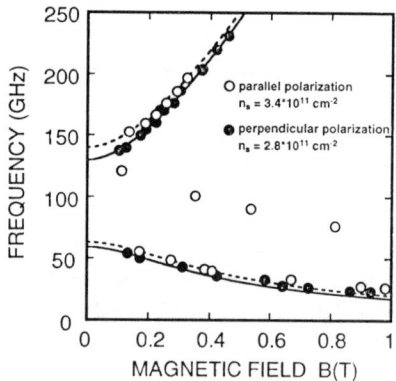

Figure 6. Measured resonance positions for an array of mesa-etched electron ellipses (dots) on GaAs for polarisation of the millimeter wave field parallel and perpendicular to the long axis. The dimensions of the long and short axes are, respectively, 116 µm and 40 µm. The lines give the calculated dispersion of the fundamental modes for parallel (----) and perpendicular (———) polarization, respectively, adjusted to fit the experimental frequencies near B = 0. (From Ref. [16].)

internal charge excitations in the confined electron systems from spectroscopic experiments. In fact initial spectroscopic experiments on quantum wires and quantum dots were motivated by the hope that they would excite internal charge oscillations in the laterally confined electron systems and would allow to spectroscopically determine subband and level spacings only little affected by collective depolarization effects. Now it is clear that this can only be achieved if the bare confining potential can be made strongly anharmonic or anisotropic. In addition it is desirable to separate the effects of intradot and interdot interactions to be able to fully understand the high frequency response in lateral superlattices. In the following we will discuss several experiments that use anisotropic as well as anharmonic confinement schemes to gain a better understanding of interelectronic interactions in lateral superlattices. Since for a parabolic confining potential there is no qualitative difference between the effects of classical confinement and quantum confinement on the infrared modes it is illuminating to first investigate the action of anisotropy and interdot interactions on classically confined electron systems as discussed below. The effects of anharmonic confinement on quantum wires and dots as well as strongly coupled superlattices will then be discussed in the subsequent paragraphs.

3. Classically Confined Dot Arrays

Classically confined electron dots with many electrons were the first systems in which the mode dispersion Eq. (1) has been experimentally observed [9]. Since it is much easier to prepare classically confined electron systems with widely varying shapes and on lattices of different geometry they are well suited to study the effects of shape anisotropy of individual dots as well interdot Coulomb interactions as illustrated in the following. When dots are no longer circular in shape it is natural that the mode degeneracy of the plasma excitation at $B = 0$ will be lifted. To study this effect quantitatively and to establish the magnetic field dispersion of the dominant infrared modes Dahl et al. [16] have studied the high frequency response of arrays of rather macroscopic ellipses defined by mesa etching in a modulation doped GaAs/AlGaAs heterostructure. At $B = 0$ two modes are observed which are polarized along the respective axes of the ellipsoids and reflect the center of mass motion of the electrons along these major axes of the bare confining potential. With increasing magnetic field the modes further separate in frequency as illustrated in Fig. 6. The lines in Fig. 6 represent the dispersion expected from a classical depolarisation model for a single ellipse with the $B = 0$ frequencies adjusted to fit the experiment. From the known dimensions of the ellipses and the two-dimensional ellectron density and with reasonable values for the electron effective mass and the background dielectric constants one calculates for a single ellipse frequencies at $B = 0$ that are within a few percent of the measured values. This implies that electronic interactions between adjacent ellipses are not apparent in the data of Fig. 6. In the ellipses one also observes an additional excitation that appears in the gap between the fundamental modes as shown in Fig. 6. This mode is excited primarily in polarization parallel to the long axis of the ellipses and is interpreted as the lowest dipole-excited harmonic mode that involves relative oscillations of the charge density along the long axis [17].

Interdot electrostatic interactions have recently achieved special attention as a possible mechanism for producing (anti-)ferroelectric phase transitions in quantum dot arrays [18–20]. Though experimental verification of such phase transitions has not yet been achieved it is valuable to devise experiments that yield a direct measure of such interdot interactions. For that purpose Dahl et al. [21] have designed a classically confined dot array consisting of circular electron disks placed on a rectangular lattice. Again the structure was defined by wet mesa etching on a modulation doped heterojunction and contains quasi-two-dimensional circular electron disks each 37 μm in diameter placed on a rectangular lattice of strongly differing periods of 40 μm and 80 μm, respectively. If there were no interdot interactions one would expect again the magnetic field dispersion of a single disk, i.e., Eq. (1). The effect of interdot interactions is directly visible in the measured dispersion displayed in Fig. 7, which is qualitatively different from Eq. (1) since it exhibits two distinct fundamental modes at $B = 0$. Near $B = 0$ the response is strongly anisotropic: depending on

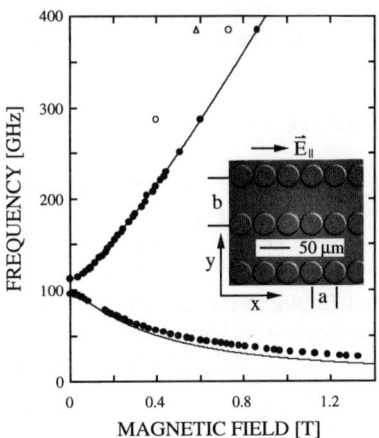

Figure 7. Measured resonance positions for an rectangular array of circular electron disks on GaAs with dimensions shown in the inset. Full dots denote the fundamental modes and open symbols are higher harmonics. The gap at B = 0 is a direct measure of the Coulomb interaction between disks. The solid line reflects the magnetic field dispersion of a dipolar model adjusted to coincide with the experimental resonance frequencies at B = 0. (From Ref. [21].)

the polarization of the incident radiation only one of the modes is visible. A detailed analysis within a dipolar coupling model yields a dispersion relation that is formally identical to the one of a single ellipsoid. If the B = 0 frequencies are adjusted to the experimental values this dispersion relation is given by the solid lines in Fig. 7. Within the accuracy with which one can determine the parameters that enter this model also the absolute prediction of the B = 0 frequencies is very good (better than 20%). The relative splitting $(\omega_x^2-\omega_y^2)/\omega_0^2$, where ω_0 denotes the frequency of the non-interacting dots is a direct measure of the ratio of interdisk to intradisk coupling. In the experiment in Fig. 7 it is 0.28 as compared to a calculated value of 0.23.

The above results demonstrate how anisotropy of the laterally confined electron system can be advantageously used to directly obtain information on electronic interactions. Though it is unlikely that the classically derived theories that have been used to explain the above experiments can be applied to quantum dot systems quantitatively one can expect them to be at least a reliable qualitative guideline. In particular should it be possible to infer from the infrared resonance frequencies of quantum dots and quantum dashes placed on various suitably chosen anisotropic lattices on the importance of intradot to interdot Coulomb interactions. A pronounced softening of a low frequency infrared mode in quantum dot arrays would then indicate the vicinity of a critical point for a ferroelectric phase transition.

4. Nonparabolic Confinement in Quantum Wires

In all studies reported on narrow quantum wires defined electrostically [5] or by deep mesa etching on modulation-doped GaAs/AlGaAs heterojunctions [22] the infrared spectra show only the fundamental center of mass mode that in a perpendicularly applied magnetic field shifts as:

$$\omega(B) = \sqrt{\omega_0^2 + \omega_c^2}.\qquad(2)$$

This implies that the bare confining potential is parabolic in very good approximation. To be able to excite internal electronic excitations it is necessary to introduce sufficiently strong non-parabolic terms in the bare confining potential [23]. In usual modulation doped heterostructures this is difficult to achieve since the modulation doped layer is separated from the high mobility electron system by a spacer such that the electrostatic gates that establish the lateral confinement are separated from the interface at which the quantum wire is formed by distances that are not much smaller than achievable wire widths. Similarly in deep mesa etched wires the lateral surfaces that cause the confinement are separated from the quantum wire by depletion regions. To define bare confing potentials that can be nonparabolic but still create highly mobile quantum wires it is therefore desirable to use devices similarly to MOSFETS in which narrow gates little separated from the heterointerface are forward biased to define quantum wires or dots. Such a device design also makes it possible to use a stacked gate technology that already has been very succesfully used on Si MOS devices to realise widely tuneable quantum wires and quantum dots [23,24].

Recently we have fabricated MISFET devices on GaAs/AlGaAs heterostructures in which narrow quantum wires can be realized by forward bias [25,26]. The layer sequence of these devices is as follows: On a semi-insulating substrate with suitable buffers a thin n^+-doped GaAs layer with typical sheet resistance of 1 kΩ is grown as a back electrode. This is followed by an undoped GaAs spacer typically 1μm thick, a short period AlAs/GaAs superlattice of total thickness of few 10 nm serving as gate insulator and a thin GaAs cap layer. On this a semitransparent grating gate is prepared which is biased with respect to the back electrode. Without bias this device contains no mobile electrons at the GaAs/AlGaAs interface. Above a positive threshold voltage electrons are injected from the back electrode to the heterointerface below the grating gate and form narrow quantum wires.

Figure 8(a) shows typical far-infrared transmission spectra of such a device with a grating gate of period a = 350 nm at T = 2.2 K with and without a magnetic field B applied perpendicularly to the interface [26]. The lowest frequency dimensional resonance reflects the center of mass motion of the laterally confined electrons and has an effective wave vector of q = $2\pi/a$. At B = 0 a higher order dimensional resonance is observed that splits in a finite magnetic field as indicated by the arrows in Fig. 8(a). This mode is the lowest dipole-active harmonic mode at effective wave vector q = $6\pi/a$ and involves a relative motion of the confined electrons [17]. Its presence shows that the bare confining potential now contains anharmonic terms. In a magnetic field this higher harmonic mode exhibits strong anticrossing with the harmonic cyclotron resonance at $2\omega_c$, as visible in the finite B

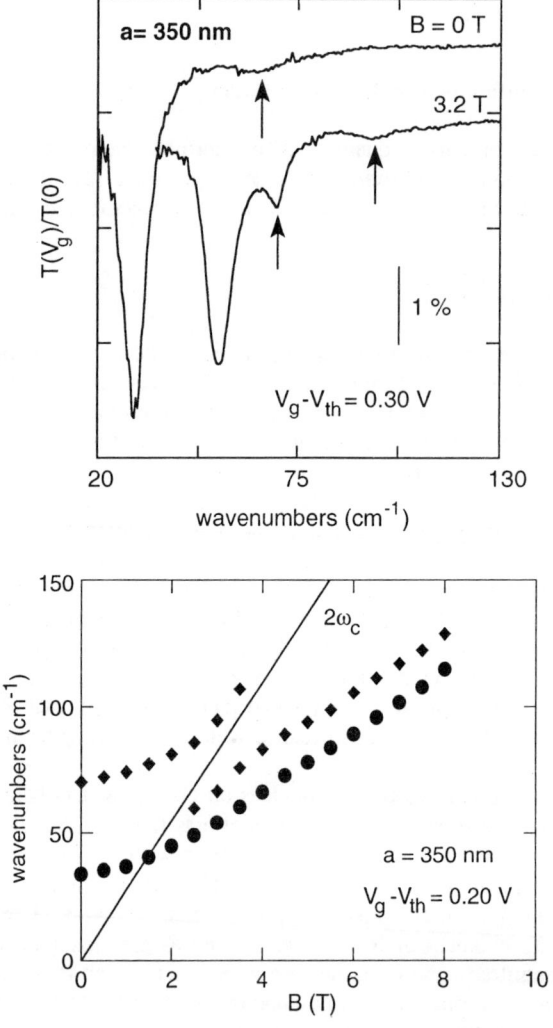

Figure 8. (a) Transmission spectra of a quantum wire array with period a without and with a magnetic field B applied perpendicularly to the plane of the array. (b) Magnetic field dispersion of the fundamental (dot) and lowest harmonic (diamond) mode. (From Ref. [26].)

spectrum in Fig. 8(a) and in the dispersion in Fig. 8(b). This anticrossing is very similar to an anticrossing that occurs in two-dimensional systems whenever the plasmon wavelength becomes comparable to the cyclotron diameter [27,28]. As expected the uniform fundamental mode does not exhibit such an anticrossing that is characteristic for nonuniform exitations. Similar anticrossing has also been observed for one-dimensional plasmons, i.e., charge density excitations with finite wave vector along the wire [22].

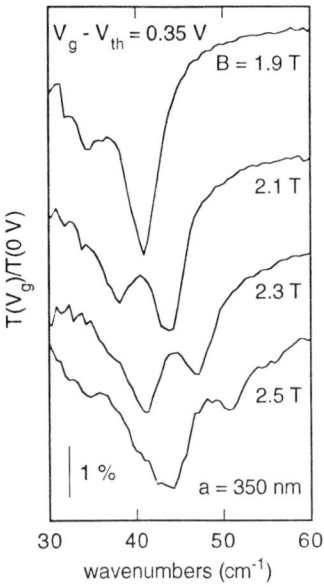

Figure 9. Transmission spectra of a quantum wire array at an electron density and magnetic fields where the fundamental electron resonance exhibits an unexpected pronounced anticrossing. (From Ref. [26].)

A completely unexpected splitting of the fundamental mode is however observed at somewhat higher electron densities in the wires, as shown in Fig. 9. This splitting which behaves as an anticrossing is consistently seen at higher electron densities in wire arrays of different periods. Since neither semiclassical [17] nor quantum mechanical [29] theories of the collective modes in electron wires have so far predicted such an anticrossing we have used the billiard model outlined above to calculate the magnetic field dependence of power spectra of the fundamental mode. Once we include nonparabolic terms in the confining potential we find an anticrossing in the power spectra that is quite similar to the experimental observations. From this model it appears that the splitting occurs when the extend of the classical motion perpendicular to the axis of the wire as calculated with harmonic confinement exceeds a critical width controlled by the strength of the anharmonic terms in the confining potential. Here the billiard model that we expect to give physical results for the fundamental mode even in non-parabolic potentials yields an insight into details of the confining potential that cannot be obtained yet from more sophisticated theories. This will be further illustrated on coupled quantum dots and antidots in the following section.

5. Excitations in Two-Dimensional Lateral Superlattices

After a fairly good understanding is achieved of the infrared electronic exciations in arrays of strongly confined electron systems, i.e., wires and dots, we now want to discuss our

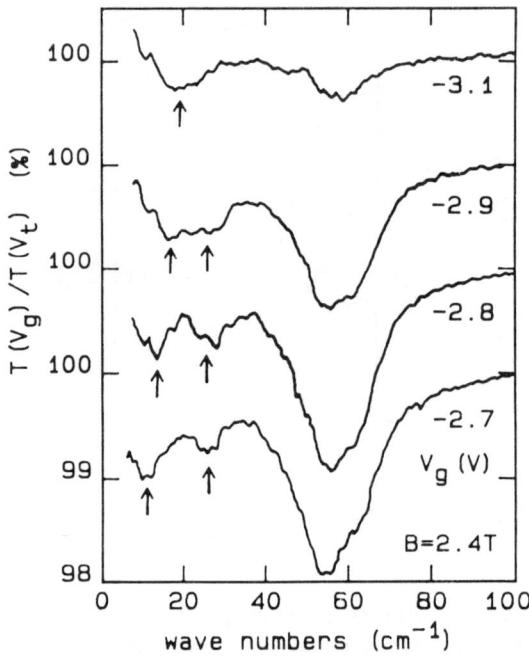

Figure 10. Transmission spectra of a square array (period a = 450 nm) of electron dots on GaAs at magnetic field B = 2.4 T and gate voltages where the dots change from being isolated (-3.1 V) to weakly coupled. Two edge-like modes (arrows) appear in the coupled case. (From Ref. [3].)

present understanding of strongly coupled situations on the basis of square superlattices as sketched in Fig. 1 spanning the whole range between weakly coupled quantum dot superlattices via antidot superlattices to the case of only moderately modulated superlattice potentials. First spectroscopic studies of such real superlattice effects were carried out by Lorke et al. [3] on electrostatically defined quantum dot superlattices in which only a small change of the gate bias resulted in changing a practically uncoupled situation into a coupled one. The uncoupled case was illustrated above in Fig. 3. On the same sample and under otherwise identical measurement conditions only a small increase in the bias voltage resulted in arrays of electrically coupled quantum dots as evidenced by the capacitance voltage trace and an essential change of the infrared spectra in a sufficiently strong magnetic field as displayed in Fig. 10. In the coupled case two additional modes appear, one with frequency below the edge mode of the individual quantum dots (see arrows in Fig. 10) and one as a shoulder on the high magnetic field side of the bulk-like mode. Figure 11 shows the magnetic field dispersion of these modes for a gate voltage close to where the coupling starts. The solid line again corresponds to Eq. (1) and describes well the two dot-like modes that persist in the weak coupling case. Within a classical model that describes the low frequency mode of the dot as an edge magnetoplasmon the new low frequency mode for the coupled case is well described as an edge magnetoplasmon with a

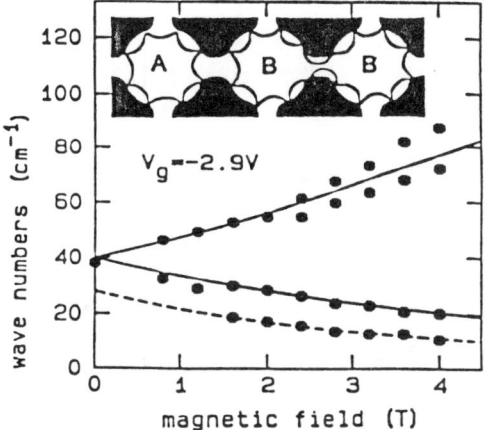

Figure 11. Magnetic field dependence of the infrared resonance positions in a coupled quantum dot array on GaAs as measured from spectra as in Fig. 10 (dots) and expected for the A-type dot modes (solid line, Eq. (1)) and the B-type edge-like mode (dashed line). (From Ref. [3].)

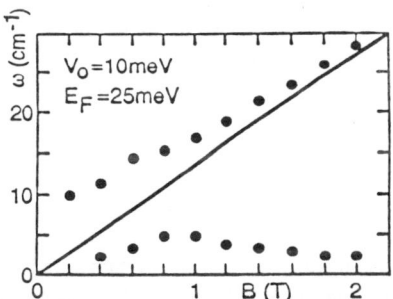

Figure 12. Eigenfrequencies calculated in a semiclassical trajectory model for a square antidot array with sinusoidal potential on GaAs (dots) in comparison to the cyclotron frequency ω_c (solid line). (From Ref. [2].)

perimeter double that of the individual dot mode as indicated in the trajectory B of the inset. The predicted dispersion for such a mode is given by the dashed line in Fig. 11 and is in surprisingly good agreement with the data. This mode involves two ballistic transmission events through the constriction between adjacent dots and is unique insofar as all other trajectories involve at least four such transmissions. With increasing gate voltage, i.e., decreasing modulation of the superlattice potential, the mode at highest field becomes the two-dimensional magnetoplasmon of wave vector $q = 2\pi/a$ and thus is a characteristic phonon-like excitation of the coupled dot lattice.

So far there exist no proper quantum theory that can correctly describe the modes that

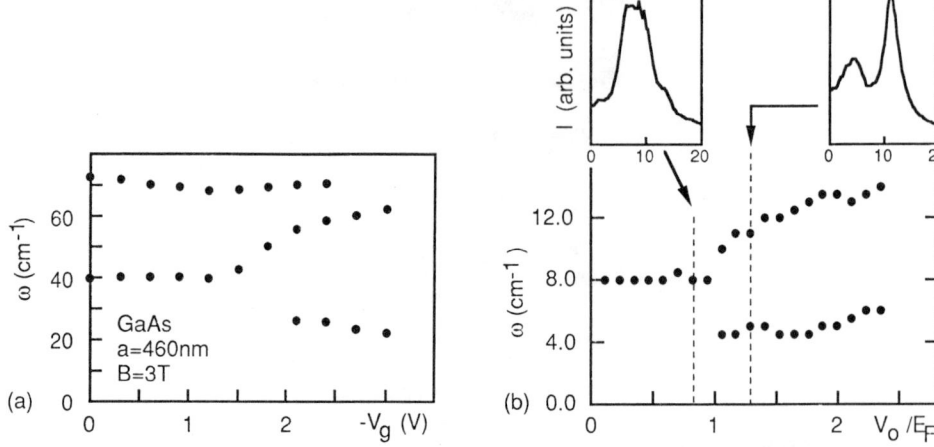

Figure 13. (a) Measured resonance positions of fundamental modes and the energetically higher magnetoplasmon mode at wave vector $2\pi/a$ in a square lateral superlattice on GaAs. (b) Calculated resonance positions of the fundamental modes for a square lateral superlattice as a function of relative potential modulation V_0/E_F. The parameters are chosen to reflect a typical superlattice on GaAs in a sufficiently strong magnetic field. A magnetic field of B = 0.6 T is chosen in the calculation to approximate the ratio of the finite B frequencies and the B = 0 frequency found in the experiment. (From Ref. [15].)

appear with the coupling whereas the semiclassical trajectory model outlined above can at least describe the observed resonance positions quite satisfactorily. Basically the same is found for other superlattice situations. In antidot superlattices two fundamental infrared modes are observed that have a magnetic field dispersion quite different from those of uncoupled or coupled dot arrays [4,30]. The only model that so far has been able to qualitatively reproduce the observed dispersion is the ballistic billiard model already used above. Figure 12 displays the resonance positions determined by Lorke [2] for an antidot superlattice within the billiard model. Qualitatively this dispersion behaves as observed by Kern et al. [30]. Since the model neglects electron-electron interactions and thus the difference between the bare and the self-consistent, partially screened confining potential we do not expect quantitative agreement with the experiment. It should be noted that in antidot superlattices where semiclassical trajectories can become chaotic [31] also the low-field magnetotransport properties [4,32–34] are surprisingly well explained by semiclassical billiard models [35,36].

In quadratic lateral superlattices and in sufficiently high magnetic fields also the experimentally observed mode dispersion in its dependence on potential modulation is surprisingly well reproduced by the semiclassical trajectory model [15]. This is illustrated in Fig. 13. Figure 13(a) displays the experimentally observed mode dispersion in a quadratic superlattice on GaAs as one tunes the gate voltage from the two-dimensional case on the right hand side through the antidot situation into the quantum dot case, essentially following the scenario sketched in Fig. 1. Note that the same qualitative behavior has been found on quadratic superlattices on Si employing two stacked gates [15]. In Fig. 13(b) the resonance positions as predicted by the trajectory model are plotted as a function of the ratio of the peak to peak amplitude of the superlattice potential to the Fermi energy. The insets also display two calculated power spectra, one just prior to the formation of antidots ($V_0/E_F < 1$) and one in the regime of antidots. Qualitatively the trajectory model again reproduces the experimental observation quite well and is at present the only model that has been used successfully for the whole range of relative superlattice modulation. It appears that the trajectory model yields a good qualitative description for many superlattice potentials whenever we consider the fundamental infrared excitations. This can be understood as that the uniform modes largely reflect the center of mass motion of the electron system even when the superlattice potential is far from parabolic. Higher order modes, however, that involve relative electron motion then are not expected to be predicted by the billiard model. This in fact is the case for all situations studied so far.

6. Conclusions

The above discussed examples show that lateral superlattices that can presently be defined electrostatically on heterojunctions exhibit electronic properties in the static as well dynamic conductivity that are largely dominated by ballistic semiclassical electron motion. Effects of lateral quantization that are clearly seen in electrostatically defined point contacts of similar lateral dimensions than the superlattices discussed here are not yet convincingly seen in lateral superlattices. In superlattices it appears that averaging over phase shifts of the electron wave that may be caused by fluctuations in the superlattice pattern definition surpresses quantization phenomena. Nevertheless progress is made in the understanding of interaction phenomena in lateral superlattices with the use of infrared spectroscopy. Dipolar interdot Coulomb interaction can now be extracted from infrared spectroscopy. Intradot excitations can be made visible by tailoring non-parabolic lateral confinig potentials. Spectroscopy of quantum dot atoms is starting to yield insight into their internal electronic structure [37]. Manifestations of ballistic superlattice transport that are clearly seen in magnetoresistance studies start to appear in infrared spectroscopy. A strong challenge for future work will be to experimentally access the transition from quasi-classical behavior in lateral superlattices to miniband transport and spectroscopy. However, one needs to further reduce periods and improve homogeneity of lateral superlattices to be able to observe some of the fascinating quantum features that have long been predicted for lateral superlattice electrons in a magnetic field [38,39].

7. Acknowledgments

This review summarizes recent work from several collaborative studies. For their most valuable contributions I wish to thank in particular Claus Dahl, Hermann Drexler, Wolfgang Hansen, and Axel Lorke. I also would like to thank Johann Alsmeier, Steve Beaumont, Frank Brinkop, Alik Chaplik, Klaus Ensslin, Theo Geisel, Detlef Heitmann, Rolf Landauer, Bert Lorenz, Ulrich Merkt, Klaus Ploog, David Wharam, Günther Weimann, Rainer Winkler, and Achim Wixforth for many stimulating discussions. The financial support of the Deutsche Forschungsgemeinschaft, the ESPRIT Basic Research action, and the Volkswagen-Stiftung is gratefully acknowledged.

References

1. W. Hansen, U. Merkt, and J.P. Kotthaus, in *Semiconductors and Semimetals*, edited by R.K. Williardson, A.C. Beer, and E.R. Weber (Academic Press, San Diego, 1992), Vol. 35, p. 279.
2. A. Lorke, Surf. Sci. **263**, 307 (1992).
3. A. Lorke, J.P. Kotthaus, and K. Ploog, Phys. Rev. Lett. **64**, 2559 (1990).
4. A. Lorke, J.P. Kotthaus, and K. Ploog, Superlattices and Microstructures **9**, 103 (1991).
5. W. Hansen, M. Horst, J.P. Kotthaus, U. Merkt, Ch. Sikorski, and K. Ploog, Phys. Rev. Lett. **58**, 2586 (1987).
6. Ch. Sikorski and U. Merkt, Phys. Rev. Lett. **62**, 2164 (1989).
7. B.A. Wilson, S.J. Allen, Jr., and D.S. Tsui, Phys. Rev. B **24**, 5887 (1981).
8. V. Fock, Z. Phys. **47**, 446 (1928).
9. S.J. Allen, Jr., H.L. Störmer, and J.C.M. Hwang, Phys. Rev. B **28**, 4875 (1983).
10. Ch. Sikorski and U. Merkt, Surf. Sci. **229**, 282 (1990).
11. L. Brey, N.F. Johnson, and B.I. Halperin, Phys. Rev. B **40**, 10647 (1989).
12. P.A. Maksym and T. Chakraborty, Phys. Rev. Lett. **65**, 108 (1990).
13. S.E. Laux, D.J. Frank, and F. Stern, Surf. Sci. **196**, 101 (1988).
14. T. Demel, D. Heitmann, P. Grambow, and K. Ploog, Phys. Rev. Lett. **64**, 788 (1990).
15. A. Lorke, I. Jejina, and J.P. Kotthaus, Phys. Rev. B **46**, 12845 (1992).
16. C. Dahl, F. Brinkop, A. Wixforth, J.P. Kotthaus, J.H. English, and M. Sundaram, Solid State Commun. **80**, 673 (1991).
17. G. Eliasson, J.-W. Wu, P. Hawrylak, and J.J. Quinn, Solid State Commun. **60**, 41 (1986).
18. K. Kempa, D.A. Broido, and P. Bakshi, Phys. Rev. B **43**, 9343 (1991).
19. P. Bakshi, D.A. Broido, and K. Kempa, J. Appl. Phys. **70**, 5150 (1991).
20. A.V. Chaplik and L. Ioriatti, Surf. Sci. **263**, 354 (1992).
21. C. Dahl, J.P. Kotthaus, H. Nickel, and W. Schlapp, Phys. Rev. B **46**, 15590 (1992).
22. T. Demel, D. Heitmann, P. Grambow, and K. Ploog, Phys. Rev. Lett. **66**, 2657 (1991).
23. J. Alsmeier, E. Batke, and J.P. Kotthaus, Phys. Rev. B **40**, 12 574 (1989).
24. J. Alsmeier, E. Batke, and J.P. Kotthaus, Phys. Rev. B **41**, 1699 (1990).
25. H. Drexler, W. Hansen, J.P. Kotthaus, M. Holland, and S.P. Beaumont, Semicond. Sci. Technol. **7**, 1008 (1992).
26. H. Drexler, W. Hansen, J.P. Kotthaus, M. Holland, and S.P. Beaumont, Phys. Rev. B **46**, 12849 (1992).
27. E. Batke, D. Heitmann, J.P. Kotthaus, and K. Ploog, Phys. Rev. Lett. **54**, 2367 (1985).
28. A.V. Chaplik and D. Heitmann, J. Phys. C **18**, 3357 (1985).
29. Q.P. Li and S. Das Sarma, Phys. Rev. B **44**, 6277 (1991).
30. K. Kern, D. Heitmann, P. Grambow, Y.H. Zhang, and K. Ploog, Phys. Rev. Lett. **66**, 1618 (1991).
31. T. Geisel, J. Wagenhuber, P. Niebauer, and G. Obermair, Phys. Rev. Lett. **64**, 1581 (1991).
32. K. Ensslin and P.M. Petroff, Phys. Rev. B **41**, 12307 (1990).
33. D. Weiss, M.L. Roukes, A. Menschig, P. Grambow, K. von Klitzing, and G. Weimann, Phys. Rev. Lett. **66**, 2790 (1991).
34. K. Ensslin, K.T. Häusler, C. Lettau, A. Lorke, J.P. Kotthaus, A. Schmeller, R. Schuster, P.M. Petroff, M. Holland, and K. Ploog, in *Low Dimensional Physics: New Concepts*,

edited by G. Bauer, F. Kuchar, and H. Heinrich (Springer, Berlin, 1993), in press.
35. A. Lorke, J.P. Kotthaus, and K. Ploog, Phys. Rev. B **44**, 3447 (1991).
36. R. Fleischmann, T. Geisel, and R. Ketzmerick, Phys. Rev. Lett. **68**, 1367 (1992).
37. B. Meurer, D. Heitmann, and K. Ploog, Phys. Rev. Lett. **68**, 1371 (1992); D. Heitmann, these proceedings.
38. D. Langbein, Phys. Rev. **180**, 633 (1969).
39. D.R. Hofstadter, Phys. Rev. B **14**, 2239 (1976).

OPTICAL SPECTROSCOPY OF QUANTUM WIRES AND DOTS

A.S. PLAUT,[1*] K. KASH,[1] E. KAPON,[1] B.P. VAN DER GAAG,[1]
A.S. GOZDZ,[1] D.M. HWANG,[1] E. COLAS,[1] J.P. HARBISON,[1] L.T. FLOREZ,[1]
H. LAGE,[2] P. GRAMBOW,[2] D. HEITMANN,[2] K. VON KLITZING,[2]
AND K. PLOOG[2]
[1]*Bellcore, Red Bank, NJ 07701-7040, USA*
[2]*Max-Planck-Institut für Festkörperforschung, 7000 Stuttgart 80, Germany*

Molecular beam epitaxy (MBE) has enabled the realization of two-dimensional (2D) structures with sharp interfaces and high electron mobilities producing much novel and unexpected physics. Recently, interest has turned to reducing the dimensionality still further in order to discover new or enhanced optical and transport phenomena.

Various different methods have been used to produce quantum wires and dots. Here we discuss optical experiments performed on a modulation-doped quantum wire system, an undoped quantum dot system, and an undoped quantum wire system. Each system has been fabricated by a different microstructuring technique: deep mesa etching, strain patterning, and growth on non-planar substrates, respectively. This therefore records the first step towards a longer term aim to compare and contrast similar systems produced by different techniques.

The energy spectra of isolated one-dimensional (1D) electron channels produced by deep mesa-etching [1] of AlGaAs–GaAs heterojunctions (where the active layers have been etched through to produce completely isolated 1D channels) have been studied using magneto-photoluminescence (PL) techniques. Electron recombination with holes bound to a δ-layer of acceptors in the GaAs produce a PL spectrum *directly* reflecting the density of states below the Fermi level. We attained the 1D quantum limit and in finite magnetic field measured the quadratic dispersion of the 1D subband. We thereby deduced typical values of 1D confinement energies, $\hbar\omega_0$ = 3–5 meV, compatible with channel widths of 30–40 nm [2].

There have been many predictions that the non-linear excitonic behaviour found in quantum well systems will be enhanced when the exciton is further confined to one or zero dimensions [3–6]. Here we report the first optical absorption saturation measurements on quantum dots. Excitonic PL from quantum dots produced by strain-patterning a AlGaAs–GaAs quantum well structure [7] was found to saturate at incident intensities more than an order of magnitude smaller than that required for quantum wells [8]. This enhanced saturation is a direct result of the reduced dimensionality of the system.

Finally, we discuss magneto-photoluminescence excitation (PLE) spectra which have been measured on AlGaAs–GaAs undoped quantum wires grown by organometallic chemical vapour deposition on V-grooved substrates [9]. In contrast to the above, these wires are formed *in situ* rather than by post-growth processing. This has the advantage of producing wires with interfaces of comparable quality to that currently available in quantum well structures. These *uniform* crescent-shaped quantum wires have subband separations up to 45 meV in the conduction band, corresponding to electronic widths of 10 nm. The PLE spectra exhibit enhanced absorption at the quantum wire subbands. We have found these PLE peaks to shift quadratically with increasing magnetic field; the size of the shift depends on the index of the subband transition, with the ground state transition appearing nearly independent of field up to 7 T.

Clearly, the study of these three systems has confirmed not only that the individual fabrication techniques are each successful in producing systems of one or zero dimensions, but that the physics of these systems does reflect this reduction in the dimensionality.

References

*   Exeter University, Exeter EX4 4QL, UK.
1. T. Demel, D. Heitmann, P. Grambow, and K. Ploog, Appl. Phys. Lett. **53**, 2176 (1988).
2. A.S. Plaut, H. Lage, P. Grambow, D. Heitmann, K. von Klitzing, and K. Ploog, Phys. Rev. Lett. **67**, 1642 (1991).
3. S. Schmitt-Rink, D.A.B. Miller, and D.S. Chemla, Phys. Rev. B **35**, 8113 (1987).
4. G.W. Bryant, Phys. Rev. B **37**, 8763 (1988).
5. H. Sakaki, K. Kato, and H. Yoshimura, Appl. Phys. Lett. **57**, 2800 (1990).
6. S. Benner and H. Haug, Europhys. Lett. **16**, 579 (1991).
7. K. Kash, D.D. Mahoney, B.P. Van der Gaag, A.S. Gozdz, J.P. Harbison, and L.T. Florez, J. Vac. Sci. Technol. **10**, 2030 (1992).
8. D.A.B. Miller, D.S. Chemla, D.J. Eilenberger, P.W. Smith, A.C. Gossard, and W.T. Tsang, Appl. Phys. Lett. **41**, 679 (1982).
9. E. Kapon, D.M. Hwang, and R. Bhat, Phys. Rev. Lett. **63**, 430 (1989).

OPTICAL PROPERTIES OF SEMICONDUCTOR QUANTUM WELL WIRES

J.M. CALLEJA,[1] A.R. GOÑI,[2] J.S. WEINER,[2] B.S. DENNIS,[2] A. PINCZUK,[2]
L.N. PFEIFFER,[2] AND K.W. WEST[2]
[1]*Dept. of Materials Physics, University Autónoma, Cantoblanco, 28049 Madrid, Spain*
[2]*AT&T Bell Laboratories, 600 Mountain Avenue, Murray Hill, N.J. 07974, USA*

ABSTRACT. Modulation doped quantum well wires have been fabricated in the one-dimensional (1D) quantum limit, where only the first 1D band is occupied. Novel properties have been found in the optical spectra. Strong Fermi edge singularities are present which markedly differ from the 2D case due to the hole recoil and phase space filling properties unique to 1D. The elementary excitations of the 1D electron gas have also been investigated, and the 1D intraband plasmon dispersion relation has been measured. Very good quantitative agreement is found with the linear dispersion expected for 1D systems.

1. Introduction

The dynamical behavior of an electron gas whose motion is restricted to one-dimensional (1D) channels is the object of increasing interest in recent years. In 1D the electron (or hole) can scatter only forwards or backwards along the channel, so that novel transport properties are expected. This, together with specific aspects of the elementary electron excitations, phase space filling, and collective phenomena in 1D are also at the origin of new and interesting optical properties [1–10]. Among them of special importance are the sharp peaks appearing at the Fermi level in the optical spectra, known as Fermi edge singularities (FES) and the behavior of 1D plasmons.

The effects of low dimensionality on the optical properties of the one-dimensional electron gas (1DEG) are expected to be more distinct in the 1D quantum limit, where only the lowest 1D band is occupied by electrons. Experiments in the 1D quantum limit have been reported recently in modulation doped GaAs quantum well wires (QWWs) 300 Å in width [8–10].

The FES appearing at low temperatures in an electron gas is produced by the collective response of the electrons to the optical excitation. It has been extensively studied theoretically [11–15] and can be described as a bound state (the Mahan exciton) between the hole created (destroyed) in the absorption (emission) process and the Fermi sea. The dynamic response of the Fermi sea to the hole potential softens the Mahan exciton, leading to the well known power-law shape of the FES [12,14] observed in the optical spectra at low temperatures where the Fermi distribution has a sharp cutoff at the Fermi energy. At higher temperatures the Fermi distribution smears out, so that the coherence of the response is lost and the FES disappears.

In two-dimensional (2D) systems FES have been observed in the photoluminescence excitation (PLE) spectra of GaAs/AlGaAs modulation-doped multiple quantum wells (MQWs) [16–23] and in InGaAs MQWs [24,25]. FES are not usually observed in photoluminescence (PL) emission because of the lack of holes at the Fermi wave vector k_F, but clear FES have been observed in PL when the hole is localized by dopant impurities in the QW [23] or by composition fluctuations in ternary compounds [24]. If an empty conduction band is close to the Fermi level the coupling by Coulomb interaction between the occupied band at k_F and the empty one at $k = 0$ also produces a sharp peak in the PL spectrum at the Fermi level [25]. From the theoretical point of view [26–29] it is clear that FES are observed in emission spectra only if holes are localized ($m_h^* \to \infty$), whereas FES are present in absorption also for mobile holes if the Fermi energy is smaller than $m_h^*/m_e^* E_M$, where m_h^* and m_e^* are the hole and electron effective masses and E_M is the binding energy of the Mahan exciton [26].

Compared to 2D, 1D systems have stronger excitonic effects due to their singular density of states, as well as specific phase space filling characteristics, because there are only two electronic states ($\pm k$) for each energy value $E(k)$. In consequence, much stronger FES are expected in 1D systems, as the hole recoil processes smear out the Fermi distribution in a 2DEG with sufficiently high electron concentration, but do not in a 1D system. On the other hand the study of carrier dynamics in semiconductor QWWs has one practical difficulty, the damage produced during the patterning process, which can produce fluctuations in the 1D potential which may localize the holes. In our recent results, reviewed here, special care was taken to minimize the patterning damage, while reducing the QWW size to as small as ~ 300 Å, where the 1D limit is attainable for reasonable electron concentrations [8]. From the theoretical view point, 1D-FES in the absorption spectrum have been studied first by exact diagonalization of a Hubbard chain using an on-site electron-hole Coulomb interaction [8]. This calculation is in qualitative agreement with experimental results and shows that, in contrast with 2D, 1D-FES can occur even if the holes are not localized and the electron concentration is high. More recent results have investigated the role played by disorder, impurity scattering and hole localization in the survival of the 1D-FES against low energy-single particle or collective excitations of the Fermi sea [31–35]. Hawrylak's calculations [32] showed that hole localization by potential fluctuations and/or impurity scattering is needed to observe the 1D-FES in emission spectra. Also impurity scattering has been invoked as the damping mechanism for low energy plasmons, which otherwise would destroy the Fermi surface and henceforth the FES [33]. Rodríguez and Tejedor [34] have shown that the use of a statically screened Coulomb interaction instead of the on site potential between the electrons and the hole, and hole wave-functions extended along the wire, results in the absence of FES unless QWWs are extremely narrow or the hole mass is very high. Finally the work by Ogawa [35] based on the Tomonaga-Luttinger model indicates that the critical exponent of the power law (i.e., the sharpness of the FES) is independent of the hole mass in 1D, which, as also stated in Refs. [8] and [31], is in striking contrast with the 2D and three-dimensional (3D) cases.

Single particle and collective excitations of the Fermi sea with finite energy can also be excited by interaction with light in low dimensional systems. Inelastic light scattering is a powerful method for the investigation of elementary excitations of the electron gas, because the energy and wave-vector dispersion of the excitations can be measured. This technique was used by Egeler et al. [6] to study plasmons in systems with many occupied 1D bands where the mode dispersions exhibit features predicted for a spatitially periodic 2DEG [37]. Clear signatures of 1D behavior are expected in the 1D quantum limit, as predicted by the few theoretical studies on the energies and dispersions of elementary excitations in QWWs in this limit [38–41]. This specific 1D behavior has been confirmed experimentally [10].

In this paper we review our results on optical measurements of the FES singularity and elementary excitations of the electron gas in the 1D quantum limit. The paper is organized as follows. In Sec. 2 we present experimental methods used to prepare QWWs supporting a 1DEG in the quantum limit. Section 3 describes the main results on absorption and emission of light by the 1DEG and their discussion in the light of the existing theoretical models. In Sec. 4 we present the energy-wavevector dispersion relations for elementary excitations of the 1DEG and their comparison with theoretical RPA predictions. Section 5 presents the main conclusions.

2. Experimental Methods and 1D Band Structure

Quantum well wires were fabricated starting from a 250 Å modulation-doped GaAs QW as described in Ref. [5]. The QW had 500 Å and 800 Å thick GaAlAs barriers below and above, respectively. Two monolayers of Si were included in the top barrier to create a 2DEG in the top interface of the QW. The initial electron concentration and low-temperature mobility of the 2DEG were $n = 3.2 \times 10^{11}$ cm^{-2} and $\mu = 1.1 \times 10^6$ cm^{2}V^{-1}s^{-1}, respectively.

The QWWs were patterned by electron beam lithography using a 25 keV, 100 pA beam on a 2000 Å thick PMMA resist, producing a periodic array of 1000 Å wide grooves separated by 1000 Å PMMA lines. The typical width fluctuation of the grooves is about 10%. Two thin layers of chromium were evaporated at ±60° angles onto the PMMA pattern, to protect it during further sample treatment, while leaving the bottom of the grooves uncovered. The pattern was finally bombarded with argon or oxygen ions at 300 V accelerating voltage and between 1 and 2×10^{-4} Acm^{-2} current density. This produces a partial depletion of the donor layers below the grooves without introducing damage at the depth of the electron gas. As a result, the electron density and therefore the electrostatic potential become modulated, and electrons and holes are confined in spatially separated 1D regions [5], as shown in Fig. 1(a). The ion bombardment produces some lateral depletion as well. As shown in Fig. 1(b), longer bombardment times increase the depletion depth and therefore the 1D confining potential, but also reduce the QWW width from L_2 to L_1. As both magnitudes cannot be controlled independently, a careful choice of the bombardment dose is needed to obtain the Fermi energy and interband spacing compatible with the 1D quantum limit. Figure 1(c) shows the band structure of QWWs and barriers parallel to the growth direction (z). Figure 2 presents the 1D band structure in the growth plane across the QWW (left) and the band dispersion along the wires (right) for a single occupied band. In the parabolic approximation, the Fermi energy E_F is related to the low energy thresholds for absorption (ε_F) and emission (ε_g) by [5]:

$$E_F = (\varepsilon_F - \varepsilon_g)\left(1 + \frac{m_e^*}{m_h^*}\right)^{-1}. \tag{1}$$

The mass ratio used in Eq. (1) is 0.15 [16].

The optical measurements were performed using an LD-700 dye laser as the

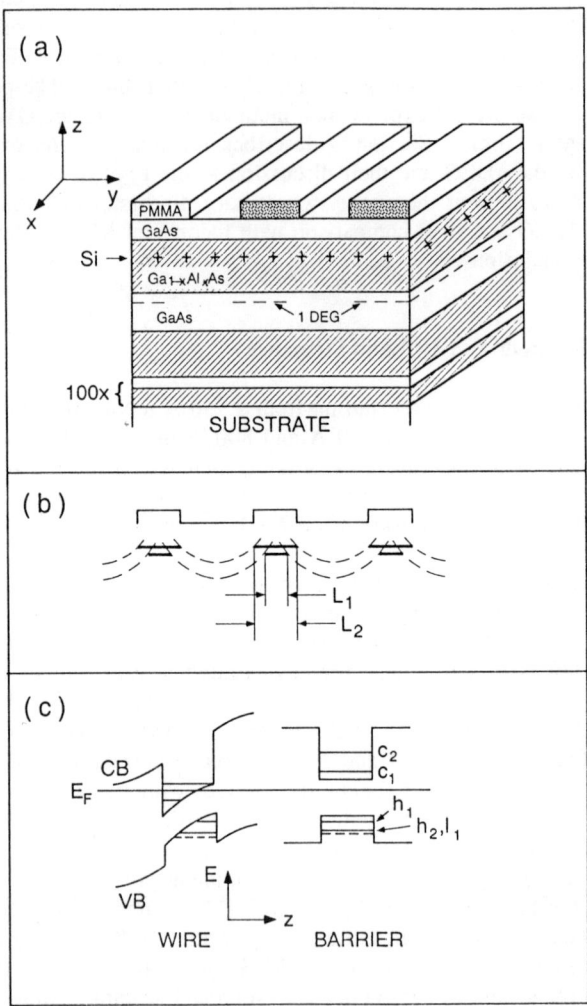

Figure 1. (a) Schematic structure of the quantum well wires showing the one-dimensional electron gas channels below the masked regions of the sample. The partial depletion is achieved by ion bombardment through the openings of the PMMA resist. (b) Depletion profiles showing how the actual wire width depends on the bombardment dose through the lateral depletion. (c) Band diagram of the one-dimensional wires and barriers along the growth direction of the quantum well.

excitation source and a Spex 1403 double monochromator with a GaAs photomultiplier tube and photon counting electronics for the detection. Typical light intensities were 0.02 W/cm^2 for PL and PLE measurements and 1W/cm^2 for inelastic light scattering, focussed into a 60 μm × 0.5 mm spot.

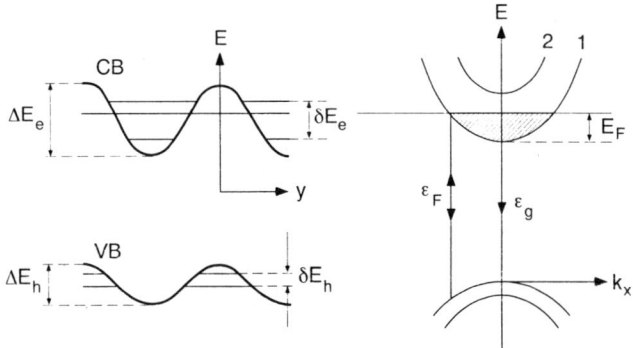

Figure 2. Band diagram of the quantum well wires across the wires showing the electron and hole intersubband spacings (left) and parabolic dispersion relations along the wires (right).

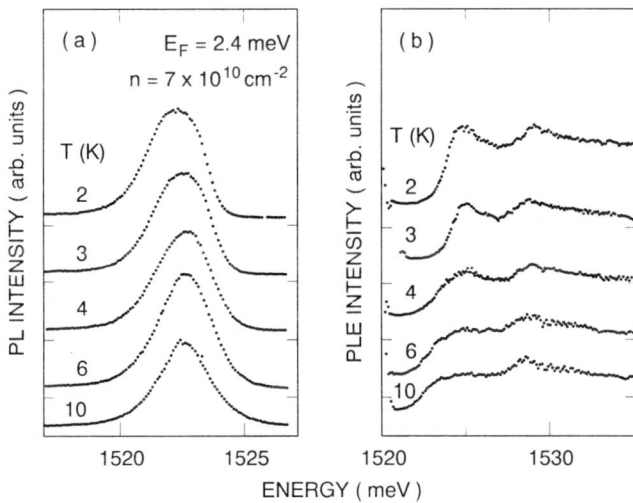

Figure 3. Temperature dependence of (a) emission (PL) and (b) absorption (PLE) spectra of a partially depleted quantum well having $E_F = 2.4$ meV and $n = 7 \times 10^{10}$ cm^{-2}.

3. PL and PLE Results and Discussion

Initially, a calibration of the bombardment dose was performed on the unpatterned 2D sample. The initial electron concentration of a 2D sample has been reduced to 7×10^{10} cm^{-2} by bombardment with argon ions at 300 V and 1.5×10^{-4} Acm^{-2} during 90 s. The corresponding PL and PLE spectra are shown in Fig. 3 for different temperatures. The

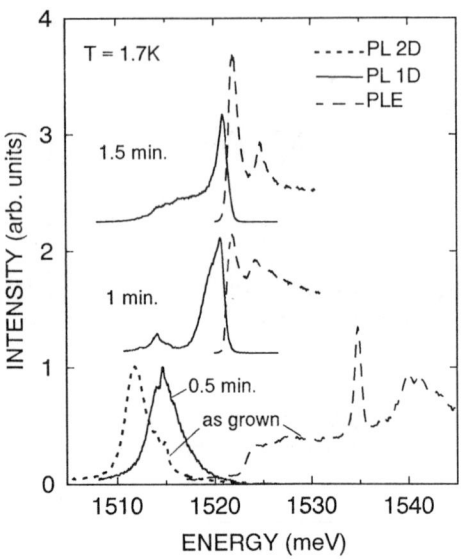

Figure 4. Photoluminescence (PL) and excitation (PLE) spectra at 1.7 K of a 1D sample as a function of ion bombardment time. One observes clearly the transition from the 2D (as-grown) shape to the 1D Fermi edge singularity.

peak observed at the PLE onset at low temperatures is a weak FES. It is absent in the initial QW at any temperature, in accordance with the expected behavior for a 2D system with mobile holes and high electron concentrations discussed in Sec. 1. A very similar FES is observed in a 2D unbombarded sample, where the initial electron concentration is reduced to 1.1×10^{11} cm^{-2} by increasing the distance between the donor layers and the QW during the growth [8]. This result proves that the bombardment process does not produce any damage in the 2DEG, and the 2D FES is therefore not due to hole localization, but to the reduction of the Fermi energy, as indicated above.

The effect of ion bombardment on a patterned sample is shown in Fig. 4, where PL and PLE spectra are present for increasing bombardment times. After short times the PL spectrum of the initial 2DEG (dotted line) has shifted to higher energies without significant changes in the PLE spectrum. As the exposure time is increased, the PL spectrum becomes asymmetric and the peak intensity shifts from the band edge to the Fermi level. At the longer times both the emission and the absorption spectra exhibit sharp peaks at the Fermi level, which correspond to the 1D-FES. The second PLE peak at 1.525 eV is the corresponding FES of the first light-hole or second heavy-hole states, which are degenerate in the initial 2D sample. Compared to the 2D case, 1D FESs are much stronger and appear also in the emission spectrum. The lack of a sharp energy onset of the emission spectrum, where the 1D density of states is singular, is discussed below.

The temperature dependence of the PL and PLE spectra of a 1D sample bombarded with oxygen for two minutes at 300 V and 10^{-4} Acm^{-2} is shown in Fig. 5. The Fermi energy obtained from Eq. (1) is $E_F = 4.3 \pm 0.5$ meV. The uncertainty comes from the location of the band edge ε_g.

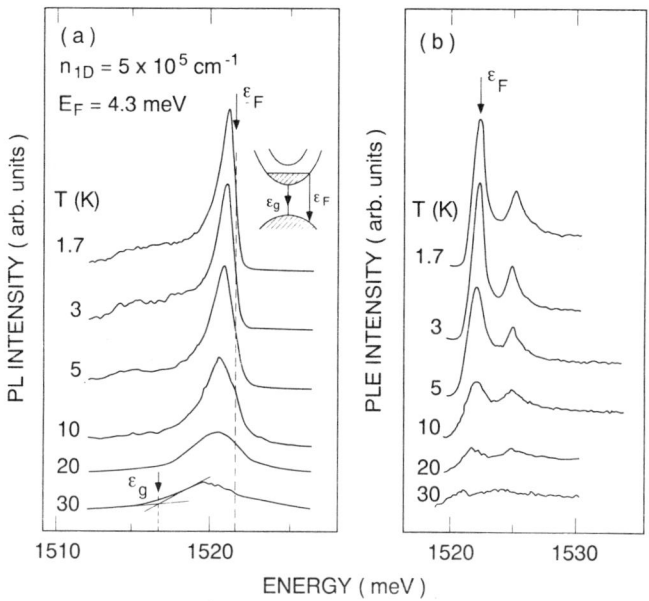

Figure 5. Temperature dependence of (a) emission (PL) and (b) absorption (PLE) spectra of a 1D sample in the quantum limit with $E_F = 4.3$ meV and $n = 6 \times 10^5$ cm^{-1}. The locations of ε_g and ε_F are indicated by arrows.

One observes the disappearance of the FES as the temperature is increased. The 1D interband spacing is measured by resonant inelastic light scattering by electronic interband excitations. The corresponding spectra are present in Fig. 6 for different laser energies and wavevector components (q) parallel to the wires [8,10]. The backscattering geometry shown in the inset of Fig. 6 permits varying q by changing the incident angle θ according to $q = (4\pi/\lambda)\sin\theta$ where λ is the wavelength of the incident light. The accessible wavevector range is 0.5–1.5×10^5 cm^{-1} (about one-tenth of the Fermi wavevector k_F). The incident and scattered light are linearly polarized parallel (H) or perpendicular (V) to the wires. One observes a peak at 5.2 meV which is mixed with a luminescence band for $q = 0$, but clear and distinct for $q > 0$. The sharpening of the peak for $q \neq 0$ is explained in the next section. As the 1D interband spacing is about 1 meV higher than E_F, the 1DEG is at the 1D quantum limit, with the second (empty) conduction band close to the Fermi level. The QWW width L_x can be estimated from the interband spacing in the parabolic well approximation [5], resulting in $L_x = 400$ Å. This is smaller than the physical line width, and indicates the importance of lateral depletion shown in Fig. 1(b). The electron density n, obtained for E_F, is $n \approx 6 \times 10^5$ cm^{-1}.

The strength of the 1D FES compared with the 2D case can be understood in terms of hole recoil arguments. In 2D systems indirect transitions from the top of the valence band to the Fermi level (see Fig. 7(a)) are possible by the excitation of low energy electron hole pairs to preserve momentum conservation. For mobile holes, if the average recoil energy is comparable to the Coulomb energy, then the absorption spectra will still have a peak at the

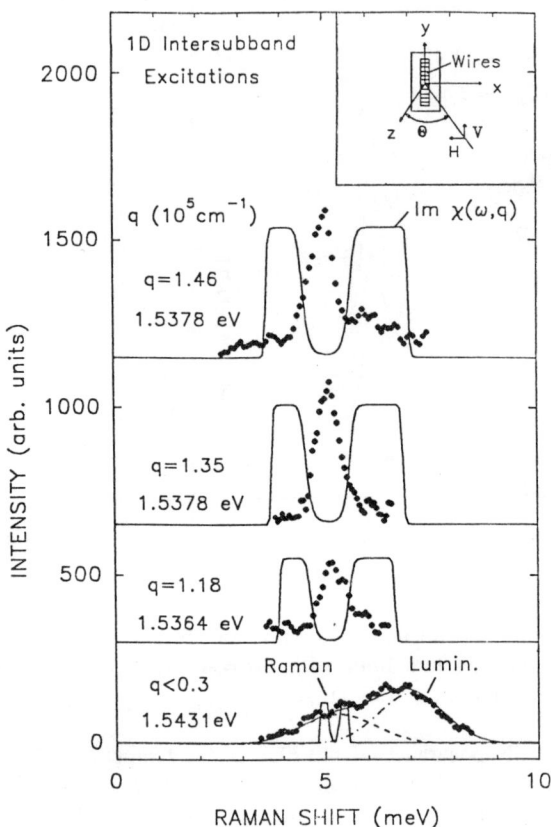

Figure 6. Raman scattering spectra of the 1D sample for different values of the wavevector component along the wires, and laser energies in the VV configuration. The interband single particle excitation is observed as a Raman peak at 5.2 meV. The peak is partially hidden by luminescence at q = 0 and is sharp and clear for q ≠ 0. The solid line represents the imaginary part of the calculated damping function (see text). The inset shows schematically the scattering geometry.

Fermi level, which disappears at higher electron concentration [26], as seen in Fig. 7(b). In 1D systems indirect transitions are not allowed, so that a broadening of the FES to the low energy side is not possible. As a result, the 1D-FES is stronger and sharper than in 2D, and will not disappear even at high electron concentrations. This also holds for the strong FES observed in the emission spectra of 1D systems, because the small Fermi energy of our samples allows some holes to be at k_F at the working temperatures. It is however difficult to understand the lack of a sharp peak at the low energy threshold of the emission spectra, corresponding to the singularity of the 1D density of states. One possible argument is the fluctuation of the band gap energy associated with the QWW width fluctuation, provided that the Fermi level is pinned to the fixed value of the 2DEG that surrounds the patterned part of the sample. On the other hand, existing calculations on direct QWWs [32,34] predict a band edge singularity in PL spectra whether the holes are completely localized [32] or not [34]. Of the possible localization sources, namely potential fluctuations or defect trapping, the latter has to be disregarded in view of the results presented in 2D systems. The former is at least dubious, as hole localization due to QWW fluctuations should present a quite small characteristic energy with similar value for all samples. Yet, the temperature dependence of the FES observed in different samples is quite different and correlates with the position of the Fermi level in the highest occupied 1D band [8]. The lack of a band edge singularity in the PL spectra has therefore to be considered as an open question, perhaps related to the spatially indirect character of our QWW (see note added in proof).

4. Inelastic Light Scattering Results and Discussion

Inelastic light scattering experiments are done with the incident and scattered light polarization directed either parallel (HH) or perpendicular (VV) to the QWW, as shown in the inset of Fig. 6. This allows the observation of intraband or interband electronic excitations, respectively.

The 1D intersubband transitions shown in Fig. 6 are recorded in the VV configruation and correspond to charge density excitations (CDEs) whose energy is δE_e [10] (see Fig. 2). At low electron densities the wire potential is nearly parabolic and the interband spacing depends weakly on density [43]. Under these circumstamces the resonance frequencies of the electron gas are independent of the electron-electron interaction [42,44]. Thus the energy of interband CDEs can be taken as δE_e. They appear dispersionless and exhibit a surprising q dependence of the bandwidths, as stated in the previous section. This can be explained in terms of the Landau damping characteristics if the second conduction 1D band is slightly occupied. This is actually the case, as the incident light power needed for the Raman experiments is 50 times higher than in PL and PLE, and the second band is less than 1 meV above the Fermi level [10]. In these conditions a gap opens in the energy spectrum of 1D interband pair excitations, as shown in Fig. 9. The solid curves in Fig. 6 represent Landau damping as given by the imaginary part of the 1D interband polarizability function $Im\chi(q,\omega)$ within the random-phase-approximation (RPA) [40]. At finite q, this function has a minimum at δE_e (see Fig. 1) and maxima at $\delta E_e \pm \hbar q v_F$, where v_F is the Fermi velocity. There, undamped 1D-CDE can be observed. For $q \approx 0$ the two maxima in the loss function merge into a single peak at δE_e and the excitations are strongly damped.

Figure 8 shows the HH light scattering spectra for different wave vectors. Peak energies increase with increasing q. The strong dispersive behavior indicates that these peaks are due to intraband excitations of the electron gas [36]. The higher energy peaks are assigned to the 1D plasmon of the wires. The low energy bands are interpreted as predominantly single-particle 1D intraband excitations (SPE) similar to those observed in the 2D case [36,45]. The solid curve in Fig. 8 again represents the loss function, which peaks at qv_F. In the crossed polarization spectra (HV, not shown) a band shape comprised of SPE and spin-density excitations (SDE) is observed [10].

In Fig. 9 we summarize the wave-vector dependence of the light-scattering peaks. The 1D intraband plasmon, which corresponds to oscillations of the charge density in the direction of the wires, exhibits an almost linear dispersion that extrapolates to a positive

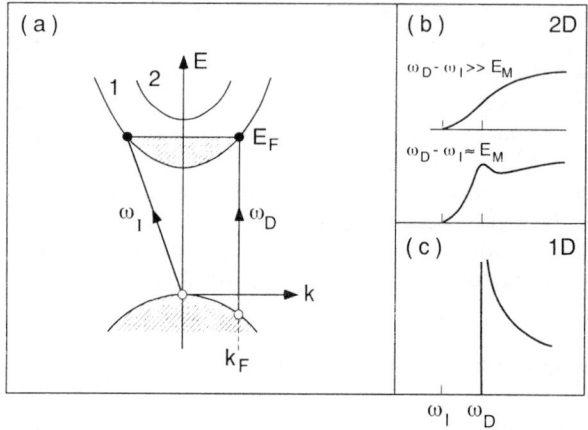

Figure 7. (a) Schematic representation of direct (ω_D) and indirect (ω_I) transitions in the light absorption process by a low dimensional electron gas. (b) The approximate shape of the absorption spectrum is shown for a 2D case, considering two different values of the average recoil energy; and (c) for the 1D case where indirect transitions cannot occur.

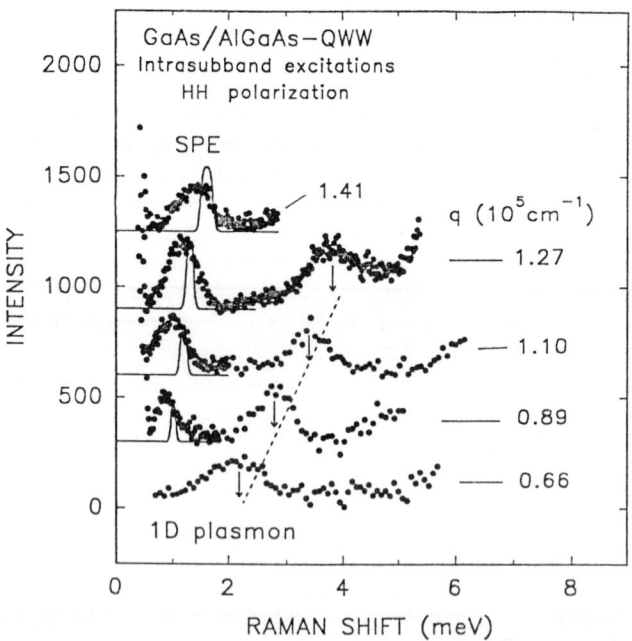

Figure 8. Polarized light-scattering spectra of a multiple QWW sample at 1.8 K for different wave vectors q along the wires. Solid curves correspond to the calculated functions Im$\chi(\omega,q)$ of intraband SPEs.

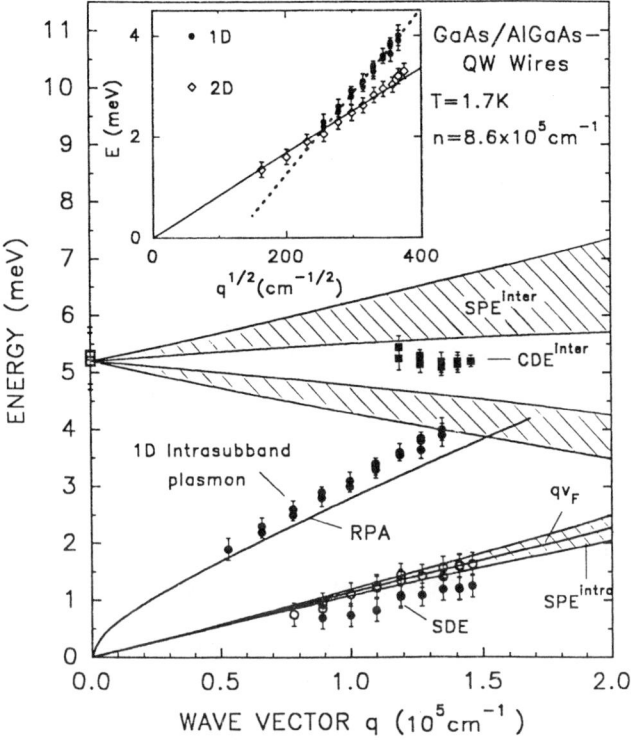

Figure 9. Wave-vector dispersions of intraband and interband excitations of a 1D electron gas in the quantum limit. Solid dots represent intraband collective CDEs and SDEs. Open circles display the position of the peak at $\hbar q v_F$ of intraband SPEs. Squares correspond to data of 1D interband CDEs measured in VV polarization. The shaded areas indicate ranges of electron-hole pair excitations (SPE) given by the condition $\text{Im}\chi(\omega,q) \neq 0$. Inset: comparison of the 1D intraband plasmon frequencies with those of a 2D electron gas with $E_F = 3.8$ meV as a function of $q^{1/2}$.

finite frequency at q = 0. The inset of Fig. 9, shows that the 1D system has markedly different behavior from the typical square root dependence of 2D plasmons. To give a quantitative interpretation of the 1D plasmon dispersion, we plot the single-band 1D plasmon frequency $\omega_+(q)$ calculated in the RPA [39,40]:

$$\omega_+(q) = \omega_0 q \left[\frac{f_{00}(q)}{2} \left[1 + \left[1 + \frac{4k_F^2}{m^2\omega_0^2 f_{00}(q)} \right]^{1/2} \right] \right]^{1/2} \quad (2)$$

with $\omega_0 = (2ne^2/m^*\varepsilon_L)^{1/2}$; ε_L is the background dielectric constant and $f_{00}(q) = (\varepsilon_L/2e^2)V_{00}(q)$, where V_{00} is the Coulomb interaction in the lowest 1D band. In the $q \to 0$ limit, Eq. (2) can be approximated by

$$\omega_+(q) \approx \omega_0 q \left[L_n\left(\frac{qL_x}{2}\right) \right]^{1/2}. \tag{3}$$

The curve labeled RPA in Fig. 9 corresponds to a fit using Eq. (2) with the electron density $n = (8.6 \pm 0.4) \times 10^5$ cm^{-1}. The matrix element $V_{00}(q)$ was calculated assuming harmonic oscillator wave functions due to the confinement in the y direction, whereas the electron gas was assumed to be of zero thickness in the z direction [40]. In these experiments, Coulomb coupling between wires [40] can be neglected because $1 < qd < 3$, d being the inter-wire distance. In spite of a small discrepancy of ~ 10%, Eq. (2) provides an excellent description of the 1D intraband dispersion. This is not surprising, in spite of the general belief that the 1DEG behaves like a Tomonaga-Luttinger liquid, because the solution for the long-wavelength plasmon dispersion has been proved to be exactly the same in the RPA and the Tomonaga-Luttinger model in 1D [40].

5. Conclusions

The 1DEG in the quantum limit exhibits optical properties markedly different from the 2D case. FESs have been observed both in absorption and emission spectra, whose strength is explained by the suppression of hole recoil effects in 1D. Electronic excitations of the 1DEG have also been studied, finding intraband plasmons that exhibit the linear dispersion predicted by RPA for 1D. Interband CDEs are dispersionless and have the interband energy as expected for a parabolic wire potential. These excitations also display the unique 1D features of Landau damping. These results provide new ideas about the collective behavior of the one-dimensional electron gas in the 1D quantum limit.

Acknowledgments

We are deeply indebted to S. Schmitt-Rink for many helpful discussions and original points of view. We also thank C. Tejedor for constructive comments.

References

1. M.A. Reed and W.P. Kirk, *Nanostructure Physics and Fabrication* (Academic Press, London, 1989).
2. W. Hansen, M. Horst, J.P. Kotthaus, U. Merkt, C. Sikorski, and K. Ploog, Phys. Rev. Lett. **58**, 2586 (1987).
3. T. Demel, D. Heitman, P. Grambow, and K. Ploog, Phys. Rev. B **38**, 12732 (1988).
4. C. Sikorski and U. Merkt, Phys. Rev. Lett. **62**, 2164 (1989).
5. J.S. Weiner, G. Danan, A. Pinczuk, J. Valladares, L.N. Pfeiffer, and K.W. West, Phys. Rev. Lett. **63**, 1641 (1989).
6. T. Egeler, G. Abstreiter, G. Weimann, T. Demel, D. Heitman, P. Grambow, and W. Schlapp, Phys. Rev. Lett. **65**, 1804 (1989).
7. T. Demel, D. Heitman, P. Grambow, and K. Ploog, Phys. Rev. Lett. **66**, 2657 (1991).
8. J.M. Calleja, A.R. Goñi, B.S. Dennis, J.S. Weiner, A. Pinczuk, S. Schmitt-Rink, L.N.

Pfeiffer, K.W. West, J.F. Müller, and A.E. Ruckenstein, Solid State Commun. **79**, 911 (1991); Surf. Science **263**, 366 (1992).
9. A.S. Plaut, H. Lage, P. Grambow, D. Heitman, K. von Klitzing, and K. Ploog, Phys. Rev. Lett. **67**, 1642 (1991).
10. A.R. Goñi, A. Pinczuk, J.S. Weiner, J.M. Calleja, B.S. Dennis, L.N. Pfeiffer, and K.W. West, Phys. Rev. Lett. **67**, 3298 (1991).
11. G.D. Mahan, Phys. Rev. **153**, 882 (1967); **163**, 612 (1967).
12. G.D. Mahan, in *Solid State Physics*, edited by H. Ehrenreich, F. Seitz, and D. Turnbull (Academic Press, New York, 1974), Vol. 29, p. 75.
13. P. Nozières and G.T. de Dominicis, Phys. Rev. **178**, 1097 (1969).
14. B. Roulet, J. Gavoret, and P. Nozières, Phys. Rev. **178**, 1072 (1969); **178**, 1084 (1969).
15. M. Combescot and P. Nozières, J. Phys. (Paris) **32**, 913 (1971).
16. A. Pinczuk, J. Shah, R.C. Miller, A.C. Gossard, and W. Wiegmann, Solid. State Commun. **50**, 735 (1984).
17. C. Delalande, G. Bastard, J. Orgonasi, J.A. Brum, H.W. Liu, M. Voos, G. Weimann, and W. Schlaap, Phys. Rev. Lett **59**, 2690 (1987).
18. G. Livescu, D.A.B. Miller, D.S. Chemla, M. Ramaswamy, T.Y. Chang, N. Sauer, A.C. Gossard, and J.H. English, IEEE J. Quantum Electron. **24**, 1677 (1988).
19. D. Huang, H.Y. Chu, Y.C. Chang, and H. Morkoc, Phys. Rev. B **38**, 1246 (1988).
20. J.S. Lee, N. Miura, and T. Ando, J. Phys. Soc. Jpn. **59**, 2254 (1990).
21. J. Wagner, A. Fischer, and K. Ploog, Appl. Phys. Lett. **59**, 428 (1991).
22. R. Cingolani, W. Stolz, and K. Ploog, Phys. Rev. B **40** 2950 (1989).
23. J. Wagner, A. Ruiz, and K. Ploog, Phys. Rev. B **43**, 12134 (1991).
24. M.S. Skolnick, J.M. Rorison, K.J. Nash, D.J. Mowbray, P.R. Tapster, S.J. Bass, and A.D. Pitt, Phys. Rev. Lett. **58**, 2130 (1987); M.S. Skolnick, D.M. Whittaker, P.E. Simmonds, T.A. Fisher, M.K. Saker, J.M. Rorison, R.S. Smith, P.B. Kirby, and C.R.H. White, Phys. Rev. B **43**, 7354 (1991).
25. W. Chen, M. Fritze, A.V. Nurmikko, D. Ackley, C. Colvard, and H. Lee, Phys. Rev. Lett. **64**, 2434 (1990).
26. A.E. Ruckenstein and S. Schmitt-Rink, Phys. Rev. B **35**, 7551 (1987).
27. S. Schmitt-Rink, D.S. Chemla, and D.A.B. Miller, Adv. in Physics **38**, 89 (1989).
28. K. Ohtaka and Y. Tanabe, Phys. Rev. B **39**, 3054 (1989).
29. T. Uenoyama and L.J. Sham, Phys. Rev. Lett **65**, 1048 (1990).
30. G.E.W. Bauer, in *Localization and Confinement in Semiconductors*, edited by G. Bauer, F. Kuchar, and H. Heinrich, 6th International Winterschool, Mauterndorf, February 1990 (Springer, Berlin, 1990), p. 27.
31. J.F. Müller, A.E. Ruckenstein, and S. Schmitt-Rink, Phys. Rev. B **46**, 8902 (1992).
32. P. Hawrylak, Solid State Commun. **81**, 525 (1992).
33. Ben Yu-Kuang Hu and S. Das Sarma, Phys. Rev. Lett. **68**, 1750 (1992).
34. F.J. Rodríguez and C. Tejedor, Phys. Rev. B **47**, 1506 (1993), and these proceedings.
35. T. Ogawa, Phys. Rev. Lett. **68**, 3638 (1992).
36. A. Pinczuk and G. Abstreiter, in *Light scattering in Solids V*, edited by M. Cardona and G. Güntherodt (Springer, Berlin, 1989), Topics in Appl. Phys., Vol. 66, p. 153.
37. G. Eliasson, J. Wu, P. Hawrylak, and J.J. Quinn, Solid. State Commun. **60**, 41 (1986).
38. F.Y. Huang, Phys. Rev. B **41**, 12957 (1990).
39. A. Gold and A. Ghazali, Phys. Rev. B **41**, 7626 (1990).
40. Q.P. Li and S. das Sarma, Phys. Rev. B **43**, 11768 (1991); Q.P. Li, S. das Sarma, and R.

Joynt, Phys. Rev. B **45**, 13713 (1992).
41. L. Wendler, R. Haupt, and R. Pechstedt, Phys. Rev. B **43**, 14669 (1991); Surf. Science **263**, 363 (1992).
42. L. Brey, N. Johnson, and B.I. Halperin, Phys. Rev. B **40**, 10647 (1989); P.A. Maksym and T. Chakraborty, Phys. Rev. Lett. **65**, 108 (1990); Q.P. Li, K. Karrai, S.K. Yip, S. Das Sarma, and H.D. Drew, Phys. Rev. B **43**, 5151 (1991).
43. S.E. Laux, D.J. Frank, and F. Stern, Surf. Sci. **196**, 101 (1988).
44. V. Shikin, T. Demel, and D. Heitman, Surf. Sci. **229**, 276 (1990).
45. M. Berz, J.F. Walker, P. von Allmen, E.F. Steigmeier, and F.K. Reinhart, Phys. Rev. B **42**, 11957 (1990).

Note Added in Proof

Very recent calculations by Rodríguez and Tejedor (to be published) on spatially *indirect* QWWs show that for sample parameters similar to ours the FES intensity grows bigger than the band edge singularity by one order of magnitude if the hole mass tends to infinity, an empty band is close to the Fermi level, and electrons are spatially separated from holes by a very critical distance (around 500 Å). Such results could lead to an interpretation of the absence of a singularity at the optical emission edge in our experiments.

OPTICAL SINGULARITIES OF THE QUASI ONE-DIMENSIONAL ELECTRON GAS

C. TEJEDOR AND F.J. RODRIGUEZ
Departamento de Fisica de la Materia Condensada
Universidad Autónoma de Madrid
Cantoblanco 28049, Madrid
Spain

ABSTRACT. We study the emission and absorption spectra of single quantum well wires doped with electrons. By solving an effective Bethe-Salpeter equation with a statically screened Coulomb interaction we analyze the necessity of hole localization in order to get bottom and Fermi edge singularities. The effect of wire width, carrier density and temperature is analized in the quantum limit in which electrons are occupying the lowest subband. Existing experimental evidence in multiwires is discussed in the light of our results.

1. Introduction

Low dimensional systems are practically produced by modulation doped semi-conductor heterojunctions being ideal for studying many-body properties of an electron gas. Since interactions and screening properties of low dimensional systems are different to those of three dimensional ones, it should be plausible to have strong optical singularities without the hole localization required in three-dimensional (3D) systems [1,2]. An excellent candidate is a quantum well wire (QWW) where singularities associated to the quasi-one dimensionality should be very strong. Different fabrication techniques have recently made feasible such systems allowing the experimental search of singularities [3–6] and provoking an increasing theoretical interest [3,7–11]. Experiments made in spatially indirect multiwires [3] clearly show Fermi edge singularities (FES) in narrow QWW. Such features have not been observed so clearly in wide direct wires [4] and multiwires [6]. We have previously shown that an on site model interaction between electrons and holes is not able to properly describe the optical singularities of QWW [11]. Here, we investigate theoretically optical singularities in QWW using a screened Coulomb interaction between electrons and holes. We pay particular attention to the necessity or not of hole localization. We consider a single wire with parabolic potentials with the same center both for the holes and the electrons at difference with some experimental situations where the two type of particles are in different spacial regions [3,5]. The holes are described in the simplest diagonal approach where the coupling between heavy and light subbands is only included by adequately varying the effective masses. By considering the actual eigenfunctions of these potentials as well as the Coulomb interaction screened by a statically consistent response function which includes electron-electron interaction by means of fluctuation effects [12], we compute the absorption and emission spectra of a quasi one-dimensional

(1D) electron gas by means of an effective Bethe-Salpeter equation.

2. Description of the Model

We are interested in describing the optical properties of a single n-doped wire in which the electron and hole states coexist in the same spatial region. Typical widths of actual wires are of the order of a few hundred Ångstroms. The states are labelled by a quantum wavenumber k along the wire direction and a subband index n. Here we will consider only one subband of electrons while the holes are described in a diagonal approach so that only the heavy-hole subband is included. Both the emission and absorption spectra are computed from the linear optical susceptibility $\chi(\omega)$ that it is related to the electron-hole Green's function $G_{n_v,n_c;n'_v,n'_c}(k,k',\omega)$ by

$$\chi(\omega) = - \sum_{n_v,n_c;n'_v,n'_c;k,k'} \langle n_v,k|e\cdot p|n_c,k\rangle \langle n'_c,k'|e\cdot p|n'_v,k'\rangle \times G_{n_v,n_c;n'_v,n'_c}(k,k',\omega), \quad (1)$$

where $|n_v,k\rangle$ and $|n_c,k'\rangle$ are valence and conduction states respectively, e is the light polarization and p is the momentum operator. The absorption intensity is directly given by the imaginary part of the susceptibility $\chi(\omega)$ while the luminescence spectrum is proportional to $g(\omega)\mathrm{Im}(\chi(\omega))$ where $g(\omega)=\{\exp[(\omega-\mu_e-\mu_h)/k_BT)]-1\}^{-1}$ is the Bose distribution function and μ_e and μ_h the electron and hole chemical potentials respectively [1,2]. The key quantity in Eq. (1) is the Green's function that can be obtained from the noninteracting electron-hole Green's function $G^0_{n_v,n_c}(k,-k,\omega)$ by means of a perturbative expansion in terms of a static electron-hole interaction potential $V_{n_v,n_c;n'_v,n'_c}(k-k')$ [1]. When the appropriate hole spectral function is consider, the ladder diagrams are the most singular and the inclusion of all the crossed diagrams can be avoided [1,2]. Then, in this ladder approximation, the Green's function is given by an effective Bethe-Salpeter equation

$$G_{n_v,n_c;n'_v,n'_c}(k,k',\omega) = G^0_{n_v,n_c}(k,-k,\omega)\delta_{n_v,n'_v}\delta_{n_c,n'_c}\delta(k+k')$$
$$+\frac{1}{L}\sum_{n''_v,n''_c,k''} G^0_{n_v,n_c}(k,-k,\omega)V_{n_v,n_c;n''_v,n''_c}(k-k'')G_{n''_v,n''_c;n'_v,n'_c}(k'',k',\omega), \quad (2)$$

L being the wire length. The ladder equation, Eq. (2), is rather easy to work with by discretizing the k space. If a mesh is defined between two cut-off values $\pm k_c$ with k_c sufficiently larger than the Fermi wavevector, the Green's function $G_{n_v,n_c;n'_v,n'_c}(k,k',\omega)$ is obtained from the inversion of a matrix $[1-G^0V]$ in the discrete indices k, k'. The results shown here have been obtained from matrices with sizes of the order of 500. Let us now discuss the ingredients G^0 and V involved in Eq. (2) starting with the interaction because it is useful in the discussion of the noninteracting electron-hole Green's function.

Since we are interested in the study of the effect of the hole localization, there are two different possibilities. First is to consider that the hole is extended to the whole wire but its effective mass can be either the usual one for perfect wires or practically infinite due to

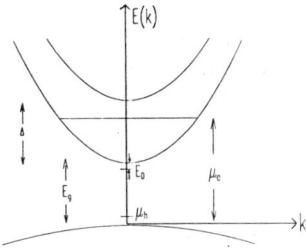

Figure 1. Model band structure for the quantum well wire. The magnitude of interest are schematically shown.

inhomogeneities in the shape of the wire. A second alternative is to have the holes extremely localized in impurities. That means different hole wave functions that are required for estimating the screened Coulomb interaction, between an electron and a hole. Firstly, we will consider that electrons and holes are completely confined in the xy plane.

For extended holes the carriers move along the x direction (the wire direction) while they are confined by parabolic potentials in the y direction as schematically shown in Fig. 1. We take the two parabolas in such a way that the ration of separations Δ between the two first levels is just given by the inverse ratio between effective masses, i.e., $\Delta_e/\Delta_h = m_h^*/m_e^*$, as it would occur for infinite well potentials. Then the wavefunction of each of those carriers is of the form

$$\psi_{n,k}(x,y,z) = \sqrt{\frac{2\delta(z)}{\pi}} \frac{e^{ikx}}{\sqrt{L}} \sqrt{\frac{1}{2^n n! \sqrt{\pi} l}} e^{-y^2/2l^2} H_n\left(\frac{y}{l}\right), \qquad (3)$$

where H_n is the Hermite polynomial corresponding to the n-th subband, and both electron and holes have the same characteristic length $l = (\hbar/m_e^*\Delta_e)^{1/2}$ (hereafter we will use Δ for the characteristic frequency of the harmonic potential confining electrons). From the wave functions of electrons and holes, it is straightforward to get the unscreened Coulomb interaction between them [12].

A second possibility is to have the holes localized in impurities. The electron-hole interaction can then be easily computed if one considers the hole wave function as absolutely localized in the centre of the wire [8]. This approach overestimates the electron-hole interaction because, in a random distribution of impurities, those situated away from the wire center involve a smaller electron-hole coupling. When one considers the hole wavefunction completely localized $\psi_h(x,y,z) = \sqrt{\delta(x)\delta(y)\delta(z)}$, only some 1/4 factors in the unscreened electron-hole interactions V^0 should be replaced by 1/8 [8,11].

Once we have the unscreened potential, we need to screen it in order to get the adequate electron-hole interaction. Two ingredients are required for the screening: the electron-electron interaction and the electron polarizability obtained from such interaction at zero temperature. Provided that l is the same for electrons and holes, the electron-electron interaction has the same shape that the interaction of electrons with extended holes with just a change of sign. The only lacking ingredient is the electron polarizability. In order to include some electron-electron effects in the polarizability, we take the form obtained from a generalized quantum Langevin approach which includes fluctuation effects by means of a diffusion constant [11,12].

The pair Green's function of noninteracting electron-hole pairs G^0 requires some attention because it contains two physical effects that are very important in the optical spectra: temperature and life time of the individual quasiparticles electron and hole. Then, taking into account that the system is n-doped and at low temperatures so that only holes at the upper subband are considered, i.e., $n_v = 0$, the expression is

$$G^0_{0,n_c}(k,-k,\omega) = \int \frac{d\omega_{n_c}}{2\pi} \int \frac{d\omega_h}{2\pi} A_{n_c}(k,\omega_{n_c}) A_h(-k,\omega_h) \times \frac{1 - f_e(\omega_{n_c}) - f_h(\omega_h)}{\omega - \omega_{n_c} - \omega_h + i\delta}, \quad (4)$$

where $\delta = 0^+$, f_e, and f_h are the Fermi distribution functions of the quasiparticles. Life time effects are included in the spectral functions of the electron A_{n_c} and the hole A_h. Such effects are not important for the electron case because its life time implies a broadening much smaller than the electron chemical potential, so that

$$A_{n_c}(k,\omega) = 2\pi\delta(\omega - E_{n_c}(k)), \quad (5)$$

$E_{n_c}(k) = E_g + \hbar^2 k^2/(2m_e^*) + n\hbar\Delta$ being the electron dispersion relation of the n-th band with E_g the energy gap and m_e^* the electron effective mass. However, the case of the hole is more complicated. The ground state wave functions of the conduction electron system, with and without the valence hole potential, are orthogonal so that just pair excitations of the conduction electron gas break the symmetry giving the possibility of observing optical transitions; this is the well known orthogonality catastrophe [1]. Such pair excitations are the essential dynamical effects responsible for a hole spectral function different from a δ-function, so that we do not include other effects. It must be stressed that this is the approach consistent with the ladder approximation made above for the correlation function [1]. In the simplest approach, the contribution of the electron-hole pairs in the first conduction subband to the spectral function for the valence band hole is obtained from a response function $R(\omega)$ within the static screening picture given by [1]:

$$R(\omega) = \frac{1}{2\pi^2} \int dk \left[\frac{v^0}{|\varepsilon(k,0)|}\right]^2 [-\text{Im}\,\chi_e(k,\omega)], \quad (6)$$

where χ_e is the electron polarizability and ε the dielectric function. From the previous expressions it is straightforward to obtain, after some algebra, $R(\omega) \to g\omega$ as the asymptotic limit for $l \to 0$ and $\omega \to 0$, where

$$g = \frac{2}{\pi^2} \frac{1}{1 + \left[\ln(1 + 1/a^2)/\pi\right]^2} \quad (7)$$

and where a is a dimensionless parameter related to the diffusion constant [11,12]. The

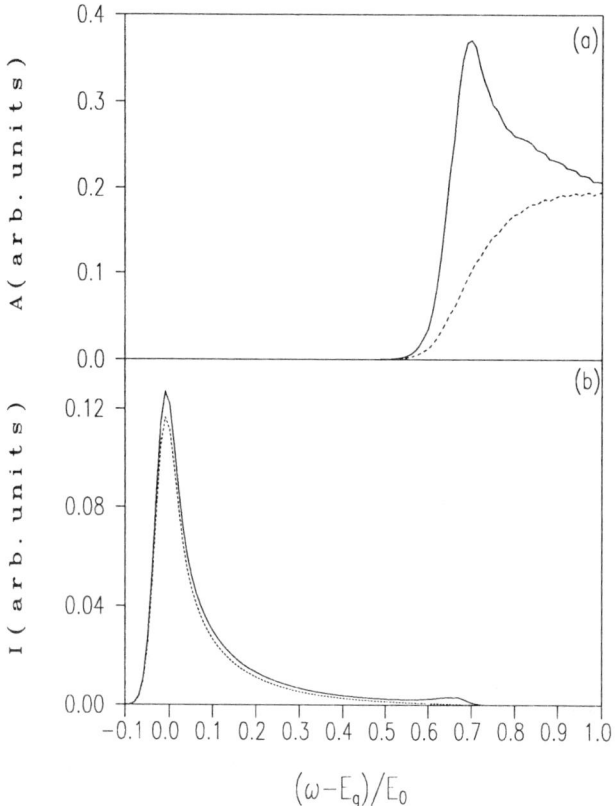

Figure 2. Absorption (a) and emission (b) spectra of a wire at $T = 0.02E_0/k_B$ with electron chemical potential $\mu_e = E_g + 0.6E_0$, $\Delta = E_0$, and $m_h^* = 8\, m_e^*$ for both screened Coulomb interacting (continuous lines) and non-interacting (dashed lines) electron-hole pairs.

quantity g does not vary very much with the value of a, which is taken to be 0.1 throughout this paper. Equation (7) leads to a hole spectral function [1]

$$A_h(-k,\omega) = \frac{2\sin(\pi g)\Gamma(1-g)}{\mu_e} \frac{e^{-|\Omega|}}{|\Omega|^{(1-g)}} \theta[\pm\Omega], \tag{8}$$

where $\Omega = (\omega - E_h(k))/\mu_e$, $E_h(k) = -\hbar^2 k^2/(2m_h^*)$ being the hole dispersion relation with m_h^* the hole effective mass. The Heaviside step function $\theta[\pm(\Omega)]$ in Eq. (8) is a consequence of the sudden creation or annihilation of a hole at the valence band in the absorption and emission processes respectively. This abrupt spectral function of the hole produces some computational instabilities when solving Eq. (2). Therefore, the numerical calculations require the introduction of some imaginary parts in the frequency and to pay some attention to the associated convergence problems.

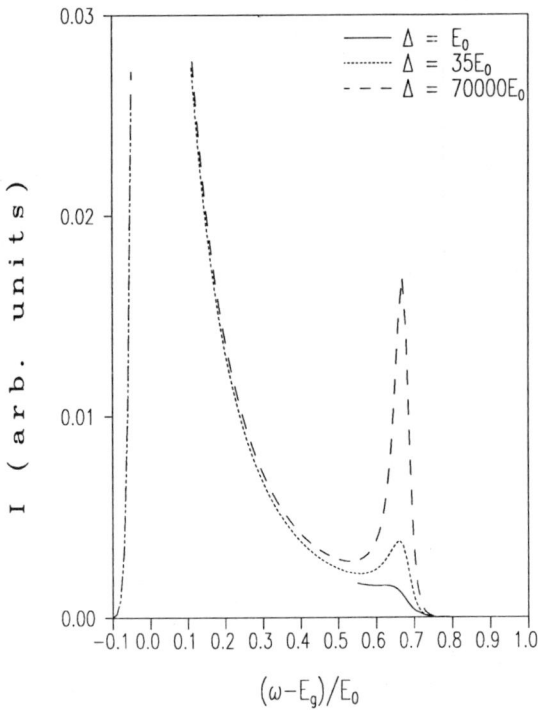

Figure 3. Emission spectra of wires with different widths (i.e., different band separations Δ) at $T = 0.02E_0/k_B$ with electron chemical potential $\mu_e = E_g + 0.6E_0$ and $m_h^* = 8m_e^*$.

3. Effect of the Hole Localization

We address ourselves in this section to the physically important question of the requirement of hole localization for having edge singularities. Such a localization is responsible for the singularities observed in the X-ray spectra of metals [1]. In two- and three-dimensional systems with mobile holes, there are not singularities due to peak broadenings produced by indirect transitions from the top of the valence band to the Fermi level accompanied by low-energy excitations of the Fermi sea to ensure momentum conservation. In 1D systems this broadening mechanism is not efficient because allowed low-energy excitations are only with either 0 or $2k_F$ momentum. Therefore, quasi-1D systems are a good candidate for presenting singularities even for cases with mobile holes [9]. So, we are going to apply our model with a screened Coulomb interaction to analyze the possibility of having singularities with mobile holes.

In order to work in the range of experimental wire widths, we start with a parabolic wire with a width $2\sqrt{2}\,l$ of the order of 400 Å. We represent the width of the wire by means of the level separation Δ that is equal to 1 (in units of E_0 as everywhere for energies in this paper). Typical electron densities correspond to Fermi energies of the order of 0.6 which is

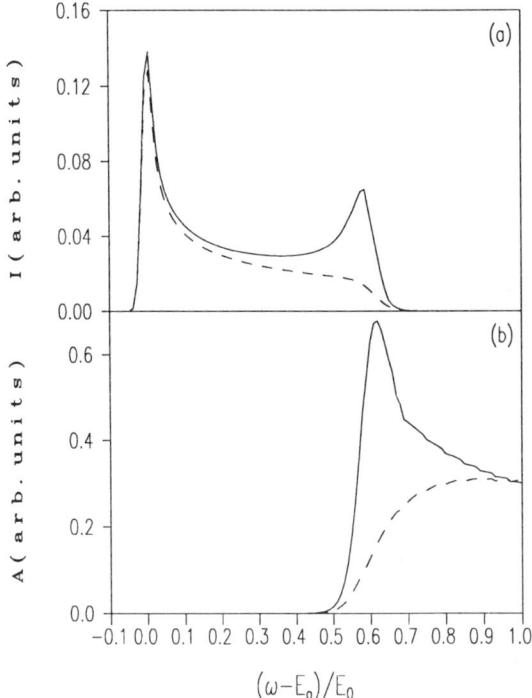

Figure 4. Emission (a) and absorption (b) spectra of a wire at $T = 0.02E_0/k_B$ with electron chemical potential $\mu_e = E_g + 0.6E_0$, $\Delta = E_0$ and $m_h^* = 1000m_e^*$ for both screened Coulomb interacting (continuous lines) and non-interacting (dashed lines) electron-hole pairs.

the value we start with. Since, for the moment, we want a system without any extrinsic holes we take $\mu_h = 1$ (i.e., well within the gap). A temperature low enough where singularities can be expected is $T = 0.02$ (in units of E_0/k_B). For describing a mobile hole we can take for its effective mass a value of 8 (in units of the electron effective mass) at most. The absorption and emission spectra obtained with those parameters is shown in Fig. 2. The continuous lines correspond to the results obtained from $G(\omega)$ while the dashed lines ones are obtained with the use of $G_0(\omega)$ so that the difference is the many-body effect produced by the screened electron-hole interaction. The FES is more clear in absorption while in emission is rather weak. However, if one just looks at the complete result (continuous lines) and try to detect singularities by comparing with background values the existence of FES is less clear. In order to get stronger singularities we reinforce the bare Coulomb attraction effect by decreasing the width of the wire, i.e., by having a more one dimensional system. This is shown in Fig. 3 where the emission spectrum is shown for different values of Δ. A FES neatly exist for values of Δ larger than 35 that means widths of the wire below 65 Å clearly far narrower than the presently available wires. The singularity saturates, and does not increase strongly for extremely narrow wires. Since FES seem to be observed in wider

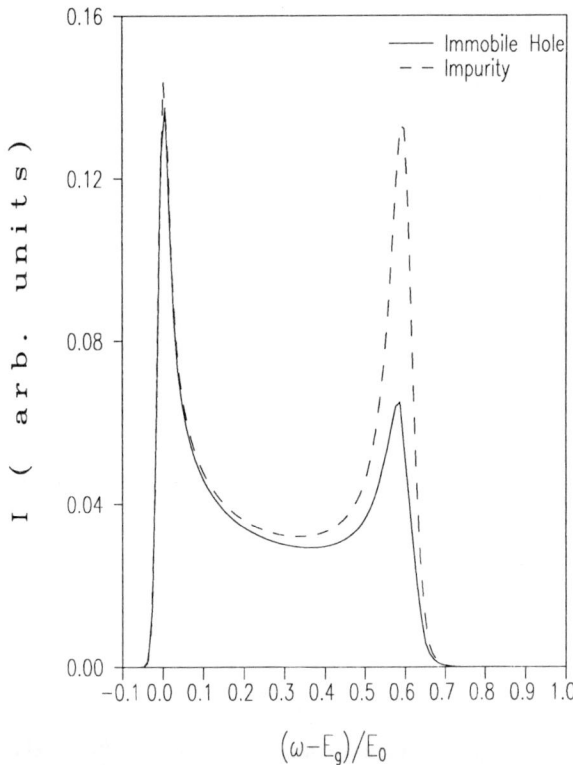

Figure 5. Emission spectra of a wire at $T = 0.02E_0/k_B$ with electron chemical potential $\mu_e = E_g + 0.6E_0$ and $\Delta = E_0$ for both an immobile extended hole with $m_h^* = 1000 m_e^*$ (continuous lines) and an impurity hole in the centre of the wire dashed lines.

multiwires [3], a much more plausible explanation is to consider a localized hole. For the case of an extended immobile hole, localization is obtained by increasing the effective mass up to infinity; that, in practice, is when $(\hbar k_F)^2/(2m_h^*) \ll k_B T$. Therefore we present in Fig. 4 the absorption and emission spectra for $m_h^* = 1000$ and $\Delta = 1$. Now the singularities are very clear not only when comparing with the non interacting case but also by comparing peaks with backgrounds. As mentioned before, another possibility is to have the holes localized in impurities. Using the model of δ functions for the hole eigenstate, the emission spectrum is that shown in Fig. 5. Now the FES is stronger than for extended immobile holes but we think this is partially due to the overestimation of electron-hole interaction involved in the model. Since all our results tend to suggest that the experimentally observed FES are connected with localized holes, we will hereafter follow the analysis of the effect of the different physical magnitudes just by using extended holes with infinite effective mass. Figure 6 shows that in this case of infinite hole mass the FES of the emission spectrum is stronger than the bottom band singularity for widths of the wire of 100 Å or smaller.

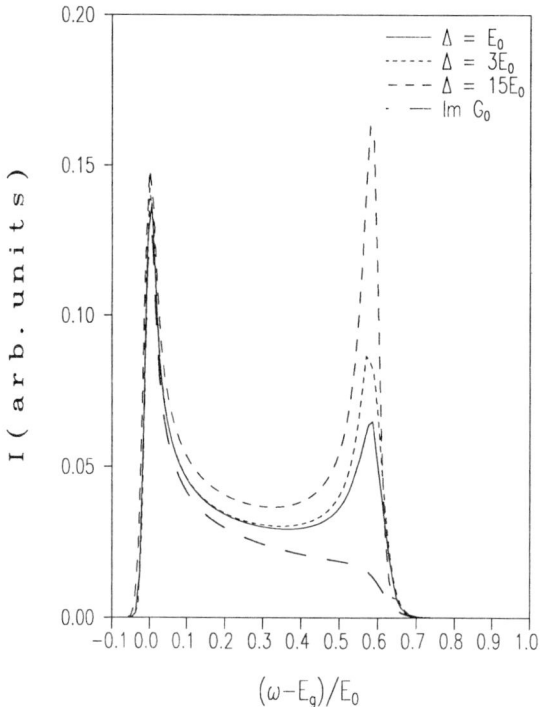

Figure 6. Emission spectra of wires with different widths (i.e., different band separations Δ) at $T = 0.02E_0/k_B$ with electron chemical potential $\mu_e = E_g + 0.6E_0$, and $m_h^* = 1000 m_e^*$.

4. Effect of the Carrier Distribution

In this section we are going to study how singularities are affected by the distribution of electrons and holes, i.e., the effect of T and μ_e. As discussed above, all the results correspond to $m_h^* = 1000$ and $\Delta = 1$, implying a width of the wire around 400 Å. For different polarizations one can have transitions from the top of the valence band to the even or to the odd parity electron bands, but they never interfere each other. Therefore we just consider the polarization which gives transitions to the first electron band [11].

In the expressions of Section 2 is not clear how the FES is affected by the variation of μ_e. In order to study this effect we compute just the emission spectrum where the singularity is analysed by comparing with the background. We take a case with low temperatures (T = 0.02) and no extrinsic holes ($\mu_h = 1$). Figure 7 shows the results for different values of the electron chemical potential. For higher temperatures everything is smoother as will be discussed later on. The FES is not extremely dependent on the chemical potential although slowly tends to disappear for high densities. When μ_e increases, the discontinuity at the Fermi level of G^0 slightly decreases, as shown in Fig. 7(b), due to the shape of 1D density of states. This effect is compensated by an increase of the interaction in the region of

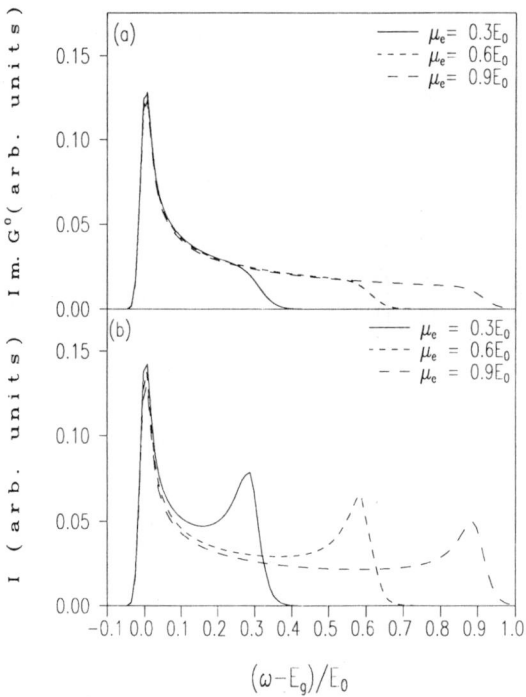

Figure 7. Emission spectra of a wire with different electron chemical potential μ_e at T = $0.02E_0/k_B$ with $\Delta = E_0$ and $m_h^* = 1000 m_e^*$. The two parts correspond to (a) non-interacting and (b) interacting electron-hole pairs, respectively.

wavevectors between 0 and $2k_F$ [11].

In order to study the effect of the temperature we concentrate again in the spectra of a wire with $\Delta = 1$ and $m_h^* = 1000$. We take $\mu_e = E_g + 0.6E_0$, which is an intermediate value among the ones considered above. Figure 8 shows the emission spectrum for different temperatures in the range between 0.02 and 0.10 (i.e., between 1.4 and 7 K). We have used a chemical potential for the holes $\mu_h = 0.01$ so that the total number of holes (and consequently the scale of the spectrum) does not change appreciably in the range of temperatures of interest as it occurs in the experiments [3]. The main result to be drawn is the disappearance of the FES for a temperature rather smaller than μ_e. In fact such a disappearance does not depend significantly on μ_e. In Fig. 9 we show the absorption spectra for the same three temperatures of Fig. 8. Once again the FES is quenched for temperatures significantly smaller than μ_e.

5. Comparison with Experiments

Our calculations have been performed for a single wire with electrons and holes in the same spatial region. Therefore they are hard to compare with experiments in multiwires with

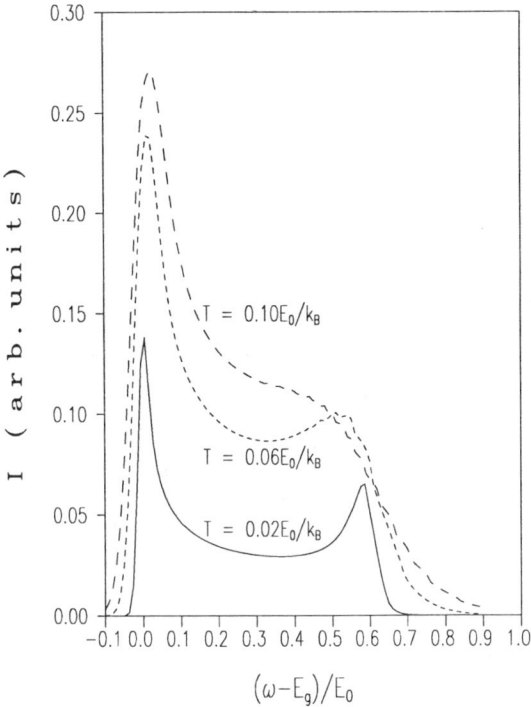

Figure 8. Temperature variation of the emission spectrum of a wire with $\mu_e = 0.6E_0$, $\Delta = E_0$, and $m_h^* = 1000 m_e^*$.

electrons and holes at different positions [3]. However, we think that some very general results should be applicable to any system made with wires. The first one is that, although formally optical singularities exist for extended holes with effective masses typical of perfect wires, in practice just with infinite m_h^* or with an impurity hole one gets clear optical singularities. The only way of getting singularities with mobile holes should be to have wires much narrower than the ones presently available. Therefore, we think that the experimentally observed singularities [3] are most probably connected with holes localized either by impurities or by some disordered variation of the wire width. An experimental fact rather difficult to understand is that the singularity of the band bottom is not observed in the emission experiments. Calleja et al. [3] suggest that this is connected with some variations in the wire width which would alter the energy position of such feature in a random way with respect to the FES which is the reference because is always at the same energy. However, we find that singularities coming from the bottom of the bands are in general much more intense than the FES. Since width fluctuations only reduce the intensity of these peaks in, at most, a factor of two [13] the lack of band bottom peaks is difficult to understand. Therefore, we think that this disappearance should be typical of systems with electrons and holes spatially separated. It should be interesting to have some experiments in a case similar to ours becaue its high symmetry would give new and interesting

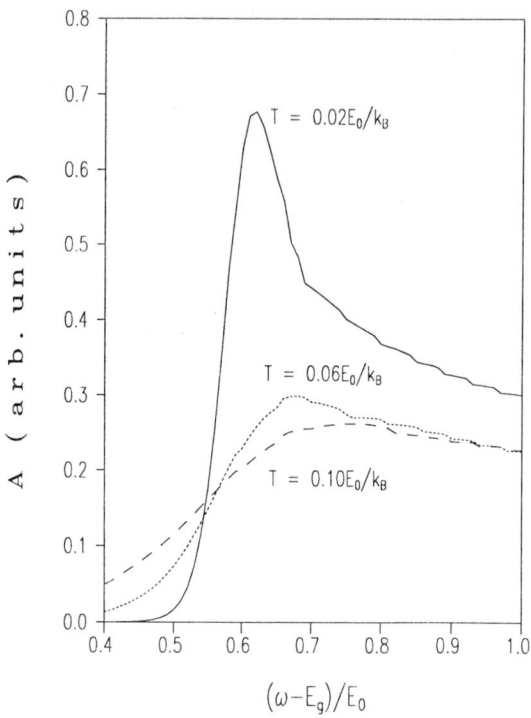

Figure 9. Temperature variation of the absorption spectrum of a wire with $\mu_e = 0.6E_0$, $\Delta = E_0$, and $m_h^* = 1000 m_e^*$.

possibilities as shown by our results.

Acknowledgments

We are indebted to L. Brey, J.M. Calleja, A. Pinczuk, and S. Schmitt-Rink for helpful discussions. This work has been supported in part by the Comision Interministerial de Ciencia y Tecnologia of Spain under contract MAT 91 0210. One of us (F.J.R.) acknowledges financial support from the World Laboratory.

References

1. G.D. Mahan, *Many Particle Physics* (Plenum, New York, 1981).
2. H. Haug and S. Schmitt-Rink, Prog. Quant. Electr. **9**, 3 (1984).
3. J.M. Calleja, A.R. Goñi, B.S. Denis, J.S. Weiner, A. Pinczuk, S. Schmitt-Rink, L.N. Pfeiffer, K.W. West, J.F. Mueller, and A.E. Ruckenstein, Solid State Commun. **79**, 911 (1991); J.M. Calleja, A.R. Goñi, B.S. Denis, J.S. Weiner, A. Pinczuk, S. Schmitt-Rink,

L.N. Pfeiffer, K.W. West, J.F. Mueller, and A.E. Ruckenstein, Surf. Sci. **263**, 346 (1992).
4. A.S. Plaut, H. Lage, P. Grambow, D. Heitmann, K. von Klitzing, and K. Ploog, Phys. Rev. Lett. **67**, 1642 (1991).
5. F. Hirler, R. Kuchler, R. Strenz, G. Abstreiter, G. Bohm, J. Smoliner, G. Trankle, and G. Weimann, Surf. Sci. **263**, 536 (1992).
6. K. Brunner, G. Abstreiter, M. Walther, G. Bohm, and G. Trankle, Surf. Sci. **267**, 218 (1992).
7. J.F. Mueller, Phys Rev. B **42**, 11189 (1990); J.F. Mueller, A.E. Ruckenstein, and S. Schmitt-Rink, Mod. Phys. Lett. **5**, 135 (1990).
8. P. Hawrylak, Solid State Commun. **81**, 525 (1992).
9. J.F. Mueller, A.E. Ruckenstein, and S. Schmitt-Rink, Phys. Rev. B **45**,8902 (1992).
10. T. Ogawa, A. Furusaki, and N. Nagaosa, Phys. Rev. Lett. **68** 3638 (1992).
11. F.J. Rodríguez and C. Tejedor, Phys. Rev. B **47**, 1506 (1993).
12. G.Y. Hu and R.F. O'Connell, J. Phys.: Condens. Matter **2**, 9381 (1990).
13. U. Bockelmann and G. Bastard, Phys. Rev. B **45**, 1688 (1992).

ACCEPTOR RELATED PHOTOLUMINESCENCE AS A PROBE OF MANY-ELECTRON STATES IN SEMICONDUCTOR NANOSTRUCTURES

P. HAWRYLAK
Institute for Microstructural Sciences
National Research Council of Canada
Ottawa, K1A OR6, Canada

ABSTRACT. We review the use of acceptor related photoluminescence as a probe of many-electron states in semiconductor nanostructures. The nonperturbative exact treatment of the Fermi edge singularity is reviewed and applied to the radiative recombination spectrum of the quasi-two-dimensional and quasi-one-dimensional electron gas.

1. Introduction

The radiative recombination of electrons in the conduction band with photo-excited holes in the valence band provides information about occupied electron states in modulation doped semiconductor nanostructures such as quantum wells, wires, and dots. To separate the effects of a valence hole from electronic effects it is useful to localise the valence hole in a well defined manner. This has been accomplished by selective doping of semiconductor microstructures with acceptors [1]. Acceptor related recombination has been used in the study of electronic properties of the electron gas [1,2], incompressible liquid [3], Wigner crystal [4], and in hot carrier relaxation [5].

Typically when acceptors are present, a well defined emission line below the band to band recombination spectrum appears [1,2]. This line corresponds to a recombination of an electron with a hole strongly bound to a negatively charged acceptor, and one can indeed focus on the electronic system. Unfortunately, even for weakly correlated electrons in the dense electron gas, the many-electron response to a localized perturbation such as recombination on an acceptor is divergent, and the line shape of the emission spectrum does not reflect the single particle density of states in a simple way. This phenomenon is known as the Fermi edge singularity (FES) [6-8]. The FES in emission spectra of modulation doped quantum wells has been observed by Skolnick et al. [9]. It manifests itself in the enhancement of the emission spectrum at high energies, i.e., in the vicinity of the Fermi surface. Hence we can think of the optical probe as inducing strong correlations among electrons even in the noninteracting electron gas. For low electron densities, low dimensions, and very high magnetic fields, the kinetic energy of electrons becomes quenched and the electron system becomes highly correlated. The radiative recombination in the correlated regime requires different methods and will be discussed elsewhere.

The enhancement of the emission spectrum of the weakly correlated electron gas at the Fermi energy is a result of three mutually compensating processes:

1. Direct scattering of the photo-excited electron and the photo-excited valence hole. This scattering process leads to bound electron-hole states, i.e., excitons. It is the only process in the absence of free carriers.
2. Exchange scattering. In this process the photo-excited electron is exchanged with the electron from the Fermi sea via virtual electron-hole pair excitations.
3. Self-energy of the hole or shake-up processes. This process describes the shake-up of the Fermi sea due to the anihilation of the valence hole. The relative contribution of each of these processes can be tuned by changing carrier density, position of acceptor with respect to the electron gas, and magnetic field, and by making use of empty subbands.

The remainder of this paper is organised as follows. In Sec. 2, I will describe the results of calculations of the recombination spectrum, which exactly sums all the three compensating processes. The method will be used to calculate the emission spectra for a quasi-two-dimensional electron gas in a single heterostructure in Sec. 3 and a quasi-one dimensional electron gas in an ideal wire in Sec. 4. For a heterostructure the effect of higher subbands and magnetic field on the emission spectrum will be discussed in Secs. 5 and 6, respectively. The comparison with experimental results will be given in Sec. 7, and the summary in Sec. 8.

2. Theory of Emission

The physical picture of the recombination process is very simple: prior to illumination (and after the recombination) our system consists of N electrons and a single negatively charged acceptor which acts as a repulsive scattering center for electrons. The single particle states and energies for N conduction electrons in a heterostructure and one valence state localised on the acceptor are denoted by $|\lambda\rangle$ and e_λ and $|h\rangle$ and $-\omega_a$, respectively. This is our final basis. It contains all the single particle effects of the confining potentials (subband structure), disorder, and magnetic field. The effect of electron-electron interactions maybe included via static self-energies. When the localised state is filled the net potential seen by electrons in a conduction band is that of a negatively charged acceptor. Upon illumination, one electron is added to the conduction band and a hole from the valence band relaxes to the state $|h\rangle$, localised by the negatively charged acceptor. The net potential seen by conduction electrons is now that of a negatively charged acceptor and a positivily charged hole, making it a neutral weak scattering center. In the absence of free carriers the recombination is due to the exciton bound to a negatively charged acceptor. The single particle states and energies of N+1 conduction electrons in a heterostructure in the presence of a neutral acceptor are denoted by $|k\rangle$ and e_k. This is our initial basis.

The emission spectrum $E(\omega)$ involves the emission of a photon with frequency ω with one of the N+1 conduction electrons making a transition to the empty level (hole) localized on the acceptor. The annihilation of the hole changes the potential seen by all electrons in the conduction subband from charge neutral to a negatively charged acceptor. This makes the transition a many-electron effect. The emission spectrum $E(\omega)$ can be derived [10] directly from the Fermi's Golden Rule (we set $\hbar=1$):

$$E(\omega) = 2\pi \sum_f |\langle \Psi_f(N+1) | \sum_{j=1}^{N+1} p_j | \Psi_i(N+1) \rangle|^2 \delta(E_i(N+1) - E_f(N+1) - \omega). \quad (1)$$

The initial state $|\Psi_i(N+1)>$ is the Slater determinant of N+1 particles occupying the N+1 lowest single particle conduction states $|k>$ in the presence of a neutral acceptor. The final states $|\Psi_f(N+1)>$ are Slater determinants of N+1 particles with one particle occupying the localized valence hole level $|h>$ and N particles occupying all possible single particle conduction states $|\lambda>$, i.e., conduction states in the presence of a repulsive acceptor potential. The main difficulty in using the Golden Rule directly is the large number of final many-particle states, for which both matrix elements and energies have to be evaluated. This problem was circumvented by Combescot and Nozieres (CN) [11], Mahan [12], and Ohtaka et al. [13]. We give here a brief derivation [10].

The total interband momentum operator couples conduction states only to the valence state localized by acceptor. Expanding final states with respect to the localized state and using the Fourier transform of a δ function allows us to rewrite Eq. (1) as:

$$E(\omega) = 2\text{Re} \int_0^\infty dt \, e^{-i(\omega-\omega_{max})t} E(t). \tag{2}$$

The maximum photon frequency is given by the difference between the ground state energies of N+1 particles before and after emission: $\omega_{max} = E_i(N+1) - (E_f^0(N) + (-\omega_a))$. Here $E_f^0(N)$ is the ground state energy of the normal (final) state, i.e., without the photoexcited electron. The time dependent emission spectrum E(t) of course assures that no frequencies larger than the maximum allowed frequency ω_{max} contribute to the frequency spectrum. From Eq. (1), E(t) is given by:

$$E(t) = \sum_f |\sum_{n=1}^{N+1} m_{k_n} (-1)^n < \Psi_f(N) | \Psi_i^{k_n}(N) > |^2 \, e^{-iE_f(N)t} e^{itE_f^0}. \tag{3}$$

In Eq. (3), transition matrix elements are expressed in terms of single particle transition matrix elements $m_k = p_{vc}<h|k>$ corresponding to transition from a conduction state $|k>$ to a localized state $|h>$; p_{vc} are conduction to valence momentum matrix elements and $<h|k>$ is the overlap of the conduction and localized electron envelope wavefunction. Each single particle matrix element is weighted by the overlap of initial and final N electron states. The initial state $|\Psi_i^{k_n}(N)>$ is formed from the N+1 lowest $|k>$ states except for the state $|k_n>$. It therefore represents a *hole in the Fermi sea*. Equation (3) can now be manipulated further to explicitly involve the summation over final states:

$$E(t) = \sum_{n=1}^{N+1} \sum_{n'=1}^{N+1} m_{k_n} (-1)^n < \Psi_i^{k_n}(N) | \sum_f | \Psi_f(N) > e^{-iE_f(N)t}$$
$$\times < \Psi_f(N) | \Psi_i^{k_{n'}}(N) > m_{k_{n'}} (-1)^{n'} e^{itE_f^0}. \tag{4}$$

The summation over a complete set of final states can now be eliminated by introducing a final state hamiltonian H_f that operates only in the space of conduction electron states. H_f is the sum of single particle hamiltonians h_f describing conduction electrons in the presence of the repulsive potential of acceptor. Using the final state Hamiltonian we can write formally:

$$E(t) = \sum_{n=1}^{N+1}\sum_{n'=1}^{N+1} m_{k_n}(-1)^n < \Psi_i^{k_n}(N)|e^{-itH_f}|\Psi_i^{k_{n'}}(N) > m_{k_{n'}}(-1)^{n'} e^{itE_f^0}. \quad (5)$$

The time dependent overlap of N electron wavefunctions of the initial basis propagated by the final state Hamiltonian $<\Psi_i^{k_n}(N)|e^{-itH_f}|\Psi_i^{k_{n'}}(N)>$ is equal to the determinant $D_{k,k'}$ of matrix Φ of order N+1 built out of matrix elements $\phi_{p,p'}$ ($\phi_{p,p'} = <p|e^{-itH_f}|p'>$, with p,p' occupied) whose nth row and n' column have been deleted. Using the relationship between the inverse of matrix Φ and determinants $D_{k,k'}$ we obtain the Combescot and Nozieres [11,12] result for emission:

$$E(t) = e^{iE_f^0 t} \text{Det}(\Phi(t)) \sum_{k,k'\leq k_F} m_k \Phi_{k,k'}^{-1}(t) m_{k'}. \quad (6)$$

The first term in Eq. (4), Det(Φ), describes the shake up of the Fermi sea due to the disappearence of the valence hole, while the last term describes vertex corrections, i.e., the scattering of the hole inside the Fermi sea by a repulsive potential in the final state Hamiltonian. This scattering process is mediated by exchange of the photo-created hole with holes (empty states) above the Fermi surface. Equation (6) involves the states of the photoexcited system while one wants to measure the states of the final basis. This is done by transforming Eq. (6) into the final basis. We define matrix $G_{\lambda,\lambda'}$ which describes the propagation of the hole in the Fermi sea in the final state basis as $G_{\lambda,\lambda'}(t) = \sum_{k,k'\leq k_F} <\lambda|k> \Phi_{k,k'}^{-1}(t) <k'|\lambda'>$, where $<k|\lambda>$ are the overlap matrix elements between the initial and final single particle states. Using the identity $\varphi_{k,k'} = \sum_\lambda <k|\lambda> e^{-ie_\lambda t} <\lambda|k'>$, the relationship between the initial matrix elements m_k and final basis matrix elements m_λ given by $m_k = \sum_\lambda m_\lambda <\lambda|k>$, and the identity $\text{Det}(\varphi) = \exp(\text{Tr}\{\ln(\phi)\}) = \exp(-iC(t))$, a set of *nonlinear* differential equations for the time evolution of the vertex (G) and self-energy (C) functions can be derived:

$$\frac{\partial}{\partial t} G_{\lambda,\lambda'}(t) = -ie_\lambda G_{\lambda,\lambda'}(t) + i\sum_{\lambda''} G_{\lambda,\lambda''}(t) e_{\lambda''} G_{\lambda'',\lambda'}(t)$$

$$\frac{\partial}{\partial t} C(t) = 2\sum_\lambda e_\lambda G_{\lambda,\lambda}(t). \quad (7)$$

The final expression for the emission function E(t) [8,10] is now given simply in terms of the vertex G and self-energy corrections C:

$$E(t) = e^{iE_f^0 t} e^{-iC(t)} \sum_{\lambda,\lambda'} m_\lambda e^{+ie_\lambda t} G_{\lambda,\lambda'}(t) m_{\lambda'}. \quad (8)$$

An important consequence of working in the final state basis is that *all* single particle states of the final basis contribute to the frequency spectrum of the emission E(t), irrespective of whether they are occupied or empty in the final ground state of the system, i.e., in the absence of the hole. The filling of phase space of initial states enters via the initial condition for the matrix $G(0)$: $G_{\lambda,\lambda'}(0) = \sum_{k \leq k_F} <\lambda|k><k|\lambda'>$. The overlap matrix elements $<k|\lambda>$ between the initial and final states are solutions of the Wannier equation:

$$e_k <k|\lambda> + \sum_{k'} V_{k,k'} <k'|\lambda> = e_\lambda <k|\lambda>. \tag{9}$$

The interaction $V_{k,k'}$ is the change in the one electron potential between initial and final bases. This change corresponds to a repulsive scattering potential due to the screened charge of the hole localized on acceptor. The matrix element is written in the initial basis, i.e., basis of the neutral acceptor.

3. Emission from Quasi-Two-Dimensional Electrons

For definiteness let us consider a single modulation doped GaAs/GaAlAs heterostructure with electron density n_s, in the range of 10^{11} cm^{-2} with typically one or two subbands occupied. The heterostructure [3] contains a delta doped layer of acceptors (Be) with sheet density of 2×10^{10} cm^{-2} at the distance $z_a = 250$ Å away from the GaAs/GaAlAs interface.

The theory of emission has been applied in Ref. [10] to a single heterojunction with only lowest subband occupied. The electron states in the absence of an acceptor can be written as a product of plane waves and the lowest subband wavefunction, which we take to be in the Stern-Howard form: $\xi_0(z) = \sqrt{b^3/2} z e^{-bz/2}$. The average position of the electron layer is given by $z_0 = 3/b$. The parameter b is obtained from variational calculations as a function of electron density. The acceptor and the hole are located away from the electron layer at position z_h. The effect of electrons and heterojunction structure on the hole is very small and we treat the acceptor as a bulk acceptor. We expand the hole wavefunction in terms of gaussians and retain only a single term in a first approximation. The hole wavefunction is therefore written as $\psi(x,y,z) = (2/\pi^2)^{3/4} \exp(-(x^2 + y^2 + z^2)/a^2)$ with parameter a playing the role of the effective hole radius. Since the hole screens the charge of the acceptor very effectively it is a very good approximation to assume a perfect screening condition, i.e., the initial basis is the basis in the absence of an impurity. In this approximation the matrix elements of the hole potential can be written as $V_{k,k'} = V(q) f(q) F(q)$, with $q = |k-k'|$ and $V(q)$ the statically screened interaction. $F(q)$ is a standard electron-hole form factor, which can be well approximated by $F(q) = \exp(-q|z_0-z_h|)$, and $f(q) = \exp(-q^2 a^2/8)$ is the Fourier transform of the in-plane hole charge density. $V(q)$ is the screened Coulomb interaction [8]. In a similar way matrix elements m_k can be effectively approximated by $m_k = m_0 (2\pi a^2 b^6/4)^{3/4} z_h e^{-bz_h/2} \exp(-k^2 a^2/4)$, where m_0 is a constant. Hence the coupling of conduction electron states with the hole is largest at the bottom of the band and decreases rapidly as the energy of the carriers increases.

In Fig. 1 we show calculated emission spectra for Fermi energy $E_F = 2$. Energies are measured in bulk Rydbergs and the Fermi energy is related to the carrier density n by $n = E_F/2\pi a_0^2$ where a_0 is the effective Bohr radius. For GaAs, $a_0 = 100$ Å and $E_F = 2$ corresponds to a carrier density $n = 3 \times 10^{11}$ cm^{-2}. The emission spectra were calculated [8,10] by propagating vertex and self-energy corrections up to a maximum time T_{max}. The emission spectra (solid lines) for two different time cutoffs $T_{max} E_F = 1.0$ and $T_{max} E_F$

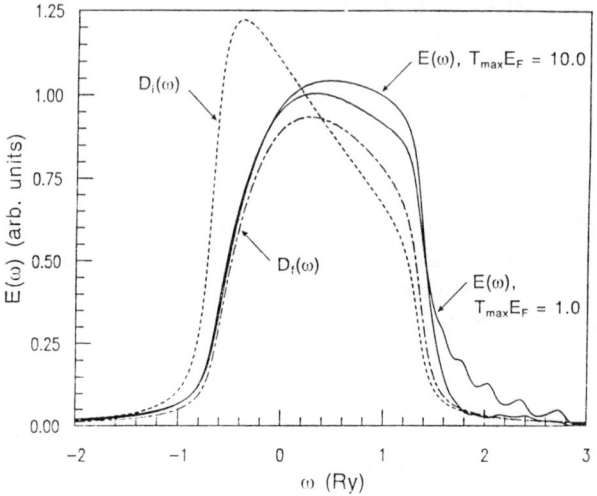

Figure 1. The emission spectrum of a heterostructure with carrier density 3×10^{11} cm^{-2} and Fermi energy $E_F = 2.0$ Ry (solid lines) for two different times of evolution. The joint density of states of the initial D_i and final D_f bases are also shown. The acceptor is located at $z_h = 15$ nm.

= 10.0 are shown. They are compared with the joint single particle density of states of conduction electrons and a hole in the absence of impurity D_i and in the presence of impurity D_f where $D_i(\omega) = \sum_{k \leq k_F} m_k^2 \delta(e_k + \omega_a - \omega)$ and $D_f(\omega) = \sum_{\lambda \leq \lambda_F} m_\lambda^2 \delta(e_\lambda + \omega_a - \omega)$, respectively. As illustrated in Fig. 1, the effect of self-energy and vertex correction is to modify the final joint density of states D_f in the vicinty of the Fermi energy. The two different times illustrate nicely the fact that as the system relaxes after emission of a photon the removal of forbidden frequencies $\omega > \omega_{max}$ corresponds to the buildup of the oscillator strength below ω_{max}, i.e., the formation of the FES. We have broadened the singularity by adding an imaginary part of 0.1–0.2 Ry to the frequency ω. The final joint density of states D_f approximates reasonable well the lineshape of the spectrum away from the FES. The joint density of initial states D_i is a poor approximation to the lineshape. However, the total integrated intensity of the emission line $E(t = 0)$ is given exactly by the total integrated intensity of the initial joint density of states D_i. In our model this is given by $E(0) = m_0^2 \xi_0^2(z_h) \sqrt{2\pi a^2} (1 - \exp(-E_F))$. This allows for a very easy interpretation of the integrated intensity as a function of carrier density, in contrast to a complicated lineshape. This should be compared with carrier densities extracted from the width of the emission line.

4. Emission from Quasi-One-Dimensional Electron Gas

The density of states $D(E)$ of a noninteracting quasi-one-dimensional electron gas diverges

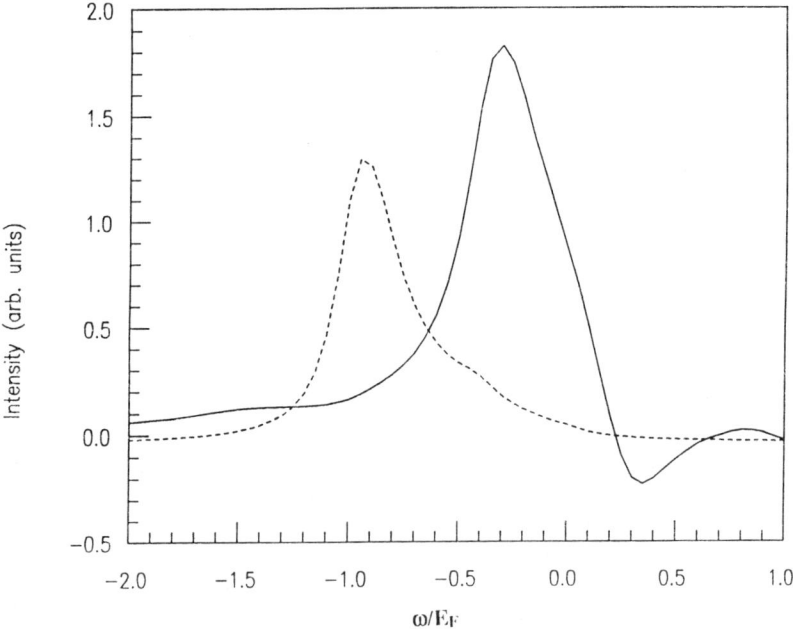

Figure 2. The emission spectrum for an ideal 1D channel of length L = 10a, repulsive impurity (model acceptor) with size a = 0.5a_0, and potential strenth V_0 = 1 Ry. There are 15 electrons present and 40 energy levels were included in the calculation. The dashed line shows the joint single particle density of states. The energy is measured in units of Fermi energy E_F.

at low energy as $E^{-1/2}$. Naively, one might therefore expect two peaks in the emission spectrum, one due to the divergent density of states at low energies, and a second peak at high energy due to FES. The acceptor related recombination of quasi-one-dimensional electron gas in quantum wires has been measured by Plaut et al. [14]. The emission spectra were quite featureless and did not resemble the single particle density of states. Similar photoluminescence experiments by Calleja et al. [15] but with free hole recombination did not reveal the anticipated single particle density of states either. One possible explanation is that the electron gas is not a Fermi liquid but a Luttinger liquid, Wigner solid, or some other highly correlated state. Another possibility is that the shake-up effects within the Fermi liquid picture are so strong that the lineshape of the emission spectrum does not correspond in a simple way to single particle density of states. Such possibility has been investigated in Ref. [16], where a simple model of an ideal one-dimensional (1D) channel of length L has been studied. In the center of the channel a repulsive impurity (a crude model of the negatively charged acceptor) in the form of a barrier of width a and height V_0 has been placed. Upon illumination a hole in the valence band is localized by a negatively charged acceptor. We assumed perfect screening, i.e., the potential of the hole screens the potential of the impurity completely. Hence the initial basis corresponds to electronic states of an ideal 1D channel of length L. With such a simple single particle basis the energy spectrum, optical matrix elements, and overlap matrix elements $<k|\lambda>$ between initial and final basis can be easily calculated. In Fig. 2 we show the emission spectrum of 15 electrons occupying 40 levels due to the recombination of one electron with the valence hole. We chose $V_0 = 4Ry$, $a = 0.5a_0$ and $L = 10a$. The dashed line shows the single particle density of states which peaks at low energies simply because the overlap of the hole and electron wavefunction is large for low energy states. However, the strong shake up effects of the 1D Fermi sea in the emission process shift the oscillator strength toward higher energies resulting in a single asymmetric peak. Similar results have been found for a parabolic confining potential and a repulsive impurity. For holes localised by potential fluctuations, calculations showed a presence of two peaks, the FES peak and a low energy peak due to divergent single particle density of states. More work is needed to understand the effect of electron correlations on emission spectra of 1D electrons.

5. Subband Spectroscopy of Electron Gas

Several groups studied the recombination process involving higher subbands [17–20]. Chen et al. [17] varied the energy of an empty subband with respect to the Fermi level and used it either to strongly enhance the recombination at the Fermi level or as an independent probe of the electron gas. By varying the position of the valence hole one can change the relative strength of optical matrix elements coupling different subbands and the effect of the hole potential on the electron gas.

The emission spectrum for several subbands can still be calculated within our general formalism [21,22]. The effect of an empty subband on the emission spectrum is two-fold: the subband mixing due to the hole potential affects the overlap matrix elements (and initial conditions) and the electron-hole pair excitation. The closer the Fermi level is to the bottom of the empty subband then the larger the mixing. In addition to this static excitonic correction, a dynamical response (shake-up) of the Fermi surface in terms of inter-subband electron-hole pair excitations leads to an enhancement of the emission at the Fermi level [21,22].

We now turn our attention to the recombination of photoexcited carriers in an otherwise empty subband. To study this separately from the recombination of equilibrium carriers, the Fermi level must be sufficiently separated from the bottom of the empty subband. We consider N carriers in the lowest subband occupying initial single particle states $k < k_F$ and one photoexcited carrier in a higher subband in some initial state $|K>$. These single particle states include the presence of the photoexcited hole. It is understood that $e_K \gg e_F$ so that we can neglect any exchange processes between the equilibrium and the photoexcited carriers, i.e., matrix elements $\Phi(t)_{k,K}$. The emission spectrum now separates into two contributions: one from equilibrium carriers and one from photoexcited carriers. The photoexcited carrier contribution can be written as:

$$E(t) = \exp\{itE_i(N) - iC(t)\} m_K^2 e^{i(e_K + \omega_a)t}. \tag{10}$$

It is a convolution of a single particle exciton-like transition superimposed with the shake-up process of the electron gas due to the annihilation of the valence hole. It is easily identified as the density of states (after Fourier transform) of the valence hole dressed by the excitations of the electron gas [16]. This is to be contrasted with the main subband emission, which is a convolution of excitonic effects (due to the creation of the hole in the Fermi sea) and shake-up processes.

The total integrated intensity I_0 of the second subband is, however, given by the single particle (exciton or free-electron–acceptor bound hole pair) oscillator strength $I_0 = m_K{}^2$. The effect of the Fermi sea is only via screening of the valence hole potential. For the acceptor, the screening of the valence hole is compensated by the screening of the repulsive acceptor potential, and the oscillator strength should be independent of the free carrier density.

6. Emission in a Magnetic Field

In a magnetic field perpendicular to the interface of a heterostructure, electrons occupy ladders of disorder broadened Landau levels l associated with different subbands n. The effect of disorder can be taken into account rather crudely by assuming a Gaussian density of states $D(E)$:

$$D(E) = \sum_n \frac{1}{\sqrt{2\pi}\Gamma \pi l_0{}^2} \sum_{l=0}^{\infty} \exp\{-(E - E_{n,l})/2\Gamma^2\}. \qquad (11)$$

Here $E_{n,l} = E_n + l\hbar\omega_c$ are subband Landau level energies, Γ is the broadening of Landau levels, $\hbar\omega_c$ is cyclotron energy, and l_0 is the magnetic length.

We consider the initial single particle basis as for a charge neutral acceptor while the final basis contains the negatively charged acceptor. Only zero angular momentum electron states contribute to the emission spectrum. Without disorder, there is only one zero angular momentum state per Landau level. However, the different angular momentum channels are mixed by the disorder and this we model by using a Gaussian density of states. The effect of localisation is neglected. We solve the Wannier equation for Gaussian broadened Landau levels with the density of states given by Eq. (11) and a short range potential for the acceptor [21,22], $V(k,k') = V_0 m_{n,l} m_{n',l'}$. The initial matrix elements are given by $m_{n,l} = m_n \exp\{-l\hbar\omega_c/E_b\}$, where E_b is the binding energy of the hole. The ratio of optical matrix elements from our self-consistent calculations is $m_2/m_1 = 3$, and we take $V_0 = 1.0$. The broadening of Landau levels due to disorder turns out to be important. When the Fermi level lies in the center of the Landau level, intra-Landau level electron-hole pair excitations lead to a FES in the emission spectrum. This is illustrated in Fig. 3(a) where the emission spectrum (solid line) corresponding to a filling factor $\nu = 4.84$ is compared with the emission spectrum without final state interactions (broken line). For this filling factor the second subband Landau level is well separated from the partially populated highest first subband Landau level. The effect of the repulsive final state acceptor potential is to shift and redistribute the oscillator strength of the emission lines corresponding to filled Landau levels to higher energies, and increase the emission intensity at the Fermi level (FES). As the magnetic field is increased and the Landau level depopulates futher, the character of response changes from a collective (dominated by the dynamical response of many electrons) to a single exciton-like particle. We illustrate this in Fig. 3(b) where the calculated emission spectrum for $\nu = 4.18$ is shown. For this filling factor, both first and second subbands are weakly populated by carriers. Without the presence of the second suband, the emission from the vicinity of the Fermi level would have been small due to

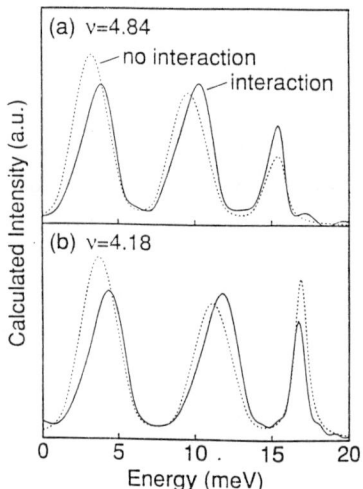

Figure 3. Calculated emission spectra for different filling factors with (solid line) and without (dashed line) electron-hole interaction for filling factor $\nu = 4.84$ (second subband empty) and filling factor $\nu = 4.18$ (second subband partially occupied).

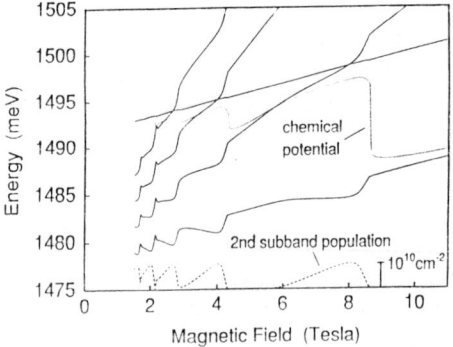

Figure 4. Calculated transition energies (solid line), chemical potential (dotted line), and occupation of the second subband (dashed line) as a function of magnetic field for a high carrier density ($n_s = 4.3 \times 10^{11}$ cm^{-2}). The curve of the second subband occupancy has been offset; in the high field limit, the second subband is unoccupied.

depopulation. The emission including the second subband, but without final state interactions, is strong due to the large matrix elements (broken line). The effect of the repulsive final state acceptor potential (solid line), however, is to reduce the intensity at the Fermi level when compared with the non-interacting emission. The decrease of intensity can be simply understood. Since the valence hole is localized on the acceptor, the repulsive finite state acceptor potential reduces the overlap of electron and hole wavefunction.

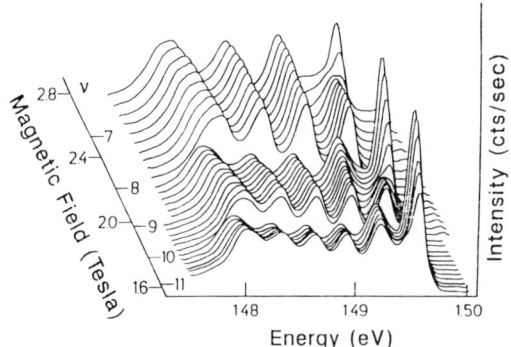

Figure 5. The measured luminescence spectra for magnetic fields in the range 1.64–2.79 T corresponding to filling factors v = 10.8–6.3 for a high carrier density $n_s = 4.3 \times 10^{11}$ cm^{-2}. The intensity of the highest Landau level is strongly enhanced towards even filling factor when the Fermi level approaches the second subband.

To understand the position of the emission line in a heterojunction in a magnetic field with Fermi level close to the bottom of the second subband we need to know the effect of carrier density on electron energies. The most important contribution comes from Hartree potentials. To separate truly many electron effects from mean field effects we perform a self-consistent calculation of subband levels in a magnetic field and for finite temperature [18,19]. The effect of disorder is taken into account by assuming a Gaussian density of states.

With the choice of Landau level broadening $\Gamma = 0.3(B)^{1/2}$ meV, other parameters include the electron density $n_s = 4.3 \times 10^{11}$ cm^{-2}, the depletion charge $N_D = 4.6 \times 10^{11}$ cm^{-2}, the acceptor sheet density $N_A = 2.0 \times 10^{10}$ cm^{-2}, and the acceptor position $z_a = 250$ Å. The actual transition energies involve energies of electrons $E_{n,l}$ and the hole energy ω_a. The hole is localized on the acceptor, and so the hole energy changes according to the changes of the self-consistent Hartree potential $V_H(z = z_a)$ at the position of the acceptor. Approximately, the energy of emitted photons is $\omega_{n,l} = \omega_a + E_{n,l} - V_H(z_a)$, where ω_a is the bare transition energy. Hence the transition energies probe both the energy levels of the electrons and the spatial distribution of the Hartree potential.

In Fig. 4, the self-consistently calculated transition energies, the chemical potential μ, and the electron density n_s in the second subband are plotted as a function of magnetic field. The pronounced oscillations in line positions as a function of magnetic field are identified with the transfer of carriers due to partially occupied first subband Landau levels crossing the lowest Landau level of the second subband. The finite width of the Landau levels means that close to the crossing point, a very small population ($n_s \approx 10^{10}$ cm^{-2}) of carriers is present in the second subband. Due to the large spatial extent of the second subband wave function, even a very small carrier density leads to changes in the self-consistent Hartree potential V_H. We find that the changes in the Hartree potential at the position of the acceptor combined with the electron confinement are responsible for the decrease of transition energies with increasing magnetic field and hence allow us to spectroscopically probe the spatial distribution of the Hartree potential.

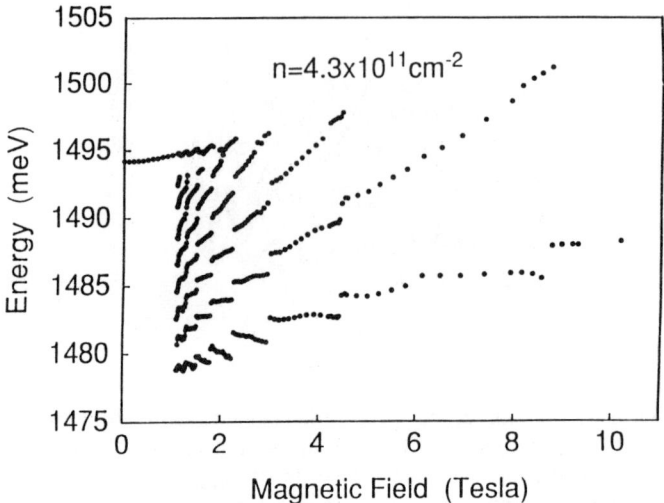

Figure 6. The measured transition energies of the Landau levels as a function of magnetic field for a high carrier density $n_s = 4.3 \times 10^{11}$ cm^{-2}. The discontinuities in level energy occur at even filling factors.

7. Experimental Results

The purpose of experiment was to test the effect of carrier density, magnetic field, and second subband on the emission spectrum of heterostructures. Photoluminescence measurements were performed by Pulsford [24] in magnetic fields up to 13 T at a temperature T = 1.5 K. The carrier concentration in the heterostructure was tuned by varying the incident laser power [3]. Typical laser intensities were in the range 10^{-5}–10^{-2} Wcm^{-2} giving electron densities in the range 1.9–4.3×10^{11} cm^{-2}. In each case, the emission spectra corresponding to the recombination of electrons with holes bound to acceptors (B-lines) were measured.

We first study the high carrier density limit $n_s = 4.3 \times 10^{11}$ cm^{-2} where the Fermi energy lies in the vicinity of the second subband. The experimentally estimated Fermi level (E_F) and subband separation (Δ) are $E_F = 15.0$ meV and $\Delta = 14.3$ meV. Luminescence emission spectra are shown in Fig. 5 for the magnetic field B increasing from 1.64 T (ν = 10.8) to 2.79 T (ν = 6.3). The data are viewed from an angle to resolve the spectral features more clearly. Transitions involving all occupied electron Landau levels are observed. The intensity of emission at the Fermi level shows striking increase in the vicinity of even filling factors, as predicted by theory (see also Fig. 4). The measured luminescence energies of the Landau levels are plotted as a function of magnetic field in Fig. 6 and should be compared with the calculated transition energies shown in Fig. 3. The energy discontinuity at even filling factors and decrease of transition energy with increasing magnetic field is in agreement with calculated transiton energies. The presence of enhancement at odd filling factors due to the FES is however not clearly visible and requires further study.

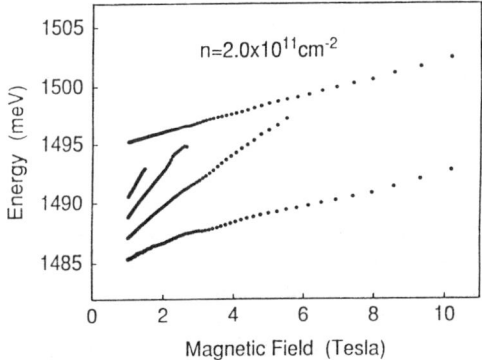

Figure 7. The measured transition energies of the Landau levels as a function of magnetic field for a low carrier density ($n_s = 2.0 \times 10^{11}$ cm^{-2}).

7.1. SECOND SUBBAND EMISSION ($E_F \ll E_2$)

We now turn our attention to the use of the second subband emission as a spectroscopic probe of the degenerate electron gas. By increasing illumination we reduce the carrier density [3], and increase the separation of the second subband energy and the Fermi energy. The lowest carrier density studied is $n_s = 2 \times 10^{11}$ cm^{-2}, with measured Fermi energy $E_F = 6.2$ meV and subband separation $\Delta = 9.5$ meV. The luminescence spectra are now dominated by emission from the occupied Landau levels in the first subband with only a weak feature corresponding to the emission from the second subband. The transition energies of the subband Landau levels are plotted in Fig. 7 as a function of magnetic field. We observe almost no oscillations of the first subband Landau level transition energies with magnetic field (or filling factor ν). This contrasts with large oscillations reported for the recombination of electrons with free holes in doped quantum wells, which is attributed to the oscillation of the valence hole self-energy with magnetic field [25].

The valence hole self-energy is due to the interaction with conduction electrons. The mobile hole drags with itself an electron cloud and lowers its energy. In the case of a free hole, the dynamic response (screening) of the electrons varies with magnetic field and leads to the oscillation in the valence hole self-energy. In comparison, a negatively charged acceptor reduces the electron density in its neighbourhood. Hence, when the hole is trapped by an acceptor, the hole and acceptor electronic clouds cancel each other. This is just another way of saying that the acceptor is charge neutral. Thus the only changes observable in acceptor samples are changes in the electron self-energy. As predicted by theory [25], the experiment shows that these changes are very small.

Let us now turn our attention to the second subband emission. The position of the emission line appears to be completely unaffected by the changes in the electron gas and follows the center of the second subband lowest Landau level. This behaviour can be understood on the basis of the discussion in Sec. 5. The valence hole potential mixes the Landau levels in both subbands and one would expect that the emission line shows an anticrossing behaviour whenever a first subband Landau level crosses the lowest Landau level of the second subband. However, the second subband recombination involves only the initial state, i.e., the electron state in the presence of a neutral acceptor. No mixing should be observed, in agreement with experiment. In contrast, an emission involving holes localized by potential fluctuations should reveal an anti-crossing behaviour, inversely periodic in Δ/B [17,22].

8. Summary

We have studied the radiative recombination processes involving electrons in a weakly correlated quasi-two-dimensional and quasi-one-dimensional electron gas. The recombination process involves the anihilation of the valence hole bound to the acceptor and the creation of the hole in the Fermi sea. This turns out to be a strong perturbation of the electron gas and carries a wealth of information about its ground and excited states. Different contributions to the emission process can be separated by varying the carrier density, applying a magnetic field, and using empty subbands.

Acknowledgments

A large part of this work was carried out in collaboration with N. Pulsford and K. Ploog. Useful discussions with A. Nurmikko are acknowledged.

References

1. I.V. Kukushkin, K. von Klitzing, K.Ploog, V.E. Kirpichev, and B.N. Shepel, Phys. Rev. B **40**, 4179 (1990); I.V. Kukushkin, K. von Klitzing, K. Ploog, and V.B. Timofev, Phys. Rev. B **40**, 7788 (1990).
2. A. Petrou, M.C. Smith, C.H. Perry, J.M. Worlock, and R.L. Aggarwal, Solid State Commun. **52**, 93 (1984).
3. H. Buhmann, W. Joss, K. von Klitzing, I.V. Kukushkin, G. Martinez, A.S. Plaut, K. Ploog, and V.B. Timofeev, Phys. Rev. Lett. **65**, 1056 (1990); H. Buhmann, W. Joss, K. von Klitzing, I.V. Kukushkin, A.S. Plaut, G. Martinez, K. Ploog, and V.B. Timofeev, Phys. Rev. Lett. **66**, 926 (1991).
4. I.V. Kukushkin, N.J. Pulsford, K. von Klitzing, K. Ploog, R.J. Haug, S.Koch, and V.B.Timofeev, Phys. Rev. B **45**, 4532 (1992).
5. R.G. Ulbrich, J.A. Kash, and J.C. Tang, Phys. Rev. Lett. **62**, 949 (1989).
6. For a review and full references, see, for example, G.D. Mahan, *Many-Particle Physics*, (Plenum Press, New York, 1981); P.Nozieres and C.T. DeDominicis, Phys. Rev. **178**, 1097 (1969); J. Gavoret, P. Nozieres, B. Roulet, and M. Combescot, J. Phys. (Paris) **30**, 987 (1969); S. Doniach and M. Sunijc, J. Phys. C **3**, 285 (1970).
7. S. Schmitt-Rink, D.S. Chemla, and D.A.B. Miller, Adv. Phys. **38**, 89 (1989).
8. P. Hawrylak, Phys. Rev. B **44**, 3821 (1991).
9. M.S. Skolnick, J.M. Rorison, K.J. Nash, D.J. Mowbray, P.R. Tapster, S.J. Bass, and A.D. Pitt, Phys. Rev. Lett. **58**, 2130 (1987)
10. P. Hawrylak, Phys. Rev. B **45**, 4237 (1992).
11. M. Combescot and P. Nozieres, J. Phys. (Paris) **32**, 913 (1972).

12. G.D. Mahan, Phys. Rev. B **21**, 1421 (1980).
13. K. Ohtaka and Y. Tanabe, Phys. Rev. B **39**, 3054 (1989).
14. A.S. Plaut, H.Lage, P. Grambow, D. Heitmann, K. von Klitzing, and K. PLoog, Phys. Rev. Lett. **67**, 1642 (1991).
15. J.M. Calleja, A.R. Goñi, B.S. Dennis, J.S. Weiner, A. Pinczuk, S. Schmitt-Rink, L.N. Pfeiffer, K.W. West, J.F. Mueller, and A.E. Ruckenstein, Solid State Commun. **79**, 911 (1991).
16. P. Hawrylak, Solid State Commun. **81**, 525 (1992).
17. W. Chen, M. Fritze, A.V. Nurmikko, D. Ackley, C. Colvard, and H. Lee, Phys. Rev. Lett. **64**, 2434 (1990); W. Chen, M. Fritze, W. Walecki, A. Nurmikko, D. Ackley, J.M. Hong, and L.L. Chang, Phys. Rev. B **45**, 8464 (1992).
18. P.E. Simmons, M.S. Skolnick, L.L. Taylor, S.J. Bass, and K.J. Nash, Solid State Commun. **67**, 1151 (1988).
19. A.J. Turberfield, S.R. Haynes, P.A. Wright, R.A. Ford, R.G. Clark, J.F. Ryan, J.J. Harris, and C.T. Foxon, Phys. Rev. Lett. **65**, 635 (1990).
20. B.B. Goldberg, D. Heiman, A. Pinczuk, L. Pfeiffer, and K. West, Surf. Sci. **263**, 9 (1992); B.B. Goldberg, D. Heiman, A. Pinczuk, L. Pfeiffer, and K.W. West, Phys. Rev. Lett. **65**, 641 (1990).
21. P. Hawrylak, Phys. Rev. B **44**, 6262 (1991).
22. P. Hawrylak, Phys. Rev. B **44**, 11236 (1991).
23. P. Hawrylak, Phys. Rev. B **42**, 8986 (1990).
24. P. Hawrylak, N. Pulsford, and K. Ploog, Phys. Rev. B **46**, 15193 (1992).
25. T. Uenoyama and L.J. Sham, Phys. Rev. B **39**, 11044 (1989).

TOWARDS FULLY QUANTIZED OPTOELECTRONIC SEMICONDUCTOR HETEROSTRUCTURES: QUANTUM BOXES OR QUANTUM MICROCAVITIES?

C. WEISBUCH
Laboratoire Central de Recherches
Thomson-CSF
F-91404 Orsay Cedex
France

ABSTRACT. It has been predicted that sharp and strong interband photoluminescence (PL) lines should be obtained in quantum box structures, allowing excellent laser performance. We explain however the usually observed decrease in interband PL efficiency with full three-dimensional quantum box quantization by the unavoidable energy relaxation bottleneck which develops in a fully quantized system. This bottleneck is conversely shown to be extremely useful in enhancing the performances of those devices based on intersubband transitions. An independent scheme can however be used to yield narrow interband PL lines (sharper than kT) through the photon quantization occuring in quantum microcavities, which selects the recombining electron-hole pair or exciton thanks to energy and momentum conservation. At exact resonance between cavity photons and excitons a normal-mode splitting appears, which can also be viewed as the vacuum-field induced Rabi splitting of the exciton.

1. Introduction

Capitalizing on the many improvements brought along by the one-dimensional (1D) quantization of layered semiconductors heterostructures (so-called 2D quantum wells (QWs)), further improvements of optical properties were predicted for structures with more fully quantized states in two or three dimensions in quantum-well wires (QWWs) or quantum boxes (QBs) [1]. Although many realizations of 1D QWWs or 0D QBs have been reported in the past few years [2], most of them reported worse, or at most as good properties as 2D QWs, at strong variance with the universal improvement brought along by QWs when compared to three-dimensional heterostructures [1]. We felt that beyond the surface and material damage brought along by the 2D and 3D manufacturing processes of 1D QWWs and 0D QBs, some intrinsic mechanism should be responsible for such a universal behaviour, independent of the many fabrication schemes used. We actually found [3,4] that whereas optical properties are predicted to improve with decreasing dimensionality thanks to the sharper optical transitions, there is conversely a price to pay for such a fuller quantization, that is reduced carrier energy relaxation rates. It is then found that energetically-injected electrons and holes in QWWs and above all in QBs most often remain in orthogonal quantum states, therefore forbidding radiative recombination in interband transitions. Whereas this relaxation bottleneck effect casts doubts on the feasibility of QB lasers, it is full of promises for intraband-based devices, which are most often limited in performance due to the fast inter-level transition rate brought along by longitudinal optical (LO) and longitudinal acoustic (LA) phonon emission.

Whereas increasing energy quantization has long been the rule of the game for electron states, it is only recently that the effect of photon quantization has been considered in the field of semiconductors. The influence of photon quantization due to optical cavity resonances has long been a matter of interest in the field of atomic physics with a prediction of the possibility of modifying the spontaneous emission rate of atoms dating back to 1946 [5]. The field has steadily developed over the years because of the information it brought about many fundamental aspects of quantum electrodynamics (QED), i.e. the interaction of atoms and electromagnetic fields [6]. This was recently revived by the observation of the deterministic, coherent behaviour of single atoms interacting with single photon modes, leading to the so-called vacuum-field Rabi oscillations [7,8]. The whole field is known as cavity QED [9] and has applications extended to fundamental quantum measurement theories ('Schrödinger cats') [10] or to the new photon squeezed states [11]. Another neighbouring field is that of photonic bandgap materials [12], where the exclusion of unwanted photon modes is provided by a periodic medium structure, instead of the phase cancellation in the multiple reflexions of optical cavities. We will show that photon quantization in microcavities actually provides an alternate scheme to reach the same goal as in QBs, i.e., narrow luminescence and gain curves. They additionally have a number of additional useful features such as controlled spontaneous emission, coherent interaction, etc.

In this progress report, we will successively describe the bottleneck effects in 1D and 0D structures and its application to interband photoluminescence (Sec. 2) and to intraband devices (Sec. 3), and finally quantum microcavity effects (Sec. 4).

2. The Phonon Relaxation Bottleneck in 0D (and 1D) Structures

In usual infinite 3D, 2D, and 1D structures electron states have a continuum density-of-states, or quasi-continuous for large enough structure dimensions. Therefore, energy relaxation downwards by emitting a phonon is always allowed. In 0D structures, the full quantization of energy levels implies that only a given phonon with energy and momentum determined by the electron (or hole) initial and final states will ensure energy and momentum conservation.

As we will see thereafter, a critical QB size can be defined below which LA phonon relaxation between two levels separated by energy ΔE is strongly diminished. In a first approximation, it is given by [13]

$$\Delta E \approx \pi^2 \hbar^2 / 2mL_\perp^2 \approx \hbar c_s 2\pi / L_z, \tag{1}$$

where $L_\perp$ and L_z are the largest (lateral) and smallest box dimensions, respectively, and c_s is the sound velocity. Due to the heavier hole mass, according to Eq. (1), the critical box size for electrons is reached before that corresponding to holes; we therefore will not consider any further hole relaxation bottleneck effects, and will only assume that holes reach a thermal distribution.

Actually, the electron relaxation of injected electrons in a real situation, for instance from a wide-bandgap confining material, involves a cascade of LO, LA and transverse acoustic (TA) phonons. The LO phonon modes are the most efficient due to the polar-optical coupling. However, LO phonons are almost dispersionless with an energy $\hbar\omega_{LO}$ in the 30 meV range for the III-V compounds most widely used in our context. Therefore, as soon as the energy difference between QB levels reaches a few meV, LO phonon relaxation can be strongly forbidden as an excited electron does not find another level situated $\hbar\omega_{LO}$ below and can therefore relax only by TA and LA phonons, which have a continuum of energies upwards from zero. In the following, we will not consider any further this LO-phonon bottleneck problem and will concentrate on the more general situation obtained whenever an electron reaches an energy level within $\hbar\omega_{LO}$ from the ground state through an LO phonon cascade, from which it can only reach the ground state through acoustic phonon emission. We also consider only LA phonons, as it is well-known that TA phonon scattering is ~ 10 times smaller than LA scattering due to its main piezoelectric coupling.

The LA phonon scattering rate between two states ψ_i and ψ_f for quantum box electrons in an infinitely deep well is given by [13]

$$w_{if} = \frac{1}{\tau_{i \to f}}$$
$$= \frac{2\pi}{\hbar} \sum_q E_q \frac{D^2}{2\rho c_s^2 \Omega} \left| \left\langle \psi_f \left| e^{\pm iqr} \right| \psi_i \right\rangle \right|^2 \delta(E_f - E_i \pm E_q) \left(n_B(E_q) + \begin{Bmatrix} 1 \\ 0 \end{Bmatrix} \right), \qquad (2)$$

where D is the deformation potential, r the crystal density, Ω a normalization volume, and q the phonon wavevector. The δ-function accounts for energy conservation. The upper (lower) sign holds for phonon emission (absorption), and $n_B(E_q)$ is the phonon occupation number given by the Bose-Einstein statistics. For acoustic phonons, considering bulk properties is a reasonable assumption due to their small reflection coefficient at well interfaces.

The square matrix element, from now on denoted as M(q), is separable, and reflects momentum conservation in the three directions

$$M(q) = M_x(q_x) M_y(q_y) M_z(q_z) \qquad (3)$$

each factor being of the form

$$M_\alpha(q_\alpha) \sim \int \sin(k_{\alpha i} r_{\alpha i}) e^{q_\alpha r_\alpha} \sin(k_{\alpha f} r_{\alpha f}) dr_\alpha, \qquad (4)$$

where α stands for x, y or z and where $k_{\alpha i(f)} = n_{i(f)} \pi / L_x$, as is the case in the infinite-well approximation. We give in the following a very simplified description of a detailed analysis which is being developed in Ref. [4].

For the matrix element M to reach its maximum value 1 implies that all $M_\alpha(q_\alpha)$ are roughly equal to 1. They do so if the q_α are such that $q_\alpha \approx (n_{\alpha i} - n_{\alpha f}) \pi / L_\alpha$ (see Fig. 1). However, at the same time, energy conservation requires $\Delta E_{i,f} = E_f - E_i = \hbar c_s q$. Therefore, we have to compare, for two levels i and f which determine the modulus of q, those wavevectors q for which all the $M_\alpha(q_\alpha)$ are ≈ 1. If we confine ourselves to such flat boxes where one dimension is much smaller than the others (typically the thickness of the QW into which QBs are etched) and for parallel kinetic energies smaller than the z-quantized level energy separation (hence $\pi/L_{x,y} < \pi/L_z$), we can determine energy and momentum conservation as represented in Fig. 1: in any α-direction, the maximum q vector of the phonons which yield $M_\alpha(q_\alpha) \sim 1$ is roughly determined by the Fourier transform of the

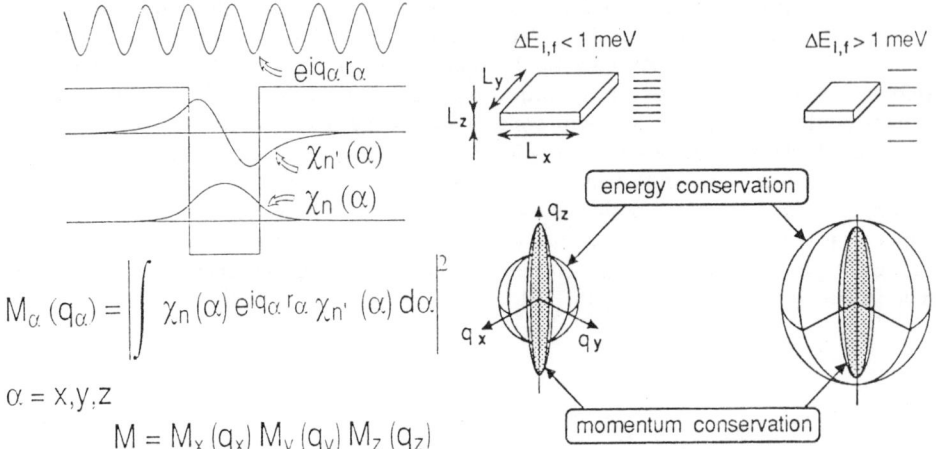

Figure 1. Schematics of the LA phonon-scattering matrix element M in a QB (left) and representation of the locus of energy and momentum conserving states (right): the ellipsoid represents the volume inside which $M \sim 1$, given approximately by the Fourier transform of the QB wavefunction, and the sphere represents the phonon momentum required for energy conservation, given by $\hbar c_s q \sim \Delta E_{n,n'}$.

electron wavefunction in that direction, i.e., $\sim \pi/L_\alpha$. The locus of q modulus ensuring the energy conservation condition is a sphere of radius $q = \Delta E_{i,f}/\hbar c_s$. Depending on $\Delta E_{i,f}$, the sphere cuts, or not, the ellipsoid which is the Fourier transform of the quantum box.

As soon as the sphere is larger than the ellipsoïd, i.e., for $\Delta E_{i,f} > \hbar c_s q_z = h c_s 2\pi/L_z$, M is smaller than 1. This defines the critical lateral box size as given by Eq. (1), remembering that in the infinite QW approximation $\Delta E_{if} \approx \pi^2 \hbar^2/2mL^2$ is a representative value. It can be shown [3,4] that $w_{i,f} = 1/\tau_{i \to f}$ decreases like $L^8_{x,y}$ beyond the critical size, very fast indeed. Then, electrons accumulate into excited states which are mostly orthogonal to the thermal hole states. A detailed calculation of the full relaxation cascade is shown in Fig. 2, assuming that besides the radiative recombination probability, an energy-independent non-radiative recombination process exists with a rate $W_{nr} = (1 \text{ ns})^{-1}$. The actual box size at the onset of diminished relaxation is only very little dependant on the precise value of W_{nr} as the relaxation rate varies as L^8_x in the region of non-k conservation, as mentioned above. From there, it is shown that the slight change from 150 nm box lateral size to 115 nm dramatically changes the occupancy factor for electrons, therefore drastically changing the radiative recombination probability. The result of the calculated quantum efficiency for varying box size is shown on Fig. 3.

A similar calculation for quantum wires leads to the result shown also in Fig. 3. It can be remarked [3] that for a rectangular wire with an aspect ratio a (i.e., with dimensions aL and L) the largest energy separation in a subset (nl) occurs for those levels (nlm) near the next subset edge (n,l+1). One can easily determine that this separation is $\pi^2 h^2/2maL^2_x$, i.e., that the value of the critical box size will vary as $a^{-1/2}$ or in other words that a rectangular box with surface area aL^2_x experiences the same bottleneck as a square box with side $\sqrt{a}L_x$ having the same area. This is well observed in Fig. 3 where the onset for diminished yield occurs at a size $\sim 1/3$ that of the square box of side the same small side for an aspect ratio of

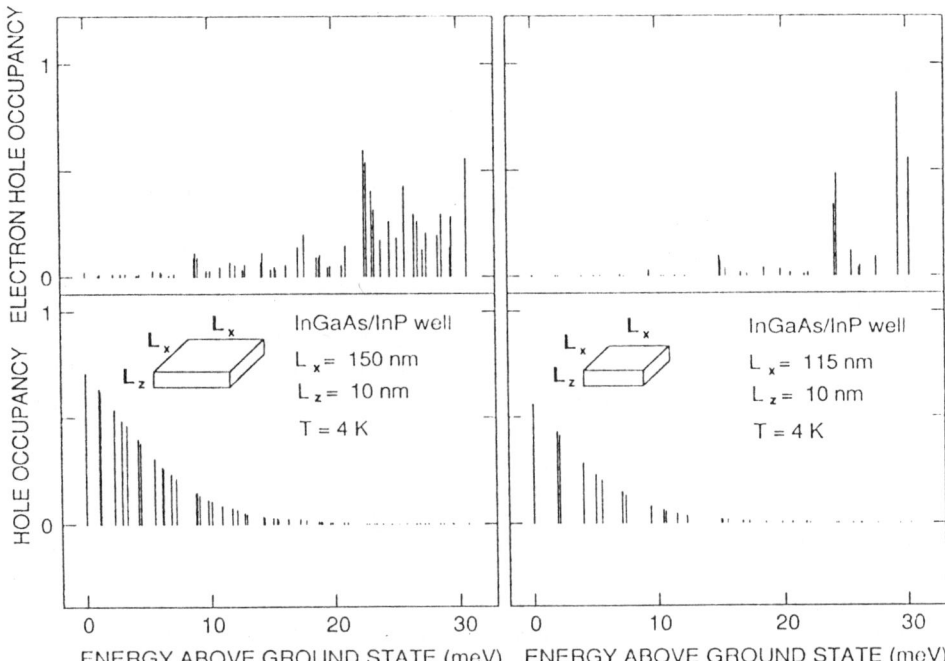

Figure 2. Calculated electron and hole occupancy factors for InGaAs/InP QBs with 10 nm thickness and varying lateral sizes. Note the already large occupancy of excited states for 150 nm. A non-radiative recombination rate $w_{NR} = (1 \text{ ns})^{-1}$ is assumed (a scale-up of 10 of the hole energies to account for the 10 times heavier hole mass has been performed in order to visualize directly the k-matched electron and holes states).

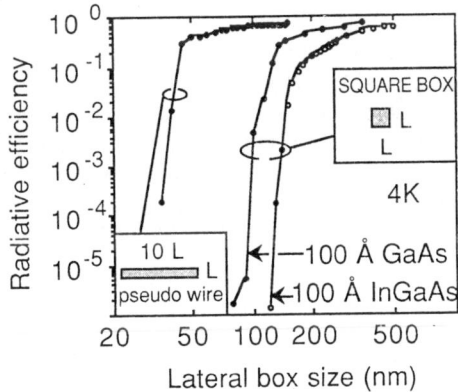

Figure 3. Calculated radiative yields at 4 K of square and elongated (approximately quantum wires) quantum boxes of dimensions $L \times L$ and $L \times 10L$ respectively, fabricated of GaInAs lattice-matched to InP or Ga(Al)As, as a function of L.

10. Of course, infinite long wires do not feel any bottleneck effect as they possess a continuum of electron states.

It is clear that this type of calculation is oversimplified: it neglects both electron-electron collisions and two-phonon transitions. The former mechanism has been shown to be very important in the special case of quasi-QBs obtained by Landau quantization of QW states in a magnetic field transverse to the QW: in that case, energy levels are equidistant and energy-conserving collisions between Landau states can promote one electron up and one down. Even in that very favorable case, strongly diminished carrier relaxation ($\sim 10^3$ factor) has been evidenced at strong quantization [14]. In the case of isolated quantum boxes where up-and-down energy spacings are unequal (we exclude the very special case of quantum boxes with parabolic confining-energy potentials), such collisions most often cannot take place because of the impossibility to conserve energy in the transition. At low temperatures, two-phonon transitions can be safely considered as negligible compared to one-phonon transitions, with the small exception of an accidental matching of the energy difference between two levels thanks to an LO phonon and an acoustical phonon emissions. However, even though such an occurrence could eventually remove the bottleneck for two given levels, another bottleneck would soon occur for some other levels for slightly smaller box sizes. Due to the very fast variation of the phonon relaxation mechanism described here, it is hard to envision any improper approximation which would remove the bottlenecks effects for the 200 Å or smaller QB required for room-temperature quantization effects, whereas the bottleneck sets in for sizes as large as 1000 Å.

Many experiments support the decrease of quantum yield with box sizes around and below 100 nm [15]. Also, a number of authors have reported the observation of non-thermal excited-states luminescence [16] or slowed-down energy relaxation rate [17]. A most convincing argument about the relevance of the intrinsic mechanism described here lies in the comparison between QWW and QB results: whereas QWW have shown good luminescence characteristics for wire width down to 20 nm [18], QBs stop at sizes ~ 100 nm. Assuming similar surface defects densities and dead layer thicknesses, one would expect any collapse of QB properties due to these two process-induced phenomena to occur at similar sizes. The fact that it does not show that something else occurs, namely the relaxation bottleneck in QBs. This is not to say of course that in many cases, some part of the decline of PL efficiency is indeed due to process-induced defects, particularly for the harsher process in the GaAs/GaAlAs materials system with no regrowth.

The conclusion of this description of the phonon relaxation bottleneck mechanism makes it hard to foresee the feasibility of QB lasers: their analysis [19] relies on the fact that electrons and holes are in the ground state where they are k-matched. Unless some clever means to inject electrons and holes into their respective ground state is found, relaxation bottleneck at the required 20 nm sizes will forbid the excited particles to reach the ground states.

3. Implications for Intersubband-Transitions Based Devices

Whereas the phonon relaxation bottleneck appears to be very detrimental to efficient interband devices, it can be put to good use in an area where the phenomenon is badly needed: intersubband-based devices have their performance limited by the very short excited state lifetime due to efficient phonon emission. Due to the strong decrease in the

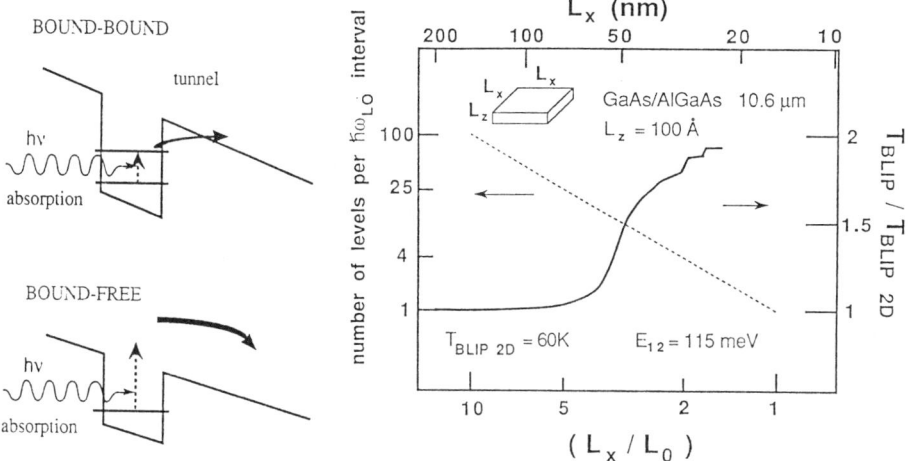

Figure 4. Schematics of intersubband-based IR QW detectors based on bound-bound or bound-unbound transitions (left) and calculation of the BLIP temperature for QB detectors as a function of box diameter (right).

phonon-limited relaxation rate in QBs, we expect a drastic improvement in device performance. We will take as examples the infrared (IR) intersubband-based detector and the IR emitter.

3.1. THE IR QB DETECTOR

The IR intersubband transition-based detector has already shown impressive results [20]. It is based on the intersubband IR absorption transition followed by the photoionization of the excited states into the barrier material surrounding the QW (bound-bound transitions), or by the direct excitation from the QW ground state to an unbound barrier state (bound-unbound transitions). However, one of its main operating parameters, namely the background limited infrared performance (BLIP) temperature, is quite low, due to the short excited state lifetime [21]: the BLIP temperature is defined by the equality of the photo-generated and of the thermally-generated excited-state population. Whereas in the usual interband detectors (such as HgCdTe) the excited state lifetime is of the order of microseconds, limited by radiative recombination, it is in the subpicosecond range in excited quantum wells (as measured by the QW IR absorption linewidth) due to LO phonon emission, leading to a much worse BLIP temperature. For QW detectors patterned as boxes, a simulation of device operation was carried out [4] taking into account the two types of quantum states [i.e., (1,1,m) and (2,1',m')] to be considered in the modelling, the result being shown in Fig. 4. The result is impressive, yielding a near doubling of the BLIP temperature (usually 60 K in 2D QWs) at lateral box sixes ~ 50 nm, well before real 0D quantization occurs and within reach of present-day fabrication techniques. It should be emphasized that the IR transition energy is mainly given by the quantum box thickness determined by the quantum well layer growth, the lateral box patterning not changing the

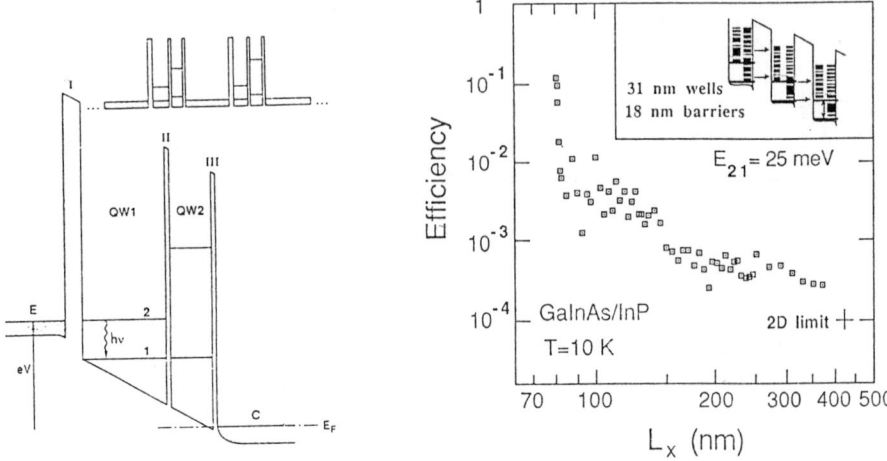

Figure 5. Schematics of an infrared IR emitting device based on excited electron injection and extraction by resonant tunneling [22] (left) and calculation of radiative yield at 10 K (right).

E_{12} energy and being only used to produce the phonon relaxation bottleneck.

3.2. THE IR EMITTER

Several schemes have been proposed for IR intersubband emitters [22]. Figure 5 shows one of these where a Multiple QW (MQW) structure with two-well period is used to inject and extract electrons by resonant tunneling. Here again the radiative recombination transitions compete with an extremely efficient phonon relaxation mechanism when using 2D QWs. The result of a model calculation [4] of the radiative quantum efficiency of boxes taking into account the various transitions is shown in Fig. 5. Here again, a major improvement is obtained (~ 3 orders of magnitude improvement) with an achievable patterning of ~ 50 nm diameter boxes.

4. Optical Microcavities

It seems so far that due to the bottlenecked phonon relaxation mechanism in QBs, one has to give up on obtaining sharp luminescence and gain curves with interband transitions. Fortunately, this is not so thanks to the new possibility offered by the photon quantization in microcavities [23-26]. The basic principle is easy to grasp: electrons and holes in energy bands recombine at the various energies allowed by carrier statistics while conserving energy and momentum ("vertical" transitions) because there exist in usual conditions a continuum of photon states which induce spontaneous emission through the so-called vacuum-field fluctuations. If we consider that the active material is placed in an optical cavity where only a single photon mode has a spectral overlap with the emission band,

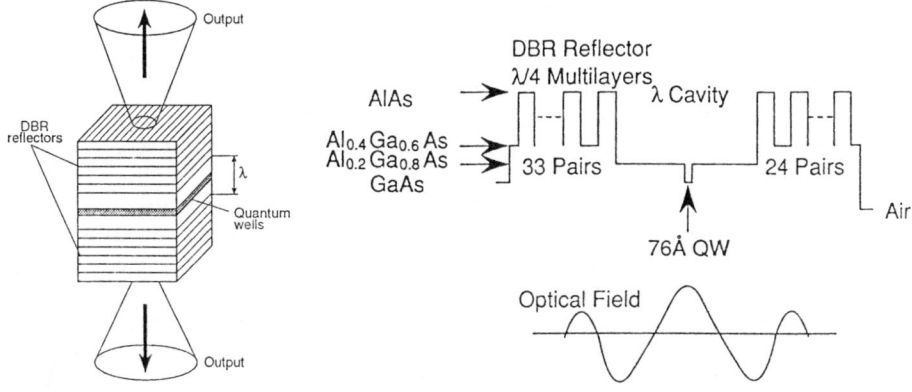

Figure 6. Schematics of a quantum microcavity (left) and its realization in the GaAlAs/GaAs multilayer materials system (right), where the band profile across the structure is schematically shown.

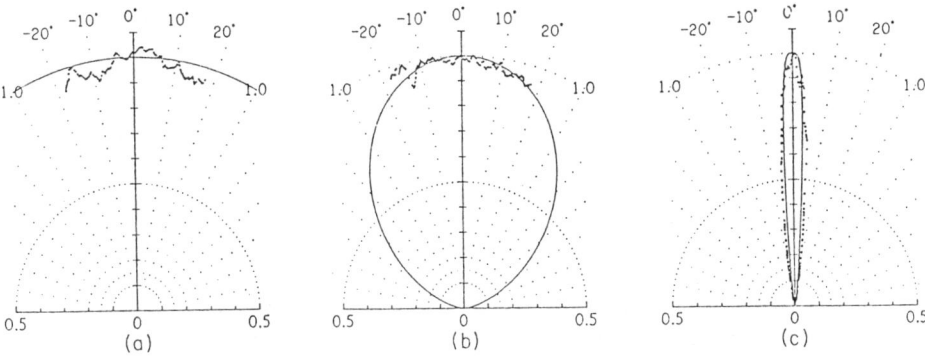

Figure 7. Angular spontaneous emission diagram for a standard double-heterostructure with no cavity for (a) s-wave and (b) p-wave, and (c) for a quantum microcavity. (From Ref. [23].)

radiative recombination can only take place in that photon mode. The emitted light has the spectral characteristics of the photon mode, i.e., can be extremely narrow as determined by the photon lifetime in the cavity. The fate of all electron-hole pairs not matching the energy and momentum of the mode photon is to either recombine non-radiatively or, in "good" samples, to reach the photon-matched states by momentum and energy relaxation before non-radiative recombination takes place.

The actual implementation (Fig. 6) of semiconductor microcavity structures improves this simple "zero-order" description: one uses planar Fabry-Perot (FP) microcavities where the mirrors are multi-quarter wave stacks distributed-Bragg reflectors (DBRs). The usual multipass interference description of the FP resonator shows that the resonant photon mode, for which all reflected waves in the cavity are in phase, has an intensity increased by a factor 4/(1-R), where R is the DBR mirror reflectivity [23]. Conversely, those non-resonant photon modes are suppressed by a factor ≈ 1-R. The cavity acts as if it concentrated the electric field from all non-resonant modes into the resonant mode. The net effect is therefore that the spontaneous lifetime is hardly changed, as long as relaxation towards the photon-coupled electron-hole state prevents a too-strong hole burning phenomenon [23–25].

Many predictions of the impact of microcavity effects have been verified. One of the most promising device-wise is the concentration of the spontaneous emitted light in the narrow angle of the cavity mode (Fig. 7), of the order of (1-R)/π inside the microcavity material [23,24]. This opens the way to high efficiency LEDs with accessible emission angles instead of the approximately Lambertian emission of cavity-less structures. It should also lead to the "thresholdless" laser [23] as the emission process will continuously switch from spontaneous to stimulated in the same single photon mode while increasing exciting power.

Recently, we have evidenced the resonant interaction of the microcavity photon mode with the QW excitons, which can be equivalently seen as the vacuum-field Rabi oscillation of the excitons [26].

Rabi oscillations can be seen as a coupled-oscillator process, by which resonantly coupled atomic and photon oscillators periodically exchange energy. In a classical mechanical oscillator description, the overall system response yields two split modes corresponding to the normal modes. In an atomic transition language, one considers the system as undergoing a coherent evolution with a photon being absorbed by an atom, which subsequently emits a photon with the same energy and wavevector k, that photon being reabsorbed, and so-on. A simple criterion for this phenomenon to spontaneously occur in a cavity can be put as [7]

$$\alpha d \gg 1-R \approx \pi/F, \tag{5}$$

where α is the absorption coefficient, d the absorbing medium length, R the cavity mirror reflection coefficient and F the cavity finesse. This inequality states that before escaping the cavity, an emitted photon is reabsorbed. To be observable, such an effect also supposes that both the atom excitation is not damped out and that it will emit in the same photon mode which led to absorption. The cavity also fulfills the latter condition, as it provides a single mode to interact with atoms (the cavity mode) and suppresses all other modes.

To obtain an order of magnitude of the expected effect is must be recalled that excitons in QWs have an oscillator strength per unit surface given by (in the exact 2D quantum limit) [27]

$$Nf = 8f/\pi\, a_B^2, \tag{6}$$

where f is an atomic oscillator strength (the one used in computing electron-hole interband transitions) and a_B the 3D exciton Bohr radius.

Our cavities, grown by metal-organic chemical vapor deposition (MOCVD) involve 24 GaAlAs/GaAs stacks on the front side (air interface) and 33 stacks on the substrate side in order to balance the front and back side reflections at ~ 98% (Fig. 6).

As is usual, the thickness variation across the wafer is such that the FP wavelength varies ~

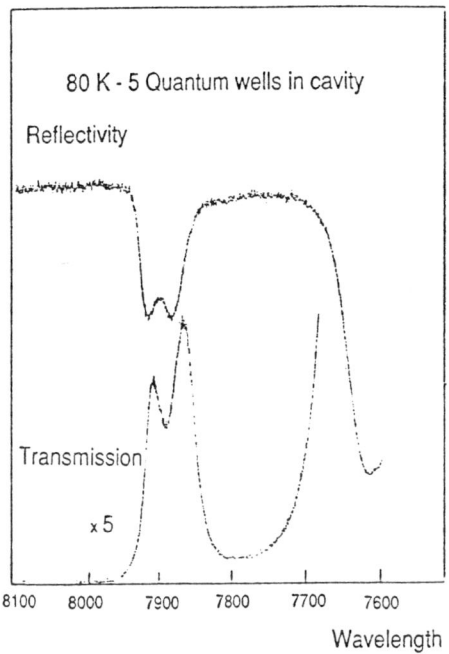

Figure 8. Reflectivity and transmission curves of a 5 QW quantum microcavity at 77 K, at a point where excitons and the photon mode of the cavity are in resonance, displaying the normal-mode splitting of resonant coupled oscillators.

300 Å from wafer center to side, while the QW exciton frequency is hardly changed, its energy corresponding mainly to the GaAs bandgap. To change the FP cavity frequency, one therefore only needs to probe different points on the wafer.

Figure 8 shows transmission and reflectivity for a microcavity with 5 QWs in the cavity. A point corresponding to exact resonance between the cavity and exciton frequencies has been selected and a clear splitting is seen. Figure 9 shows the crossing behaviour observed when scanning through the resonance.

A simple theoretical treatment consists in considering the DFB reflectors as simple, wideband mirrors of constant reflectivity and transmission coefficients. Then, the standard multi-beam Fabry-Perot analysis can be carried out, which yields the transmission coefficients as [7]

$$T(\nu) = \frac{T^2 e^{-\alpha d}}{\left(1 - R e^{-\alpha d}\right)^2 + 4 R e^{-\alpha d} \sin^2(\Phi/2)}, \quad (7)$$

where R and T are reflection and transmission coefficients of the front and back mirrors assumed identical, α the absorption coefficient of the cavity QWs with total length d, and Φ the dephasing over the various materials of the cavity.

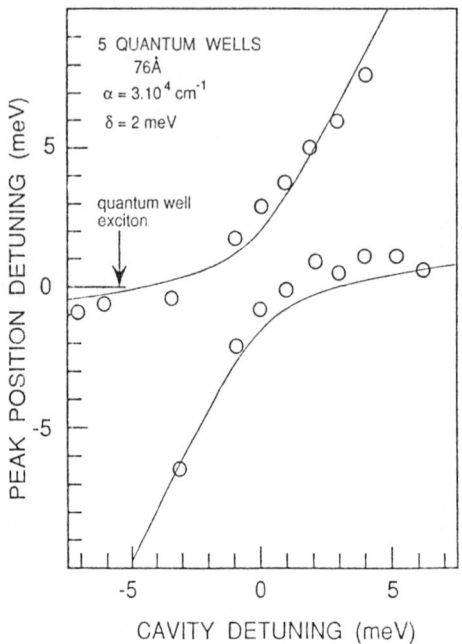

Figure 9. Variation of the reflectivity peak positions as various point on the sample are tested, allowing various values of the detuning between excitons and cavity photons.

$$\Phi(\omega) = 2\pi(\Delta - \Delta_{res})/\Delta_{FSR} + 4\pi(n - n_b)dv/c, \qquad (8)$$

where $\Delta = \omega - \omega_{exc}$, $\Delta_{res} = \omega_{cav} - \omega_{exc}$, $\Delta_{FSR} = c/2L$ is the cavity free spectral range, n and n_b are respectively the resonant and background index of refraction of the cavity quantum wells. We use for the exciton dielectric function a simple two-level approximation with Lorentzian linewidth $\delta = 2$ meV. The peak absorption α is adjusted to the standard value of 3×10^4 cm^{-1}. Such a description classical linear-dispersion model of the system then yields two peak or dips in the transmission or reflectivity curves respectively, separated by, when the oscillator frequencies are equal ($\omega_{exc} = \omega_{cav}$) [7]

$$\Omega = \left(\frac{\alpha \, d \, \delta \, \Delta_{FSR}}{\pi}\right)^{1/2} \qquad (9)$$

The theoretical fit is obtained from the above simplified analysis, or a standard multiple-interference analysis of the DBR-FP interferometer using a matrix formulation for optical propagation [28]. From the fit with $\alpha = 3 \times 10^4$ cm^{-1} and $\delta = 2$ meV, we can deduce Nf = 4×10^{12} cm^{-2}, in excellent agreement with theoretical evaluations [29].

The quantum microcavity concepts should influence various devices: the structure is of course that of vertical cavity surface emitting lasers (VCSELs) [30]. It is also similar to the Fabry-Perot structures used as electro-optic modulators [31] and to the non-linear Fabry-

Perot etalons used as optical-optical modulators [32]. So far, to our knowledge, no account of coupled-moding was done in such systems.

In VCSELs, a main impact will occur if laser action is based on exciton recombination. However, it is well-known that, at least at room temperature, exciton dissociation and carrier interactions are much faster than the exciton radiative lifetime (in the nanosecond range), so that excitons are usually wiped out into electron-hole pairs [27]. The existence of the rapid Rabi oscillation might change drastically this state of affair, as the oscillation could be faster than the dissociation time, leading to efficient radiative recombination whenever a coupled atom-photon mode escapes the cavity.

In order to obtain thresholdless lasers, it is often assumed that 3D quantum microcavities are required to have coupling in only one photon mode as many other modes exist in DBR planar 1D cavities [23]. Our experiment clearly shows that coupling to the main cavity mode is strongly dominant, at least 70%, as otherwise damping would wipe out the splitting.

For those devices which rely on unrelaxed QW excitations, excitons are still very important at room temperature [27] and are at the root of the unsurpassed performance of QW heterostructures-based systems. So far, these structures involved many quantum wells as active materials, typically 30–100 [31,32]. The present experiments evidence that, due to coupled-moding, strong modifications of cavity properties can be obtained with as few as 3 QWs . This indicates that modification of the QW properties by the usual external actions such as an electric field (quantum confined Stark effect) or carrier-band filling can be exerted on much less material when in a strong coupled-moding situation (i.e., a high Q cavity). Under such conditions, one can expect figures of merit of such devices to be improved by a factor 10–30. An additional bonus is the very high sensitivity of the coupled-mode to any perturbation of one of its components, the exciton or cavity mode, be it by an applied electric field or a non-linear saturation intensity for electro-optic or non-linear devices respectively. One might wonder why the coupled-modes were not observed so far in these widely researched structures: in our view, it is most certainly due to the fact that the usual active media investigated use too many QWs, which lead to a poor cavity quality factor. On the other hand, under such conditions the cavity width is so large that a good operating point is always obtained, which is not the case in our high-Q system unless the thickness is very precisely controlled.

5. Acknowledgments

I wish to thank Henri Benisty for the very fruitful collaboration we had developing the concept of phonon relaxation bottleneck. I also would like to thank the NTT Corporation for support through an endowed chair at the University of Tokyo, which was made available in the summer and fall of 1991 during which the microcavity work was performed. I also thank RCAST for an excellent hospitality during that time. At Thomson, M. Papuchon and J.P. Pocholle provided useful discussions, and E. Costard and N. Laurent experimental help. Recently, M. Potin and R. Aymard developed further luminescence measurements.

References

1. For a general background, see, for example, C. Weisbuch and Borge Vinter, *Quantum*

Semiconductor Devices (Academic, Boston, 1991).
2. See, for example, the contributions by A. Plaut, D. Heitmann, G. Abstreiter, and J. Kotthaus, in these proceedings.
3. H. Benisty, C.M. Sotomayor-Torres and C. Weisbuch, Phys. Rev. B **44**, 10495 (1991).
4. H. Benisty and C. Weisbuch, to be published.
5. E.M. Purcell, Phys. Rev. **69**, 681 (1946).
6. S. Haroche and D. Kleppner, Physics Today **42**, 24 (January 1989).
7. Y. Zhu, D.J. Gauthier, S.E. Morin, Q. Wu, H.J. Carmichael, and T.W. Mossberg, Phys. Rev. Lett. **64**, 2499 (1990).
8. R.J. Thompson, G. Rempe, and H.J. Kimble, Phys. Rev. Lett. **68**, 1132, (1992).
9. For recent excellent reviews, see, for example, S. Haroche, in *Fundamental Systems in Quantum Optics*, edited by J. Dalibard, J.M. Raimond, and J. Zinn-Justin (Elsevier, Amsterdam, 1991); E.A. Hinds, in *Advances in Atomic and Molecular Physics* Vol. 20, edited by D. Bates and B. Bederson (Academic, Boston, 1991), p. 237.
10. M. Brune, S. Haroche, J.M. Raimond, L. Davidovich, and N. Zagwry, Phys. Rev. A **45**, 5193 (1992).
11. Y. Yamanoto, S. Machida, K. Igeta, and H. Horikoshi, in *Coherence and Quantum Optics VI*, edited by L. Mandel, E. Wolf, and J.H. Eberly (Plenum, New York, 1990), p. 1249.
12, See, for example, S. John, Physics Today **45**, 32 (May 1991).
13. U. Böckelmann and G. Bastard, Phys. Rev. B **42**, 8947 (1990).
14. J. Stark, W.H. Knox, and D.S. Chemla, Phys. Rev. Lett. **68**, 3080 (1992); D.S. Chemla, in these proceedings.
15. See, for example, E.M. Clausen, H.G. Craighead, J.M. Worlock, J.P. Harbison, L.M. Schiavone, L. Florez, and B. Van der Gaag, Appl. Phys. Lett. **55**, 1427 (1989); B.E. Maile, A. Forchel, R. Germann, D. Grützmacher, H.P. Meier, and J.P. Reithmaier, J. Vac. Sci. Technol. **B7**, 20301 (1989) and references therein.
16. M.A. Reed, R.T. Bate, K. Bradshow, W.M. Duncan, W.R. Rensley, J.W. Lee, and H.D. Shih, J. Vac. Sci. Technol. B **4**, 358 (1986); J.N. Patillon, R. Gamonal, M. Iost, J.P. André, B. Soucail, C. Delalande, and M. Voos, J. Appl. Phys. **68**, 3789 (1980); G. Abstreiter, in these proceedings.
17. G. Mayer, H. Leier, B.E. Maile, A. Forchel, H. Schweizer, G. Weimann, and W. Schlapp, in *Proc. 20th Int. Conf. on the Physics of Semiconductors*, edited by E.M. Anastassakis and J.D. Joannopoulos (World Scientific, Singapore, 1990), p. 2415.
18. See these proceedings.
19. Y. Arakawa and H. Sakaki, Appl. Phys. Lett. **40**, 939 (1982); M. Asada, Y. Miyamoto, and Y. Suematsu, IEEE J. Quantum Electron. **QE-22**, 1915 (1986); Y. Miyamoto, Y. Miyake, M. Asada, and Y. Suematsu, IEEE J. Quantum Electron. **QE-25**, 2001 (1989).
20. For a general and recent reference, see E. Rosencher, B. Levine, and B. Vinter, *Intersubband Transitions in Quantum Wells*, (Plenum, New York, 1992).
21. M.A. Kinch and A. Yariv, Appl. Phys. Lett. **55**, 2093 (1989).
22. See, for example, A. Kastalsky, V.J. Goldman, and J.H. Abeles, Appl. Phys. Lett. **59**, 2636 (1991) and references therein.
23. Y. Yamamoto, S. Machida, K. Igeta, and G. Björk, in *Coherence, Amplification and Quantum Effects in Semiconductor Lasers*, edited by Y. Yamanoto (Wiley, New York, 1991), p. 561; G. Björk, S. Machida, Y. Yamamoto, and K. Igeta, Phys. Rev. A **44**, 669 (1992).

24. H. Yokohama, Science **256**, 66 (1992).
25. T. Baba , S. Hamano, F. Koyama, and K. Iga, IEEE J. Quantum Electron. **QE-27**, 1347 (1991).
26. C. Weisbuch, M. Nishioka, A. Ishikawa, and Y. Arakawa, Phys. Rev. Lett. **69**, 3314 (1992).
27. D.A.B. Miller and D.S. Chemla, J. Opt. Soc. Am. B **2**, 1155 (1985).
28. See, for example, M. Born and E. Wolf, *Principles of Optics*, 6th Ed. (Pergamon, Oxford, 1986).
29. L. Andreani and A. Pasquarello, Phys. Rev. B **42**, 8928 (1990).
30. See, for example, J. Jewell, J.P. Harbison, A. Scherer, Y.H. Lee, and L.T. Florez, IEEE J. Quantum Electron. **QE-27**, 1332 (1991) and references therein.
31. G.D. Boyd and G. Livescu, Optics and Quantum Electronics **24**, 147 (1992); M. Whitehead, G. Parry, and P. Wheatley, Physica Scripta T **35**, 210 (1991).
32. N. Peyghambarian and S.W. Koch, in *Non-Linear Photonics*, edited by H.M. Gibbs (Springer, Berlin, 1990), p. 7.

LUMINESCENCE PROPERTIES OF GaAs QUANTUM WELLS, WIRES, DOTS, AND ANTIDOTS

G. ABSTREITER, G. BÖHM, K. BRUNNER, F. HIRLER, R. STRENZ, AND G. WEIMANN
Walter Schottky Institut
Technische Universität München
D-8046 Garching, Germany

ABSTRACT. Luminescence properties of doped and undoped GaAs/(AlGa)As quantum wells, wires, dots, and antidots are discussed. The complex lineshape of high quality modulation doped single quantum wells can be understood in terms of wave vector conserving and nonconserving transitions. A shallow etching technique together with holographic optical lithography results in a spatial separation of electrons and photogenerated holes. The luminescence features of wires, dots, and antidots show both direct and indirect character in real space. Isolated single quantum dots have been fabricated by local interdiffusion of undoped GaAs/AlGaAs quantum wells. Due to the inherent exclusion of inhomogeneous broadening in single dot structures very sharp luminescence lines are observed in originally 3 nm wide quantum wells.

1. Introduction

In recent years extensive research activity has been focused on the fabrication and characterization of one- and zero-dimensional semiconductor quantum structures. The quality of such structures, however, often suffers from the poor properties of the side walls with a high density of nonradiative recombination centers. Apart from optical and electron beam lithography together with deep etching or ion implantation and subsequent annealing, also special growth methods and strain patterning techniques have been applied [1–5]. Shallow etching or local interdiffusion by focused laser beam are different approaches to realize lateral potential or energy gap modulations. Results obtained with both techniques are presented in this contribution.

2. Shallow Etched Doped Quantum Well Structures

To achieve a lateral band edge modulation we apply a shallow wet chemical etching technique. A remote doped single quantum well (SQW) serves as starting point. A typical layer sequence (Fig. 1) consists of a 60 Å GaAs quantum well grown on a thick $Al_{0.35}Ga_{0.65}As$ buffer, followed by a 200 Å undoped $Al_{0.35}Ga_{0.65}As$ spacer layer, a 500 Å Si doped $Al_{0.35}Ga_{0.65}As$ layer, and a 100 Å GaAs cap layer. The corresponding electron density and the mobility of the two-dimensional electron gas (2DEG) at 4.2 K under illumination are 3.8×10^{11} cm^{-2} and 56000 cm^2/Vs, respectively. A wire, antidot, or dot photoresist pattern with a period of typically 500 nm is produced by means of laser holography on top of the sample. To create a lateral potential modulation (see schematic

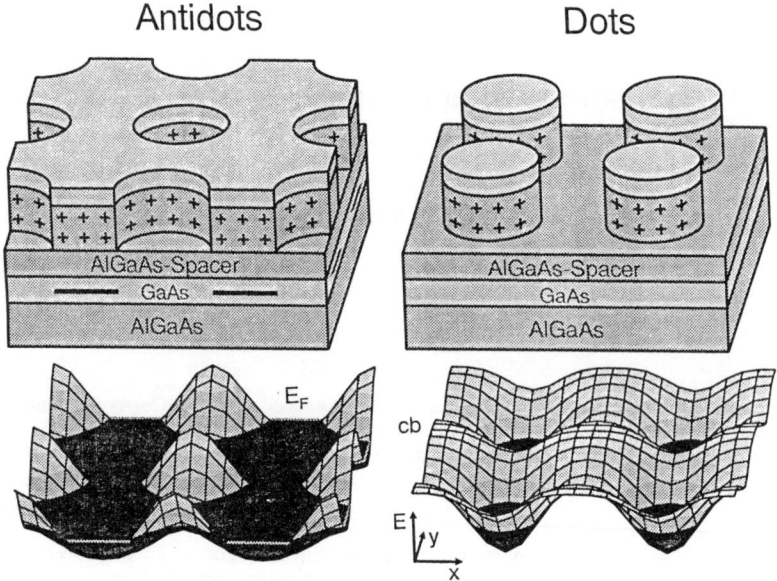

Figure 1. Schematic view of shallow etched antidot and dot samples with corresponding lateral potential modulation. The black areas indicate the electron gas in the conduction bands.

conduction band structure in Fig. 1) the pattern is transferred into the sample using a slow wet chemical etching technique. The etching process is stopped 100 to 200 Å above the quantum well in order to avoid nonradiative recombination centers within the well. The depletion of electrons during etching is monitored by measuring the resistance of the SQW via ohmic contacts. As a consequence of the potential modulation, the electrons and laser induced holes are separated into adjacent regions and form a lateral type II superlattice.

Figure 2 shows photoluminescence (PL) as well as photoluminescence excitation (PLE) spectra of an as grown, as well as an unstructured, homogeneously etched reference sample, and an antidot and a dot sample at liquid-He temperature. Results for wires have been published elsewhere [6]. The as grown SQW exhibits the well known PL lineshape of an electron plasma of a modulation doped quantum well with a low energy maximum, a high energy tail and a cut off at the Fermi energy [7]. A careful analysis of the spectrum allows a separation of wave vector conserving and nonconserving transitions up to the Fermi wave vector k_F. We find no indication of additional Fermi edge enhancement in the luminescence of high quality quantum wells [8]. The PLE spectrum reveals a sharp onset at the Fermi energy with a heavy hole and a light hole peak being separated by 19 meV. The spacing is smaller as compared to the homogeneously etched sample due to a smaller curvature of the light hole inplane energy dispersion with respect to the heavy hole one. The excitonic luminescence of the homogeneously etched well is much narrower (FWHM 3 meV) and shifted to higher energies due to reduced bandgap renormalisation [9]. Its PLE spectrum is Stokes shifted by 2.5 meV and shows an energy spacing of heavy hole peak to light hole peak of 21.3 meV. Depending on the remaining carrier concentration

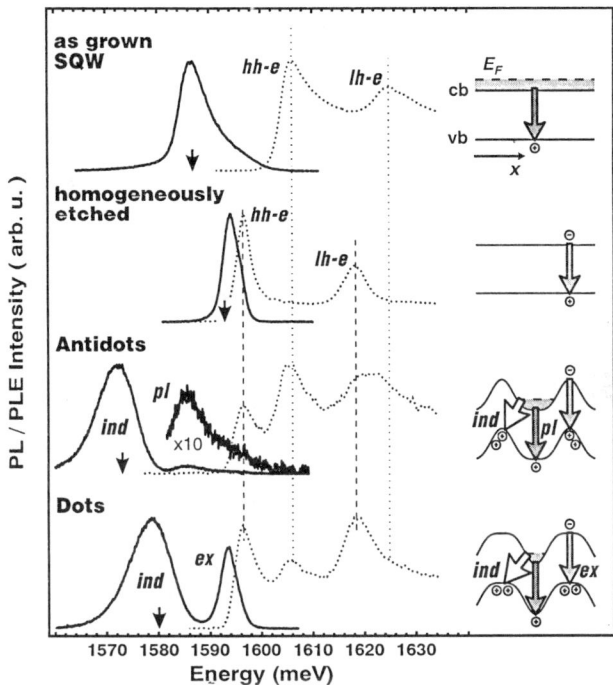

Figure 2. PL (solid lines) and PLE (dotted lines) spectra of various samples at T = 5 K. The solid arrows indicate the PLE detection energies.

we observe a FWHM down to 1.7 meV and a Stokes shift as small as 0.5 meV for 80 Å wells, showing that the excellent sample quality does not suffer from the etching procedure.

The PL of antidot and dot samples is dominated by an additional broad feature below the fundamental gap of the quantum well which is due to *spatially indirect* transitions, similar to wires [6] and to bulk nipi structures [10], whose transition energies are strongly dependent on etch depth and laser illumination density. We also observe a high energy luminescence that is very similar to either the luminescence of the 2D electron plasma or the excitonic peak of the homogeneously etched sample. It is identified to be due to *spatially direct* transitions. The PLE spectra are a superposition of the PLE spectra of the as grown as well as the homogeneously etched SQW. In the case of dots, where the direct PL is excitonic, also a stronger exciton contribution to PLE is observed.

The origin of the different PL contributions is sketched in the insets of Fig. 2. The broad indirect luminescence (*ind*) is due to carriers accumulated in the respective energy minima. But as the overlap is many orders of magnitude smaller with respect to direct transitions, indirect transitions are not visible in PLE. The high energy luminescence is emitted from regions with high electron density (*plasma*) below unetched areas as well as low carrier density (*excitonic*) below etched areas. Though the density of one type of

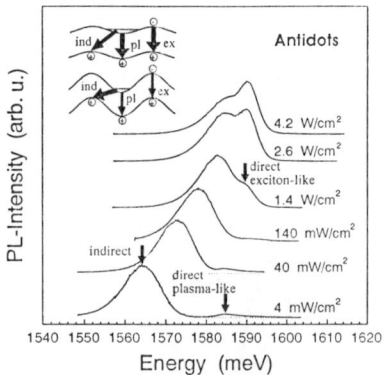

Figure 3. PL spectra of an antidot sample at various laser excitation densities (T = 1.5 K).

carriers is very low for both kinds of direct recombinations, their luminescence intensity can be comparable to the indirect PL because of the large overlap of the carrier wavefunctions. The striking difference between direct antidot and dot PL is not clear so far, because both, etched and unetched areas contribute to PLE and the Fermi energies are nearly identical as can be seen from the position of the PLE peaks. From the ratio of etched to unetched areas a plasma contribution about 16 times larger is expected for antidots with respect to dots. On the other hand, the different potential shapes could influence the carrier relaxation and the reduction of the dimensionality in quantum dots may affect the selection rules for plasma contributions to PL.

A variation of the laser illumination density (Fig. 3 for antidots) reveals the different characters of direct and indirect transitions. With increasing excitation intensity, the indirect luminescence shifts continuously towards the direct one, until they finally merge. The direct plasma luminescence can be observed only at the lowest excitation densities. Above 1.4 W/cm^2 the excitonic luminescence increases in intensity relative to the indirect one, until it finally dominates. The transition energies of both kinds of direct recombinations are almost independent of the excitation density. This behaviour is due to the change of the lateral potential with illumination. An increasing illumination density generates additional electrons and holes that are separated by the intrinsic, lateral field. The separation of carriers flattens the band edge modulation and the effective bandgap for indirect transitions increases. Direct transition energies are determined by the fundamental gap and remain almost unaffected. Concomitant with a flattening of the modulation and a heating of the carriers at the highest excitation densities, the carrier separation becomes less effective and the occupation probability at the potential maxima increases. Consequently, the excitonic direct recombinations with their large oscillator strength gain in intensity at the expense of the indirect ones. In addition, the spatially indirect character of the involved recombinations turns into a more direct one which yields an increase of the absolute intensities normalised to the excitation density.

The maximum of the indirect luminescence is due to transitions of electrons from the Fermi energy, which have the largest overlap with the spatially separated holes near the valence band maximum. The energy separation of this peak to the direct transitions at k_F therefore allows to estimate the total amplitude of the potential modulation. Values up to

50 meV are achieved corresponding to an electron subband spacing of about 2 meV.

In conclusion, we fabricated lateral type II superlattices with potential modulations up to 50 meV and high optical quality. The PL spectra are dominated by spatially indirect transitions and the spatially direct transitions are either plasma like or excitonic depending on the shape and dimensionality of the electron gas. The spatial separation of electrons and holes explains the observed features. There is no indication of a strong additional Fermi edge enhancement in our structures as discussed by Weiner et al. [11] for wires with much lower potential modulation and smaller carrier concentration.

3. Single GaAs Quantum Dots

In the last part we present results on the PL of spatially direct quantum dot structures. One very special issue of our structures is the optical isolation of single quantum dots. Microscopic spectroscopy of a single zero-dimensional (0D) system eliminates inhomogeneous line broadening which seems inevitable in macroscopic spectroscopy on arrays of dots or wires.

A schematic view of a single quantum dot structure is shown in the inset of Fig. 4. Local interdiffusion of a GaAs/$Al_{0.35}Ga_{0.65}As$ single quantum well structure (L = 30 Å) grown by molecular beam epitaxy is used to get a spatially direct (type I) quantum dot. Thermal interdiffusion is induced by a focused laser beam (FLB) of about 0.5 μm in diameter, which heats up the sample to a local temperature of about 1000 °C [12]. The extremely nonlinear increase of Al/Ga diffusion with temperature reduces the lateral width of the diffusion profile by about a factor of five compared to the laser spot profile. This improvement in patterning resolution allows direct writing of various quantum structures just by scanning the sample underneath the laser spot. Well defined lateral structures require a high precision xyz-translation stage of 10 nm accuracy, an intensity stabilized Ar^+-laser and an autofocus system. A quantum dot is produced by drawing a square frame of size w with the FLB. Optical isolation of the dot is achieved by interdiffusion of the surrounding region. The intended lateral bandgap modulation is sketched in the lower part of the inset of Fig. 4.

In Fig. 4 some typical PL spectra from a series of quantum dots with different size w = 300 nm up to 2000 nm are shown. Using a cold-finger He flow cryostat at the microscope stage and a pinhole at the PL image plane, we get a spatial resolution of about 1.5 μm and a sample temperature T = 5 K. In large dots (w > 700 nm) the PL resembles the wide smooth luminescence line of the as-grown quantum well. The PL exhibits a more and more pronounced peak structure and shifts to slightly higher energy when reducing w to a value of 450 nm. At still smaller dot size the PL is stronger blue shifted but the peak separation is decreased. Within the homogeneously interdiffused region around the dots we observe weak PL blue shifted up to about 70 meV. These recombinations are attributed to local minima of band edge energy and defects. The high efficiency and the small blue shift of PL from the dots compared to the surrounding region allows spectroscopy of the isolated dots.

The monotonic increase of PL peak energy and the energy splitting of the predominant PL peaks as a function of dot size are summarized in Fig. 5. The shift and splitting of the PL line can be qualitatively understood considering the lateral modulation of quantum well interdiffusion. In a simple model we calculate the electron and heavy hole confinement energies in an interdiffused quantum well. Evaluating the lateral modulation of

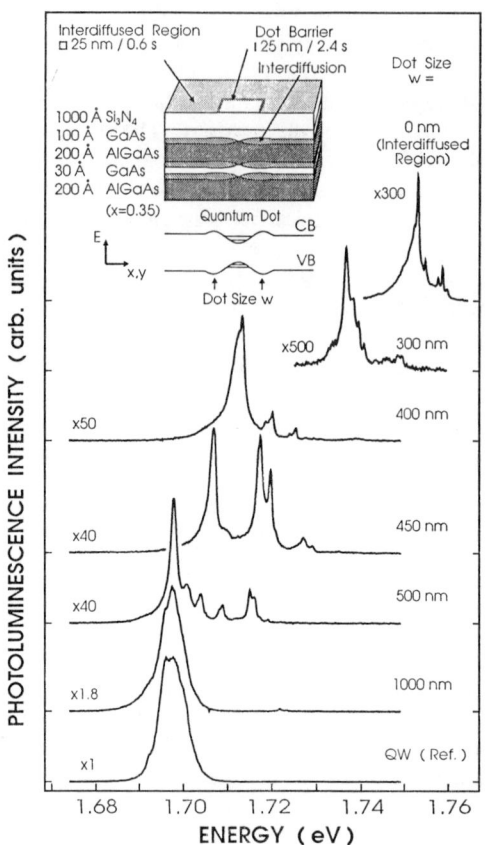

Figure 4. PL spectra of single dot structures of various size w. Laser excitation is E_{exc} = 1.96 eV with P = 1 µW focused to a spot of size 1.5 µm. The inset shows a schematic diagram of a dot structure fabricated by FLB induced interdiffusion of a 30 Å thick GaAs/AlGaAs single quantum well.

interdiffusion from the equation for heat conduction and from the thermal Al/Ga diffusion relation we obtain a lateral modulation of the effective band edge energies. Using reasonable laser processing parameters and sample properties we get similar lateral potential barriers for electrons and holes. They are nearly parabolic and isotropic near the center of the dot with w = 450 nm and they result in electron and hole confinement energies af about 6 meV and 4 meV, respectively. So the nearly equidistant energy separation of the main PL peaks ΔE = 10 meV in the 450 nm dot compares well to the energy of quantum number conserving transitions $n_e = n_h$ = 1,2,3. An energy shift of only a few meV originates in the bare quantum well interdiffusion at the dot center, while the groundstate quantization energy is about 10 meV due to the steep lateral potential barriers. At larger dot size the confinement energies decrease due to a wide, flat bottom of the potentials. At smaller dot size the large blue shift indicates quantum well interdiffusion at the dot center. The

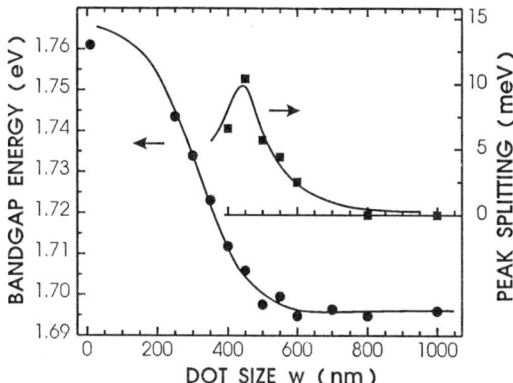

Figure 5. Energy splitting of predominant PL peaks and minimum PL peak energy versus the geometrical dot size w. The solid lines are just a guide to the eye.

decrease of quantization energy is attributed to a smaller potential modulation height in these more homogeneously interdiffused dots.

Concomitant with the observation of maximum peak splitting at maximum lateral potential height strongest PL intensity of the higher energy levels is obtained from the dot w = 450 nm. The population of excited levels has been shown to be an intrinsic property of 0D semiconductor systems due to the drastically slowed down energy relaxation of photogenerated carriers [13,14]. Theory reveals that the carrier relaxation in 0D systems by longitudinal acoustic phonons is strongly decreased when the level separation is beyond a threshold energy $E_t = hc_s 2\pi/L$. This is caused by the momentum mismatch of 0D electronic states and acoustical phonons with linear dispersion. With the sound velocity of GaAs c_s = 3700 m/s and the smallest quantum dot dimension L = 30 Å we obtain E_t = 5.1 meV. Quantitatively, an electron relaxation rate of about 15 ns for the w = 450 nm sample with an electron quantization energy beyond E_t was calculated [15]. The electrons in the weaker quantized dots as well as the holes in all dots are expected to relax relatively fast. The intensive excited level transitions at maximum peak splitting are attributed to a small relaxation rate of 0D electrons. The systematic dependence of peak splitting and blue shift on dot size and the appearance of a slowed electron relaxation expected in 0D systems with large peak splitting strongly indicate that the observed PL splitting is caused by the discrete levels in a 0D system and is not caused by any potential fluctuation or introduction of defects.

In Fig. 6 PL spectra from the dot with maximum peak separation ΔE = 10 meV are shown for various excitation powers. The spectra are normalized by the laser power to represent the PL efficiency. The photon energy E_{exc} = 1.77 eV used in the upper spectra corresponds to excitation near the edge of the confining potential. At low excitation the PL reveals a single peak of the groundstate transition and a set of distinct peaks at about 1.72 eV, which are attributed to transitions from the first excited electron level. With increasing power the efficiency of these two main peaks remains nearly constant and no saturation of level occupation is observed. This indicates that the dot excited by about 1 μW still has a relatively low average occupation with electron hole pairs. The transitions at higher

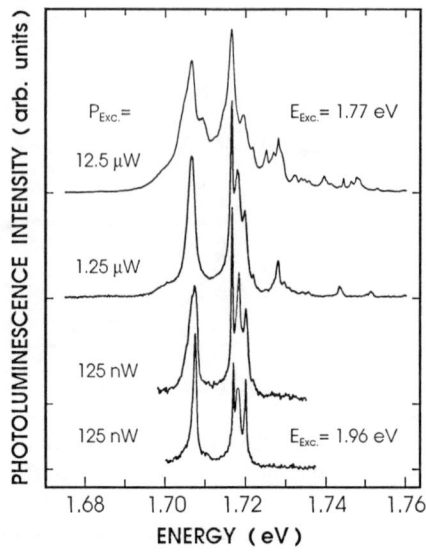

Figure 6. PL spectra from the dot of size w = 450 nm on an intensity scale normalized by laser excitation power. The photon energy was E_{exc} = 1.77 eV, except for the lowermost spectrum where E_{exc} = 1.96 eV.

energies gain intensity and the peaks broaden with increasing excitation power. The spectral width of individual lines is below 0.5 meV at low excitation density. The line width is further reduced by exciting the dot in the continuum E_{exc} = 1.96 eV. Line broadening in our spectra may be caused by interlevel scattering of photogenerated carriers and by Coulombic shifts of transition energies induced by a fluctuating occupation of dot levels. The fine structure as well as the shoulders in between the predominant PL lines imply quantum number nonconserving electron-hole transitions. The selection rules of optical transitions in a symmetric potential can be weakened by a nonuniform lateral potential. Such transitions in between the main peaks are also observed in PL excitation spectroscopy and in PL at increased sample temperatures [15].

We have presented PL spectra from single GaAs/AlGaAs quantum dots fabricated by FLB induced interdiffusion. The splitting of the PL with peaks separated by up to 10 meV is attributed to transitions of electrons and holes in consecutive levels. The high optical quality allows spectroscopy at moderate excitation. Individual PL lines reveal widths below 0.5 meV attributed to intrinsic mechanisms because inhomogeneous broadening is inherently excluded in a single 0D structure. The observation of transitions from higher energy levels and the decrease of the groundstate PL efficiency indicates slow relaxation of quantized electrons in small dots.

Acknowledgments

We want to thank U. Bockelmann for stimulating discussions and theoretical help. This work was supported financially by the Deutsche Forschungsgemeinschaft via SFB 348 and by the Bundesministerium für Forschung und Technologie via DFE Verbundprojekt.

References

1. K. Kash, J.M. Worlock, A.S. Gozdz, B.P. van der Gaag, J.P. Harbison, P.S.D. Lin, and L.T. Florez, Surf. Sci. **229**, 245 (1990).
2. D. Gershoni, J.S. Weiner, S.N.G. Chu, G.A. Baraff, J.M. Vandenberg, L.N. Pfeiffer, K. West, R.A. Logan, and T. Tanbun-Ek, Phys. Rev. Lett. **65**, 1631 (1990).
3. M. Tsuchiya, J.M. Gaines, R.H. Yan, R.J. Simes, P.O. Holtz, L.A. Coldren, and P.M. Petroff, Phys. Rev. Lett. **62**, 466 (1989).
4. Y.J. Li, M. Tsuchiya, and P.M. Petroff, Appl. Phys. Lett. **57**, 472 (1990).
5. A.S. Plaut, H. Lage, P. Grambow, D. Heitmann, K. von Klitzing, and K. Ploog, Phys. Rev. Lett. **67**, 1642 (1991).
6. F. Hirler, R. Küchler, R. Strenz, G. Abstreiter, G. Böhm, J. Smoliner, G. Tränkle, and G. Weimann, Surf. Sci. **263**, 536 (1992).
7. A. Pinczuk, J. Shah, H.L. Störmer, R.C. Miller, A.C. Gossard, and W. Wiegmann, Surf. Sci. **142**, 492 (1984).
8. R. Küchler, G. Abstreiter, G. Böhm, and G. Weimann, Semicond. Science and Technol. **8**, 88 (1993).
9. C. Delalande, G. Bastard, J. Orgonasi, J. A. Brum, H.W. Liu, and M. Voos, Phys. Rev. Lett. **59**, 2690 (1987).
10. G.H. Döhler, H. Künzel, D. Olego, K. Ploog, P. Ruden, H.J. Stolz, and G. Abstreiter, Phys. Rev. Lett. **47**, 864 (1981).
11. J.S. Weiner, G. Danan, A. Pinczuk, J. Valladares, L.N. Pfeiffer, and K. West, Phys. Rev. Lett. **63**, 1641 (1989).
12. K. Brunner, G. Abstreiter, M. Walther, G. Böhm, and G. Tränkle, Surf. Sci. **267**, 218 (1992).
13. U. Bockelmann and G. Bastard, Phys. Rev. B **42**, 8947 (1990).
14. H. Benisty, C.M. Sotomayor-Torres, and C. Weisbuch, Phys. Rev. B **44**, 10945 (1991).
15. K. Brunner, U. Bockelmann, G. Abstreiter, M. Walther, G. Böhm, G. Tränkle, and G. Weimann, Phys. Rev. Lett. **69**, 3216 (1992).

OPTICAL PROPERTIES OF STRAIN-INDUCED NANOMETER SCALE QUANTUM WIRES

D. GERSHONI,[1] J.S. WEINER,[2] E.A. FITZGERALD,[2] L.N. PFEIFFER,[2] AND N. CHAND[2]
[1]Physics Department and the Solid State Institute, Technion, Haifa 32000, Israel
[2]AT&T Bell Laboratories, Murray Hill, New Jersey 07974, USA

ABSTRACT. We have fabricated single quantum wires and quantum wire arrays of nanometer scale lateral dimensions. The wires are produced by two steps of epitaxial growth in orthogonal directions. We utilize the concept of pseudomorphic growth of lattice mismatched epitaxial layers to generate strain modulation within the plane of a conventional quantum well. The strain modulation, as large as 2%, modulates the electronic band structure for carrier motion parallel to the quantum well plane. These potential modulations which are of order 0.1 eV, confine carriers laterally to regions comparable in size to the size of a quantum well. We present in this work optical studies of these one dimensional systems by means of low temperature cathodoluminescence, photoluminescence, photoluminescence excitation and time resolved spectroscopies. We analyse our observations using a new multi-band effective mass and deformation potential model. The measured optical transitions and polarization selection rules agree well with the model. In particular, we find that electrons and light holes are laterally confined to the same region while heavy holes are spatially separate from them. Finally, we demonstrate a novel p-i-n linear junction which provides a means to electrically contact these one-dimensional quantum structures.

1. Introduction

Ever since the first quantum well was fabricated [1], demonstrating the ability of epitaxial growth with control on a monolayer level, there has been a search for ways to realize such control in more than one dimension. In particular, there has been much interest in one- and zero-dimensional semiconductor systems, both for their possible contribution to our understanding of solid state physics and for their potential device applications [2].

We have recently demonstrated a new approach to realize quantum wires [3], in which the size of the two confined directions is comparable and can be made as narrow as the typical width of an epitaxially grown quantum well (QW). Our method utilizes two steps of epitaxial growth in orthogonal directions. It allows for a perfect control over the size in two dimensions, while at the same time it maintains the high quality and perfect crystallinity of the material. This is in contrast to conventional approaches based on lithographic techniques. Here the in-plane lattice constant of a QW is spatially modulated by growing it on the cleaved side of a strained layer (SL) superlattice (SLS) or QW (SLQW), which had been grown initially. Thus, the concept of pseudomorphic growth is utilized here twice.

We report here on our observations of strain-induced confinement in single, isolated quantum wires, grown over single strained layer quantum wells, and in quantum wire arrays, grown over strained layer superlattices. In addition, we describe a new method which

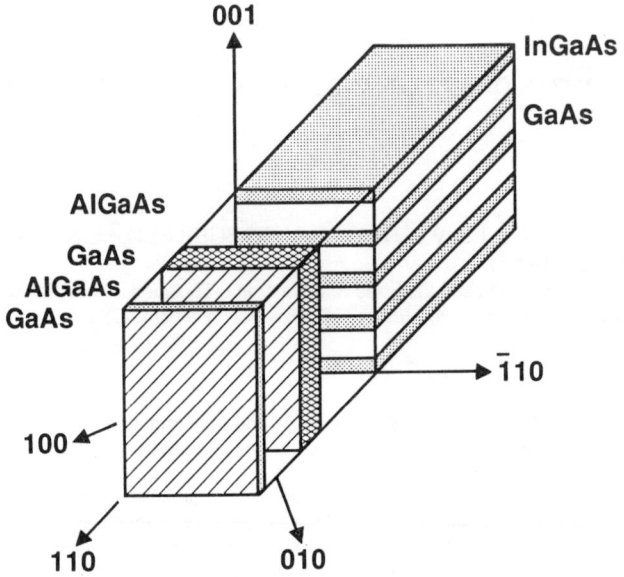

Figure 1. Schematic of the strained quantum wire structure.

utilizes the two direction orthogonal growth technique to create a linear p-n junction. The linear junction allows for selective carrier injection into the quantum wires.

The paper is organized as follows: in Sec. 2 we briefly discuss the two direction epitaxial growth. In Sec. 3 we present cathodoluminescence studies of single quantum wires. Studies of quantum wire arrays by means of optical excitations are presented in Sec. 4. In Sec. 5 we discuss the experimental data and compare it with the theoretical model. Section 6 describes the linear p-i-n junction and a short summary is contained in Sec. 7.

2. Growth

In the first growth step we prepare the substrate which contains strained layers of $In_xGa_{1-x}As/GaAs$. The widths of and the strain in the layers are measured using transmission electron microscopy (TEM) and high resolution x-ray diffraction (HRXRD) [4]. The strained layers are grown using either molecular beam epitaxy (MBE) or metal-organic chemical vapor deposition (MOCVD) on a (001) oriented GaAs substrate. This growth stage is concluded by a 2 μm thick GaAs cap layer in order to prevent edge effects during the next growth stages. The samples are then inserted into another MBE machine, for the second growth step. They are cleaved in situ under high vacuum conditions and GaAs/AlGaAs quantum structures are grown on the (110) cleaved facets [5]. The (110) oriented quantum structures reported on here consisted of two GaAs QWs of 80 and 34 Å width, respectively, separated from each other and from the substrate by 200 Å thick $Al_{0.3}Ga_{0.7}As$ barriers. The structure was capped by 200 Å $Al_{0.3}Ga_{0.7}As$ and 50 Å GaAs layers. A typical structure is schematically described in Fig. 1.

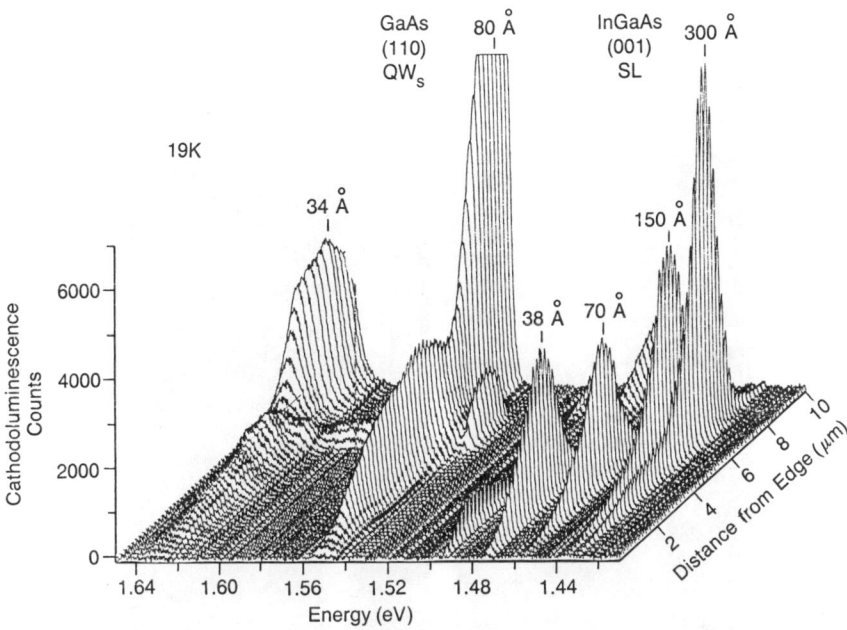

Figure 2. Cathodoluminescence spectra of the single strained quantum wire sample for various beam positions along the (001) crystallographic direction. The beam is incident normal to the regrown cleaved edge.

3. Single Strained Quantum Wires

Cathodoluminescence (CL) spectra obtained at different electron beam positions over the edge of a single strained quantum wire sample are shown in Fig. 2. The sample under study here contains in its substrate five different strained layers of $In_{0.06}Ga_{0.94}As$. The layers of 300, 150, 75, 40, and 20 Å width, respectively, are 1 µm apart. The CL spectra were obtained at a sample temperature of 19 K with a 2 KeV electron beam incident normal to the (110) regrown facet of the sample. The beam current was 250 pA. The CL was detected by a silicon diode array mounted on a 0.25 m monochromator. The beam position was varied in intervals of 2000 Å along the (001) direction (growth direction of the SL substrate) starting from the sample edge. Five peaks are discernible at the lower energy side of the spectra. These peaks result from the radiative recombination of carriers within each (001) SLQW. Each peak reaches its maximum intensity when the electron beam is at a different distance from the edge. That distance corresponds to the location of the SLQW in which the recombination occurs. We note that both the magnitude and the spatial width of the peaks scale with the SLQW widths, or the cross sections that they expose to carrier collection. The higher energy side of the spectra is dominated by two peaks which result from radiative recombination of carriers within the GaAs (110) QWs. Thus, by correlating the spatial dependence of this CL with that of the SLQWs it is possible

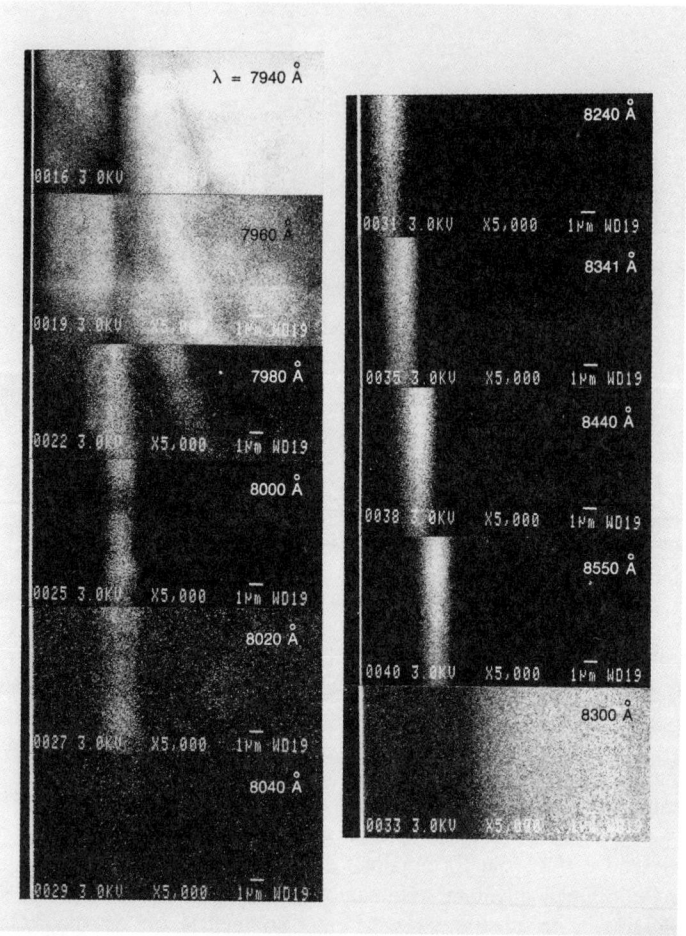

Figure 3. Fixed wavelength CL images of the regrown cleaved edge of the sample containing single strained quantum wires. The right set shows emission lines originating in the substrate. The left set shows emission lines originating in the layers deposited on the cleaved edge.

to determine how each strained layer in the substrate affects the emission from the quantum well above it. There are clear lower energy satellite peaks associated with each main peak. These satellites are due to carrier recombination within the regions of GaAs (110) QWs which lie over the strained layer regions in the substrate, i.e., the strained quantum wires (SQWRs). Four satellites can be resolved for each main peak. The energy shifts of the satellites are 4.9, 7.4, 10.2, and 14.5 meV from the 80 Å QW peak and 6.7, 8.1, 17, and 35 meV from the 34 Å QW peak, respectively.

It is important to note that the CL from both the closer 80 Å thick and the farther 34 Å thick quantum wells is red shifted considerably. Clearly, this indicates that the strain fields reach the second QW 500 Å away from the strained substrate. The distance that the strain fields propagate away from the substrate is surprisingly large. It cannot be understood in terms of the theory of elasticity which predicts complete decay of the fields within a distance comparable in size to the width of the strained layer that initiates it [6]. We believe that the low temperature pseudomorphic epitaxial growth involved in fabrication of the SQWRs requires a more detailed theoretical treatment, taking into account the strain distribution in these extreme conditions, energy terms that are associated with surfaces and interfaces, and the possible formation of metastable states. Although, such a treatment has not yet been carried out, the experimental evidence for the field propagation as presented in Figs. 2 and 3 below cannot be disputed.

In Fig. 3 we present two sets of fixed wavelength CL images of the (110) regrown edge of the sample, close to the SLQWs. The images were obtained by scanning the electron beam over the area, while monitoring the emitted light at a fixed wavelength. The signal was detected with a cooled GaAs photomultiplier mounted on a 0.85 m double monochromator. The four uppermost images in the right side of Fig 3. were obtained at wavelengths corresponding to the emission from the four narrowest SLQWs. The fifth image was obtained at the wavelength of a corresponding extrinsic emission from the GaAs substrate. The emission of the widest SLQW was at too long a wavelength to be detected by our photomultiplier. Its position can be determined, however, by comparison of the emission from the narrower SLQWs with that of the substrate. The left side of Fig. 2 presents images taken at fixed wavelengths corresponding to emission from the 80 Å (110) GaAs QW (at 7940 Å) and its lower energy satellites. By comparison between the two series one can clearly identify the spatial origin of the two lowest energy satellites as emission from the 80 Å times 150 Å and the 80 Å times 300 Å SQWRs. The two higher-energy satellites observed in Fig. 2, are not resolved here, due to their weaker intensities (which scale with the SLQW width) and smaller energy shifts.

4. Strained Quantum Wire Arrays

For the studies of the strained quantum wires by optical excitation we have prepared a 150 period strained layer superlattice (SLS) of 71 Å of $In_{0.063}Ga_{0.937}As$ followed by 240 Å of GaAs in the first growth stage. The following steps were performed exactly the same as described above. Thus, a quantum wire array is formed, extending laterally over a width of roughly 4 μm. This lateral extension permits optical probing using low temperature photoluminescence (PL) and photoluminescence excitation (PLE) techniques, as described below.

Figure 4 shows a dark field TEM micrograph of the cross-sectional view of the sample using the ($\bar{2}\bar{2}0$) reflection. Since the sample in Fig. 4 is tilted away from the ($00\bar{1}$) zone axis, the InGaAs/GaAs SLS grown along the (001) crystallographic direction does not show clear contrast. The SLS was found, however, to be of high quality and sharp interfaces using TEM of the ($00\bar{2}$) reflection (not shown here) and HRXRD. A very low density of misfit dislocations was observed only in the first interface of the SLS. This is expected for such a thick SLS. It is clear from the micrograph that the two QWs are of high quality and that they form a perfect crystallographic structure with the SLS beneath them. The in-plane

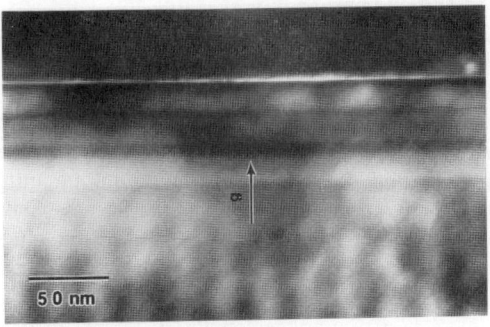

Figure 4. Dark field ($\overline{2}20$) reflection TEM micrographs of the (110) cleaved-edge grown quantum wire array.

lattice constant of the quantum well is thus modulated by the mismatch strain associated with this two directional pseudomorphic structure.

Figure 5 displays low temperature CL, PL, and PLE spectra of the sample. The spectra in Fig. 5(a) were obtained from QWs grown over the unstrained, cleaved GaAs substrate. Those in Fig 5(b) were obtained from QWs grown over the SLS. The CL spectra (lowest curve, solid line) were obtained at 19 K using a 10 pA, 5 kV electron beam incident normal to the (110) plane. Here too, we used fixed wavelength CL images (not shown), to spatially resolve the different regions of the sample, and to verify the source of each spectral feature. Two important points can be readily learned from the CL spectra. First, the luminescence peaks of the SQWRs associated with both the 80 and 34 Å QWs are shifted ~ 21 meV to lower energy relative to the unstrained QWs. The uniform shift indicates again that both QWs experience similar strain, although they are spatially separated by 200 Å. This result is compatible with our studies of the single quantum wire as observed in Fig. 2. Second, almost no broadening of the CL from the SQWRs relative to that from the QWs is observed, indicative of the high quality of the crystalline structure. We did observe, however, increased shifts when the electron beam was directed close to the SLS-buffer interface. We attribute these shifts to strain variations in this slightly dislocated region. Third, the CL from the SQWRs is as efficient as that from the QWs. This is indicative of the increased carrier collection efficiency of the SQWR and the absence of nonradiative channels.

For the PL and PLE measurements, light from a cw dye laser was focused at normal incidence onto the (110) overgrown facet of the sample using a × 60 microscope objective mounted in the He-flow cryostat. The emitted light was collected from the sample side and analyzed by a 0.25 m double monochromator and cooled photomultiplier. The lowest dashed curves in Fig. 5(a) and (b) are the PL spectra. Two PLE spectra are also shown for each QW and each SQWR. The dashed and solid PLE curves were obtained with incident polarization parallel (1$\overline{1}$0 crystallographic direction) and perpendicular to the wires (001 direction), respectively. The PL line widths of the unstrained QWs are comparable to those measured by CL. However, the line widths of the SQWRs are broader. This is because PL also probes the dislocated areas between the buffer and the SLS.

The assignments of excitonic transitions in the PLE spectra are marked in Fig 5(a). The

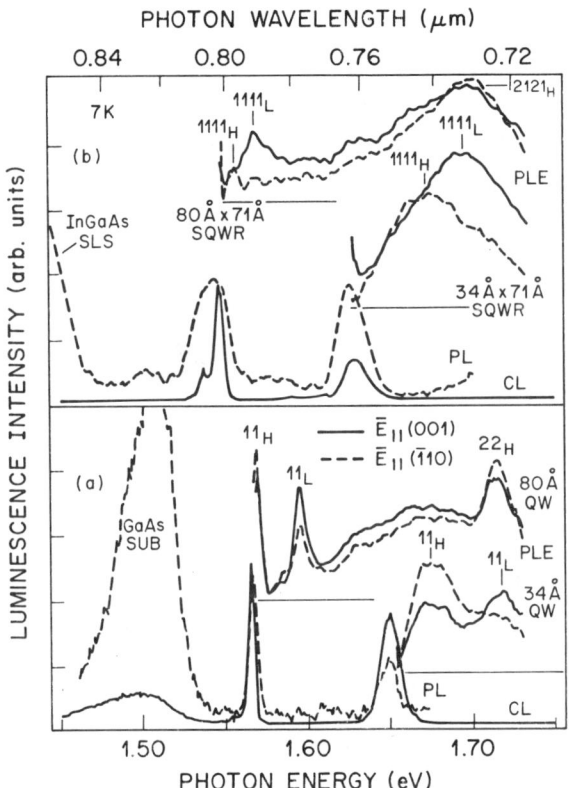

Figure 5. Cathodoluminescence (CL), photoluminescence (PL), and PL excitation (PLE) spectra of (a) (110) cleaved-edge grown quantum wells and (b) (110) side-grown strained quantum wire arrays. The PLE spectra are shifted vertically for clarity, and the zeros are indicated by horizontal lines.

numbers refer to the conduction and valence band sublevels respectively. The letter H(L) indicates that this is a heavy (light) hole subband. These sublevels result from composition-induced-confinement in the (110) direction. The observed transitions agree to within 3 meV with the calculated energies [7]. In marked contrast to the unpolarized PLE of a conventional (100) QW, the (110) QWs show clear polarization dependence. The light hole transitions are preferentially polarized in the (001) direction and heavy hole transitions are polarized in the ($1\bar{1}0$) direction. We have observed this effect in (110) oriented QWs grown over planar, uncleaved substrates [7]. We have shown in Ref. [7] that this anisotropy in the linear polarization of the absorption of the QWs reflects the anisotropy of the GaAs valence band structure, which is revealed in this case by the reduced symmetry of the quantum confinement direction (110). In Fig. 5(b) additional quantum numbers are added to the conduction and valence sublevels to indicate the confinement in the (001) direction. The heavy hole to electron transitions in the PLE of the SQWR are red shifted ~ 17 meV

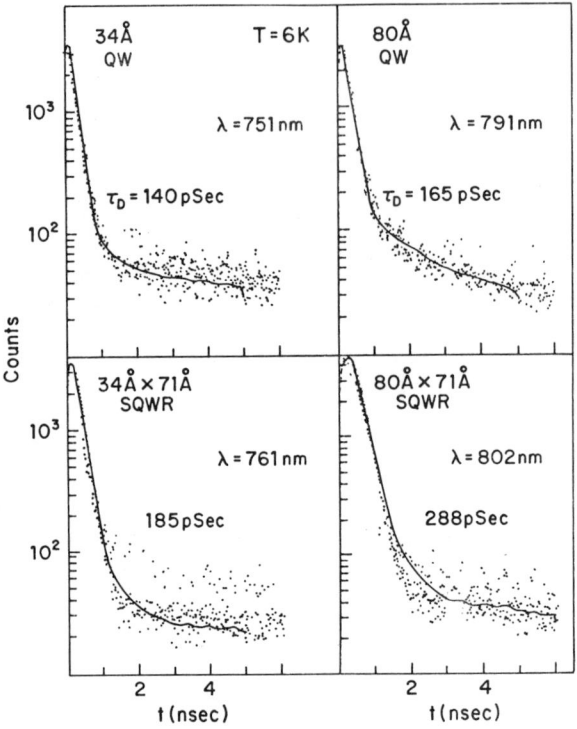

Figure 6. Photoluminescence transients from the (110) oriented quantum wells and strained quantum wires.

relative to the corresponding QW transitions, while the light hole to electron transitions are red shifted by as much as 28 meV. The shifts are almost the same for both SQWRs and for both $n_{(110)} = 1$ and $n_{(110)} = 2$ transitions. Optical transition intensities are measured by integrating under the excitonic peaks after background subtractions. Extreme changes are observed in the transition intensities and polarization anisotropy of the SQWRs compared to those of the unstrained QWs. We first note that the light hole transitions are stronger than the heavy-hole ones, in clear contrast with the unstrained QWs. Second, the light hole transitions are polarized more than 5:1 perpendicular to the wires (001 direction) and the $n_{(110)} = 1$ heavy hole transitions are polarized roughly 3:1 parallel to the wires ($1\bar{1}0$ direction), whereas the $n_{(110)} = 2$ heavy hole transition is almost unpolarized.

In Fig. 6. PL decay transients from the QWs and from the SQWRs are presented. For these measurements the sample was excited by 1.5 ps long pulses from an actively mode locked Ti:Sapphire laser (Spectra Physics Tsunami) at an energy of 1.717 eV. The transients were detected by a micro channel plate photomultiplier and conventional time correlated single photon counting electronics with a temporal resolution of about 60 ps. The solid lines represent best fits obtained by convoluting an exponential decay with the system response function. It is clear from Fig. 6. that the decay times of the PL from the SQWRs are longer than those of the corresponding QWs.

5. Discussion

In order to analyze the data the strain in the system should be evaluated. This was done by solving Hook's tensorial equation for the side-grown epitaxial layer, assuming that no strain relaxation takes place. The stress applied by the substrate to the layer is biaxial with unequal components in the $(1\bar{1}0)$ and (001) directions. For our system, the displacement in the $(1\bar{1}0)$ direction is zero and the displacement in the (001) direction is given by the difference between the GaAs lattice constant and the strained InGaAs lattice constant. With these constraints we find:

$$\varepsilon_{xx} = \varepsilon_{yy} = \frac{\varepsilon_{xy}}{2} = \frac{-C_{12}}{C_{11} + C_{12} + 2C_{44}} \varepsilon_{zz}. \qquad (1)$$

The ε_{ij} are the components of the strain tensor, the C_{ij} are the components of the stiffness tensor and ε_{zz} is given by: $(a_{SLS} - a_{GaAs})/(a_{GaAs})$. Here a_{SLS} is the strained $In_xGa_{1-x}As$ lattice constant which is measured by HRXRD [4]. This gives $\varepsilon_{zz} = 0.87\%$.

To model the structure within the effective mass approximation, it is necessary to solve eight coupled differential equations [8], which contain both first- and second-order k•p terms [9] and strain terms [10]. We have recently developed an eigenfunction expansion method for solving such problems [11]. By applying our method to the strained quantum wires, we find that the electron has only one discrete level associated with confinement in the lateral (001) direction. This results in two laterally confined eigenstates ($n_{(110)} = 1$; $n_{(001)} = 1$) and ($n_{(110)} = 2$; $n_{(001)} = 1$) with probability distributions as described in Fig. 7. The energy associated with this lateral confinement is roughly 20 meV, for both the $n_{(110)} = 1$ and $n_{(110)} = 2$ states of the 80 Å QW and for the $n_{(110)} = 1$ state of the 34 Å QW.

Thus, roughly speaking, the electrons have one discrete level associated with the strain-induced lateral confinement, and the level is energetically located in the middle of the lateral potential well, which amounts to some 40 meV below the unstrained bulk GaAs conduction band energy.

Light holes are also marginally confined laterally to the strained regions with calculated confinement energies of less than 10 meV, while the heavy hole wave functions reside preferentially in the unstrained GaAs regions. The energies associated with the lateral confinement of holes are thus smaller or equal to the excitonic binding energy. Our model, which does not take into account the Coloumb interaction between the electron and the hole, is therefore inaccurate. Nevertheless, the trends are in perfect agreement with the experimental observations: first, the 10 meV increase in the relative energy red-shifts of the light-hole to electron transitions as compared to those of the heavy hole to electron; second, the enhancement of the light hole to electron transitions relative to those from the heavy hole in the SQWRs, as shown in Fig. 5(b); and third, the longer decay times of the luminescence from the SQWRs relative to that of the QWs, as is shown in Fig 6. These trends can be intuitively understood in terms of the smaller overlap between the heavy hole and electron envelope wave functions due to the lateral confinement to different spatial regions.

The increased anisotropy in the optical absorption of the SQWR as revealed by the polarized PLE spectra of Fig. 5 is also in agreement with our model, where the optical transition intensities are calculated from the eigenfunctions resulting from the diagonalization procedure [7]. We emphasize here that the analysis of the polarization

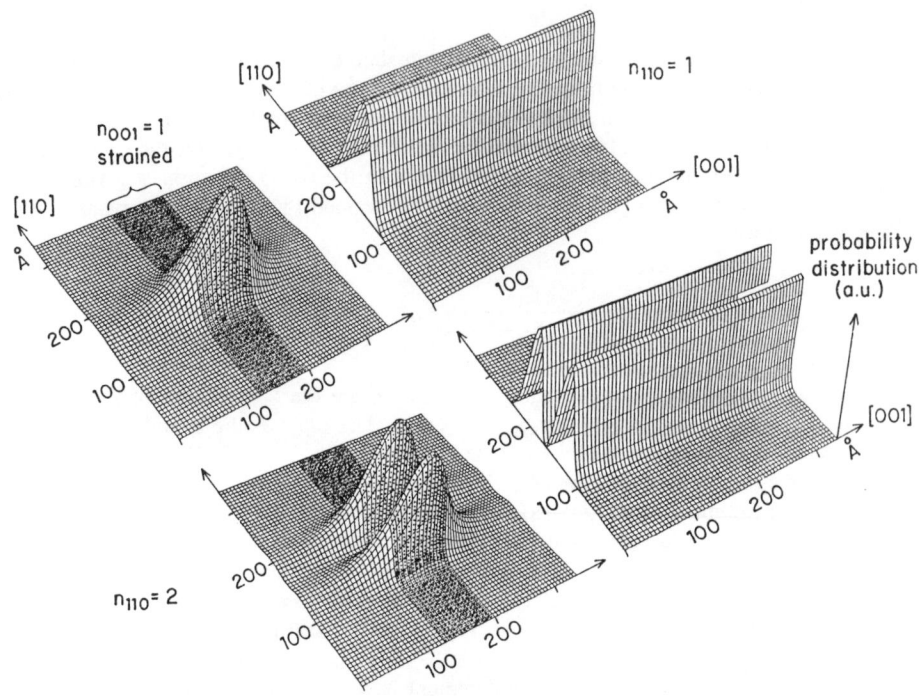

Figure 7. Calculated probability distribution of the discrete electronic levels of the 80 Å (110) oriented quantum wells and of the 80 Å × 71 Å strained quantum wires.

selection rules in this case is quite complicated since the anisotropy results from the reduced symmetry of the composition induced confinement direction [7], the strain [12], and the strain-induced lateral confinement. These effects are considered by our model. In addition we use the QWs, which are grown over the unstrained regions of the cleaved edge, as perfect control samples to isolate the component of the polarization induced by the reduced symmetry of the (110) growth axis.

A particularly strong indication of lateral quantum confinement is found in the different polarization anisotropy of the 1111h and the 2121h transitions. This mainly reflects the different symmetries of the envelope wavefunctions. A similar strong indication of the 1D nature of the system has already been observed in wider quantum wires [13].

6. Linear p-i-n Junction

Many possible applications of quantum wires [2] require electrical contacts. We have used doping in the two orthogonal growth directions involved in the fabrication of the SQWRs to prepare a novel linear p-i-n junction in which the SQWRs are embedded. For this purpose we have grown a four-period Be-doped $In_{0.12}Ga_{0.88}As$/GaAs SLS on a semi-insulating GaAs substrate. SQWRs have been overgrown on the (110) oriented cleaved

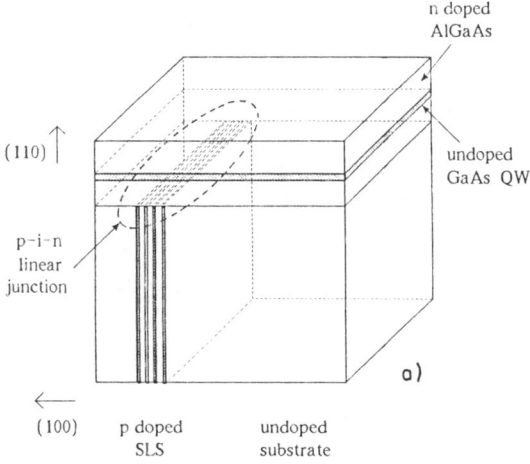

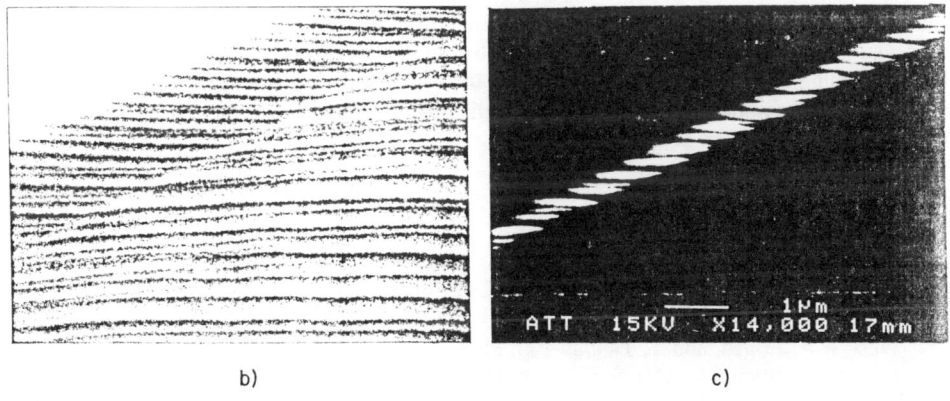

Figure 8. (a) Schematic description of the linear p-i-n junction. (b) Secondary electron imaging of the p-i-n regrown cleaved edge. The arrow indicates the same position as the arrow in (a). (c) Electron beam induced current (EBIC) imaging of the region as in (b). The scale bar applies to both (b) and (c). White areas indicate places of high induced currents.

edge of this sample. The overgrowth procedure was similar to that described above, but with the following difference: on top of the two quantum wells an additional 1000 Å thick layer of Si doped n-type $Al_{0.3}Ga_{0.7}As$ was deposited. The resulting structure is schematically depicted in Fig. 8(a). It can be seen that a p-i-n junction will be formed along the line of intersection between the two orthogonal and differently doped films.

Ohmic contacts were then made to the doped layers and the sample I-V characteristics were measured. It exhibited a typical diode curve with turn on voltage of about 1 V and differential resistance in forward bias of a few hundred ohms. The diode could be reverse biased up to ~ 25 V before breakdown occurred.

Figure 8(b) is a secondary electron microscope (SEM) image of the overgrown cleaved edge of the sample. Figure 8(c) is a room temperature electron beam induced current (EBIC) image of the same area of the sample. The white areas in Fig. 8(c) indicate regions of high induced currents. It is clear that the electrons are collected most efficiently from the region where the p- and n-doped films intersect. Since the resolution of the EBIC mapping is limited by the diffusion length of the electrons (~ 1 μm), it is safe to assume that the width of the linear junction is comparable to the width of the p-doped SLS (~ 1000 Å). Carriers can therefore be either injected into or collected from the SQWRs via this junction.

7. Summary

We have fabricated GaAs/GaAlAs quantum wires which exhibit optical properties due to nanometer scale lateral confinement. The confinement is produced by modulating the in-plane lattice constant of a quantum well grown on an InGaAs/GaAs strained layer superlattice. The strain induced band gap modulation and lateral confinement of carriers manifest themselves in a drastic change of the optical polarization selection rules, large red shifts of the PL and PLE spectra, and increase in the PL decay times. The experimental observations are in very good agreement with a theoretical model based on an eight band k•p approximation and deformation potential theory. These structures provide a new range for studies of the optical and transport properties of 1D systems.

References

1. R. Dingle, W. Wiegmann, and C.H. Henry, Phys. Rev. Lett. **33**, 827 (1974).
2. Y. Arakawa and H. Sakaki, Appl. Phys. Lett. **40**, 939 (1982).
3. D. Gershoni, J.S. Weiner, S.N.G. Chu, G.A. Baraff, J.M. Vandenberg L.N. Pfeiffer, K. West, R.A. Logan, and T. Tanbun-Ek, Phys. Rev. Lett. **65**, 1631 (1990).
4. J.M. Vandenberg, D. Gershoni, R.A. Hamm, M.B. Panish, and H. Temkin, J. Appl. Phys. **66**, 3635 (1989).
5. L.N. Pfeiffer, K.W. West, H.L. Stormer, J. Eisenstein, K.W. Baldwin, D. Gershoni, and J. Spector, Appl. Phys. Lett. **56**, 1697 (1990).
6. See, for example, S.P. Timoshenko and J.N. Goodier, *Theory of Elasticity*, 3rd Edition, (McGraw-Hill, New York, 1970), pp. 53-60.
7. D. Gershoni, I. Brener, G A. Baraff, S.N.G. Chu, L.N. Pfeiffer, and K. West, Phys. Rev. B **44**, 1930 (1991).
8. G. Bastard, Phys. Rev. B **25**, 7584 (1982).
9. E.O. Kane, in *Physics of III-V Compounds*, Vol. 1 of Semiconductor and Semimetals, edited by R.K. Willardson and A.C. Beer (Academic Press, New York, 1966), p. 75.
10. G.E. Pikus and G.L. Bir, Fiz. Tverd. Tela **1**, 154 (1959) [Soviet Phys.-Solid State **1**, 136 (1959)].

11. G.A. Baraff and D. Gershoni, Phys. Rev B **43**, 4011 (1991).
12. F.H. Pollak and M. Cardona, Phys. Rev. **172**, 816 (1968).
13. M. Kohl, D. Heitmann, D. Grambow, and K. Ploog, Phys. Rev. Lett. **63**, 2124 (1989).

SPECTROSCOPY OF QUANTUM-DOT ATOMS

D. HEITMANN, B. MEURER, AND K. PLOOG*
Max-Planck-Institut für Festkörperforschung
Heisenbergstraße 1
D-7000 Stuttgart 80
Germany

ABSTRACT. Progress in submicron technology makes it possible to realize man-made low-dimensional electronic systems with quantum confined energy states, i.e., quantum wires and quantum dots. In this review we will address experiments where we approach the quantum limit, i.e., quantum dots with small numbers of electrons per dot. We have prepared arrays of field-effect confined quantum dots with diameters smaller than 100 nm starting from $Al_xGa_{1-x}As$–GaAs heterostructures. In far-infrared spectroscopy, we observe discrete steps in the gate voltage dependence of the integrated absorption strength indicating directly the incremental occupation of each dot with $N = 1, 2, 3$, and 4 electrons. From the experiments we can deduce that this discrete number is stabilized by the high Coulomb charging energy of about 15 meV in the very small dots. With this well defined small number of electrons per dot, it becomes possible to perform a kind of 'atomic' spectroscopy on these systems. We will show that the dynamic excitations exhibit an interesting complex interplay of atomic-like single-particle and collective many-body effects.

1. Introduction

Many novel and unique properties of layered two-dimensional (2D) semiconductor structures with quantum confined energy states have challenged scientists to prepare and study systems with further reduced dimensionality, specifically quantum wires and quantum dots. In these systems the original free dispersions of the electrons in the lateral directions are also quantized due to an additional lateral confinement. One ultimate limit is a quantum dot, where, induced by a confining potential in both the x- and y-directions, artificial 'atoms' with a totally discrete energy spectrum, are formed [1–10]. (The growth direction is labeled z in the following.) Typical confinement energies are in the few-meV regime. Thus the most direct information on the quantum confined energy levels in these low-dimensional systems should be obtained by investigation of optical transitions with far-infrared (FIR) spectroscopy. It turns out that the dynamic response of these systems exhibits an interesting complex interplay of atomic-like single-particle and many-body effects which will be discussed by reviewing some recent experiments on quantum dots. We will in particular address experiments [9,11] on periodic arrays of quantum dots where electrons are confined by the field effect of a laterally structured gate electrode, which have been prepared, starting from two-dimensional electron systems (2DES) in $Al_xGa_{1-x}As$–GaAs heterostructures. In FIR spectroscopy we observe discrete steps in the gate voltage (V_g) dependence of the integrated absorption strength. Since the integrated absorption

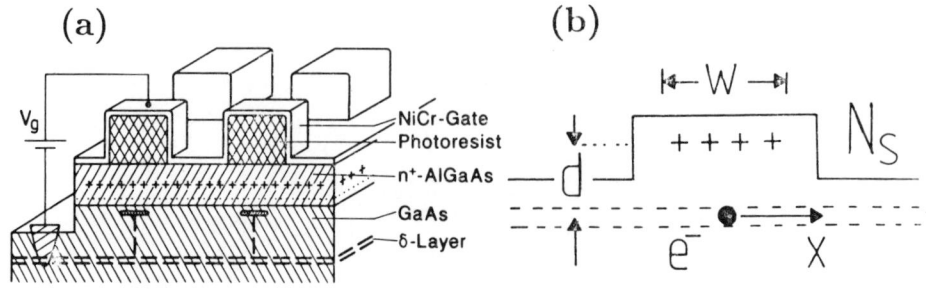

Figure 1. (a) Sketch of field-effect confined quantum-dots in an AlGaAs/GaAs heterostructure. (b) A model to calculate the lateral confinement potential for electrons in a quantum dot.

strength is proportional to the number of electrons per dot, this indicates directly the stepwise, discrete charging of each dot with 1, 2, 3, and 4 electrons. This links our experiments to another interesting topic, i.e., single-electron charging effects, which are extensively studied in both small metallic and semiconductor systems [12–17]. From the gate voltage dependence of the steps, we can estimate the Coulomb charging energy to be about 15 meV. This large value stabilizes, at a given V_g, a well defined number of electrons in each individual dot of the array and allows us to study the spectral fine structure of quantum-dot atoms.

2. Sample Preparation and Experimental Details

The quantum-dot samples are sketched in Fig. 1(a). They have been prepared starting from $Al_{0.32}Ga_{0.68}As$–GaAs heterostructures grown by molecular beam epitaxy. The thickness of the spacer layer was 36 nm, of the doped AlGaAs layer was 56 nm and of the cap GaAs layer was 9 nm. A Si δ-doped layer in the GaAs, deposited at a distance of 330 nm from the AlGaAs-GaAs interface, acts as a back contact to charge the dots. The doping density of 2×10^{12} cm^{-2} was optimized to have enough conductivity for charging the dots but still being semitransparent for FIR radiation. The 2D density was about $N_s = 2 \times 10^{11}$ cm^{-2} and the mobility about 900,000 cm^2/Vs (at 2.2 K). On top of the heterostructure we prepared a periodic photoresist dot array by holographic lithography. The periods ranged from a = 500 nm down to 200 nm and the lateral photoresist dot sizes were about half the period with a height of about 100 nm. An 8 nm thick semitransparent NiCr gate of 4 mm diameter was evaporated onto the photoresist structure. Outside the gate area the electron gas was removed by etching the AlGaAs. Contacts were alloyed to the δ-doped back contact, so that with a negative gate voltage we could confine the electrons under the photoresist dots and vary the number of electrons [2–4,7,8]. FIR transmission spectroscopy was carried out with a Fourier transform spectrometer and with FIR lasers in perpendicular magnetic fields B. We recorded the normalized transmission of unpolarized radiation, $T(V_g)/T(V_t)$, where V_t is the threshold voltage at which the dots are totally depleted.

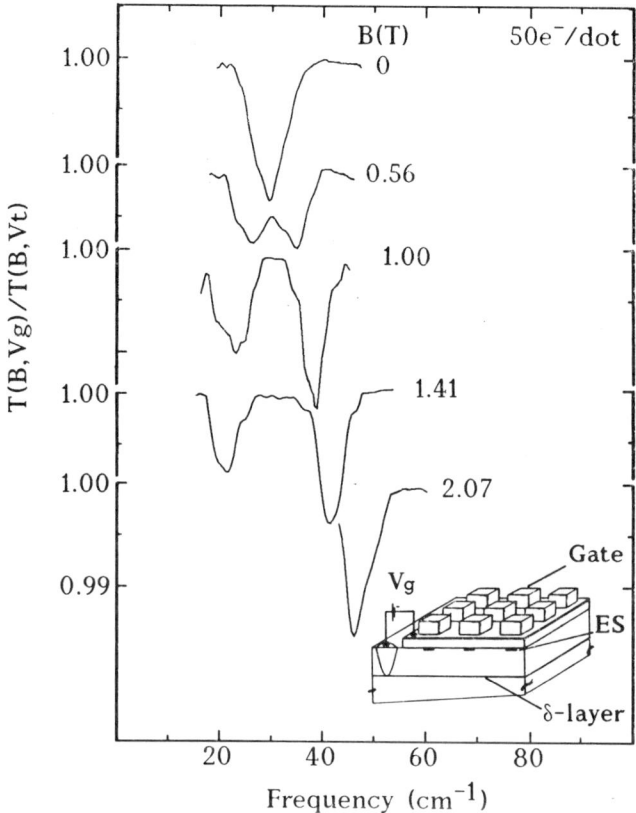

Figure 2. FIR transmission measured at various magnetic fields for a quantum-dot array with a period of a = 500 nm and a gate voltage corresponding to N = 50 electrons per dot. (From Ref. [11].)

3. Experiments

Experimental FIR transmission spectra for a sample with period a = 500 nm are displayed in Fig. 2. At a gate voltage of $V_g = -0.8$ V there are about N = 50 electrons per dot (for the determination of N, see below). At B = 0 we observe one resonance at about 30 cm^{-1} (= 3.7 meV). For B > 0 this resonance splits into two resonances. The dispersion for these two resonances is depicted in Fig. 3(a), the lower branch decreases with increasing magnetic field, the upper branch increases with increasing B and approaches the cyclotron resonance frequency $\omega_c = eB/m^*$. In Fig. 3 we have also plotted the dispersions for two other gate voltages, $V_g = -0.9$ V and $V_g = -1.05$ V, where we have, respectively, N = 25 and N = 12 electrons per dot. Figure 3 shows that these dispersions do not depend significantly on the number of electrons per dot. There is only a small shift of the B = 0 frequency to higher energies with decreasing N.

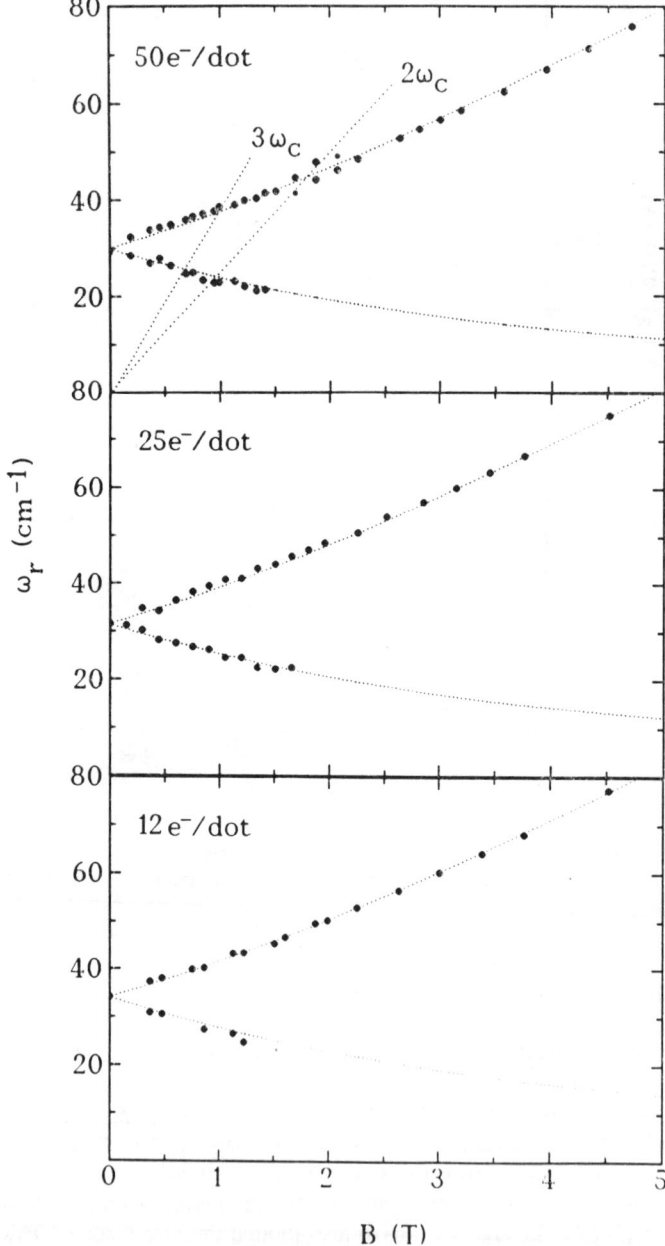

Figure 3. Magnetic field dispersion of the resonance frequency in a quantum-dot array with a period of a = 500 nm and, from the top, N = 50, 25, and 12 electrons per dot. (From Ref. [11].)

4. Theoretical Models

The dispersions in Fig. 3 have been observed before, not only for quantum dots with a small number of electrons [4,6,7,9], but similarly also in finite-sized 2DES with much larger dimensions and electron numbers [18–20]. (These latter experiments have been interpreted as depolarization, or, equivalently, as edge magnetoplasmon modes [21–24]) To explain the experimentally observed behavior in an "atomic" model we need information about the confining potential. Self-consistent bandstructure calculations show that the external confinement potential for electrons in a field-effect confined quantum dot has a nearly parabolic shape [25]. In a simplified way we can see this from the model sketched in Fig. 1(b). Due to the strong original 2D confinement the electron can only move in the x–y directions. A positively charged layer of 2D density N_S and width w of donors in the AlGaAs/GaAs layer or, in a gated structure, from a gate voltage, holds the electron in an equilibrium position. Moving the electron in the x direction produces a force approximately linear in the elongation, F = Kx, which gives rise to a parabolic confinement potential $V(x) = Kx^2/2$. We can calculate the force constant K and thus the eigenfrequency by integrating the electric fields from the charged layer at the position of the electron and find approximately $\Omega_0^2 = N_S e^2 \pi/(2m^* \bar{\varepsilon} \varepsilon_0 w)$, where we have assumed that the distance d in Fig. 1(b) is small and the dielectric surrounding is described by an effective dielectric function $\bar{\varepsilon}$.

Now we consider the one-electron Schrödinger equation in a magnetic field B for the potential $V(r) = m^* \Omega_0^2 r^2/2$. The energy eigenvalues have been calculated by Fock [26]

$$E_{n,\ell} = (2n+|\ell|+1)\sqrt{(\hbar\Omega_0)^2 + (\hbar\omega_c/2)^2} + \ell\hbar\omega_c/2, \tag{1}$$

where n = 0,1,2,... and ℓ = 0,±1,±2,... are, respectively, the radial and azimuthal quantum numbers. From calculations of the matrix elements one finds that dipole allowed transitions have transition energies

$$\Delta E^\pm = \sqrt{\hbar^2\Omega_0^2 + (\hbar\omega_c/2)^2} \pm \hbar\omega_c/2. \tag{2}$$

This dispersion is exactly observed in our experiments as demonstrated by the fit in Fig. 3.

However, we do not have only *one* electron in a quantum dot, but several. In this case electron-electron (ee) interaction should strongly influence the energy spectrum and the dynamic response. For a large number of electrons one usually calculates the static effective one-electron energy spectrum within the Hartree-Fock approximation and then applies the random phase approximation (RPA) to determine the dielectric response (see, e.g., Ref. [27]). For a small number of electrons per dot, however, it is possible to calculate many-body-electron wavefunctions and energy states directly [28–30]. In Fig. 3 we compare the one-particle energy spectrum with the two-particle energy spectrum in a parabolic confinement [29,30]. The many-body energy spectrum is govened by two energies (at B = 0): the quantum confinement energy, $E_q = \hbar^2/(m^* l_0^2)$, and the Coulomb energy, $E_c = e^2/(4\pi\varepsilon\varepsilon_0 l_0)$. Here $l_0 = \sqrt{\hbar/m^*\Omega_0}$ is the confinement length of the harmonic oscillator. With increasing l_0 the Coulomb energy becomes increasingly important with respect to the confinement energy. The scale is set by the effective Bohr radius a* which is 10 nm in GaAs. In Fig. 3 with E_q = 3.4 meV and correspondingly l_0 = 20 nm, the Coulomb energy is E_c = 7 meV. E_c reflects the energy which is required to "squeeze" the second electron

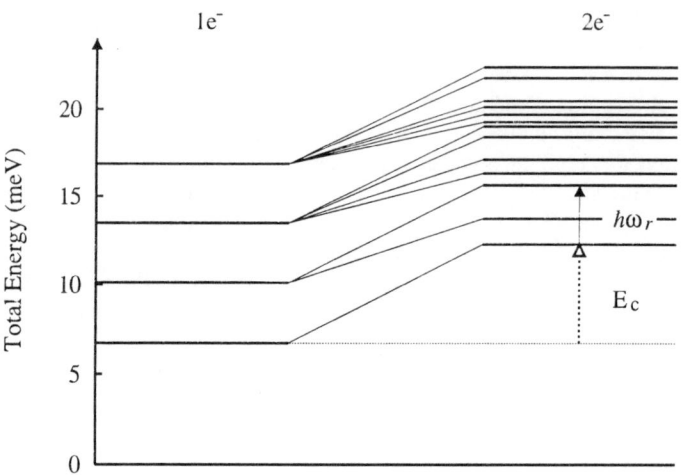

Figure 4. The one-particle and the two-particle excitation spectrum ('quantum-dot-He') in a parabolic confinement [30]. The indicated dipole allowed transition ($\hbar\omega_r$) for quantum-dot-He has exactly the same energy as for quantum-dot-H. (From Daniela Pfannkuche, with many thanks.)

into the quantum dot. This energy is represented in Fig. 4 in the increase of the lowest energy level of the two-particle spectrum with respect to the one-particle spectrum. This "Coulomb charging energy" of the atom will become important for our experiments as discussed below. Due to the ee-interaction some of the degeneracies of the one-particle spectrum are now lifted giving rise to a complex excitation spectrum.

The unique and important point of a parabolic confinement is, however, that the only allowed dipole transition ($\hbar\omega_r$ in Fig. 4) for two electrons has exactly the same energy as in the one-particle case. Moreover, this result also holds for an arbitrary number of electrons, N. It has been shown for quantum wells [31,32] and quantum dots [33] with parabolic confinement, that the Hamiltonian for N electrons, i.e., including ee-interaction, separates into a center-of-mass (CM) motion and into relative internal motions. The CM motion solves exactly the one-electron Hamiltonian and is the only allowed optical dipole excitation. Thus, the optical dipole response of a quantum dot with parabolic confinement represents a rigid collective CM excitation at the frequency of the bare external potential, i.e., for B = 0, at $\hbar\Omega_0$. This also means that for a parabolic confinement, dipole excitation frequencies are insensitive to ee-interactions and independent on the electron density in the dot. This result can be considered as a generalization of Kohn's famous theorem [34], which says that the cyclotron frequency in a translationally invariant system is not influenced by ee-interactions. Experimentally this theorem is demonstrated by the small effect of the gate voltage and related number of electrons on the resonance position in Fig. 3. Sikorski and Merkt [4] found that for their InSb quantum dots the resonance frequency was independent on the gate voltage. Theoretically the same conclusion has also been drawn from a classical description of an electron system in a parabolic confinement by Shikin et al. [21]. The quantum mechanical CM motion corresponds exactly to the classical dipole plasma mode of the quantum dot.

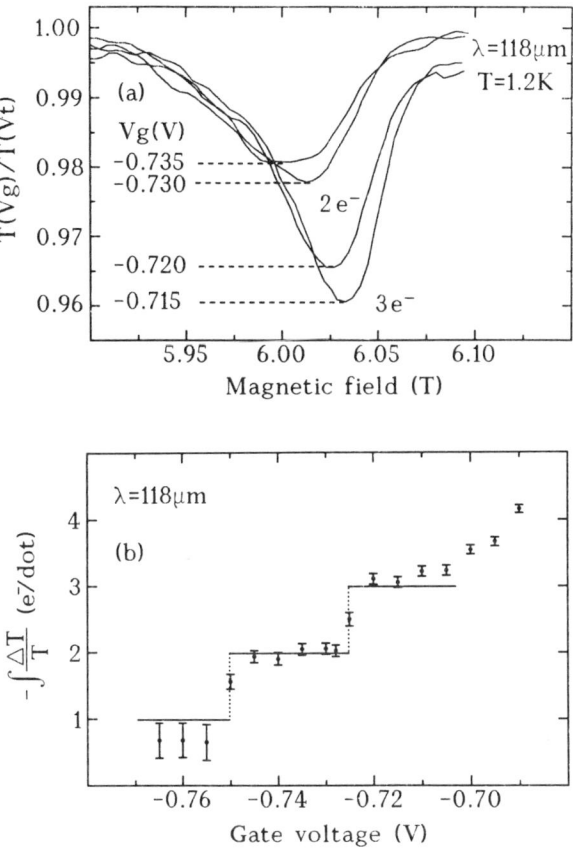

Figure 5. (a) FIR transmission measured at a fixed laser frequency of 10.5 meV at various gate voltages V_g for a quantum-dot array with period a = 200 nm. Spectra for 3 electrons per dot (V_g = -0.715 V and -0.720 V) and for 2 electrons per dot (V_g = -0.730 V and -0.735 V) are shown. (b) Integrated absorption strength versus gate voltage V_g for a series of spectra (including (a)). The step-wise increase of the absorption strength indicates the incremental occupation of the dots with N = 1, 2, and 3 electrons per dot with increasing V_g. (Error bars mark the accuracy of the fits.) The temperature is T = 1.2 K. (From Ref. [9].)

5. Coulomb Charging of Quantum Dots

The generalized Kohn's theorem tells us that for a parabolic confinement we cannot determine the number of electrons from the excitation frequency of the quantum-dot atoms. However, FIR spectroscopy does not only give us the resonance frequency, but also the absorption strength, which we can measure in absolute units. This allows us to determine the electron number per dot. Experimental FIR transmission spectra for a sample with period a = 200 nm are displayed in Fig. 5(a). At the laser frequency of 10.5 meV (84 cm^{-1}) we observe the resonance at about B = 6 T. The interesting observation in Fig. 5(a) is that the absorption amplitude does not increase in proportion to $V_g - V_t$. As we will elaborate in the following, this directly reflects that the quantum dots are occupied with a well defined number of electrons in different V_g intervals. A closer look at the experimental spectra shows that not only the absorption amplitude but also the lineshape varies with V_g. In certain V_g and B regimes the resonance consist clearly of two overlapping resonances, as shown, for example, in Fig. 6(a). To understand this behavior, we have performed a large series of measurements on different samples and in different B regimes. In Fig. 6(a) we show spectra for a fixed B = 3 T with two slightly different values of V_g. We observe that the resonance shape changes significantly even on this very fine V_g scale.

To evaluate these spectra we apply a classical model for the FIR response of electrons in a parabolic potential which have a high frequency conductivity of [35]

$$\sigma(\omega) = Ne^2\tau/\{2m^*a^2[1+i(\omega^2-\omega_r^2)\tau/\omega]\}, \qquad (3)$$

where $\hbar\omega_r = E^+$ is the high frequency resonance in Eq. (2), N the number of electrons per dot, a the period, and t a phenomenological scattering time that describes the linewidth. The relative change in transmission $\Delta T/T = (T(V_g) - T(V_t))/T(V_t)$ is then given by the well known expression [4,6,35]

$$\Delta T/T = -2Re(\sigma(\omega))/[(1+\sqrt{\varepsilon}+r_v/r_g)\varepsilon_0 c], \qquad (4)$$

where ε = 12 is the dielectric function of GaAs, r_v = 377 Ω the vacuum impedance, and $r_g \approx$ 1 kΩ, the combined impedance of front gate and back contact. Note that the integrated classical absorption strength is identical with the quantum mechanical result [36]. The result that the integrated absorption strength is proportional to the number of electrons per dot is a special case of the well known sum rule in atomic spectroscopy, the special behavior here is that, for the parabolic confinement, all the absorption is in only one line, whereas in "natural" atoms with a 1/r potential the absorption is distributed over several lines. We can check the determination of N via Eq. (2) (and all the the prefactors in this equation) if we tune the gate voltage to zero. Then, without confining potential, we have a 2DES and can determine the 2D density N_S from the observed cyclotron resonance absorption. For a 2DES we can determine N_S independently in magneto-capacitance measurements from Shubnikov-de Haas (SdH) oscillations and find that, as has been demonstrated before [37], the optical determination of the carrier density agrees with in better than 5% with the density determined from SdH.

We can thus determine directly the absolute number of electrons per dot from the experimental signal strength $\Delta T/T$. We have fitted the spectra in Figs. 5(a) and 6(a) with the ansatz Eqs. (2) and (3) using a superposition of two resonances with frequencies ω_{r1} and ω_{r2}. The results are plotted in Figs. 5(b) and 6(b), respectively, and demonstrate that we can directly determine the number of electrons from the FIR absorption strength. The accuracy of this line profile fits is demonstrated, for example, in Fig. 6(b). In Fig. 5(b) we find that, above V_t up to -0.64 V, the dots contain N = 1 electron; between -0.63 V and -0.61 V, N = 2 electrons; between -0.60 V and -0.58 V, N = 3 electrons; and between -0.57 V and -0.555 V N = 4 electrons. (For N > 4 the electron wavefunctions of the dots begin to

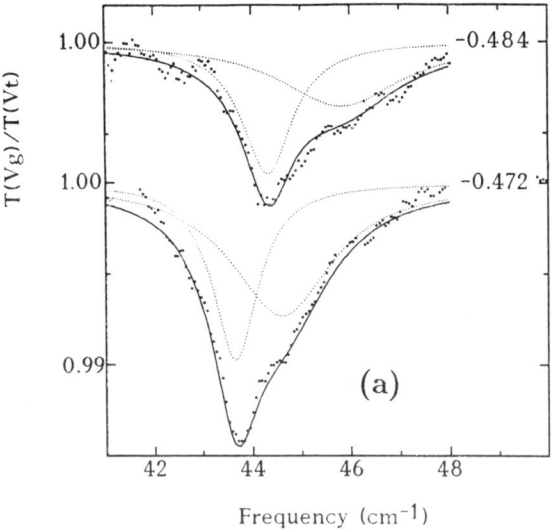

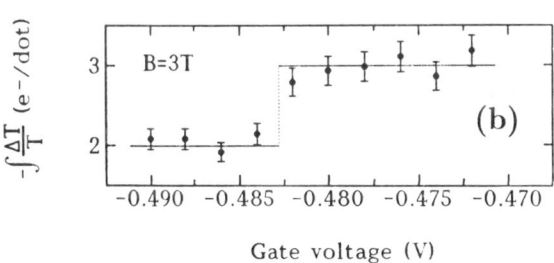

Figure 6. (a) Experimental FIR transmission spectra (symbols) and fits (solid lines) with two resonances (dotted lines) at B = 3 T and two gate voltages V_g as indicated. Different spectra are shifted vertically. For this V_g and B a clear splitting into two resonances is observed. The spectral resolution is 0.1 cm^{-1} and the temperature is T = 2.2 K. (b) Variation of the gate voltage V_g in very small increments to demonstrate the step wise transition from N = 2 to N = 3 electrons from the integrated absorption strength. (From Ref. [11].)

overlap with neighbouring dots and the spectra become more complex [7,38].) In Fig. 6 we have varied V_g in very small increments to demonstrate the sharp onset of the absorption strength which is in particular observed for the gate voltage transition from N = 2 to N = 3 electrons per dot. In Fig. 7 we show the same stepwise increase of the absorption strength for a sample with a different period a = 250 nm.

It is surprising that for our large number of about 10^8 dots we can charge all the dots simultaneously with the same number of electrons (within the error bars given in Figs. 5(b)

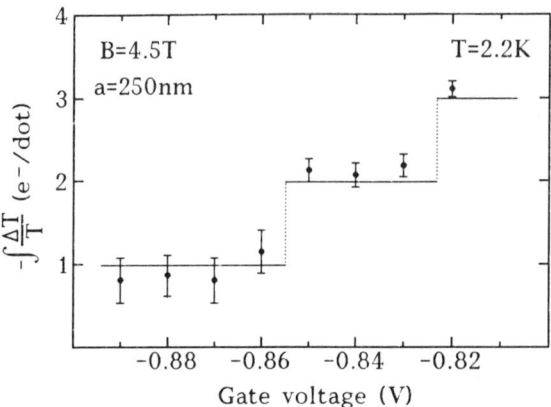

Figure 7. Integrated absorption strength versus gate voltage V_g for a sample with a different period of a = 250 nm. The step-wise increase of the absorption strength indicates again the incremental charging of the dots with N = 1, 2, and 3 electrons per dot with increasing V_g. (From Ref. [11].)

and 6(b)). What is the reason for this well defined charging behavior? The reason is the Coulomb energy of the dots! We have already seen in the manybody calculation of the 2-electron quantum dot ("quantum-dot-He") in Fig. 4 that it takes of the order of 10 meV for the dimensions here to "squeeze" the next electron into the dot. How can we estimate this Coulomb energy for our structure? A rough estimate can be obtained if we try to calculate the "capacitance" of the dot. It is not easy to calculate this for our complicated geometry. However, if we use the model of a parallel-plate capacitor, describing the effective dot area by the extent of the wavefunction in a harmonic oscillator (54 nm for N = 2) and take the distance to both the front gate and the back contact layer we find roughly a value of $C = 5 \times 10^{-18}$ F and a related "Coulomb charging energy", defined as $E_c = e^2/2C$, of 15 meV. This value is indeed significantly larger than kT and ensures that Coulomb effects are not smeared out either by temperature or by potential fluctuations from impurities and related local variations of the threshold voltage in our sample. Note that the definition of a capacitance and the related Coulomb charging energy is necessarily somewhat different from that in classical metallic systems where the capacitance is fixed [12,13]. When we charge the quantum dot with one additional electron, we increase the electronic diameter of the dot and the capacitance. Hence we vary the Coulomb energy with varying V_g. The estimation of a capacitance in this atomic regime is of course very rough. (Does for example the harmonic oscillator radius determine the effective capacitance area?)

However, we have a much more accurate information in our experiments. From Fig. 5(b) we can see that it requires an increase in the voltage of $\Delta V_g = 30$ mV to increment the number of electrons from N = 2 to N = 3. This corresponds to a capacitance of $C = e/\Delta V_g$ = 5.3×10^{-18} F and a related Coulomb charging energy $E_c = e^2/2C = 15$ meV. This value is similar to the value estimated above and confirms the model.

From the very beginning it is not directly clear that this gate voltage interval determines the Coulomb charging energy. However we can demonstrate that this determination is correct by comparing self-consistent bandstructure calculations with the Coulomb energy in the two-electron calculation. Self-consistent Hartree calculation for quantum-dot structures in GaAs with nearly the same dimensions as discussed here have been performed by Kumar et al. [25]. From the two-particle calculations [29,30] of quantum dots in a parabolic confinement one can deduce that the increase of the two-electron ground state energy with respect to the one-electron ground state, due to the Coulomb interaction, is $E_c \approx (e^2/\varepsilon l_0)$. ($l_0 = \sqrt{\hbar/m^*\omega_0}$ is the harmonic oscillator radius.) Using this "atomic model" we can now demonstrate the reliability of determining Coulomb energies from ΔV_g. For the parameters in the calculations in Ref. [25] we have for the external confinement potential 5 meV and thus $l_0 = 15$ nm which gives $E_c = e^2/\varepsilon l_0 = 7.9$ meV. This value agrees excellently with the determination of the Coulomb energy $E_c = e\Delta V_g/2 = 7.5$ meV from the calculated gate voltage interval in the same paper and demonstrates that this is a reliable and an independent method to determine E_c.

We like to note that Coulomb charging of quantum dots with few electrons has been discussed, for example, by Kumar et al. [25], Silsbee and Ashoori [15], and also very recently by Ashoori et al. [16].

6. Spectral Fine Structure and Relative Motions

Let us now come back to the spectral fine structure for the quantum-dot atoms. As discussed above, we expect for a fixed external potential of parabolic shape *one* resonance at a frequency which is *independent* of the number of electrons. For our experiments however, with a well defined number of electrons per dot, we observe a small but well resolved frequency shift (see Fig. 6(b)) which oscillates in phase with the number of electrons [9]. This shows that the real potential is not exactly parabolic. We find linewidth oscillations and, in a magnetic field, splitting into *two* resonances. We attribute this behavior to the fact that, due to the Coulomb charging effect, the next electron has to be "squeezed" into the dot. This influences the confining potential in a complex way, in particular, induces deviations from a perfect parabolic shape. These deviations vary in strength as we tune V_g within the interval of a fixed number of electrons. A possible explanation for this splitting is the following. Calculations for a parabolic potential and N = 2 show that, with increasing B, the ground state switches at a certain B_S to another state with a different angular momentum [29,33]. For a parabolic confinement this cannot be observed in FIR experiments, since the energy difference between the new ground state and the corresponding excited state is still the same. However, it has been calculated recently [30] that the degeneracy of the excited state is lifted for $B > B_S$ if the confining potential is not parabolic. This might account for our observation.

There is another interesting observation in Fig. 3(a) that the upper branch shows a resonant anti-crossing behavior with the energies $2\omega_c$ and $3\omega_c$. This interaction is again not allowed in a strictly parabolic confinement potential and indicates deviations from this form. It resembles a similar interaction of 2D plasmons with the harmonics of the cyclotron resonance, the so called "Bernstein" modes [39]. We find a similar interaction also in quantum wires [11,40] and such interactions were previously also observed in other types of microstructures, for example, antidots [38] and for the mixed-mode plasmon in wires [41]. We can explain this interaction theoretically in RPA calculations on dots with nonparabolic confinement. The unique feature of the interaction here is that, depending on the strength of the nonparabolicity, this interaction does not occur directly at $2\omega_C$ but can also be shifted to significantly smaller energies, somewhere in the regime between ω_C and $2\omega_C$! These results will be published separately in more detail [42].

7. Summary

In summary, we have realized field-effect confined quantum-dot atoms were a small discrete number of electrons is stabilized by a high Coulomb charging energy of 15 meV which is directly reflected in a step-like FIR absorption strength. For these artificial atoms it is possible to observe various types of spectral fine structures in a magnetic field.

Acknowledgments

We thank R.R. Gerhardts, V. Gudmundsson, and D. Pfannkuche for many discussions on the theory of quantum dots, R. Nötzel for the MBE growth of our samples, H. Lage and C. Lange for expert help with the preparation of the dots and the Bundesministerium für Forschung und Technologie for financial support.

References

* Present address: Paul-Drude-Institut für Festkörperelektronik, Hausvogteiplatz 5-7, O-1086 Berlin, Germany.
1. M.A. Reed, J.N. Randall, R.J. Aggarwal, R.J. Matyi, T.M. Moore, and A.E. Wetsel, Phys. Rev. Lett. **60**, 535 (1988).
2. T.P. Smith, III, K.Y. Lee, C.M. Knoedler, J.M. Hong, and D.P. Kern, Phys. Rev. B **38**, 2172 (1988).
3. W. Hansen, T.P. Smith, III, K.Y. Lee, J.A. Brum, C.M. Knoedler, J.M. Hong, and D.P. Kern, Phys. Rev. Lett. **62**, 2168 (1989).
4. Ch. Sikorski and U. Merkt, Phys. Rev. Lett. **62**, 2164 (1989).
5. C.T. Liu, K. Nakamura, D.C. Tsui, K. Ismail, D.A. Antoniadis, and H.I. Smith, Appl. Phys. Lett. **55**, 168 (1989).
6. T. Demel, D. Heitmann, P. Grambow, and K. Ploog, Phys. Rev. Lett. **64**, 788 (1990).
7. A. Lorke, J.P. Kotthaus, and K. Ploog, Phys. Rev. Lett. **64**, 2559 (1990).
8. J. Alsmeier, E. Batke, and J.P. Kotthaus, Phys. Rev. B **41**, 1699 (1990).
9. B. Meurer, D. Heitmann, and K. Ploog, Phys. Rev. Lett. **68**, 1371 (1992).
10. D. Heitmann, K. Kern, T. Demel, P. Grambow, K. Ploog, and Y.H. Zhang, Surface Sci. **267**, 245 (1992).
11. B. Meurer, PhD-Thesis, Stuttgart, Germany (1992).
12. I. Giaver and H.R. Zeller, Phys. Rev. Lett. **20**, 1504 (1968).
13. For recent work see, for example, R. Wilkens, E. Ben-Jacob, and R.C. Jaklevic, Phys. Rev. Lett. **63**, 801 (1989) and references therein.
14. H. van Houten and C. W. J. Beenaker, Phys. Rev. Lett. **63**, 1893 (1989).
15. R.H. Silsbee and R.C. Ashoori, Phys. Rev. Lett. **64**, 1991 (1990).
16. R.C. Ashoori, R.H. Silsbee, L.N. Pfeiffer, and K. West, in *Nanostructures and*

Mesoscopic Systems, edited by W. Kirk and M. Reed (Academic, New York, 1991), p. 323; R.C. Ashoori, H.L. Störmer, J.S. Weiner, L.N. Pfeiffer, S.J. Pearton, K.W. Baldwin, and K.W. West, Phys. Rev. Lett. **68**, 3088 (1992).
17. U. Meirav, M.A. Kastner, and S.J. Wind, Phys. Rev. Lett. **65**, 771 (1990).
18. S.J. Allen, Jr., H.L. Störmer, and J.C. Hwang, Phys. Rev. B **28**, 4875 (1983).
19. D.C. Glattli, E.Y. Andrei, G. Deville, J. Poitrenaud, and F.I.B. Williams, Phys. Rev. Lett. **54**, 1710 (1985).
20. D.B. Mast, A.J. Dahm, and A.L. Fetter, Phys. Rev. Lett. **54**, 1706 (1985).
21. V. Shikin, T. Demel, and D. Heitmann, Surf. Sci. **229**, 276 (1990).
22. A.L. Fetter, Phys. Rev. B **32**, 7676 (1985); Phys. Rev. B **33**, 5221 (1986).
23. V.B. Sandomirskii, V.A. Volkov, G.R. Aizin, and S.A. Mikhailov, Electrochimica Acta **34**, 3 (1989).
24. V. Shikin, S. Nazin, D. Heitmann, and T. Demel, Phys. Rev. B **43**, 11903 (1991).
25. A. Kumar, S.E. Laux, and F. Stern, Phys. Rev. B **42**, 5166 (1990).
26. V. Fock, Z. Phys. **47**, 446 (1928).
27. V. Gudmundsson and R.R. Gerhardts, Phys. Rev. B **43**, 12098 (1991).
28. G.W. Bryant, Phys. Rev. B Lett. **59**, 1140 (1987).
29. U. Merkt, J. Huser, and M. Wagner, Phys. Rev. B **43**, 7320 (1991).
30. D. Pfannkuche and R.R. Gerhardts, Phys. Rev. B **44**, 13132 (1991).
31. P. Ruden and G.H. Döhler, Phys. Rev. B **27**, 3547 (1983).
32. L. Brey, N. Johnson, and B. Halperin, Phys. Rev. B **40**, 10647 (1989).
33. P. Maksym and T. Chakraborty, Phys. Rev. Lett. **65**, 108 (1990).
34. W. Kohn, Phys. Rev. B **123**, 1242 (1961).
35. B.A. Wilson, S.J. Allen, Jr., and D.C. Tsui, Phys. Rev. B **24**, 5887 (1981).
36. U. Merkt, Ch. Sikorski, and J. Alsmeier, in *Spectroscopy of Semiconductor Microstructures*, edited by G. Fasol, A. Fasolino, and P. Lugli (Plenum, New York, 1989), p. 89.
37. K. Ensslin, D. Heitmann, H. Sigg, and K. Ploog, Phys. Rev. B **36**, 8177 (1987).
38. K. Kern, D. Heitmann, P. Grambow, Y.H. Zhang, and K. Ploog, Phys. Rev. Lett. **66**, 1618 (1991).
39. E. Batke, D. Heitmann, J.P. Kotthaus, and K. Ploog, Phys. Rev. Lett. **54**, 2367 (1985).
40. P. Grambow, PhD-Thesis, Darmstadt, Germany (1992).
41. T. Demel, D. Heitmann, P. Grambow, and K. Ploog, Phys. Rev. Lett. **66**, 2657 (1991).
42. P. Grambow, V. Gudmundson, B. Meurer, R.R. Gerhardts, D. Heitmann, D. Pfannkuche, and K. Ploog, to be published.

TRANSMISSIONS THROUGH BARRIERS AND RESONANT TUNNELING IN A LUTTINGER LIQUID

C.L. KANE
Department of Physics
University of Pennsylvania
Philadelphia, PA 19104
USA

ABSTRACT. We present a review of the theory of the single channel Luttinger liquid applied to transport in nanostructure devices. Our main focus will be on the effects of barriers, or constrictions in a single channel wire. In particular, we show that repulsively interacting electrons incident upon a single barrier are completely reflected at zero temperature (T). At finite T, the conductance vanishes as a power of T, and at T = 0, power law current-voltage characteristics are predicted. We next discuss resonant tunneling through a double barrier structure and show that the resonance peaks have a width which vanishes as a power of T as T → 0. Moreover, the T dependence of the resonance lineshapes are described by a universal scaling function, which has non Lorentzian tails. Possible experimental realizations of this are discussed, including a connection to edge transport in the fractional quantum Hall regime.

1. Introduction

In recent years nano-fabrication technology has advanced to the stage where it is possible to fabricate nanostructures in which electrons are confined to one dimension. A number of new interesting issues have been raised by such structures. For instance, quantization of the conductance in units of e^2/h has been observed in narrow point contacts [1]. Moreover, interesting charging effects have been observed in one dimensional (1D) wires associated with the Coulomb blockage [2,3].

From a theoretical point of view, one dimension is particularly interesting. Whereas in higher dimensions Fermi liquid theory is generally believed to be an appropriate paradigm for an interacting electron gas, in one dimension this is not the case. Interactions destroy the Fermi surface. Though interacting many body problems are in general notoriously difficult, much is known about the interacting problem in one dimension [4–8]. Away from charge density wave and superconducting instabilities there exists a simple description which Haldane has named Luttinger liquid theory [7]. As we shall see, the properties of a Luttinger liquid can be qualitatively different from that of a Fermi Liquid.

The purpose of this article is to review the physics of the Luttinger liquid and to show how such ideas can be applied to nanostructure devices. Particular emphasis will be paid to transport properties. We will be interested in the effects of a barrier, or constriction in the one dimensional channel on the conductance. The situation which we have in mind is illustrated in Fig. 1. Imagine a GaAs-AlGaAs interface which has been gated in such a way to create a narrow channel of electrons. Provided the channel is narrow enough and the density is low

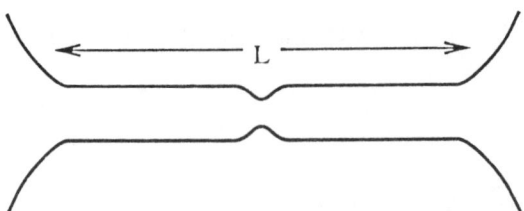

Figure 1. Schematic picture of a one dimensional channel connected at each end to wide leads.

enough, the electrons in the channel can be confined to a single 1D subband. In this case, the system is truly 1D in the sense that all motion transverse to the "wire" is quantum mechanically frozen out. In order to measure transport properties of such a device, the 1D wire must be attached to wide leads (as indicated in Fig. 1). We must be sure that we are probing the 1D physics of the wire rather than the three dimensional physics of the leads. We shall see that this will be the case provided the length L of the wire is longer than a thermal coherence length $L_T = \hbar v_F/kT$. Alternatively, for a fixed length wire, the temperature must be larger than $T_L = \hbar v_F/kL$. For a 1 μm GaAs wire with a typical electron separation of 10 nm, $T_L \sim 10$ mK. Our final (and probably most severe) assumption is that the wire is sufficiently clean that transport is ballistic on the scale of the length of the wire, except for scattering by a single barrier or constriction in the wire.

For the purposes of this article, we will confine our interest to spinless electrons. This will be applicable to a system in a magnetic field which spin polarizes the electrons. Such a situation is actually desirable, since a magnetic field will tend to specially separate the left and right moving electrons, which will tend to reduce the unwanted backscattering in the 1D leads. At the end of this article we shall also point out a connection between transport in a one dimensional Luttinger liquid and edge transport in the Fractional quantum Hall regime of a two dimensional electron gas. We shall see that a number of our predictions may be directly applied to such systems.

2. The Luttinger Liquid

For non interacting electrons, it is well known that the two terminal conductance of a single channel wire is equal to e^2/h. As a first clue to the difference between a Luttinger liquid and a Fermi liquid, let us see how this is modified by interactions. To do so, recall the argument given by Landauer and Buttiker [9]. Imagine that the left and right leads of the 1D wire are in equilibrium with reservoirs at chemical potentials separated by eV. Then there will be an excess of $\kappa eV/2$ states carrying current away from the reservoir at higher chemical potential. Here $\kappa = \partial n/\partial \mu$ is the compressibility. The current carried by each state is determined by the velocity ev_F. The remarkable quantization of the conductance $G = e^2/h$ then follows from the special relationship between the Fermi velocity and the compressibility for 1D non-interacting electrons, $\kappa v_F = 1/(\pi \hbar)$. For interacting electrons, both the compressibility and the velocity are renormalized, and this cancellation need no longer precisely occur. Therefore, if we define $g = \pi \hbar \kappa v$, then the conductance is given by

$$G = g\frac{e^2}{h}. \tag{1}$$

The dimensionless parameter g is a measure of the strength of the interactions and plays a central role in our theory. For non interacting electrons, g = 1, whereas for repulsive interactions (which decrease the compressibility), g is less than one and can be estimated roughly as $g^2 \sim (1 + U/2E_F)^{-1}$. Here U is the Coulomb interaction between neighboring electrons and E_F is the Fermi energy. On the other hand, for attractive interactions, g > 1. Equation (1) should be viewed with some caution, because at present we have no theory for the interface between the one dimensional Luttinger liquid and the two or three dimensional (Fermi liquid) leads. The contact resistance between the 1D sample and the leads may well be very important in practice.

In order to gain more insight into the nature of the 1D interacting electron gas, a number of powerful techniques have been developed over the years [4-8]. Electron interactions may be treated in a perturbative renormalization group scheme, which leads to the so-called "g-ology" theory [6]. A central result of this theory is that in the absence of Umklapp scattering, a system of spinless electrons can be described at low energies and long length scales by a line of renormalization group fixed points characterized by a single parameter, g (which turns out to be the same as the g defined above). A particularly simple way of describing the physics near these "Luttinger liquid" fixed points is via the method of "bosonization". We will not attempt to explain the technical details of the bosonization method here, though below we will present a heuristic discussion of the physics which is implicite in it.

The central feature of a Luttinger liquid which we would like to emphasize is that a Luttinger liquid is "almost" a charge density wave (CDW) at zero temperature. By this we mean that the density-density correlation function contains an oscillatory component at wave vector $2k_F$ (corresponding to the mean interelectron spacing), which decays only algebraically at large distances,

$$<n(x)n(0)> \approx \frac{\sin 2k_F x}{x^g}. \qquad (2)$$

For non-interacting electrons (g = 1), this is the well known result which is a consequence of the sharp Fermi surface. For repulsive interactions (g < 1), the exponent is smaller, so that the CDW correlations are longer range. Indeed, in the strong interaction limit, g → 0, true long range order is approached.

The very strong interaction limit is a useful starting point for developing intuition regarding the nature of the Luttinger liquid. In this limit, it is appropriate to describe the system as a Wigner crystal plus fluctuations [10]. The low energy excitations about this crystal will be the long wavelength phonons. If we define a phonon coordinate θ, which describes the displacement of a particle from its original lattice position (in units of the lattice spacing), then the long wavelength phonon modes may be described by the following continuum Lagrangian,

$$L = \frac{\pi}{2g\upsilon} \left[\left(\frac{\partial \theta}{\partial t}\right)^2 - \upsilon^2 (\nabla \theta)^2 \right]. \qquad (3)$$

The first term in Eq. (3) is the kinetic energy, $m\dot{r}^2/2$, and the second term is the energy of a long wavelength lattice distortion, while g and υ in Eq. (3) are defined consistently with the above definitions. In one dimension, quantum fluctuations of these long wavelength phonon modes destroys the long range crystalline order, and one is left with algebraically decaying positional correlations at T = 0, as described in Eq. (2). This may be directly seen from the logarithmic divergence of the Debye-Waller factor.

It is remarkable that the description of the electron gas in Eq. (3), which was motivated from the strong interaction (Wigner crystal) limit, is actually valid for weak interactions, no interactions and even attractive interactions. In these situations, the low energy excitations of the system continue to be phonon like collective density fluctuations. The bosonization technique provides a rigorous formulation of this, and provides a formal means of computing the electronic properties of such systems, such as Eq. (2).

3. Single Barrier

We now turn to the effects which a barrier in an otherwise clean single channel electron gas will have. It is worthwhile first to remind the reader of what is expected for non interacting electrons. In that case, it is well known that an electron incident on a localized scattering potential will be partially reflected and partially transmitted, with probabilities which depend on the size of the barrier [9]. The conductance is then given by the transmission probability times e^2/h. We will now demonstrate that the presence of electron electron interactions has a profound and qualitative effect on such scattering. Our approach is to treat the effects of the barrier perturbatively. This can be done in two limits: a very weak barrier or a very strong barrier (or equivalent weak tunneling).

To compute the effect of a weak scattering potential, we must determine the form of the appropriate correction to Eq. (3) due to the scatterer. The bosonization technique provides a formal recipe for translating perturbations into the boson language. However, here we shall be content with a heuristic motivation based on the Wigner crystal picture. In this case, the effect of a weak barrier is to act as a pinning site for the Wigner crystal. Such a term must be invariant under translating all of the particles by one lattice space, but provide a barrier towards doing so. The simplest such term has the form [11,12],

$$\delta L = \upsilon \cos 2\pi\theta(x=0), \tag{5}$$

The coefficient υ is related to the Fourier component of the scattering potential at wavevector $2k_F$, which corresponds to the interparticle spacing. In general there can be higher Fourier components to the pinning potential, such as $\cos 4\pi\theta$ or $\cos 6\pi\theta$. However, these terms turn out to be "irrelevant" in the sense that their contribution is negligible at low temperatures. Equation (5) may also be interpreted as a term which backscatters electrons, changing their momentum by $2k_F$.

By performing an analysis of this problem perturbative in υ, we find that the conductance is reduced in the following way at finite temperature T [11,12]:

$$G(T) \approx g\frac{e^2}{h} - a\upsilon^2 T^{2(g-1)}, \tag{6}$$

where a is a constant depending on the Fermi energy. For non interacting electrons, $g = 1$, we obtain the expected temperature independent reduction of the conductance. For repulsive interactions, $g < 1$, however, the correction diverges at low temperatures, so that perturbation theory breaks down. Expressed in the language of the renormalization group, this says that the

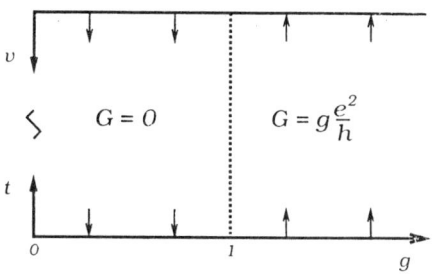

Figure 2. The phase diagram for a one-channel gas of spinless interacting electrons incident upon a single barrier. Here g is the dimensionless conductance of the gas in the absence of the barrier, and G is the conductance in the presence of the barrier. The small arrows indicate the renormalization group flows. Notice that for repulsive interactions (g < 1) a single barrier is completely reflecting. The dashed line at g = 1, which corresponds to a non-interacting electron gas, indicates a fixed line along which the conductance through the barrier varies continuously.

scattering potential υ is a "relevant" perturbation. For attractive interactions, however, the correction vanishes at low temperature, which means that υ is "irrelevant". For non-interacting electrons, the perturbation is "marginal".

Since for repulsive interactions the barrier is relevant, as one goes to lower temperatures, its effective strength increases, and one leaves the regime where perturbation theory is valid. In this case it is useful to consider the opposite limit of a very strong barrier, which may be described as two semi-infinite wires connected by a very weak hopping matrix element, t. By performing an analysis similar to the above, perturbative in t, we find that

$$G(T) \approx t^2 T^{2(1/g-1)}. \tag{7}$$

Now, the situation is exactly the reverse as before. For repulsive interactions, g < 1, the conductance vanishes as T → 0, which means that t is "irrelevant". For attractive interactions, on the other hand the perturbation theory diverges, so that t is "relevant". Again, however, for non-interacting electrons t is "marginal".

By piecing these two limits together, we obtain the phase diagram displayed in Fig. 2. For repulsive interactions, g < 1, it is very plausible that we may connect the renormalization group flows (indicated as small arrows Fig. 2), which means that in the limit of low temperatures, an arbitrarily weak scatterer will cause the wire to be perfectly insulating. For g = 1/2 an exact solution is available from which the matching of the two limits can explicitly be seen [12]. On the other hand, when the electron electron interactions are attractive, an arbitrarily strong barrier will be transparent at low temperatures. This behavior of barriers at low energies is one of the central results of this work, and it is in striking contrast to the conventional Landauer result. At finite temperatures there will be power law corrections as in Eq. (7). Moreover, at low temperatures, the I-V characteristic will have a similar power law behavior,

$$I \approx V^{2/g-1}. \tag{8}$$

At low temperatures and voltages, the power law behavior is cut off in a finite length wire by T_L defined in the introduction. Thus, Eqs. (7) and (8) will be valid only for energy scales larger than T_L. Below that, there will be ohmic conductance with T replaced by T_L in Eq. (7).

Physically, we may interpret the insulating behavior of a single barrier with repulsive interactions in the following way. As described above, the Luttinger liquid has a tendency towards CDW order. For $g < 1$, these CDW correlations are long range enough that a single barrier will pin the incipient charge density wave at low energies.

4. Resonant Tunneling

We now turn to the problem of resonant tunneling through a double barrier structure. This situation is particularly interesting in light of recent experiments in which oscillation in the conductance through a double constriction have been observed in the conductance as a function of the chemical potential [2]. These oscillations are believed to correspond to the addition of single electrons to the "island" between the two constrictions, and to be a result of the Coulomb blockade which forbids tunneling onto the island unless the chemical potential is tuned to a value where another electron can be added with no cost in energy [13,14].

For non-interacting electrons, as the incident energy of the electron is varied, the Schrodinger equation typically gives resonances, or peaks in the transmission. For a symmetric double barrier potential, the transmission at the peaks is perfect, and the conductance is $G^* = e^2/h$. The width of the resonance is finite, as $T \to 0$, and is determined by the height and width of the barriers and the density of states of the electrons in the leads. What is the effect of electron-electron interactions on such resonances? There are two important effects. The first arises due to the electron-electron interactions on the island between the constrictions. This gives rise to a capacitive charging energy $e^2/2C$ for adding a single electron to the island. This charging energy can be as large as ~ 1 K, and gives rise to the "Coulomb blockade". However, if the chemical potential on the island is adjusted to the point where another electron is just about to be added, then an electron may hop onto the island at no energy cost. In this situation, the Coulomb blockade does not impede the tunneling of electrons. This is believed to be the physics responsible for the oscillations in the conductance observed as a function of gate voltage, which occur as single electrons are added to the island. The other important effect of electron electron interactions comes from the interactions among the electrons in the leads. We saw for a single barrier that such interactions played a crucial role, and it is these effects which we will focus on below.

Again, we can perform a perturbation theory and renormalization group analysis in two limits: the limit of very strong barriers and the limit of very weak barriers. These two limits may then be pieced together to give the low temperature phase diagram for resonant tunneling. Rather than go through the details here, we will simply summarize our results [15,12]. We find that there can be perfect transmission on resonance for a symmetric double barrier, provided the electron electron interactions are not too strongly repulsive (that is $g > 1/4$). In this case, however, we find that in contrast to the non interacting case, the resonances become sharper as the temperature is lowered and have a width which vanishes as a power of the temperature. Moreover, at low temperatures, the lineshape of the resonance is predicted to have a universal shape described by the scaling function:

$$G(T,\delta) = \tilde{G}(\delta/T^{1-g}). \tag{9}$$

Here δ is the distance" to resonance, $V_G - V_G^*$. The scaling function $\tilde{G}(X)$ is universal in the

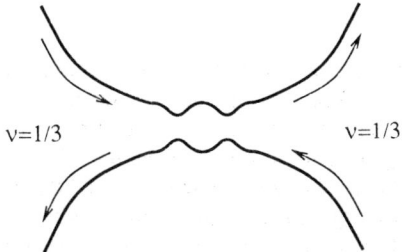

Figure 3. Schematic picture of a "double barrier" with two dimensional leads in a fractional quantum Hall state (for instance, $\nu = 1/3$). The arrows represent the direction of propagation of the edge states, which can be backscattered at the constrictions.

sense that it depends only on the lead conductance g and not on the details of the shapes of the barriers. The width of the resonance thus vanishes at low temperatures as T^{1-g}. Moreover, with a suitable rescaling of the axes, plots from different temperatures should collapse onto a single scaling curve. The scaling function $\tilde{G}(X)$ has non-Lorentzian tails, which decay as $X^{-2/g}$.

5. Connection to Fractional Quantum Hall Effect

There is a deep connection between the 1D transport which we have studied here and edge transport in the fractional quantum Hall (FQH) regime of a two-dimensional electron gas. Indeed the FQH regime may well be the best place to observe the phenomena which we havepredicted. Specifically, Wen has shown in a recent series of articles [16,17] that the edge excitations of the edge excitations of a two dimensional electron gas in the FQH regime may be described as a "chiral Luttinger liquid". A chiral Luttinger liquid is simply a Luttinger liquid consisting only of electrons moving in one direction. (It is in a sense "half" of an ordinary Luttinger liquid). A quantum Hall bar which is long, but not one dimensional will consist of chiral Luttinger liquids on each edge which move in opposite directions. This is formally identical to a usual Luttinger liquid. Moreover, as Wen has shown, the dimensionless conductance of this Luttinger liquid is simply given by the filling factor of the bulk fractional quantum Hall state, at least for the odd integer states, $g = \nu = 1/m$. Since the two edges are assumed well separated from each other, tunneling between the edges is strongly suppressed. Therefore, the problem of backscattering in the bulk Luttinger liquid is eliminated. Moreover, since it is much easier to fabricate large two dimensional electron gasses, the cutoff due to finite system size (L) should no longer be a limiting factor.

Imagine now a constriction in this quantum Hall "wire" which allows tunneling between the two edges, as depicted in Fig. 3. This is precisely the analogue of the single barrier problem described above, and in a $\nu = 1/3$ state, for instance, the two terminal conductance should vanish at low temperatures as T^{-4} [17]. In the presence of two such constructions, we have precisely the resonant tunneling problem. At low temperatures the resonance lineshapes should be described by a truly universal scaling function with no undetermined free parameters g. It is also clear that there will be a qualitative difference between integer and fractional Hall effects, with only the latter having resonant widths which vanish in the zero temperature limit.

6. Conclusion

In conclusion, we have discussed the effects of electron electron interactions on transport in one dimension. The most notable feature is the fact that for repulsive interactions an arbitrarily weak barrier in an otherwise pure Luttinger liquid will cause total reflection of an incident current and hense insulating behavior at zero temperature. The experimental signature of this Luttinger liquid physics will be the power law scaling of the conductance as a function of temperature for temperatures larger than the finite length cutoff T_L. In the resonant tunneling problem, the signature will be a sharpening of the resonance linewidth as the temperature is lowered. Due to the difficulties of fabricating long, clean wires, it appears that the most promising system to observe these effects is in the quantum Hall regime.

Acknowledgment

This work has been a product of collaboration with Matthew P.A. Fisher.

References

1. G. Timp, in *Mesoscopic Phenomena in Solids*, edited by B.L. Altshuler, P.A. Lee, and R.A. Webb (Elsevier, Amsterdam, 1990).
2. U. Meirav, M.A. Kastner, M. Heiblum, and S.J. Wind, Phys. Rev. B **40**, 5871 (1989).
3. For a recent review of the Coulomb blockade and related phenomena, see D.V. Averin and K.K. Likharev, in *Mesoscopic Phenomena in Solids*, edited by B.L. Altshuler, P.A. Lee, and R.A. Webb (Elsevier, Amsterdam, 1990).
4. J.M. Luttinger, J. Math. Phys. **15**, 609 (1963).
5. A. Luther and L.J. Peschel, Phys. Rev. B **9**, 2911 (1974); Phys. Rev. Lett. **32**, 992 (1974); A. Luther and V.J. Emery, Phys. Rev. Lett. **33**, 589 (1974).
6. J. Solyom, Advances in Physics **28**, 201 (1970); V.J. Emery, in *Highly Conducting One-Dimensional Solids,* edited by J.T. Devreese (Plenum, New York, 1979).
7. F.D.M. Haldane, J. Phys. C **14**, 2585 (1981).
8. F.D.M. Haldane, Phys. Rev. Lett. **47**, 1840 (1981).
9. R. Landauer, Phil. Mag. **21**, 863 (1970); M. Buttiker, Phys. Rev. Lett. **57**, 1761 (1986).
10. L.I. Glazman, I.M. Ruzin, and B.I. Shklovskii, Phys. Rev. B **45**, 8454 (1992).
11. C.L. Kane and M.P.A. Fisher, Phys. Rev. Lett. **68**, 1220 (1992).
12. C.L. Kane and M.P.A. Fisher, Phys. Rev. B **46**, 15233 (1992).
13. H. van Houten and C.W.J. Beenakker, Phys. Rev. Lett. **63**, 1893 (1989).
14. Y. Meir, N.S. Wingreen, and P.A. Lee, Phys. Rev. Lett. **66**, 3048 (1991).
15. C.L. Kane and M.P.A. Fisher, Phys. Rev. B **46**, 7268 (1992).
16. X.G. Wen, Phys. Rev. B **43**, 11025 (1991); Phys. Rev. Lett. **64**, 2206 (1990).
17. X.G. Wen, Phys. Rev. B **44**, 5708 (1991).

NOVEL ELECTRO-OPTICAL DEVICE STRUCTURES BASED ON QUANTUM WELLS

E.E. MENDEZ
IBM Research Division
T.J. Watson Research Center
Yorktown Heights, NY 10598
USA

The variety of material combinations available for epitaxial semiconductor heterostructures has made it possible to envision optical devices with new and desirable characteristics. Here I describe two examples proposed recently, in one of which tunable stimulated emission is achieved from a pair of type I coupled quantum wells subject to an electric field. In the other example, a type II heterostructure is presented that could lead to intersubband laser emission.

The application of an electric field along the growth direction of a coupled quantum-well type I heterostructure (as in the GaAs-GaAlAs material system) produces two significant effects on its optical properties: an increase of the recombination time of electrons and holes in their respective ground states, and a linear (red) shift of the corresponding interband optical transition.

This shift is readily observed in both absorption and luminescence experiments, which, on the other hand, show a very different behavior in relation to the strength of the transitions. While the intensity of the absorption line decreases drastically with increasing field as a result of the decreasing overlap between electron and hole wavefunctions, the overall strength of the photoluminescence spectrum remains unchanged for a very large range of fields. The constant intensity of the emission spectrum that accompanies the field-induced spectral shift has led to the proposal and experimental demonstration of tunable stimulated emission in coupled quantum wells [1].

Two tightly coupled wells, formed by a 50 Å – 20 Å – 50 Å GaAs–GaAlAs–GaAs heterostructure, were incorporated in the active region of an otherwise conventional GRINSCH laser structure with p and n regions on its ends. These heavily doped regions, normally used for carrier injection in lasers, were utilized here to apply an electric field to the coupled wells. A laser cavity was made by cleaving 300 μm bars and depositing a high-reflectivity facet coating. Population inversion in the wells was achieved by optical excitation at 7600 Å, selected to generate electron-hole pairs exclusively inside the well region.

Measurements at low temperatures (5 K) have shown lasing action that can be tuned by more than 14 meV using moderate electric fields, as illustrated in Fig. 1. This result is doubly significant. First, the ability to achieve lasing in a system in which electrons and holes are spatially separated is remarkable because of the small optical absorption inherent to this kind of transition. In addition, the degree of tunability is relatively large, in view of the considerable electric field created by the electron-hole dipoles that tend to screen the external field. That screening is still appreciable is shown by comparing the shifts at low (spontaneous) and high

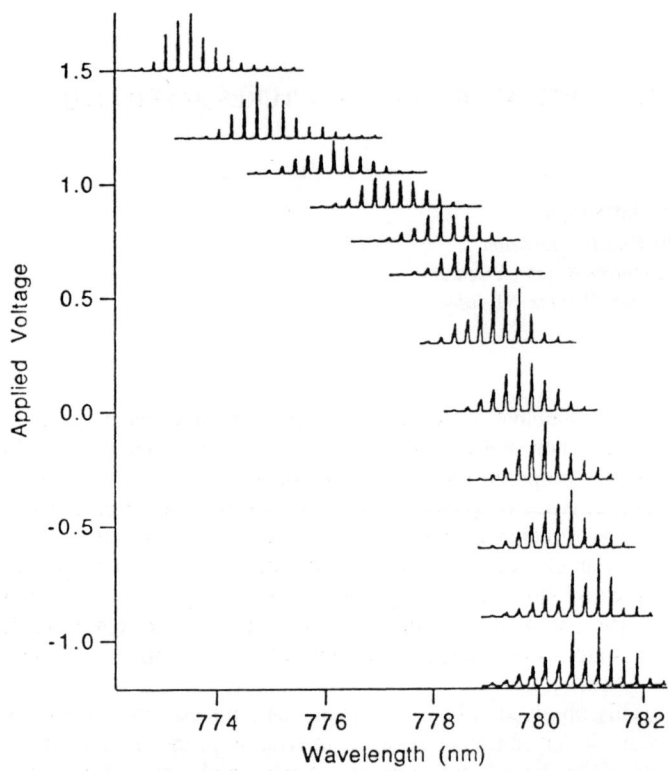

Figure 1. Stimulated emission of spectra of a 50 Å – 20 Å – 50 Å GaAs–GaAlAs–GaAs coupled quantum well imbedded in a GRINSCH heterostructure, for various biases applied between n and p GaAs electrodes. As the voltage is decreased from flat band toward reverse bias the laser spectra shift to the red, first linearly with voltage and then more gradually until finally the shift saturates at about -1 V. The total shift observed at T = 5 K is approximately 14 meV. Stimulated emission was achieved by optical pumping at λ = 760 nm a 300 μm device with coated facets. (After Ref. [1].)

excitation (stimulated) emission for the same applied bias. From such a comparison it can be concluded that the carrier screening reduces the electric field to about 25% of its external value. Thus, although screening is quite important, its consequences are less severe than in single quantum wells because of the linearity of field effects in coupled wells. For practical devices, room-temperature operation and electrical injection are essential. These requirements will entail a design that minimizes leakage current and non-radiative recombination and that at the same time allows for lateral injection of carriers.

While all present semiconductor lasers are based on interband optical transitions between the conduction and the valence bands, there have been several proposals for lasers that rely on intersubband transitions within the conduction (or valence) band of a quantum well. This would permit easy access to wavelengths (in the infrared) not available with conventional

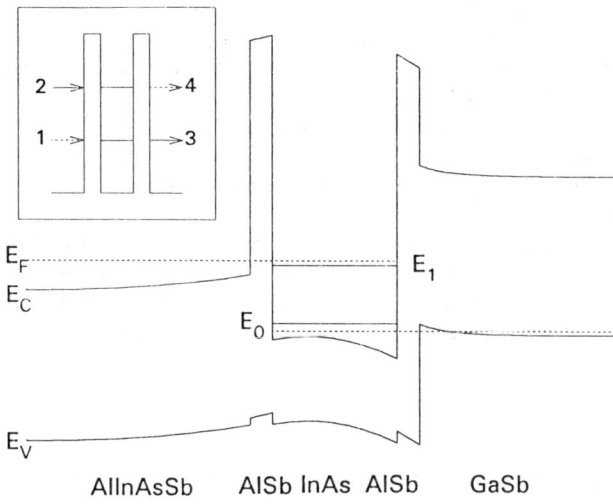

Figure 2. Band diagram for a proposed intersubband laser based on polytype type II heterostructures. Electrons are injected from an n-type AlInAsSb electrode into the first excited state of an InAs quantum well and then they decay radiatively to the ground state of the well. From there, electrons tunnel into the partially filled valence band of the p-type GaSb electrode. Leakage of electrons from the excited state is prevented by the blocking action of the energy gap of GaSb. The inset of the figure sketches the flow of carriers in a generic device that utilizes an intersubband transition for stimulated light emission. (After Ref. [2].)

lasers. The main obstacle is the difficulty of injecting carriers efficiently into an upper subband and extracting them from the lower subband. Here we describe briefly a proposal [2] to overcome this problem by using a type II polytype heterostructure, InAs-GaSb-AlSb. This material system is unique in that the top of the valence band of GaSb is higher in energy than the bottom of the conduction band of InAs, while the band gap of AlSb covers the GaSb and most of the InAs bandgap regions.

This system has been studied in depth recently in relation to resonant interband tunneling between electron and hole states in InAs and GaSb, respectively. Magnetotunneling experiments have shed light on the tunneling mechanism and have shown ways of using resonant tunneling to probe the two-dimensional (2D) electron gas formed in the InAs quantum well [3]. The most significant results have been the observation of zero-conductance regions when the Fermi energy is between two Landau levels, in close analogy with the quantum Hall effect observed in two-dimensional transport measurements, and the unambiguous determination of magnetic-field-induced oscillations of the Landé factor g of 2D electrons.

The principle of operation of the proposed device can be understood with the help of Fig. 2. An InAs quantum well that has at least two quantum states is clad by AlSb barriers and then by doped AlInAsSb and GaSb electrodes on each side. The quaternary-alloy electrode serves to inject electrons into the excited state, E_1, of the quantum well. The GaSb electrode acts as a collector of carriers in its valence band. The band alignment between InAs and GaSb should make possible an inversion of the population in E_1 and the ground level E_0 by eliminating the leakage detrimental to other proposed schemes. Electrons injected into E_1 cannot tunnel out directly because of the "blocking" bandgap of GaSb; they can only escape out of the well by a transition to E_0 and subsequent tunneling to the GaSb valence band.

By selecting properly the thickness of the various layers, it would be possible to control emission wavelength in a broad range. For instance, emission in the 5 μm band could be achieved with well and barrier thicknesses of 110 Å and 20 Å, respectively, and an external bias of 0.3 V. For this configuration, the lifetime of the quasibound state E_0 is estimated to be 0.7 ps, which is much shorter than the expected radiative recombination lifetime, calculated to be almost two orders of magnitude longer.

Using the same principle it is straightforward to design a light detector in the same range of wavelengths. In this case, E_0 should lie in the allowed region of the emitter and in the energy of the collector, while E_1 should be in the forbidden gap of the emitter and in the allowed region of the collector when a reverse bias is applied to the structure.

Acknowledgments

The work summarized here has been sponsored in part by the Army Research Office and has been done in collaboration with L. Esaki, L.Y. Liu, H. Meier, and H. Ohno.

References

1. L.Y. Liu, E.E. Mendez, and H. Meier, Appl. Phys. Lett. **60**, 2971 (1992).
2. H. Ohno, L. Esaki, and E.E. Mendez, Appl. Phys. Lett. **60**, 3153 (1992).
3. E.E. Mendez, Surf. Sci. **267**, 370 (1992).

EXCITED STATE POPULATIONS OF THE QUANTUM WELLS OF DOUBLE BARRIER RESONANT TUNNELING STRUCTURES

P.D. BUCKLE,[1] J.W. COCKBURN,[1] M.S. SKOLNICK,[1] D.M. WHITTAKER,[1] W.I.E. TAGG,[1] R. GREY,[2] G. HILL,[2] AND M.A. PATE[2]
[1]*Department of Physics,* [2]*Department of Electronic and Electrical Engineering*
University of Sheffield
Sheffield S3 7RH, UK

ABSTRACT. The observation of electroluminescence recombination from electrons in both ground (E1) and excited states (E2) of the quantum wells of double barrier resonant tunneling structures is reported. Analysis of the relative intensities of luminescence from the E1 and E2 levels permits deduction of their relative populations. The population ratios of the E2 to E1 levels are shown to be controlled by the ratio of the inter-sub-band scattering rate to the tunneling-out rate from the quantum wells. Deliberate engineering of these ratios by variation of the collector barrier width is reported.

1. Introduction

The use of photoluminescence (PL) spectroscopy to study double barrier resonant tunneling structures was first reported by Young and co-workers in 1988 [1]. The basic principle of the technique is that minority carrier, photo-injected holes act as a probe of the charge transport of the majority carrier electrons in the quantum well active region of the n-i-n devices. In this way spectroscopic information has been obtained on key properties of double barrier resonant tunneling structures (DBRTS) such as charge build-up in the quantum well (QW) [1,2] the deduction of the nature of the tunneling processes [2], measurement of electron temperatures [2] under device operation, the occurrence of impact ionization at high applied bias [3,4], minority carrier hole tunneling [5-7], etc.

More recently electroluminescence (EL) from DBRTS embedded in p-i-n structures has also been studied, where both electrons and holes are injected into the QW active region from the doped contact layers [8-10]. The use of p-i-n structures has the advantage that no optical excitation is necessary. As a result small (< 25 µm) diameter devices can be easily studied at low temperature. In addition there are potential improvements in sensitivity for detection of weak signals at high energy which might otherwise be obscured by stray light from the laser beam used for excitation in a PL experiment. A possible disadvantage in the use of p-i-n structures is that neither the electrons nor the holes are minority carriers in the QW region. It may no longer be justifiable to consider one carrier as a non-invasive probe of the charge accumulation of the other carrier type. For example, at low bias electron-hole recombination may be the dominant loss mechanism for carriers which accumulate in the QW, as opposed to carrier escape by tunneling in more conventional n-i-n structures.

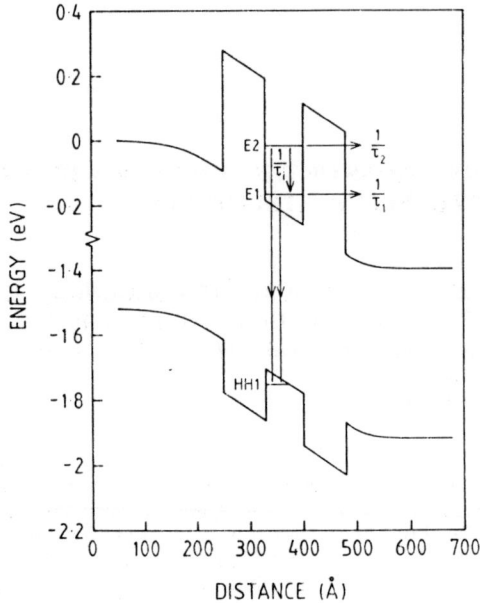

Figure 1. Band diagram calculated from solution of Poisson's equation for p-i-n double barrier structure with 70 Å GaAs quantum well, 80 Å $Al_{0.4}Ga_{0.6}As$ barriers. The tunneling-out rates from the E2 and E1 quantum well levels, $1/\tau_2$ and $1/\tau_1$, and the inter-sub-band scattering rate $1/\tau_i$ are indicated. The structure is biased for electron tunneling into the E2 quasi-confined state.

2. Results and Discussion

The present paper describes recent work at Sheffield on the study of EL recombination from both excited and ground electronic states of double barrier resonant tunneling structures embedded in p-n junctions. The structures investigated are all of the general design: n^+ GaAs substrate, 1 μm n = 1×10^{18} cm^{-3} GaAs, 1000 Å n = 1×10^{17} cm^{-3} GaAs, 50 Å undoped GaAs spacer, the $Al_{0.4}Ga_{0.6}As$-GaAs-$Al_{0.4}Ga_{0.6}As$ double barrier region, 50 Å undoped GaAs spacer, 1000 Å p = 5×10^{17} cm^{-3} GaAs, 1 μm p = 1×10^{18} cm^{-3} GaAs top contact. Structures with AlGaAs-GaAs-AlGaAs double barrier regions of widths 80-70-80 Å and 80-80-50 Å will be discussed here. For these GaAs QW widths, there are two quasi-confined electron states (E1 and E2) in the QW regions.

A calculated band diagram of the 80-70-80 Å (barrier-well-barrier widths) structure is shown in Fig. 1, at an applied forward bias of 2.0 V. The bias of 2.0 V is equivalent to 0.5 V on an n-i-n structure, since the built-in voltage of 1.5 V must first be overcome before forward bias, current flow can occur. The structure is shown biased at the second electron tunneling resonance. Electrons which tunnel into E2 can either tunnel out through the collector barrier at a rate $1/\tau_2$, or undergo inter-sub-band scattering to E1 at a rate $1/\tau_i$, and then tunnel out of the well from E1 at a rate $1/\tau_1$. The study of EL recombination from

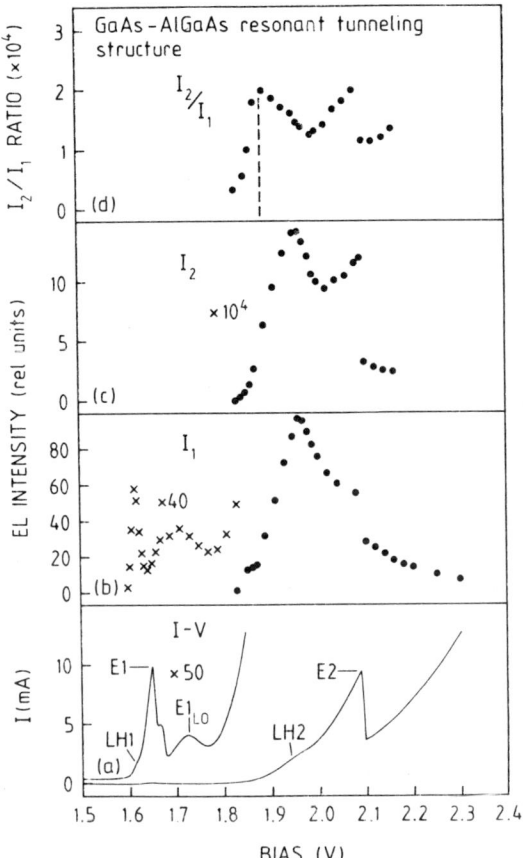

Figure 2. Current, I_1 (E_{11h}), I_2 (E_{21h}) luminescence intensities and I_2/I_1 ratio plotted as a function of applied bias for 80 Å barrier, 70 Å quantum well structure (80-70-80 Å structure). All the data are obtained at 4.2 K. The dashed vertical line in Fig. 2(d) indicates the bias voltage, at the onset of the E2 resonance, to the left of which the E2 population is very low.

electrons in E2 and E1 with holes in the n = 1 heavy hole level (E_{21h} and E_{11h} recombination respectively) provides a direct measure of the ratio of the E2 to E1 populations. This ratio can then be analyzed in terms of the $1/\tau_i$ and $1/\tau_1$ rates discussed above.

The I-V characteristics of the 80-70-80 Å structure are shown in Fig. 2(a). The principal resonances at 1.65 and 2.08 V are attributed to electron tunneling into the n = 1 and n = 2 levels of the QW. These attributions are obtained from comparison with the I-V characteristics of an n-i-n DBRTS of very similar design, and from device simulations of the sort shown in Fig. 1. Weak features at 1.62 and 1.95 V arise from light hole tunneling resonances.

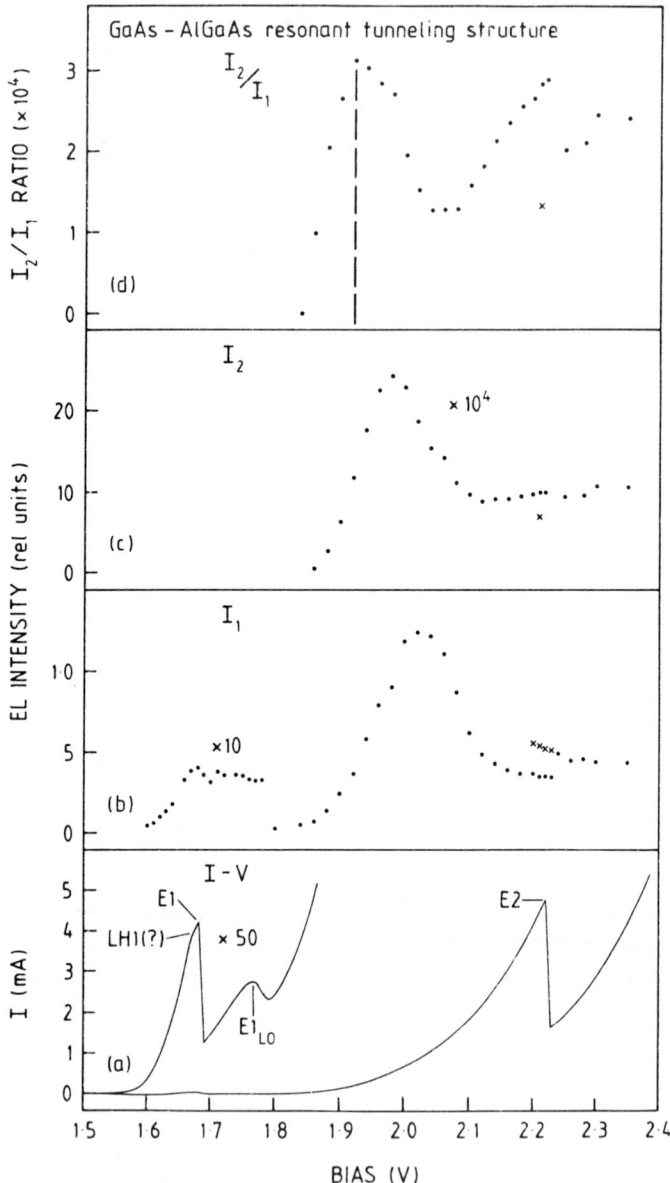

Figure 3. As Fig. 2, but for a structure with 80 Å barriers, 80 Å well (80-80-80 Å structure).

For comparison, the I-V characteristics of the 80-80-80 Å structure are shown in Fig. 3(a). Very similar I-V characteristics are observed, as expected. The most significant difference from Fig. 2(a) is that the LH1 feature is no longer clearly visible. However, it is believed that the distortion of the E1 resonance peak at ~ 1.67 V, marked LH1(?), is due to the LH1 resonance. On going from a 70 Å to an 80 Å QW (from Fig. 2 to Fig. 3), the difference in E1 and LH1 confinement energies decreases from 13 to 10 meV. It is thus reasonable that the closely spaced E1 and LH1 resonances should be less well resolved in Fig. 3(a) than in Fig. 2(a).

Representative spectra, taken at 4.2 K, from the 80-70-80 Å structure, as a function of applied bias, are shown in Fig. 4. EL emission is first observed at ~ 1.60 V, close to the onsets of the E1 and LH1 resonances, as shown in Fig. 4(i). The emission peak at 1.579 eV (at 1.60 V) arises from E1-HH1 (E_{11h}) recombination. As the bias is increased the EL intensity increases very rapidly (Fig. 4(ii) and 2(b)) and peaks in the region of the LH1 resonance. The EL intensity is controlled by both the electron and hole densities in the QW [11]. In general, features in I_{EL} are expected at both electron and hole resonances [5,7]. However, the occurrence of a very sharp resonance, such as LH1 in Fig. 2(a), may dominate the variation of I_{EL}; the subsequent very rapid decrease of the hole population in the region of E1, beyond LH1, will then be expected to suppress any feature in I_{EL} at E1, as observed. Comparison with the results for the 80-80-80 Å structure in Fig. 4(b) confirm this picture. In this case, the E1 and LH1 resonances are more closely spaced in bias, with the result that I_{EL} peaks close to the E1 resonance.

As the bias is increased further to the onset of the E2 resonance, an additional EL peak at 1.715 eV is observed, arising from E2-HH1 (E_{21h}) recombination, as shown in the right hand side of Fig. 4(iv). The E_{21h} recombination is ~ 10^4 times weaker than that due to E_{11h}. As expected the initial observation of the E_{21h} luminescence correlates very well with the onset of tunneling into the E2 state, as seen by comparison of the E_{21h} intensity (I_2) variation in Fig. 2(c), with the I-V characteristics of Fig. 2(a). A very similar correlation of the initial increase of the I_2 intensity with the onset of the E2 resonance in I-V is seen for the 80-80-80 Å sample in Figs. 3(a) and (c).

The energetic separation of the E_{21h} and E_{11h} recombination peaks of Fig. 4 of 143 meV (at 1.90 V) is very close to that calculated from finite potential well calculations of 147 meV, further confirming the attribution of the 1.715 eV EL peak. The E_{21h} - E_{11h} separation for the 80-80-80 Å sample of Fig. 3 is 127 meV, again close to the calculated E2-E1 separation of 122 meV.

The ratios of the E_{21h} to E_{11h} EL intensities (I_2/I_1) are plotted as a function of bias for the 80-70-80 Å and 80-80-80 Å structures in Figs. 2(d) and 3(d). Very similar results are found for the two structures. The rapid increase in the ratio at the onset of the E2 resonance (to the left of the dashed vertical lines on Figs. 2(d), 3(d)) arises since the E2 population is rapidly increasing from zero through this region, whereas the E1 level has significant non-zero population due to non-resonant tunneling processes. Once the devices are biased beyond the onsets of the E2 resonances (~ 1.9 V on Figs. 3, 4), the I_2/I_1 ratios vary by about a factor of two over the bias ranges investigated.

In the main region of interest from 1.9 to 2.2 V, the absolute values for the I_2/I_1 ratios for the two structures are in very good agreement (2.0 to 1.2×10^{-4} for the 80-70-80 Å structure, 3 to 1.4×10^{-4} for the 80-80-80 Å structure). Since both E2 and E1 electrons recombine with HH1 holes, the ratio I_2/I_1 is independent of the hole population, and is given by

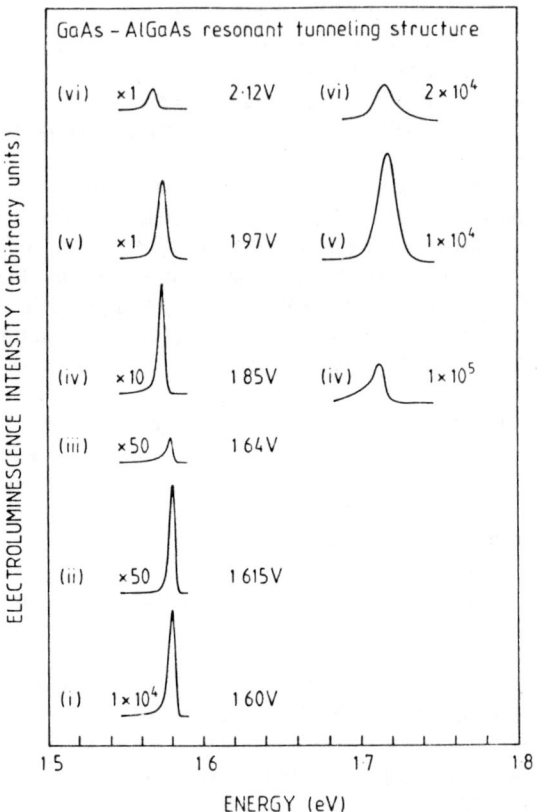

Figure 4. EL spectra as a function of applied bias for the 80-70-80 Å structure. E_{11h} spectra are plotted in the left hand side and E_{21h} spectra on the right hand side. In the bias range up to 1.85 V, only E_{11h} recombination is observed (Figs. 4(i) to (iii)). Beyond the onset of the E2 resonance at 1.85 V, E_{21h} emission is also seen (Figs. 4(iv) to (vi)).

$$\frac{I_2}{I_1} = \frac{n_2}{n_1} \times \frac{f_2}{f_1}, \qquad (1)$$

where n_2 and n_1 are the E2 and E1 populations respectively, and f_2/f_1 is the ratio of oscillator strengths for E_{21h} to E_{11h} transitions [12]. The f_2/f_1 ratio was calculated from a solution of Schrödinger's equation for the QW, using electric field values obtained from solution of Poisson's equation. At 2.0 V bias f_2/f_1 is calculated to be 0.3 [13]. Some absorption of the EL will occur in the top GaAs contacts. This is a small effect, but will lead to attenuation of I_2 relative to I_1 by a factor of about two [14]. Using the average 2.0 V values of $I_2/I_1 = 1.3 \times 10^{-4}$ ($\times$ 2 for GaAs attenuation) for the two structures, and $f_2/f_1 = 0.3$, n_2/n_1 at 2.0 V is obtained from Eq. (1) to be 8.7×10^{-4}.

When the structure is biased at the E2 resonance, n_1 and n_2 are determined by the following rate equations:

$$\frac{dn_2}{dt} = G - \frac{n_2}{\tau_2} - \frac{n_2}{\tau_i} - \frac{n_2}{\tau_{R2}} \qquad (2)$$

$$\frac{dn_1}{dt} = \frac{n_2}{\tau_i} - \frac{n_1}{\tau_1} - \frac{n_1}{\tau_{R1}}, \qquad (3)$$

where G is a "generation" rate for n_2 electrons determined by tunneling in from the emitter contact, τ_2, τ_1, and τ_i were defined earlier (see Fig. 1), and $(\tau_{R2})^{-1}$ and $(\tau_{R1})^{-1}$ are radiative recombination rates. In the steady state, from Eq. (3)

$$\frac{n_2}{\tau_i} - \frac{n_1}{\tau_1} - \frac{n_1}{\tau_{R1}} = 0 \qquad (4)$$

and so $n_2/n_1 = \tau_i \left(\dfrac{1}{\tau_1} + \dfrac{1}{\tau_{R1}}\right)$. Electrons relax from E2 to E1 by rapid longitudinal optical (LO) phonon inter-sub-band scattering at a rate $1/\tau_i \sim 2 \times 10^{12}$ s^{-1} [15]. Using the deduced value of n_2/n_1 of 9×10^{-4}, $\left(\dfrac{1}{\tau_1} + \dfrac{1}{\tau_{R1}}\right)$ is thus found to be 1.8×10^9 s^{-1}. At the probable carrier densities in the well ($< 10^{10}$ cm^{-2} for at least one of the carriers away from the peak of resonance) τ_{R1}^{-1} will very likely be $< 10^8$ s^{-1}, and can be neglected relative to τ_1^{-1} [16]. As a result τ_1^{-1} is found to be 1.8×10^9 s^{-1}. Calculated values for τ_1^{-1} are extremely sensitive to the barrier width and height employed. For example, for d = 80 Å, we obtain a value for τ_1^{-1} of 0.73×10^9 s^{-1}, from the calculated width of the E$_1$ resonance at 2.0 V applied bias, increasing to 3.2×10^9 s^{-1} for d = 70 Å. Bearing in mind this sensitivity to the input parameters we conclude that the experimental value of τ_1^{-1} of 1.8×10^9 s^{-1} is in very satisfactory agreement with the calculated values, and provides confidence in the analysis of I_2/I_1 intensities.

The extreme sensitivity of the $1/\tau_1$ rates to the width (w) of the collector barrier raises the possibility of increasing the I_2/I_1 ratio (= τ_i/τ_1) by decreasing w. In order to achieve this, a structure with w = 50 Å (a 50-80-50 Å structure) was grown. The I-V characteristics of the structure were very similar to those of Fig. 2, with E1 and E2 resonances at 1.67 and 2.03 V respectively. In accordance with the increased $1/\tau_1$ rates (decreased τ_1) expected at w = 50 Å, the I_2/I_1 ratio was found to have increased to 5×10^{-3} (at 1.96 V) for the 50-80-50 Å structure. This I_2/I_1 ratio corresponds to an n_2/n_1 population ratio of 3.3×10^{-2}, compared to 8.7×10^{-4} at w = 80 Å. The observed increase of n_2/n_1 of a factor of ~ 40 on decreasing the collector barrier width from 80 to 50 Å compares with a factor of 60 increase in $1/\tau_1$ predicted from calculations of the change in width of the E1 resonance on going from w = 80 to 50 Å. In view of the sensitivity to the input parameters discussed above, the observed increase in I_2/I_1 ratio between 80 and 50 Å is considered to be in very fair agreement with theoretical expectation. Experiments to further increase the n_2/n_1 ratio by additional decrease of the collector barrier width are in progress at the present time. In this context, it is worth noting that $n_2/n_1 = 1$ is predicted for w ~ 30 Å [17].

It was pointed out in Ref. [10] that although the n_2/n_1 ratio is determined by the τ_i/τ_1 ratio, the absolute values of n_2 and n_1 are controlled by the E2 tunneling-in (G) and tunneling-out ($1/\tau_2$) rates. As the collector width w is reduced to increase the n_2/n_1 ratio the absolute value of n_2 (and hence n_1) will be reduced as the tunneling-out rate $1/\tau_2$ is increased [18]. However, this reduction in n_2 can be compensated by a corresponding reduction of the emitter barrier width, and hence increase of the tunneling-in rate G, albeit at the cost of an increase in overall current density through the device.

3. Conclusion

In conclusion, electroluminescence recombination from the ground and excited states of double barrier resonant tunneling structures has been reported. It has been shown how these results can be analyzed to provide a measure of the relative populations of the two quasi-confined electronic states in the quantum well active region. Initial steps towards the engineering of the ratio of the populations of the two states have been reported.

References

1. J.F. Young, B.W. Wood, G.C. Aers, R.L.S. Devine, H.C. Liu, D. Landheer, M. Buchanan, A.J. SpringThorpe, and P. Mandeville, Phys. Rev. Lett. **60**, 2085 (1988).
2. M.S. Skolnick, D.G. Hayes, P.E. Simmonds, A.W. Higgs, L. Eaves, M. Henini, O.H. Hughes, G.W. Smith, C.R. Whitehouse, H.J. Hutchinson, M.L. Leadbeater, and D.P. Halliday, Phys. Rev. B **41**, 10754 (1990).
3. C.R.H. White, M.S. Skolnick, L. Eaves, and M.L. Leadbeater, Appl. Phys. Lett. **58**, 1164 (1991).
4. J.W. Cockburn, M.S. Skolnick, J.P.R. David, R. Grey, G. Hill, and M.A. Pate, Appl. Phys. Lett. **61**, 825 (1992).
5. N. Vodjdani, D. Cote, D. Thomas, B. Sermage, P. Bois, E. Costard, and J. Nagle, Appl. Phys. Lett. **56**, 33 (1990).
6. S. Charbonneau, J.F. Young, and A.J. SpringThorpe, Appl. Phys. Lett. **57**, 264 (1990).
7. C.R.H. White, M.S. Skolnick, P.E. Simmonds, L. Eaves, M. Henini, D.H. Hughes, G. Hill, and M.A. Pate, Superlattices and Microstructures **8**, 195 (1990).
8. C. van Hoof, J. Genoe, R. Merlens, G. Borghs, and E. Goovaerts, Appl. Phys. Lett. **60**, 77 (1991).
9. C.R.H. White, H.B. Evans, L. Eaves, P.M. Martin, M. Henini, G. Hill, and M.A. Pate, Phys. Rev. B **45**, 9513 (1992).
10. J.W. Cockburn, P.D. Buckle, M.S. Skolnick, D.M. Whittaker, W.I.E. Tagg, R.A. Hogg, R. Grey, G. Hill, and M.A. Pate, Phys. Rev. B **45**, 13757 (1992).
11. M.S. Skolnick, P.E. Simmonds, D.G. Hayes, A.W. Higgs, G.W. Smith, A.D. Pitt, C.R. Whitehouse, H.J. Hutchinson, C.R.H. White, L. Eaves, M. Henini, and O.H. Hughes, Phys. Rev. B **42**, 3069 (1990).
12. H.T. Grahn, H. Schneider, W.W. Ruhle, K. von Klitzing, and K. Ploog, Phys. Rev. Lett. **64**, 2426 (1990).
13. This value for f_2/f_1 at 2.0 V bias is about a factor of four greater than that quoted in Ref. [10].
14. M.D. Sturge, Phys. Rev. **127**, 768 (1962).
15. M.C. Tatham, J.F. Ryan, and C.T. Foxon, Phys. Rev. Lett. **63**, 1637 (1989).

16. T. Matsusue and H. Sakaki, Appl. Phys. Lett. **50**, 1429 (1987).
17. Inverted light and heavy hole populations have been reported recently by White et al. in Ref. [9], in a DBRTS with sub-band spacing less than the LO phonon energy, biased for tunneling into the n = 1 light hole level. Similar behaviour is also found at higher hole resonances.
18. $1/\tau_2$ is calculated to be 5×10^{11} s^{-1} at 2.0 V bias and w = 80 Å, and 70 Å quantum well width.

QUANTUM WELL LUMINESCENCE BY RESONANT TUNNELING INJECTION OF ELECTRONS AND HOLES

H.B. EVANS,[1] L. EAVES,[1] C.R.H. WHITE,[1] M. HENINI,[1] P.D. BUCKLE,[2]
T.A. FISHER,[2] D.J. MOWBRAY,[2] AND M.S. SKOLNICK[2]
[1]*Department of Physics, University of Nottingham, NG7 2RD, UK*
[2]*Department of Physics, University of Sheffield, S3 7RH, UK*

ABSTRACT. The electroluminescence and current-voltage characteristics of two forward biased p-i-n double barrier structures based on GaAs/AlAs are investigated. Electroluminescence lines due to carrier recombination in the GaAs contact layers and in the quantum well are observed. The bias dependence of the intensity of these lines exhibits pronounced peaks which are also seen in the I(V) characteristics and which are due to electron and hole resonant tunneling into the quantum well. Two quantum well emission lines are observed in the electroluminescence spectrum. Their relative intensity is strongly dependent on the concentration of electrons in the quantum well, which can be varied by a large amount by voltage tuning on and off the electron resonance. Photoluminescence excitation spectroscopy measurements indicate that the two quantum well emission lines involve recombination of electrons with holes in the lowest heavy hole state (HH1) and holes bound to acceptors in the quantum well.

1. Introduction

The energy relaxation and recombination processes of electrons and holes in quantum confined semiconductor heterostructures are not only of fundamental interest in condensed matter physics but also determine the properties of a variety of optoelectronic devices including quantum well lasers and novel far infrared sources. These processes have been extensively studied by means of optical techniques, including time-resolved measurements and resonant tuning of quantum well (QW) levels with an external bias [1–10]. Photoluminescence spectroscopy has also been used to study the carrier transport and relaxation in unipolar double barrier resonant tunneling devices [7,8].

In this paper, we describe how a double barrier tunnel structure incorporated inside a p-i-n diode and based on a GaAs/AlAs heterostructure can be used to resonantly inject both electrons and holes into the confined states of the conduction and valence band quantum wells. Electroluminescence (EL) is observed from the QW due to recombination of electrons and holes which have tunnelled in from the respective n- and p-doped regions. In addition, near band edge (bulk GaAs) luminescence is observed due to recombination of electrons (holes) which tunnel through both barriers to recombine as minority carriers in the p-doped (n-doped) layer. The intensities of the EL lines arising from the QW and from the bulk GaAs show strongly resonant features as a function of bias that can be associated with the resonant peaks in the I(V) characteristics. We will be particularly concerned with the behaviour of the two EL peaks associated with the QW recombination. We describe how photoluminescence excitation spectroscopy is used to identify the two transitions giving rise

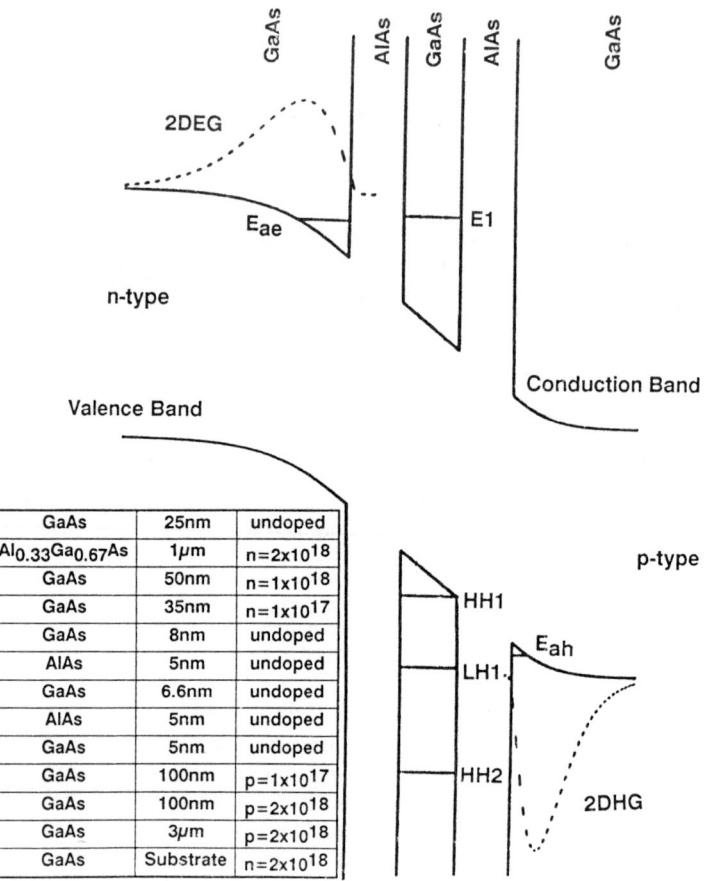

Figure 1. A schematic diagram of the conduction and valence band profiles in the p-i-n double barrier device at a forward bias of about 1.65 V. The inset shows the layer structure of the 6.6 nm wide QW device; the doping densities are quoted in cm^{-3}.

to these two peaks.

2. Details of the Devices

The composition of one of the devices used in this investigation is shown in the inset of Fig. 1. The central region is a resonant tunneling device with two 5-nm thick AlAs barriers and a GaAs quantum well of width 6.6 nm. The layer was grown by molecular beam epitaxy on a (100) substrate and was processed into 100 μm diameter mesas with a 20 μm thick outer annular metallic top contact. This allows efficient collection of the EL. The heavily doped n$^+$(AlGa)As top layer has a sufficiently wide bandgap to transmit the EL from the QW. A similar device with a QW width of 8.2 nm was also investigated.

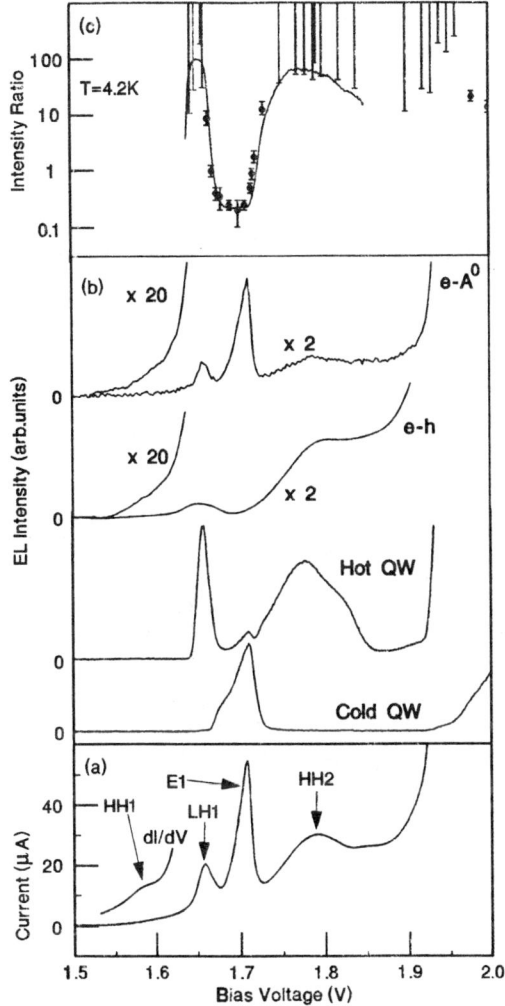

Figure 2. (a) The current-voltage characteristics of the device with the 6.6 nm wide quantum well measured at 4.2 K. The HH1 resonance is observed as a weak feature in the conductance, dI/dV. (b) The variation with bias of the electroluminescence intensity of the four emission lines observed. The spectra were taken with the spectrometer set at the peak of the emission lines, as described in the text, with the device at 4.2 K. The intensities of the e-A^0 and e-h lines corresponding to recombination in the GaAs contact layers are weaker than the two QW lines (E1-HH1 and E1-LH1). For this reason, these plots are shown at higher gains (×2, ×20). (c) The ratio of the intensities of the hot and cold quantum well lines as a function of bias. The continuous curve is obtained by ratioing the E1-LH1 and E1-HH1 intensity shown in Fig. 2(b). The closed circles are obtained directly from the EL spectra at various biases.

Figure 1 shows a schematic band diagram of the device at a forward bias of around 1.65 V. Electron and hole accumulation layers form in the lightly doped spacer layers on the n-type and p-type sides of the two barriers. The excess negative (positive) charge in the electron (hole) accumulation layer is in the form of a quasi-two-dimensional electron (hole) gas with bound state energy E_{ae} (E_{ah}) as shown in the figure. As the voltage is increased, both electrons and holes can resonantly tunnel from their respective accumulation layers into the quasi-bound states of the conduction and valence band QWs. The notation E1 refers to the conduction band QW; HH1, LH1, HH2 refer to the heavy (HH) and light (LH) hole states in the valence band QW; the quantum number (n = 1, 2) refers to the envelope function. This assignment is strictly applicable only for in-plane momentum $k_{\parallel} = 0$ when there is no light-heavy hole admixing.

3. Experimental Results

3.1. CURRENT-VOLTAGE CHARACTERISTICS AND ELECTROLUMINESCENCE SPECTRA

The forward bias current-voltage characteristic, I(V), measured at 4.2 K, is shown in Fig. 2(a) for the device with the 6.6 nm wide QW. Three strong resonant peaks are seen at 1.66, 1.71, and 1.79 V. An additional weak resonance at 1.59 V is observed in the conductance, dI/dV. A model which takes into account the quantum confinement energy of the conduction and valence band states in the QW and in the two accumulation layers gives the following ordering of these resonant features in I(V) with increasing voltage: HH1, LH1, E1, HH2. This assignment is confirmed in separate magneto-tunneling measurements of I(V) at high magnetic field, which map out the energy-momentum dispersion curves of the conduction and valence band quantum well states [11].

The EL occurs in two distinct spectral regions as shown in the plots in Figs. 3 and 4, taken at 4.2 K and at various biases for the 6.6 nm QW device. The first at a photon energy of around 1.6 eV arises from recombination of electrons and holes in the QW; the second at around 1.5 eV from recombination in the bulk GaAs doping layers. The QW signal corresponds to two distinct lines: a higher energy line or *hot line* at around 1.61 eV and a *cold line* at around 1.59 eV. The intensities of these two lines are strongly bias-dependent and show resonant features that correspond to those seen in the I(V) characteristics. The presence of EL from the QW at *all* biases above 1.64 eV implies the simultaneous presence of both electrons and holes in the QW. This in turn implies that both electrons and holes can enter the QW by non-resonant processes even when the device is biased well away from a specific electron or hole resonance in I(V).

The EL signal due to bulk GaAs also consists of two distinct emission features at photon energies of 1.49–1.50 eV and of 1.518 eV. Their intensities are also strongly bias-dependent. The lower energy feature corresponds to electron recombination at neutral acceptors (e-A^0), whilst the higher energy feature is due to recombination of more weakly bound carriers (e-h, excitons, or free hole recombination with electrons bound to shallow donors). The photon energies of the e-A^0 and e-h lines have a weak dependence on voltage. The hot line from the QW shows a pronounced red shift with increasing bias, due to the quantum confined Stark effect [12]. The cold line from the QW is seen only over a small range of bias and has only a small red shift.

Although complete spectra were taken at a large number of fixed voltages (e.g., Figs. 3

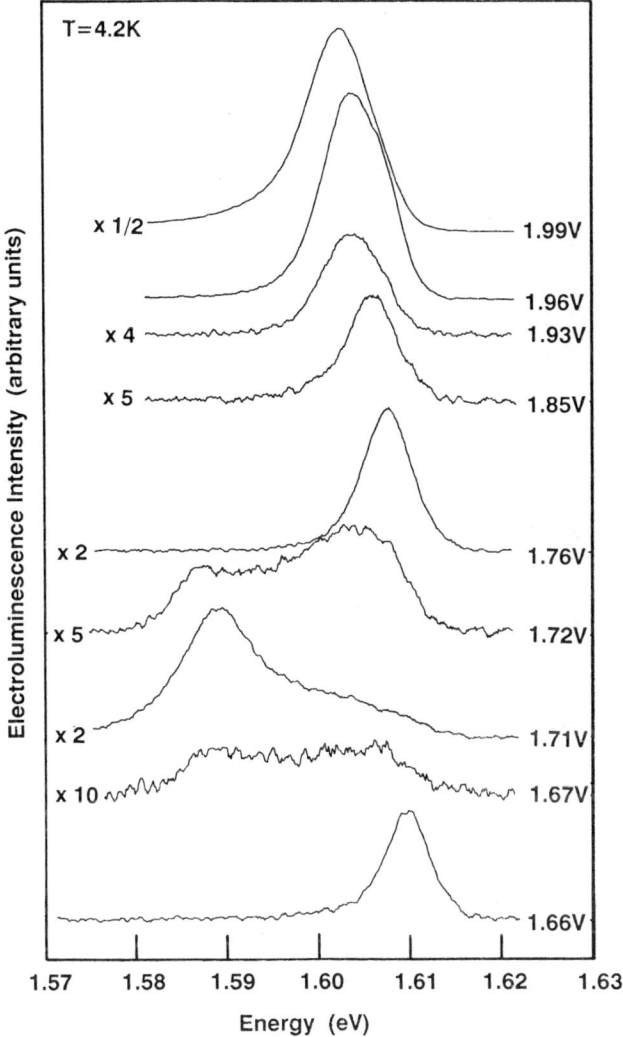

Figure 3. Spectra of the electroluminescence arising from electron-hole recombination in the QW for the device with the 6.6 nm wide quantum well (T = 4.2 K) at various forward bias voltages.

and 4), a good indication of the intensity variation of the four lines can be obtained by setting the spectrometer at the peak of each line and sweeping the voltage. This approach is particularly suited to the two band edge lines of the GaAs doping layers, because of their weak sensitivity to bias voltage. The resulting curves are compared with the I(V) characteristics in Fig. 2(b). Note that the intensity curves for the e-A^0 and e-h lines are plotted at higher gains (×2, ×20) because these lines are weaker than the QW recombination

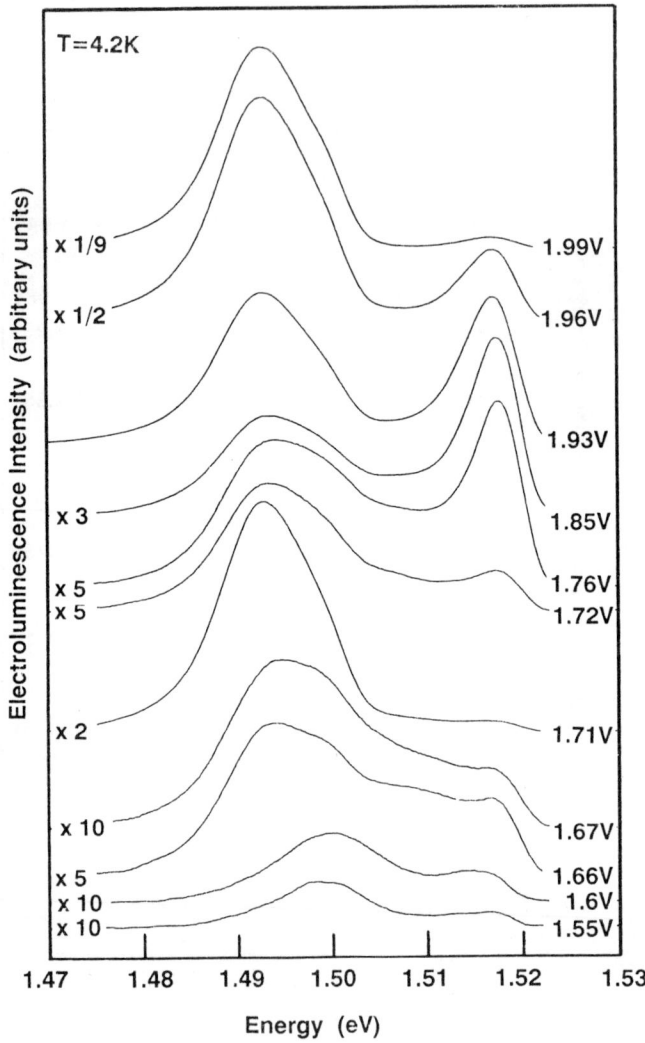

Figure 4. Spectra of the electroluminescence arising from carrier recombination in the GaAs contact regions for the device with the 6.6 nm wide quantum well (T = 4.2 K) at various forward bias voltages of the device.

lines. The e-A⁰ signal (spectrometer set at 1.490 eV) exhibits a sharp peak at 1.71 V, corresponding to the E1 resonance in I(V), and two smaller peaks at 1.66 V and 1.80 V due to the LH1 and HH2 resonances. In contrast, the e-h signal (1.518 eV) shows the two hole resonances, with only a very weak feature corresponding to the strong E1 resonance in I(V). Both e-A⁰ and e-h intensity plots show weak shoulders at around the HH1 resonance in dI/dV. We interpret their voltage dependence as follows. The strength of the E1 resonance feature in the e-A⁰ intensity plot and its absence from the e-h intensity plot indicates that those resonantly tunneling electrons that do not recombine in the QW undergo recombination with holes, which are bound to neutral acceptors in the p-doped contact regions. The holes that pass through the QW in the regions of the LH1 and HH2 resonances either recombine as essentially free particles (e-h) in the n-type region or else bind to the residual acceptor impurities (e-A⁰) prior to recombination. The e-A⁰ recombination is relatively weaker than e-h recombination for holes tunneling through the HH2 resonance than for the LH1 and HH1 features. This indicates that on the higher voltage HH2 resonance, many of the hot holes injected into the n⁻ region are unable to relax all of their energy by shallow acceptor capture prior to recombination.

From Fig. 2(b) we see that the hot QW line (1.610 eV) dominates at voltages corresponding to the LH1 and HH2 resonances of the I(V) curve. At a lattice temperature of 4 K, the cold QW line (1.589 eV) is only observed in the voltage range around the E1 resonance, where a large electron density is expected in the QW. It cannot be detected above the background noise level over wide ranges of applied voltage on either side of the E1 resonance. Note that in Fig. 2(b), the gain settings for the hot and cold QW lines are the same so that the cold line maximum on E1 is comparable in intensity to the hot line maxima at LH1 and HH2. The background noise level permits us to measure the ratio of the intensities of the hot and cold lines up to a value of approximately 100:1; this ratio is plotted in Fig. 2(c). The continuous curve is obtained from the ratio of the EL intensity plots in Fig. 2(b); the discrete points (closed circles) are obtained directly from analysis of the EL spectra at fixed voltages. Above 1.9 V, when the Stark shift of the hot line becomes significant, the continuous curve is unreliable. The error bars in Fig. 2(c) indicate the lower limit of the intensity ratio, corresponding to the background noise level. The intensities of the hot and cold QW lines were investigated at biases corresponding to the E1 resonance at lattice temperatures between 4 and 30 K. The relative intensity of the QW cold line increased with increasing temperature. The possible reason for this behaviour is discussed later.

Figures 5 and 6 show the I(V) characteristics, EL spectra, and spectral intensity plots for the second device with the 8.2 nm wide quantum well. An interesting feature of the I(V) plot in Fig. 5(a) is the appearance of the strong resonance in I(V) at around 1.9 V, which we assign to electron resonant tunneling into the first excited state (E2) of the conduction band QW. The two EL lines from the QW occur at around 1.56 eV with a separation of ~ 16 meV (see the 1.615 V plot in Fig. 6). The hot line from the QW shows a strong peak on the LH1 resonance in I(V) and again peaks strongly on the LH2 resonance at 1.88 V. Both of these light hole resonances occur at voltages quite close to the electron resonances (E1 and E2 respectively). The HH2 resonance in I(V) gives rise to no discernible peak in the quantum well luminescence. This may be due to the fact that it is well away from any electron resonance so that the electron density in the quantum well in the region of HH2 is low. A weak, broad peak occurs in the intensity of the hot quantum well line at ~ 1.7 eV. This does not correspond directly to either an electron or a hole resonance in I(V), but may be due to an increase of the electron density in the quantum well, due possibly to the LO-phonon assisted tunneling satellite of the E1 resonance [11]. The cold line that we attribute to E1 → HH1 (see Fig. 6(b)) increases strongly on both the E1 and E2 resonance in I(V), on the peaks of which it has an intensity comparable to that of the hot line.

In the region of the E2 resonance of I(V), we also see a weak higher-energy luminescence line from the quantum well, which corresponds to recombination of electrons

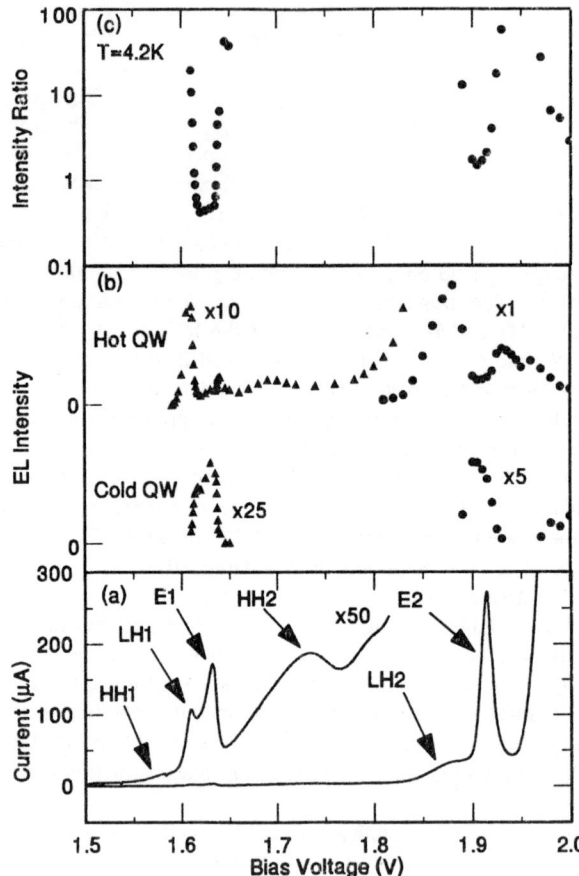

Figure 5. (a) The current-voltage characteristics of the device with the 8.2 nm wide quantum well, measured at 4.2 K. (b) The variation with bias of the electroluminescence intensity of the two quantum well emission lines. The gains ×5, ×10, and ×25 are used to clearly reveal the behaviour of the EL features at low intensity. (c) The ratio of the intensities of the hot and cold quantum well lines as a function of bias. The cold line is not detectable in the range of bias below the E1 resonance and in the ranges 1.65–1.87 V and 1.935–1.960 V.

in the upper E2 state with holes. The intensity of this line is much weaker (by a factor of 10^4) than that of the corresponding E1 transition, indicating a very small electron population in E2. This behaviour is very similar to work that was recently reported by Cockburn et al. [13] in a p-i-n $GaAs/Al_{0.4}Ga_{0.6}As$ resonant tunneling device. The E2 line is weak because of the rapid inter-subband transition rate of electrons from E2 to E1 via the emission of polar-mode optic phonons. In our device, we see evidence of a splitting of the E2 electroluminescence feature. The intensity dependence of the two components is similar to that of the hot and cold lines associated with the E1 recombination.

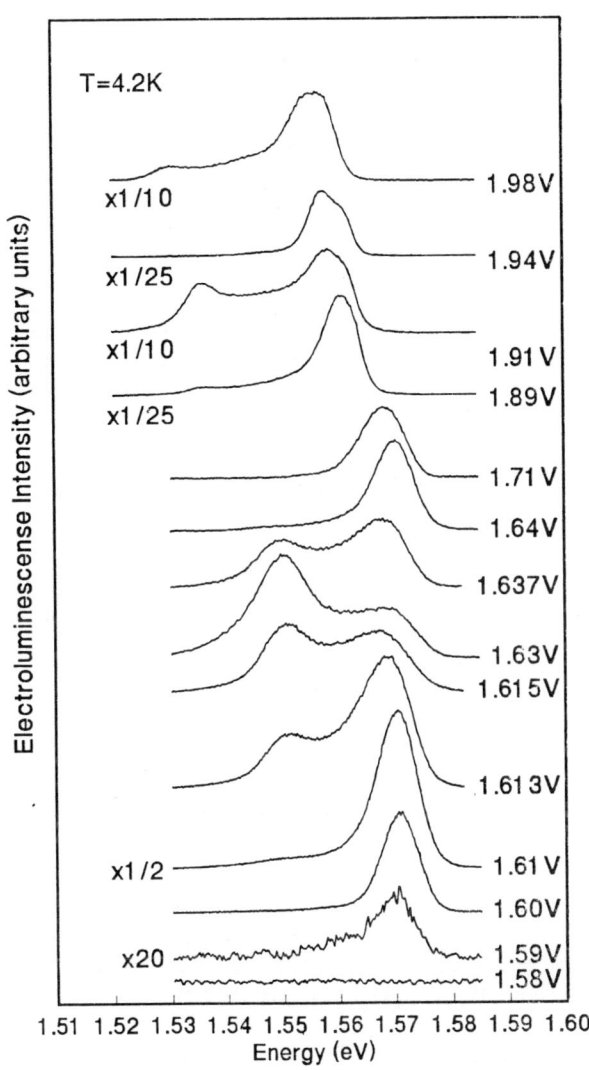

Figure 6. EL spectra due to QW recombination for the device with the 8.2 nm wide QW (T = 4.2 K) at various forward bias voltages.

3.2. PHOTOLUMINESCENCE EXCITATION SPECTRA

To help identify the origin of the hot and cold QW lines we performed photoluminescence excitation (PLE) measurements on the device with the 6.6 nm wide quantum well. The experiments were carried out at 4.2 K with a tunable Ti-sapphire laser pumped by an Ar-ion laser. A typical photoluminescence (PL) spectrum is shown in the lower part (dashed curve) of Fig. 7. It was taken at a photon excitation energy of 1.65 eV and incident power

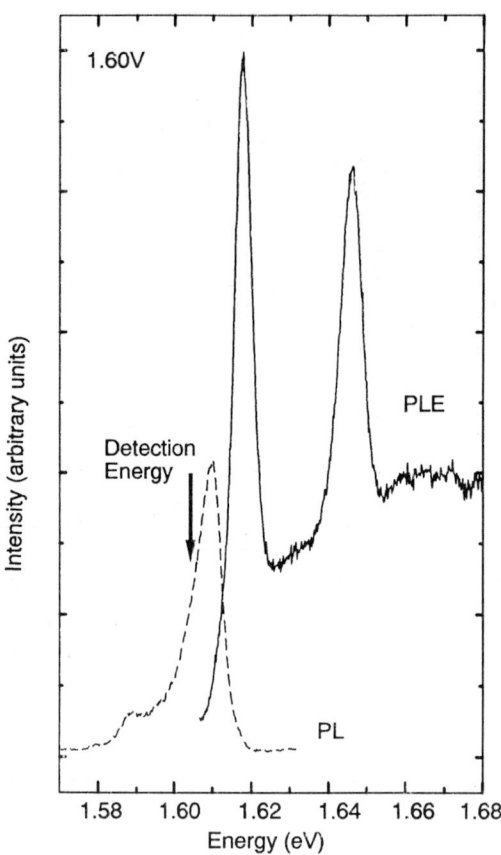

Figure 7. The photoluminescence excitation spectrum (solid curve) taken at a forward bias of 1.6 V and at a laser intensity of a few watts per square centimetre. The spectrum was taken at a detection energy of 1.604 eV, as indicated by the arrow in the lower part of the figure, which also shows the photoluminescence spectrum (dashed line).

of a few watts per square centimetre. The spectrum reveals both the hot line at 1.61 eV and the cold line at 1.59 eV, in agreement with the EL spectra. The PL spectra were obtained at or below bias voltages of 1.6 V, for which the EL emission is negligible. The hot line dominates the spectrum and the cold line shows evidence of saturation at high laser intensity.

The PLE spectrum (solid curve in Fig. 7) was obtained by detecting at 1.604 eV, on the low energy side of the hot line. It reveals two strong peaks at 1.618 and 1.646 eV and its overall form is typical of a GaAs quantum well. The same two lines were observed in PLE when the detection energy was moved to the cold line at 1.59 eV. We ascribe the two lines in PLE to the HH1 → E1 and LH1 → E1 transitions of the quantum well, respectively. The energy and position of these lines are consistent with our calculated values of these transitions for a 6.6 nm wide quantum well. The lower energy line has the larger intensity, as expected from the oscillator strengths for these transitions. The HH1 → E1 line shows a Stokes shift (~ 8 meV) from the corresponding high energy peak (the hot line) in the PL spectrum.

4. Discussion

In a previous article [14], the hot and cold EL lines of the quantum well were ascribed to electron-hole recombinations involving the E1 → LH1 and E1 → HH1 transitions respectively. The new PLE measurements described in the previous section indicate clearly that the hot EL line from the QW corresponds to the E1 → HH1 transition, since its Stokes shift from HH1 → E1 line observed in PLE is small. We must therefore assign the lower energy cold line to a transition in which one of the recombining particles is bound (localised) on a defect or impurity. It is more likely that the bound particle is a hole since the energy separation between the hot and cold lines indicates a fairly large binding energy of around 20 meV for both quantum well widths (6.6 and 8.2 nm). The binding energy of a donor at the centre of a quantum well varies with well width, reaching a maximum of around 13 meV for narrow quantum wells [15]. The binding energy is lower for a donor displaced from the centre of the QW. We therefore propose that the cold line involves recombination of an electron with a hole bound at an acceptor impurity in the quantum well (i.e., E1 → A^0_{QW}). In the samples under investigation, the spacer layer between the Be-doped (1×10^{17} cm^{-3}) p-type contact layer and the tunnel barrier is only 5 nm. At the growth temperature of 650° C, it is possible that Be-acceptors have diffused into the quantum well from the contact layer. Such an effect has been identified recently in the case of donor diffusion from n-type contact layers [16]. The binding energy of an isolated Be-acceptor is 28 meV, though this value increases in a quantum well due to the confinement of the wavefunction [17]. The diffusing Be impurities are likely to spread over the entire width of the quantum well, giving a range of binding energies below the maximum value. The higher energy line corresponding to E1 → HH1 may be excitonic in nature. Hence the measured difference in energy (21 meV) between the hot and cold EL emission lines would correspond to the difference between the mean acceptor binding energy in the well (~ 30 meV) and the exciton binding energy (~ 9 meV). It is also worth noting that the separation between the two peaks in the EL from the quantum well is similar to that of the two lines from the GaAs band-edge spectral region, the lower energy component of which is also acceptor-related.

The apparently anomalous temperature dependence of the intensity of the cold line in the electroluminescence spectrum is also consistent with previously reported temperature dependence of impurity related luminescence in quantum wells. Skolnick et al. [18] have observed that the intensity of the lower-energy bound hole quantum-well recombination line increases with increasing temperature due to the increased mobility of holes for capture at impurities, similar to the behaviour found in our double barrier resonant tunneling structures.

The notable and unusual feature of the electroluminescence spectra described above is that the cold QW line due to the bound hole electroluminescence is present only in the voltage region of the electron resonant tunnelling peaks (E1 and E2), at which a large electron density in the well can be expected. Under these conditions, the hole density in the quantum well will be relatively low, because the device is biased well away from a hole resonance. In addition, the large number of electrons will increase the recombination rate for holes and hence will decrease the hole density. The comparable intensities of the hot and cold lines in the region of the electron resonances indicate that the number of unbound holes available for E1 → HH1 recombination is comparable with the number of holes recombining at neutral acceptors. In the voltage regions away from the electron resonances, the hole density in the QW is much higher. Hence there is a much larger density of unbound holes in HH1 which dominates the EL spectra, whilst the QW recombination involving the acceptor bound holes is saturated. Saturation of acceptor recombination in the luminescence from a resonant tunneling device at high hole densities was suggested recently by Cockburn et al. [19]. It is, however, surprising that the cold line is completely undetectable in the voltage ranges away from the electron resonances. We postulate that the capture of holes at the acceptors is relatively slow in general, making the E1 → A^0_{QW} recombination process relatively inefficient when holes predominate in the QW. The strength of this process on the E1 and E2 resonances may be due to Auger-like processes in which electrons aid the capture of holes onto acceptors, from which they can subsequently recombine radiatively.

Acknowledgments

This work is supported by the Science and Engineering Research Council (UK). We are grateful to Messrs. T.S. Turner and P.M. Martin for helpful discussions. We are grateful to Drs. G. Hill and M.A. Pate (University of Sheffield) for processing the layers.

References

1. A. Pinczuk, J. Shah, R.C. Miller, A.C. Gossard, and W. Wiegmann, Solid State Commun. **50**, 735 (1984).
2. J. Shah, Solid State Electronics **32**, 1051 (1989) and references therein.
3. J.F. Ryan, Physica **134B**, 403 (1985).
4. W.W. Rühle, M.G.W. Alexander, and M. Nido, in *Proc. 20th Int. Conf. on the Physics of Semiconductors* (Thessaloniki, Greece, 1990), edited by E.M. Anastassakis and J.D. Joannopoulos (World Scientific, Singapore, 1990), p. 1226.
5. H.-J. Polland, W.W. Rühle, J. Kuhl, K. Ploog, K. Fujiwara, and T. Nakayama, Phys. Rev. B **35**, 8273 (1987).
6. D. Bimberg, J. Christen, A. Steckenborn, G. Weimann, and W. Schlapp, J. Lumin. **30**, 562 (1985).
7. J.F. Young, B.M. Wood, H.C. Liu, M. Buchanan, D. Landheer, A.J. Springthorpe, and P. Mandeville, Appl. Phys. Lett. **52**, 1398 (1988).
8. M.S. Skolnick, D.G. Hayes, P.E. Simmonds, A.W. Higgs, G.W. Smith, H.J. Hutchinson, C.R. Whitehouse, L. Eaves, M. Henini, O.H. Hughes, M.L. Leadbeater, and D.P. Halliday, Phys. Rev. B **41**, 10754 (1990).
9. H.T. Grahn, H. Schneider, W.W. Rühle, K. von Klitzing, and K. Ploog, Phys. Rev. Lett. **64**, 2426 (1990).
10. M. Helm, P. England, E. Colas, F. DeRosa, and S.J. Allen Jr., Phys. Rev. Lett. **63**, 74 (1989).

11. P.M. Martin, R.K. Hayden, C.R.H. White, M. Henini, L. Eaves, D.K. Maude, J.C. Portal, G. Hill and M.A. Pate, in *Proc. Int. Conf. on Hot Carriers in Semiconductors* (July, 1991), Semicond. Sci. Technol. 7 (special issue), 456 (1992).
12. D.A.B. Miller, D.S. Chemla, T.C. Damen, A.C. Gossard, W. Wiegmann, T.H. Wood, and C.A. Burrus, Phys. Rev. B **32**, 1043 (1985).
13. J.W. Cockburn, P.D. Buckle, M.S. Skolnick, D.M. Whittaker, W.I.E. Tagg, R.A. Hogg, R. Grey, G. Hill, and M.A. Pate, Phys. Rev. B **45**, 13757 (1992).
14. C.R.H. White, H.B. Evans, L. Eaves, P.M. Martin, M. Henini, G. Hill, and M.A. Pate, Phys. Rev. B **45**, 9513 (1992).
15. R.L. Greene and K.K. Bajaj, Solid State Commun. **45**, 825 (1983).
16. M.W. Dellow, P.H. Beton, C.J.G.M. Langerak, T.J. Foster, P.C. Main, L. Eaves, M. Henini, S.P. Beaumont, and C.D. W. Wilkinson, Phys. Rev. Lett. **68**, 1754 (1992).
17. P.O. Holtz, M. Sundaram, J.L. Merz, and A.C. Gossard, Phys. Rev. B **40**, 10021 (1989).
18. M.S. Skolnick, P.R. Tapster, S.J. Bass, A.D. Pitt, N. Apsley, and S.P. Aldred, Semicond. Sci. Technol. **1**, 29 (1986).
19. J.W. Cockburn, P.D. Buckle, M.S. Skolnick, D.M. Whittaker, W.I.E. Tagg, R. Grey, G. Hill, and M.A. Pate, Superlattices and Microstructures **12**, 413 (1992).

ELECTRONIC EXCITATIONS AND OPTICAL PROPERTIES OF C_{60} MOLECULES

E. BURSTEIN AND M.Y. JIANG
Physics Department
University of Pennsylvania,
Philadelphia, PA 19104, USA

ABSTRACT. We present here a brief summary of key theoretical and experimental investigations of the electronic excitations and optical properties of C_{60} molecules and solid C_{60}, reviewing, in particular, the optical absorption by vibronic transitions, Raman scattering, photoluminescence, resonant two-photon photoemission, nonlinear optical processes, the charge state of C_{60} molecules adsorbed on metal and semiconductor surfaces, and surface enhanced Raman scattering, and also a brief comment about the C_{60} molecule as a quantum dot.

1. Electronic Structure of C_{60}

The C_{60} molecule has a truncated icosahedral structure, consisting of twelve pentagon rings and twenty hexagon rings which have two types of carbon-carbon bonds: single bonds that form the pentagon rings and doublebonds that occur on alternate sites of the hexagon rings (see Fig. 1) [1]. The resonant bonding structure normally exhibited by hexagon rings with conjugated bonds is disrupted by the presence of the single bonded pentagons and, as a consequence, the reaction of the double bonds of the C_{60} molecule is similar to that of arenes and alkenes, rather than to that of aromatic molecules [2]. Molecular orbital calculations show that the molecule has a closed pi-orbital shell singlet ground state, and a relatively large energy gap between the HOMO and LUMO levels (see Fig. 2) [3]. The electron affinity and the ionization energy of the molecule are 2.65 and 7.75 eV, respectively. The relatively large magnitude of the electron affinity is due in part to a sizable "chemical energy" (~ 4.5 eV) and the relatively small magnitude of the energy to electrostically charge the molecule (~ 1.8 eV). The truncated-icosahedral structure of C_{60} in which all of the carbon atoms have identical bonding, has been verified by the observation of the expected number of infrared active vibration modes, by the observation of the Raman active vibration modes, by the observation of a single NMR line for the $^{13}C_{60}$ molecule, and by X-ray diffraction.

The HOMO level, which has h_u symmetry, is 5-fold degenerate. The LUMO level, which has t_{1u} symmetry, is 3-fold degenerate and can accommodate six extra electrons. The next higher unoccupied level, which has t_{1g} symmetry, is also 3-fold degenerate. Due to strong vibronic interactions, the transitions between the molecular energy levels correspond to vibronic transitions [4]. The electronic transitions between the h_u level and the t_{1u} level are electric dipole forbidden since the two levels have the same parity. However, vibronic

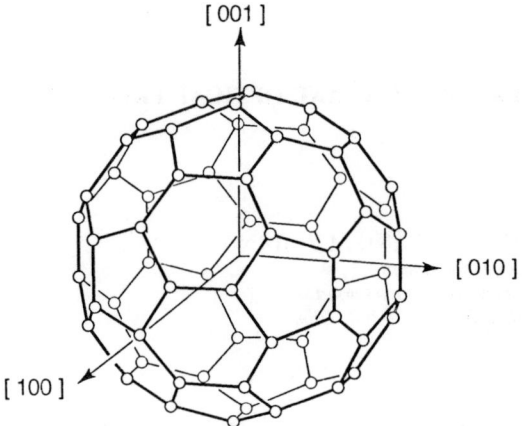

Figure 1. Geometry of the C_{60} molecule. There are two different types of C–C bonds. One is on the pentagons and the other is shared by neighboring hexagons.

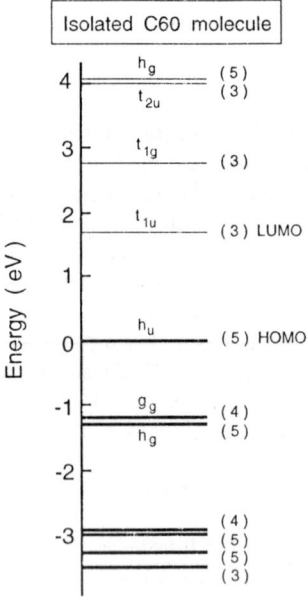

Figure 2. Energy Levels of a free C_{60} molecule. The numbers in parenthesis indicate the degeneracies of the levels.

transitions between the h_u level and the t_{1u} level can occur as a result of coupling of the t_{1u} and t_{1g} levels by non-totally symmetric vibration modes (Herzberg-Teller mechanism). Optical transitions between the h_u and t_{1g} levels and between the h_g and t_{1u} levels, which are electric dipole allowed, correspond to vibronic transitions that involve changes in the quantum numbers of totally symmetric vibration modes, and whose strengths depend on the deformation of the molecule in the excited state relative to the ground state (Franck-Condon mechanism). Because of the high symmetry and consequent large degeneracy of the C_{60} molecule and its relatively large size, the optical matrix elements for the h_u–t_{1g}, h_g–t_{1u} and t_{1u}–t_{1g} electronic transitions are quite large. The double bonds of the hexagon rings form a polyene type sequence of conjugated bonds that circumscribe the C_{60} molecule. The electronic excitations of the molecule are therefore believed to lead to the formation of exciton polarons [5]. On this basis, it has been suggested that the observed luminescence of C_{60} molecules excited by a green laser originates from self-trapped polaron excitons. Calculations (based on the Su-Schrieffer-Heeger model for polyacetylene) indicate that in the C_{60}^{1-} anion, the added electron in the ground state of the ion corresponds to a polaron whose geometry corresponds to a loop circling the molecule, and in the C_{60}^{2-} anion, the two additional electrons form a bipolaron with a similar geometry [6].

The C_{60} molecules form a van der Waals "molecular" crystal and at room temperature it has the fcc structure in which the C_{60} molecules are rotating. The crystal undergoes an orientational ordering phase transition to a sc structure at 249 K in which the C_{60} molecules are still rotating [7]. At temperatures below 100 K the molecules cease to rotate and their orientation become locked in with a double bond of a C_{60} molecule facing the pentagonal ring of the neighboring molecule, yielding a superstructure corresponding to a $2a_0$-fcc crystal structure in which there are two groups of molecules that are rotated by an angle ϕ around the <111> directions [8]. The interactions between the pi-electron systems of neighboring molecules modifies the sharp molecular levels into bands. Band structure calculation for the fcc crystal indicate that the energy bands have a close correspondence to the energy levels of the free C_{60} molecule and are relatively narrow (i.e., widths of about 0.5 eV) [9].

2. Linear and Nonlinear Optical Properties

The optical absorption spectrum of solid C_{60} and that of (free) C_{60} molecules in solution are quite similar, exhibiting strong absorption bands at 3.6, 4.6, and 5.8 eV and a weaker band with an onset at about 1.5 eV [10]. On the basis of a value of 2.3 eV for the bandgap of solid C_{60} (derived from photoemission and inverse photoemission measurements) and a value of 1.6 eV for the on-site molecular C_{60} Coulomb interaction (derived from Auger spectroscopic data) the excitation energy of Frenkel-type singlet excitons in the C_{60} molecule is estimated to be in the range of 1.5 to 2 eV [11]. A loss peak at 1.55 eV in the electron energy loss spectrum (EELS) for solid C_{60} is the lowest electronic excitation and is attributed to the excitation of intramolecular Frenkel triplet excitons via exchange processes [12]. The lowest frequency absorption band is correspondingly attributed to a singlet Frenkel exciton resonance at 1.79 eV. The optical and EELS data thus indicate a singlet-triplet splitting of about 0.2 eV.

Raman scattering by the intramolecular (internal) vibration modes of molecules are two-step processes which involve virtual vibronic excitations between the ground and the excited (singlet exciton) states [13]. In the case of dipole-allowed electronic transition, one observes Raman scattering by totally-symmetric vibration modes via the Franck-Condon mechanism. In the case of dipole-forbidden electronic transitions, one observes Raman scattering by non-totally symmetric vibration modes via the Herzberg-Teller mechanism. Photoluminescence is similarly a two-step process, but involves real vibronic transitions between the ground and excited (singlet exciton) states. When there is an intersystem crossing of the singlet exciton to a long lived triplet exciton, the radiative recombination of the triplet exciton corresponds to a phosphorescence at longer wavelengths, unless the recombination of the triplet exciton is non-radiative. Because of the very high symmetry of the C_{60} molecules, the radiative recombination of the h_u–t_{1u} triplet exciton is strongly forbidden in the free molecule.

The long lifetime of the triplet state of free C_{60} molecules has been utilized in two-photon (pump-probe) photoemission measurements to obtain (optical absorption) data on the optically-induced vibronic transition of isolated low temperature C_{60} molecules in a supersonic jet [14]. The excited triplet states, formed by the rapid intersystem crossing from the singlet states that are created by the visible and near-ultraviolet (UV) dye laser photon-induced vibronic transitions, were one-photon ionized by 6.4 eV photons of an ArF laser to produce positive C_{60} ions that were then pulse extracted into a time-of-flight mass spectrometer. Direct absorption spectrophotometry of a solid solution of C_{60} in methylcyclohexane at 77 K was also carried out. The strongest bands of the solid solution and the resonant two-photon ionization (R2PI) spectra in the 375–415 cm^{-1} range correspond quite closely, apart from a solvent shift of about 351 cm^{-1}. The R2PI spectrum of C_{60} in the 595–640 nm region exhibits a complex pattern of very sharp lines, some of less than 5 cm^{-1} full width at half maximum (FWHM) corresponding to optically induced vibronic excitations.

Measurements of the transient absorption spectra of C_{60} molecules in hexane solution have been carried out following the excitation of the ground state molecules by 355 nm and by 532 nm pump pulses [15]. The 30 ns absorption spectra exhibit a band near 740 nm corresponding to the optical absorption by the C_{60} triplet exciton to create a higher excited triplet exciton state. Kinetic measurements at frequencies in the vicinity of the absorption band yield a singlet exciton lifetime of 650 ps confirming the fast intersystem singlet-triplet crossing rate absorption band for the C_{60} molecules. The quantum yield for the triplet state formation is near unity and the absorption cross-section for the photoexcitation of the triplet state is larger than that of the ground state [16]. The latter has been the basis for using a C_{60} film as an optical-limiting material [17], whose saturation threshold is found to be equal to or lower than those reported for optical-limiting materials now in use.

Measurements of Raman scattering and luminescence excitation-profile spectra enable one to obtain information about the optical absorption by vibronic excitations of the C_{60} molecule. The excitation-profile spectrum for Raman scattering by the $A_g(2)$ symmetry 1469 cm^{-1} vibration mode for a C_{60} film is found to exhibit a resonance peak near 2.4 eV (the h_u–t_{1u} electronic excitation) [18]. The corresponding excitation profile spectrum for C_{60} dissolved in CS_2 does not exhibits a resonance peak at 2.4 eV, but rather exhibits increasing strength toward the UV (i.e., toward the resonance corresponding to the h_u–t_{1g} transition). The observation of the 2.4 eV peak for the C_{60} film, and its absence for C_{60} in CS_2, is attributed to reduced symmetry of the C_{60} molecules in the film. Since C_{60} films are invariably polycrystalline with crystal grains less than 10 nm in size, unless great care is taken to prepare epitaxial films on lattice matched substrates at elevated temperatures [19], we surmise that the 2.4 eV peak may rather be due to Raman scattering by C_{60} molecules with modified symmetry at crystal grain boundaries. The Raman scattering spectra for single crystal C_{60} obtained under resonance enhanced conditions (using 541 nm radiation) at 40 K exhibit narrow bands that are split by crystal field effects and that are consistent with selection rules for the $2a_0$-fcc phase of solid C_{60} [20].

The photoluminescence spectrum of solid C_{60} for low excitation and an incident photon energy of 1.9 eV (i.e., above the singlet exciton resonance) exhibits a broad emission band in the spectral range 1.3–1.8 eV with clearly resolvable structure corresponding to red shifted vibration mode replica [21]. The corresponding photoluminescence excitation (PLE) spectra exhibit peaks which are shifted by 45 meV to higher energies with respect to the detected photon energy. This energy separation between the vibronic recombination and the vibronic excitation is consistent with the energy of the "squashing" vibration mode.

The C_{60} molecule and solid C_{60} possess inversion symmetry, and therefore second-order nonlinear optical processes (i.e., second harmonic generation) are dipole forbidden. However, third-order nonlinear optical processes are allowed. Measurements of degenerate four wave mixing which involve four virtual vibronic transition steps (i.e., h_u–t_{1g}, t_{1g}–t_{1u}, t_{1u}–t_{1g} and t_{1g}–h_u), yield relatively large values for the third-order electric susceptibilities [22]. The presence of a surface, or of an interface with another medium, removes the inversion symmetry of the molecules in the surface region, and solid C_{60} is found to exhibit a strong surface-induced second harmonic generation, indicating a large surface-induced second order nonlinear susceptibility [23]. The second harmonic generation (SHG) by thin C_{60} films on a hydrogen terminated Si(111) substrate is found to be isotropic and to exhibit interference between contributions from the surface of the C_{60} film and from the interface with the substrate [24]. The surface-induced SHG is found to exhibit a strong and sharp resonance at 1.80 eV, which corresponds to an output resonance at 3.60 eV, i.e., a resonance of the output 3.60 eV photons with the surface modified t_{1g}–h_u vibronic transition. The input 1.80 eV photons are at the same time reasonably close to resonance with the surface modified h_u–t_{1u} vibronic transition.

The singly charged C_{60}^{1-} ion is stable in the gas phase in accord with the relatively large electron affinity of the C_{60} molecule. Long lived doubly charged C_{60}^{2-} ions have been also observed in the gas phase and, in cyclic voltametry experiments, C_{60} molecules are found to exhibit one-electron reduction steps to C_{60}^{2-} anions [28]. The stability of the C_{60}^{2-} anions is due in part to dielectric screening of the anions by the solvent and to the "donicity" character of the solvent. In the superconducting alkali fullerides M_3C_{60}, in which there are three alkali metal ions in the primitive unit cell, the C_{60} molecule has a -3e charge and the t_{1u} level is half-filled [25]. In the insulating alkali fullerides M_6C_{60}, in which there are six alkali metal ions in the unit cell, the C_{60} molecule has a -6e charge and the t_{1u} level is completely filled. The attachment of as many as six electrons by the C_{60} molecules in the alkali fullerides is stabilized by the screening of the electrostatic potential of the negatively charged C_{60} ions by the surrounding alkali ions [26].

In the case of C_{60}^{n-} ions adsorbed on a metal substrate, the electrostatic potential is screened by the interaction of the anion with its image charge in the substrate [27]. The charge state depends on the bonding to the substrate, the work function of the substrate, and on the screening of the electrostatic potential and, in the case of a low work function (i.e., alkali) metal, theoretical calculations indicate that at most two electrons can be transferred from the substrate to the C_{60} molecule [28]. The screening of the C_{60} anions should be greater for the case of C_{60} molecules adsorbed on a metal monolayer or embedded within a metal film.

The broadening of the LUMO level of the adsorbed molecule due to resonance with the energy levels of the metal substrate will allow a partial (i.e., fractional) charge-transfer even for a metal substrate having a work function as high as 4.5 eV. Moreover, the symmetry of the adsorbed molecules is reduced and the degeneracies of the energy levels are lifted. The broadening and shifting of the h_u, t_{1u}, and t_{1g} levels is expected to be different due to differences in spatial symmetry and resonances with the electronic levels of the substrate. The energy separation between the levels will therefore be different from those in the free molecule. Also, as a result of the reduced symmetry (i.e., removal of the center of inversion, etc.) the selection rules for optical transitions between the levels are modified (i.e., the normally forbidden transitions between the h_u and t_{1u} levels become electric dipole allowed in the adsorbed C_{60} molecule and can be induced by p-polarized photons), and the selection rules for infrared absorption and Raman scattering by vibration modes are relaxed. Moreover, the adsorbate-substrate (i.e., metal-C_{60}) complex will exhibit charge-transfer (intermolecular) vibronic excitations [29] as well as intramolecular optical transitions between the occupied $t_{1u}1$ level and the t_{1g} level.

The surface enhanced Raman spectra for C_{60} molecules adsorbed on a roughened Au electrode exhibit new features that are attributed to the presence of C_{60} anions that undergo a Jahn-Teller distortion [30]. Moreover, the high frequency bands that arise from vibration modes involving bond stretching appear at lower frequencies than in bulk C_{60}^{1-} and are further downshifted at more negative potentials. The downshifting of the bands is attributed to the weakening of the bonds by the electrons that are transferred to the adsorbed molecule from the metal substrate. The spectra also exhibit a band at 340 cm^{-1}, which was tentatively attributed to a Au-C_{60}^{1-} vibration mode. The Raman scattering spectrum of C_{60} molecules adsorbed on a Ag island film obtained by our group also exhibited a feature at 350 cm^{-1} [31], and we have suggested that it was due to the silent octupolar vibration mode whose Raman scattering becomes allowed as a result of the reduced symmetry of the adsorbed molecules.

The surface-enhanced Raman spectra of monolayer C_{60} deposited on noble metals indicate substantial downshifts of the high frequency $A_g(2)$ mode, which are attributed in part to the transfer of electrons from the metal substrate to the adsorbed C_{60} molecules [32]. The shifts, which increase in the series Au, Cu, and Ag, are consistent with the decrease in the metal work function in this sequence. The shifts for Ag and Cu are similar to those found for the alkali-intercalated M_3C_{60} films. X-ray photoemission and inverse X-ray photoemission measurements for C_{60} molecules adsorbed on various metal substrates (Ag, Cu, Cr, and Mg) and GaAs(110) show that the t_{1u} levels are aligned with the Fermi level of the substrate for all of the metals, indicating charge-transfer from the metal to the C_{60} molecules and LUMO-metal state mixing [33]. In the case of GaAs, a charge-transfer and substrate band bending was observed for n-type GaAs, but not for p-type GaAs.

3. C_{60} as a Quantum Dot

Finally, we note that the C_{60} molecule is, in effect, a quantum dot. It has discrete electronic and vibrational levels, and the electron-electron correlation and exchange interaction plays an important role in electronic transitions, as in the case of semiconductor quantum dots. However, it differs from a semiconductor quantum dot in having strong electron level-vibrational mode coupling (i.e., the transitions between the electronic levels correspond to vibronic transitions) and a rapid intersystem crossing from singlet to triplet states, which reduces the lifetime of the singlet states, and, thereby, the intensities of Raman scattering, luminescence, and nonlinear optical processes. Moreover, the lowest electronic state is optically allowed in direct gap semiconductors (i.e., III-V compounds, CdS, etc.), whereas it is forbidden in the C_{60} molecule. The C_{60} molecule is similar to a Cu_2O quantum dot in this respect, and one should be able to observe similar two-photon process, such as two-photon absorption and inverse (electronic) Raman scattering [34]. In the case of negatively charged C_{60}^{1-} molecules, and in photoexcited C_{60} molecules, in which the t_{1u} level is occupied by an electron, it should be possible to observe resonant inelastic scattering by t_{1u}–t_{1g} excitations. The electrostatic charging potential of the C_{60} molecule is quite high (~ 1.8 eV) due to its small size. This will limit the number of electrons that can be transferred stepwise to the C_{60} quantum dot in electron tunneling (i.e., Coulomb blockade effect). One should, however, be able to observe features corresponding to the excitation of vibrational modes by tunneling electrons in metal/insulator/metal structures having C_{60} dots embedded within the insulating film.

Acknowledgments

We wish to acknowledge valuable discussions with S.C. Erwin, R.P Messmer, H. Talaat, and I. Yurchenko. This work was supported in part by the National Science Foundation MRL Program under Grant No. DMR-8519059 at the University of Pennsylvania Laboratory for Research on the Structure of Matter.

References

The literature on C_{60} is quite extensive and we have only cited some of the references that have made key contributions to the subject of this overview.

1. H.W. Kroto, J.R. Heath, S.S. O'Brien, R.F. Curl, and R.E. Smalley, Nature **318**, 162 (1985); D.R. Huffman, Physics Today **44** (11), 22 (1991).
2. P.J. Fagan, J.C. Calabrese, and P. Malone, Science **252**, 1160 (1991); R.C. Haddon, L.F. Schneemeyer, J.V. Waszczak, S.H. Glarum, R. Tycko, G. Dabbagh, A.R. Kortan. A.J. Muller, A.M. Mujsce, M.J. Roseinsky, S.M. Zahurak, A.V. Makhija, F.A. Thiel, K. Raghavachari, E. Cockayne, and V. Elser, Nature **350**, 46 (1991).
3. R.C. Haddon, L.E. Brus, and H. Raghavachari, Chem. Phys. Lett. **125**, 459 (1986).
4. J. B. Birks, *Photophysics of Aromatic Molecules* (Wiley, New York, 1972).
5. M. Matus, H. Kuzmany, and E. Sohmen, Phys. Rev. Lett. **68**, 2822 (1992).
6. B. Feldman, Phys. Rev. B **45**, 1455 (1992).
7. P. Heiney, J.F. Fischer, A.R. McGhie, W.J. Romanow, A.M Denenstein, J.P. McCauley, Jr., and A.B. Smith III, Phys. Rev. Lett. **66**, 2911 (1991); R.D. Johnson, C.S. Yannoni, H.C. Dorn, J.R. Salem, and D.S. Bethune, Science **255**, 1235 (1992).
8. G. Van Tendeloo, S. Amelinckx, M.A. Verheijen, P.H.M. van Loosdrecht, and G. Meijer, Phys. Rev. Lett. **69**, 1065 (1992).
9. S. Saito and A. Oshiyama, Phys. Rev. Lett. **66**, 2673 (1991).

10. S.L. Ren, Y. Wang, A.M. Rao, E. McRae, J.M. Holden, T. Hager, K.-A. Wang, W.-T. Lee, H.F. Ni, J. Seleque, and P.C. Eklund, Appl. Phys. Lett. **59**, 2678 (1991); W.Y. Ching, M.-Z. Huang, Y.-N. Xu, W.G. Harter, and F.T. Chan, Phys. Rev. Lett. **67**, 2045 (1991).
11. R.W. Lof, M.A. Veenendaal, B. Koopmans, H.T. Jonkman, and G.A. Sawatzky, Phys. Rev. Lett. **68**, 3294 (1992).
12. G. Gensterblum, J.J. Pireaux, P.A. Thiry, R. Caudano, J.P. Vigneron, Ph. Lambin, A.A. Lucas, and W. Krätschmer, Phys. Rev. Lett. **67**, 2171 (1991).
13. A. Brotman and E. Burstein, Physica Scripta **32**, 385 (1985); A.C. Albrecht, J. Chem. Phys. **34**, 1476 (1961).
14. R.E. Haufler, Y. Chai, L.P.F. Chibante, M.R. Fraelich, R.B. Weisman, R.F. Curl, and R.E. Smalley, J. Chem. Phys. **95**, 2197 (1991).
15. R.J. Sension, C.M. Phillips, A.Z. Szarka, W.J. Romanow, A.R. McGhie, J.P. McCauley, Jr., A.B. Smith III, and R.M. Hochstrasser, J. Phys. Chem. **95**, 6075 (1991).
16. J.W. Arbogast, A.P. Darmanyan, C.S. Foote, Y. Rubin, F.N. Diederich, M.M. Alvarez, S.J. Anz, and R.L. Whetten, J. Phys. Chem. **95**, 11 (1991).
17. L.W. Tutt and A. Kost, Nature **356**, 225 (1992).
18. K. Sinha, J. Menendez, R.C. Hanson, G.B. Adams, J.B. Page, O.F. Sankey, L.D. Lamb, and D.R. Huffman, Chem. Phys. Lett. **186**, 287 (1991).
19. X.-D. Xiang, J.G. Hon, G. Briceno, W.A. Vareka, R. Mostovoy, A. Zettl, V.H. Crespi, and M.L. Cohen, Science **256**, 1190 (1992).
20. P.H.M. van Loosdrecht, P.J.M. van Bantum, M.A. Verheijen, and G. Meyer, Chem. Phys. Lett. **198**, 587 (1992).
21. J. Feldman, R. Fischer, W. Guss, E.O. Gobel, S. Schmitt-Rink, and W. Kratschmer, Europhys. Lett. **20**, 553 (1992).
22. Z.H. Kafafi, J.R. Linde, R.G.S. Pong, F.J. Bartok, L.J. Leug, and J. Millikan, Chem. Phys. Lett. **188**, 492 (1992).
23. X.K. Wang, T.G. Zhang, W.P. Lin, S.Z. Liu, G.K. Wong, M.M. Kappes, R.P.H. Chang, and J.B. Ketterson, Appl. Phys. Lett. **60,** 810 (1992).
24. B. Koopmans, A. Anema, J.T. Jonkman, G.A. Sawatzky, and F. van der Woude, Phys. Rev. B, in press.
25. S. Erwin and W.E. Pickett, Science **254**, 842 (1991).
26. S. Erwin and M.R. Pederson, Phys. Rev. Lett. **67**, 16 (1991).
27. E. Burstein, A. Brotman, and P. Apell, J. de Physique C **10** Suppl., 429 (1983).
28. E. Burstein, S.C. Erwin, M.Y. Jiang, and R.P. Messmer, Physica Scripta T **42**, 207 (1992).
29. J.D. Jiang, E. Burstein, and H. Kobayashi, Phys. Rev. Lett. **57**, 1793 (1986).
30. R.G. Garrell, T.M. Herne, C.S. Szafranski, F. Diederich, F. Ertl, and R.L. Whetten, J. Am. Chem. Soc. **92**, 6302 (1991).
31. M.Y. Jiang and H. Talaat, *Workshop on Fullerites and Solid State Derivatives*, Philadelphia PA, August (1991), unpublished.
32. S.J. Chase, W.S. Bacsa, M.G. Mitch, L.J. Pilione, and J.S. Lannin, Phys. Rev. B **46**, 7863 (1992).
33. T.R. Ohno, Y. Chen, S.E. Harvey, S.E. Kroll, G.H. Weaver, J.H. Haufler, and R.E. Smalley, Phys. Rev. B **44**, 13747 (1992).
34. D. Froelich, A. Nothe, and R. Wille, Phys. Rev. Lett. **55** 1355 (1985); D. Froehlich, K. Reimann, and R. Wille, J. Lumin. **38**, 235 (1987).

OPTICAL PROPERTIES OF POROUS SILICON

D.J. LOCKWOOD
Institute for Microstructural Sciences
National Research Council
Ottawa, Ontario K1A 0R6
Canada

ABSTRACT. The optical properties of porous Si films produced by electrochemical dissolution of Si are reviewed. From measurements of the optical absorption spectra of a number of samples, an inverse relationship between the optical gap energy and the average nanoparticle size is obtained demonstrating for the first time the quantum confinement of electron-hole pairs in porous Si. Two different sources of photoluminescence in anodized Si have been identified.

1. Introduction

Silicon is the dominant material in the microelectronics industry and is likely to remain so for many years to come [1]. The need for optical interconnects in the emerging technology of optoelectronics (or photonics) has lead recently to the search for Si-based structures that emit light with a higher quantum efficiency than the $\sim 10^{-4}\%$ of bulk Si: bulk Si emits band to band luminescence at 1.1 eV that is typically observed only at low temperatures, because it involves excitons with a small binding energy (~ 15 meV) that thermally dissociate well below room temperature. Such optoelectronic devices would then allow monolithic integration of mature Si technology with optical signal processing. Optical sources investigated to date include $Si_{1-x}Ge_x$ strained-layer emission, Si/Ge atomic layer superlattices, radiative centers in bulk Si, Si-based molecules (such as siloxene) and polymers, SiH_x, and porous Si (p-Si). Of these, p-Si has received the most attention, because of its visible light emission with high quantum efficiency (1–10%) at room temperature [2,3].

Porous silicon was discovered over 35 years ago by Uhlir at Bell Telephone Laboratories, USA [4]. The porous material is created by electrochemical dissolution in HF-based electrolytes. Hydrofluoric acid, on its own, etches single-crystal Si extremely slowly, at a rate of only nanometers per hour. However, passing an electric current between the acid electrolyte and the Si sample speeds up the process considerably, leaving an array of deep narrow pores that generally run perpendicular to the Si surface. Pores measuring only nanometers across, but micrometers deep, have been achieved under specific etching conditions [5,6]. The evenly-spaced pore structure was explained in terms of available paths for the etching current, with each pore surrounded by an insulating layer of material depleted of electrons. However, despite many years of basic research on the chemical and physical properties of p-Si, the formation mechanism of the network of pores is still not completely understood and the resulting porous microstructure depends critically not only on the HF acid concentration and anodization parameters (current density or voltage) but also on the type and doping level of the Si crystal [7,8]. The Si skeleton left behind has been shown to retain the substrate crystallinity, but with a small lattice expansion compared to bulk Si [9]. In the mid 1970s, p-Si was already being considered for use as a dielectric in the development of Si-on-insulator technology for very-large-scale integrated circuits [7,10,11]. It is only recently that the optical characteristics of p-Si have been explored in detail. In 1984, Pickering et al. [12] reported on low temperature luminescence in p-Si anodized under a range of conditions and noted a band at higher energy (above the bulk-Si band gap) that shifted up in energy with decreasing p-Si density, but they did not investigate the emission properties at room temperature.

In July 1989, Canham conceived the idea of fabricating Si quantum wires in p-Si by reverting to the much slower chemical HF etch after the electrochemical etch [13]. In this way Canham proposed to join up the pores leaving behind an irregular array of undulating free standing pillars of Si only nanometers wide. In 1990, Canham [14] described for the first time the observation of intense *visible* (red) light *emission at room temperature* from p-Si that had been etched under carefully controlled conditions. Visible luminescence ranging from green to red in colour was soon reported by Canham et al. for other p-Si samples [15] and ascribed to quantum size effects in wires of width ~ 3 nm [14,16]. Independently, Lehmann and Gösele [17] reported on the optical absorption properties of p-Si. They observed a shift in the bulk Si absorption edge to values as high as 1.76 eV that they also attributed to quantum wire formation. Visible photoluminescence (PL) in p-Si at room temperature was soon also reported by Bsiesy et al. [18], Koshida and Koyama [19], and Gardelis et al. [20], while visible electroluminescence was observed by Halimaoui et al. [21] during the anodic oxidation of p-Si and, later, by Koshida and Koyama [22] with a diode cell.

Considerable activity on research into the physical and chemical characteristics of p-Si has ensured from these early reports [2,3]. From all this work there is now a general consensus that quantum size effects play a key role in producing visible PL in p-Si [23], but there is still vigorous debate on the precise origin(s) of the PL. Other proposals include the formation of SiH_x [24–26] (infrared absorption studies have shown that the inner surface of the as-prepared p-Si is chemisorbed almost perfectly by H atoms [27]) or siloxene derivates [28] during the etching process, amorphous Si [2], radiative surface states [3,29,30], and localized rather than delocalized states [3,31]. The PL quantum efficiency, wavelength dependence, and lifetime have been found to be sensitive to oxidation of the p-Si and to surface adsorbates (solvents) [2,24,25,32–35]. To obtain intense luminescence, it has been shown that surface passivation is important—either by H [25,36,37] or by O [30,38]—to eliminate dangling Si bonds. Fully oxidized p-Si, when oxidized at high temperature, produces a stable PL intensity, while in the case of freshly prepared hydrogenated samples the PL intensity is very sensitive to the external environment [2,3,25,30,39]. Theoretical calculations ranging from a simple infinite-potential-well model [40] through tight binding [41,42] and envelope function [43–45] analyses to sophisticated first-principles approaches based on the local density approximation [46–49] have all produced results indicating that quantum confinement effects in Si wires (or dots) can produce an intense PL like that observed in p-Si. Other supporting experimental evidence for quantum size effects in p-Si has come from measurements of a blue shift in the x-ray absorption in the vicinity of the Si L edge in freshly prepared material [50] and from observation of bright orange-red light emission from crystalline Si nanoparticles produced by gas phase synthesis in a microwave plasma [51–53]. Nevertheless, direct experimental proof of the correlation between the blue shift in the optical properties with a decrease in the wire/particle size in p-Si is still lacking [54]. One of the major problems in this regard is the lack of uniformity in the wire/particle size in p-Si samples.

In this paper we review the optical properties of p-Si [55] using, as an example, National Research Council Canada results, which are representative of the work of many other groups, and provide optical absorption evidence for quantum confinement effects in p-Si.

2. Optical Properties of p-Si

2.1. SAMPLE PREPARATION

Porous silicon samples were made in standard fashion by anodizing boron doped, (100) oriented, and 10-35 Ω cm resistivity p-type silicon wafers [40]. To make the anodic current distribution uniform, an ohmic back contact was made of Al by vacuum deposition followed by annealing at 450 °C. The anodization process was carried out in a single teflon container at room temperature, with continuous stirring of the solution using an ultrasonic bath to prevent the sticking of gas bubbles to the surface of the sample. The electrochemical treatment was performed in 20% HF aqueous (or alcohol) solution in the dark using Pt as a counterelectrode and supplying constant current density in the 10–100 mA/cm^2 range. In order to increase the porosity of the layers, extended chemical dissolution was applied to some samples. Previously fabricated p-Si samples were kept in 40% HF aqueous solution for up to 120 h in the dark at room temperature. The dissolution rate was increased in some cases by using 20% HF in absolute alcohol solution in the presence of light for up to 15 min. To obtain free p-Si membranes [56], the current was raised to $\sim$ 350 mA/cm^2 for several seconds at the end of the electrochemical process, which detaches the p-Si layer from the Si wafer.

2.2. STRUCTURAL CHARACTERISTICS

The structural characteristics of some representative p-Si samples were investigated by electron microscopy and all samples were studied by Raman spectroscopy.

2.2.1. *Transmission Electron Microscopy*.
Cross-sectional transmission electron microscope (TEM) specimens were prepared by the small-angle cleavage technique [57]. This method was chosen to avoid the surface amorphization produced by ion milling that obscures the delicate structures in this material. The TEM work [40] was performed on a Philips EM430T operating at 250 keV.

Lattice images of most samples showed a $\sim$ 10 µm thick layer of partially amorphized silicon, with no other larger structure such as wires evident. Selected area diffraction patterns (SADPs) showed the single crystal nature of the layer, but the Kikuchi lines were obscured and the diffraction spots were slightly fuzzy, indicating the partial amorphization of the layer. Randomly spaced striations ran parallel to the surface throughout the reacted layer, but were usually most heavily concentrated in the region 1/3 to 1/2 way through the layer as measured from the substrate side. These striations were at consistent depths in the

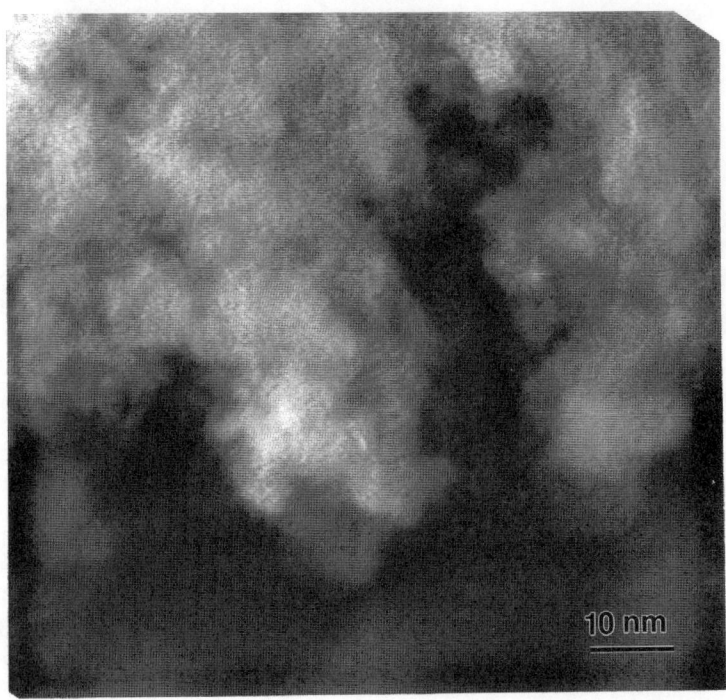

Figure 1. High-resolution transmission electron micrograph of a p-Si sample anodized for 300 s at a current density of 20 mA/cm^2 in a 20% HF aqueous solution. The crystallinity near the base of the reacted area is evident from the lattice fringes. (From Ref. [40].)

layer in all TEM samples made from the material, and were thus not a TEM sample preparation-induced artifact. They are most likely a result of the anodization process. In some samples, a sponge-like columnar structure perpendicular to the surface was evident in the region between 1/3 and 1/2 way through the layer as measured from the substrate side. The columns passed apparently unaltered through the randomly spaced striations. The columns varied in diameter along their length, as the approximately 10–17 nm wide intervening etched areas were irregular in shape: the column diameter ranged from approximately 3–8 nm, as can be seen in Fig. 1. Again, the SADPs showed a consistent single-crystal pattern throughout the layer, with the Kikuchi lines obscured and with some fuzziness of the diffraction spots. These results are consistent with those obtained from other TEM studies of p-Si [2,6–8,16,58,59], but, as noted above, they observed a p-Si

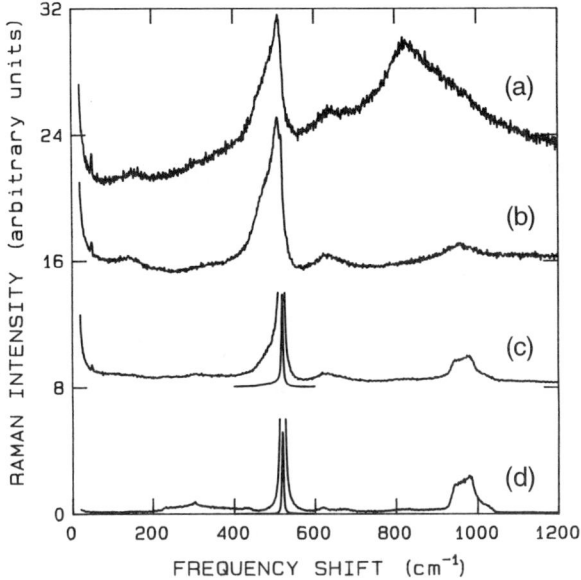

Figure 2. Room temperature Raman spectra of p-Si samples prepared using (a) 10 mA/cm^2 for 600 s in 20% HF aqueous solution, (b) 20 mA/cm^2 for 300 s in 20% HF aqueous solution, and (c) 15 mA/cm^2 for 40 s in 20% HF alcohol solution. Curve (d) is a bulk Si reference spectrum recorded under the same conditions. (From Ref. [40].)

microstructure that is Si-material and sample-preparation dependent.

In a p-Si sample anodized for a short time (40 s) in 20% HF alcohol solution, TEM showed a featureless thin layer. No columnar structure nor crystalline structure of any kind was apparent and the SADPs confirmed that the 0.6 μm thick layer was completely amorphous.

2.2.2. *Raman Spectroscopy.* Raman scattering spectroscopy was used very early on for the characterization of p-Si layers [2,60–62] and a feature in the spectrum just below the bulk Si optical-mode frequency of 520 cm^{-1} has been assigned to nanometer sized structures in p-Si [60,61,63,64].

The Raman spectra of our p-Si samples immersed in He gas were excited with 220 mW of 457.9 nm argon laser light using a quasi-backscattering geometry [40]. Representative Raman spectra of the various p-Si samples after a few months storage in air, together with a reference spectrum of bulk Si recorded under the same conditions, are shown in Fig. 2. A number of broad peaks are seen in the frequency range 10–1200 cm^{-1} together, in some cases, with a sharp line at 520 cm^{-1} due to the Si substrate. In all samples, a broad peak is observed near 480 cm^{-1} and weaker ones near 155 and 635 cm^{-1}. In most of the samples, a broad strong peak appears just below the 520 cm^{-1} bulk-Si line position, while another broad peak appears near 950 cm^{-1}, which lies close to the pure Si second-order feature near 960 cm^{-1} (see Fig. 2). The very broad feature appearing near 830 cm^{-1} in Fig. 2 curve (a) was not evident in other samples.

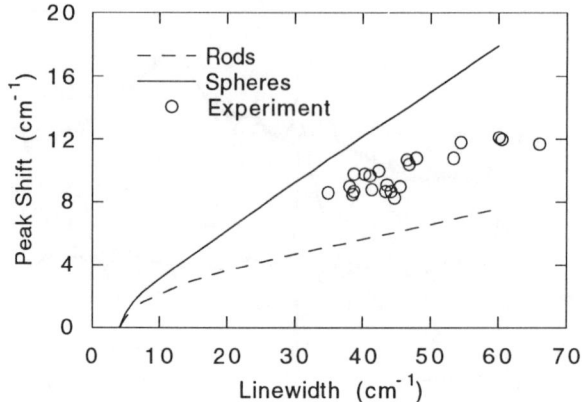

Figure 3. The distribution of room temperature Raman results for the linewidth versus frequency shift (from 520 cm^{-1}) for the broad line near 520 cm^{-1} observed in p-Si samples.

In earlier reports on Raman studies of p-Si, the broad asymmetric feature ~ 10 cm^{-1} below the Si substrate line has been attributed to crystalline Si within p-Si [60,61,63,64]. The small size of the Si crystallites in p-Si results in a shift to lower frequency and a broadening of what was the bulk Si line at 520 cm^{-1}. The confinement effect on the Si phonons has been investigated in detail in earlier studies on nanocrystals of Si [65–68], where the shift in frequency $\Delta\omega$ from 520 cm^{-1} and the line width Γ have been modeled [66–69] for the ideal cases of spheres and rods (or wires). These models have recently been applied to p-Si [63], and have shown that in that particular case the p-Si samples comprised spherical nanocrystals of Si with diameters in the range 2.5–3 nm. In our case, from the experimental data of $\Delta\omega = 9$ cm^{-1} and $\Gamma = 30$ cm^{-1} for Fig. 2 curve (a) and $\Delta\omega = 9$ cm^{-1} and $\Gamma = 37$ cm^{-1} for Fig. 2 curve (b), the models [63] indicate that the curve (a) sample is mostly comprised of 3 nm diameter spheres, and the curve (b) sample contains a mixture of 3 nm diameter spheres and other more anisotropic (wire-like) nanocrystallites. Although these numbers represent only the "average" of a distribution of particle sizes, they are in good agreement with Si nanocrystallite sizes and shapes determined directly by TEM in these same samples. Thus the parameters Δw and Γ for this line in the Raman spectrum provide a good estimate of the crystalline Si size and shape in p-Si. In many p-Si samples, the Raman results for this line lie between the theory curves for spheres and rods (see Fig. 3) indicating the Si nanocrystals are more in the shape of interconnected irregular spheres, which we shall call spherites, rather than free standing wires. This is in accord with the findings of other groups [29,70,71], who consider p-Si to comprise interconnected "dots" rather than undulating "wires".

Amorphous Si has a dominant Raman line at 480 cm^{-1}, a weaker line at 170 cm^{-1}, and a very weak broad line near 950 cm^{-1} [72]. The Raman spectrum of Fig. 2 curve (c) is a close match to the amorphous Si spectrum when allowance is made for overlap with the substrate bulk-Si spectrum. As it is known from TEM that this sample layer is amorphous, it is clear that these features can be used as an indicator of amorphous Si in p-Si samples. The presence of the 480 cm^{-1} peak in the Raman spectrum of the samples shows that they all contain amorphous Si to varying extents, again in accord with TEM analysis. The other

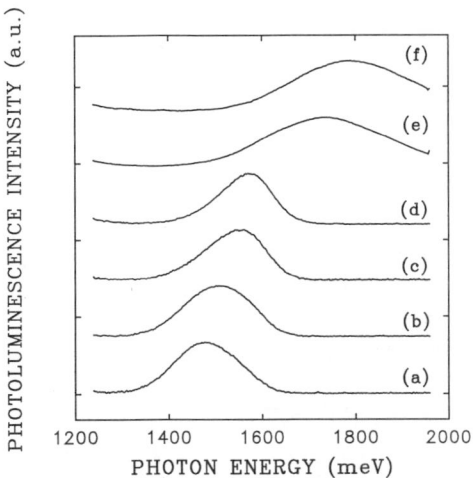

Figure 4. Normalized PL spectra of p-Si samples recorded at 5 K. The sample preparation conditions were (a) anodization for 600 s at 10 mA/cm^2, (b) as for the previous sample plus 17 h chemical dissolution in the dark, (c) 120 s at 50 mA/cm^2, (d) as for the previous sample plus 20 h dissolution in the dark, (e) 600 s at 20 mA/cm^2 plus 3 h dissolution in the dark, and (f) 300 s at 20 mA/cm^2 plus 1 min dissolution in the light. All anodization steps were in 20% HF aqueous solution. (From Ref. [40].)

weak peak seen near 635 cm^{-1} in most samples is likely due to SiH$_x$ [73] formed during the chemical treatment.

2.3. PHOTOLUMINESCENCE

Investigations of the visible light emission in p-Si have been reported now by many groups [2,14,16,18–20,24,25,29,31,33–35,70,74–83] including samples where the porosity was increased up to 85%. A strong PL signal and a progressive shift of the PL spectrum from the near infrared towards the visible as the porosity increases has been observed.

In our work, temperature dependent steady-state and time-resolved PL have been used in the optical characterization of a series of p-Si samples. The steady state PL spectra were obtained by exciting the samples with 100 mW of unfocused 514.5 nm argon laser light and detected using a Bomem DA 3.02 Fourier transform infrared interferometer [40]. For the time-resolved experiments [40], the excitation source was a mode-locked Nd^{3+}:YAG laser synchronously pumping a Rodamine 640 dye laser system (pulse duration ~ 6 ps). A Pockels cell modulator was used to reduce the repetition rate to the operating frequency of 4 MHz and the detection instrumental response time to the dye-laser excitation pulses was 100 ps.

Each of the anodized samples produced a broad PL spectrum covering a range of energies from 1.3 to 2.0 eV and typical 5 K PL spectra are shown in Fig. 4. The PL spectra obtained here are similar to those obtained very early on by other groups [14,18–20]. The spectra show a blue shift with increasing current density, as shown by curves (a) and (c) in Fig. 4, and when extended chemical dissolution in the dark was also used (compare curves

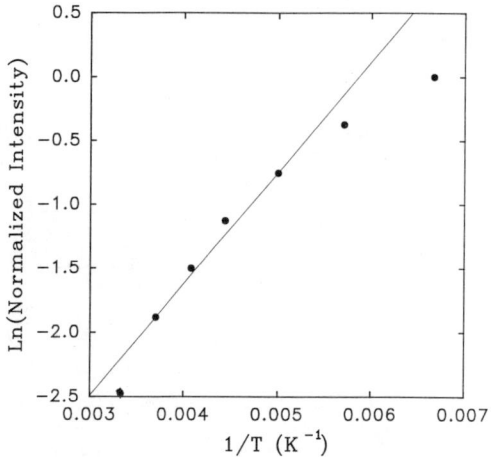

Figure 5. Temperature dependence of the PL intensity for a p-Si sample anodized for 120 s at 50 mA/cm^2 followed by dissolution for 20 h in the dark. The solid line represents an activation energy of 75 meV. (After Ref. [40].)

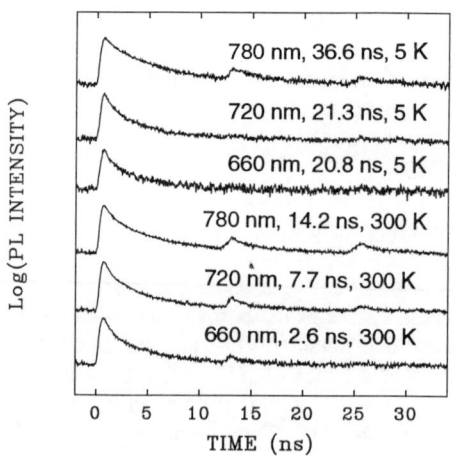

Figure 6. Luminescence decay curves for a p-Si sample anodized for 120 s at 50 mA/cm^2 plus dissolution for 20 h in the dark. The decay is non-exponential, and only the initial decay times (ns) at various emission wavelengths (nm) are given at sample temperatures of 5 and 300 K. The small trailing pulses ~ 13 ns apart are "spurions" owing to leakage from the Pockels cell. (After Ref. [40].)

(a) and (b) and curves (c) and (d)). It was also observed that much smaller immersion times were required to produce noticeable blue shifts when the chemical dissolution was carried out in the presence of light (compare curves (e) and (f)).

According to Beale et al. [8], the porosity of p-Si increases with increasing anodization current density. Therefore, the behavior of the PL spectra in Fig. 4 reflects the differences in sample porosity and hence in the dimensions of Si nanocrystallites.

The effect of temperature change was also investigated for some p-Si samples. A small red-shift and broadening of the PL peak was observed when the temperature was raised, consistent with the bandgap shift of bulk Si with temperature and some broadening associated with the increase in thermal energy. The PL was found to quench as the sample temperature was increased from ~ 120 to 295 K, as shown, for example, in Fig. 5. From such plots, activation energies in the range 25–75 meV were obtained for the samples studied. Gardelis et al. [20] found a similar temperature dependence of the integrated p-Si PL signal and deduced activation energies in the range 34–73 meV. The integrated PL intensity was found to vary linearly with the excitation intensity at both low temperature and room temperature, again in agreement with Gardelis et al. [20] and Barbour et al. [74]. These results indicate that confined localized excitons may be involved in the radiative process [20].

The PL lifetime was measured on a number of p-Si samples at low temperature and at room temperature. The results obtained for one sample are shown in Fig. 6. It can be seen that the lifetime depends strongly on the PL emission energy and on the temperature. Also, the decay is clearly non-exponential and longer-scale lifetimes (up to milliseconds) have been observed by others [18,29,37,70,84].

In a few samples, a very different behavior in both time-resolved and temperature-dependent PL was observed, as shown in Fig. 7. The PL peak is very broad and a low energy shoulder is observed at ~ 1.1 eV. Such structure was not observed in the PL spectra of the other samples discussed above. The PL signal vanishes at temperatures less than 80 K, whereas in all of the "normal" p-Si samples the luminescence persists for all temperatures in the range 5–300 K. The PL showed almost exponential decay (see Fig. 7) with a time constant of 1.5 ns independent of emission energy, and no long-lived component could be observed. Raman and TEM analysis showed that these samples comprised thin layers of amorphous Si only.

2.4. OPTICAL ABSORPTION

A typical optical transmission curve obtained at room temperature from a p-Si membrane sample is shown in Fig. 8 together with the corresponding low temperature PL spectrum. The transmission data are similar to results reported previously by Lehmann and Gösele [17] on p and p^+ type p-Si layers. For all samples, the absorption in the thin film rises fairly sharply at a higher energy than the peak in the PL spectrum.

3. Evidence for Quantum Confinement in p-Si

The most remarkable feature about p-Si is that the anodized layers exhibit very efficient PL, many orders of magnitude larger than that observed in bulk Si, and that it can be observed by eye even at room temperature. As noted previously [14,18–20], the PL shift from the infrared to the visible can be associated with the increase in the effective energy gap due to quantum confinement of the recombining electrons and holes in a one-dimensional quantum wire or zero-dimensional quantum dot. The increase in PL quantum efficiency

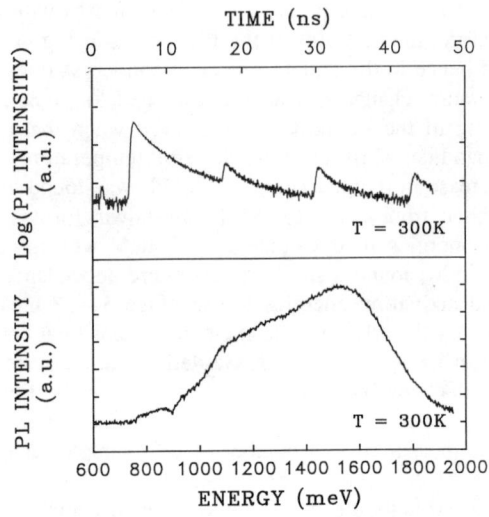

Figure 7. The room-temperature PL decay curve at an emission energy of 720 nm and PL spectrum for a sample anodized for 40 s at 15 mA/cm^2 in 20% HF alcohol solution. (From Ref. [40].)

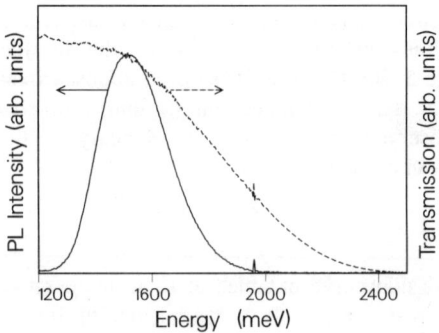

Figure 8. Membrane transmittance spectrum measured at 300 K (broken line) and the corresponding PL spectrum measured at 5 K (solid line) for a p-Si sample anodized in 20% HF aqueous solution for 240 s at 100 mA/cm^2. There is a noticeable red shift of the PL from the p-Si absorption edge. (From Ref. [40].)

from bulk Si to p-Si is also consistent with the quantum confinement interpretation. Calculations have shown that as the free-standing Si pillars decrease in size, to reach a few nanometers in diameter, p-Si can behave like a direct bandgap material due to the quantum confinement [41–49]. In order to observe dramatic quantum size effects, the characteristic diameters of the Si nanocrystals need to be significantly less than the dimensions of the free-exciton Bohr radius of bulk Si (~ 5 nm).

To demonstrate the effects of confinement, Lockwood et al. [40] have considered the

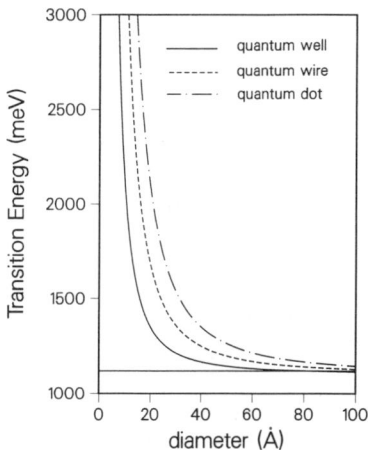

Figure 9. Theoretical PL transition energies as a function of system diameter for quantum wells, wires, and dots assuming infinite confining potentials for the electron and hole. The solid horizontal line is the bulk room temperature band gap. (From Ref. [40].)

simplest model of an infinite confining potential and have calculated the PL transition energies from the lowest electron and heavy-hole eigenergies and the room temperature bandgap, ignoring exciton energies and bulk band symmetry. Figure 9 shows the system size dependence of the PL transition energy for a quantum well, wire, and dot. It can be seen that transition energies consistent with experiment are obtained with system sizes of a few nanometers for the "effective diameter" of the p-Si nanocrystals. Using this model, the wire diameter distribution corresponding to the PL spectrum can be calculated. Such a calculation is illustrated in Fig. 10. It is important to note that since the geometry of the p-Si layer is more complex than that assumed in the calculations, the calculated transition energies have to be considered as order-of-magnitude estimates. The calculation shown in Fig. 10 indicates, for example, that the corresponding "effective" diameter of a Si quantum wire having a transition energy of ~ 1.5 eV is ~ 2.5 nm. This value is in reasonable agreement with the nanoparticle sizes obtained by TEM and Raman spectroscopy. In fact, the diameters obtained directly by TEM or inferred from Raman spectra should be larger than those calculated here, because these techniques measure the physical diameter of the nanoparticles whereas the size obtained from the PL peak energy is an electronic diameter that is reduced from the real physical diameter by the total depletion of the walls of the wires.

As noted earlier, the PL spectrum red shifts with respect to the absorption spectrum (see Fig. 8) and therefore narrows the nanoparticle size distribution towards larger diameter values. This PL red shift is probably due to the tendency of the confined excitons to diffuse towards regions of lowest bandgap in the sample, which correspond to the largest diameter silicon nanoparticles, before the electrons and holes recombine. Simple calculations of the optical absorption using a size distribution based on the PL spectrum would predict a sharp drop in the transmission curve near the centre of the PL peak and with a substantially narrower energy width than that experimentally observed. This analysis

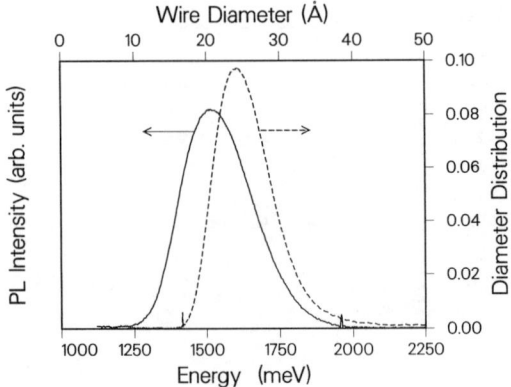

Figure 10. The calculated quantum wire diameter distribution (broken line) corresponding to the 300 K PL spectrum (solid line) of a p-Si sample anodized in 20% HF aqueous solution for 1200 s at 20 mA/cm^2. The diameter distribution was calculated ignoring matrix element dependencies. (From Ref. [40].)

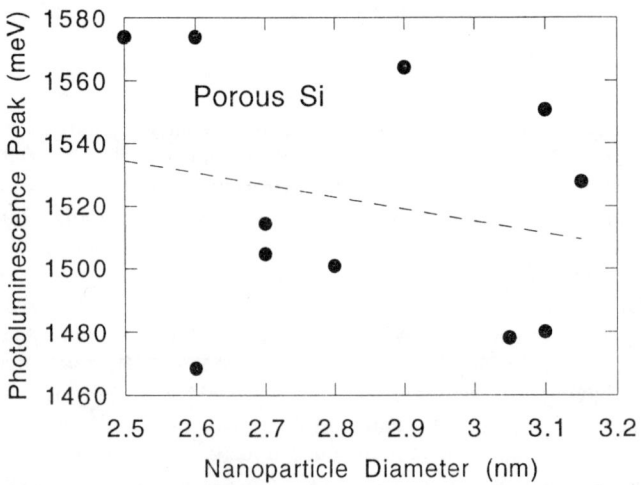

Figure 11. The dependence of the peak of the PL in p-Si samples at 5 K on the average spherite diameter, as determined by Raman spectroscopy. Only a weak correlation is found (shown by the broken straight line obtained from a least-squares fit to the data).

indicates that care must be taken in interpreting the PL spectra and in deducing the energy (or size) distribution of the recombination centres in p-Si samples. Other evidence supporting this conclusion comes from the plot of peak PL energy versus nanoparticle diameter shown in Fig. 11. Here, the Raman scattering $\Delta\omega$ and Γ data for the broad peak near 510 cm^{-1} have been used to estimate the "effective" nanocrystal diameter according to the theory curve for spheres [63]. The plot shows that there is very little correlation between the PL energy and the spherite size.

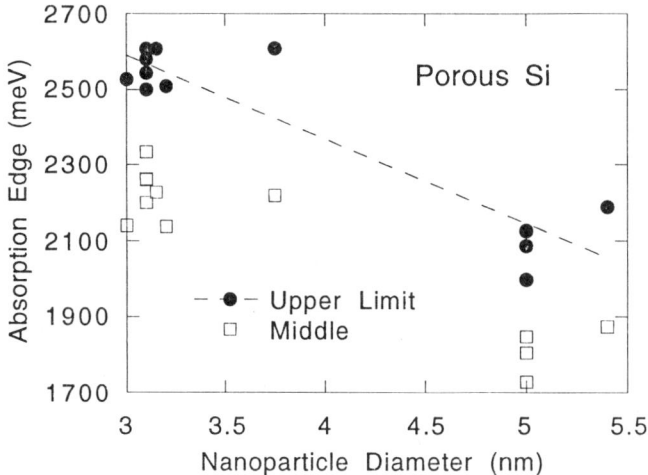

Figure 12. The dependence of the optical absorption in p-Si samples at 300 K on the average spherite diameter, as determined by Raman spectroscopy. A strong correlation is found (shown by the broken straight line obtained from a least-squares fit to the data).

Since the PL red shifts, the only reliable indicators of quantum confinement effects are direct measurements of the p-Si band gap such as optical absorption or PL excitation spectra. The results of a study of the optical absorption and nanoparticle size in a number of p-Si samples are shown in Fig. 12. The samples used in this study were shown by TEM to be of the type where nanocrystals of Si exist in uniform fashion throughout the layer with no evidence of extended wire formation. In these circumstances, the $\Delta\omega$ and Γ parameters of the Raman peak near 510 cm^{-1} provide a reasonable estimate for the average size of the nanocrystals using the theory for spheres [63]. From the theory, the average spherite diameter can be found from the Raman data to within 0.05 nm. Figure 12 shows that there is a *strong inverse correlation* between the spherite diameter and either the high-energy absorption cut-off or the middle of the absorption edge—giving, for the first time, convincing experimental evidence for quantum confinement effects in p-Si. It is difficult to make the analysis more precise, because the nanoparticle size estimated from the Raman data is not likely to be the "active" optical diameter, as discussed above, and because the spread of the absorption edge indicates the existence of a range of nanoparticle sizes in a given sample. Nevertheless, the overall trend is clearly demonstrated from these estimates of the "average" optical properties and spherite diameters.

The decrease in PL lifetimes at shorter emitted wavelength is also consistent with the quantum confinement interpretation, as the change in lifetime can be related to the sizes of the emitting spherites in the porous layer [18]. The shortest lifetimes correspond to transitions with the highest energies, which are associated with the smallest nanoparticle sizes. The resulting extra confinement increases the overlap between the electron and hole wavefunctions, decreasing the radiative lifetime. Moreover, due to the tendency of the excitons to diffuse to regions of lowest bandgap energy (largest diameter nanoparticles), the emission at higher energy will be faster since it has at least one extra decay channel in addition to those available to the lower energy excitations. Carrier tunneling through SiO_2

barriers [70] and carrier hopping in a low-dimensional walk [84] have also been invoked to explain the PL decay behavior. Both local-density-approximation [46–49] and effective-mass-theory [43–45] calculations have demonstrated that the radiative lifetimes resulting from quantum confinement are strongly size (and symmetry) dependent. The thermalization data of Fig. 5 is consistent with the dissociation of confined excitons and the thermal quenching is then a measure of the binding energy. Previous [20,85] and more recent [45] theoretical estimates predict a binding energy of 50–60 meV for such excitons in a 5 nm diameter Si quantum wire or dot, which is in remarkable agreement with the p-Si values of 25–75 meV [20,40]. The thermally activated decrease in the PL intensity and the observed shortening of the PL lifetime, as the temperature is increased from 5 to 300 K, are both consistent with exciton dissociation coupled to a strong non-radiative shunt path [20].

For the thin-layer amorphous Si samples the quantum confinement model cannot be considered in explaining the PL. For these samples the anodization time in alcohol was quite short allowing only the first chemical steps to occur, which is mostly the formation of Si...O...H complexes [80]. Thus the very different PL observed in the amorphous Si samples is likely due to the presence of silicon oxyhydride species or other adsorbates [86]. It is not due to the formation of siloxene or siloxene derivates, as the PL in these compounds does not disappear at low temperatures [87]. Although many aspects of the vibrational and PL properties of siloxene resemble those of p-Si [28], the vibrational spectrum of brightly-luminescing p-Si is quite different from that of siloxene if care is taken to avoid exposing the p-Si to air [88]. Further, siloxene or its derivates would not withstand the high temperature annealing in oxygen that has been shown to produce strong PL in p-Si [30,38]. Thus siloxene is now an unlikely explanation for the optical properties of fully H- or O-passivated p-Si. Nevertheless, there is a sufficient diversity of PL results in the literature to indicate that other light emitting centers can and do exist in anodized or chemically etched Si.

4. Conclusions

From these and other studies [2,3] it is apparent that photoluminescence in p-Si is very sensitive to the chemistry of p-Si production. Crystalline Si wires, crystalline Si spherites, and amorphous (disordered) Si material, or any combination of them, may be formed in a given sample. The p-Si layers thus formed may be far from uniform, which adds to the difficulties in analyzing their optical properties. Other light emitting species may be formed on the surfaces of the anodized and/or chemically etched Si.

We have found *two* distinct sources of photoluminescence in our samples. One can be explained very well by *quantum confinement effects* in nanocrystalline Si. The other is likely associated with silicon oxyhydride species in amorphous Si. In general, there could be, and it is likely that there are, other photoluminescence sources too in p-Si, which may be the reason for the current confusion in the literature. The studies of photoluminescence and optical absorption in p-Si reported here have demonstrated that the photoluminescence is an unreliable indicator of the optical centers involved and that direct absorption or photoluminescence excitation techniques are preferred for analysis.

Definitive explanations of the various light emitting sources in p-Si will only come from further study of the surface chemistry in the pores and from new methods of producing *uniform* Si wires or dots.

Acknowledgments

The results reviewed here based on work performed at the National Research Council Canada have been obtained in collaboration with G.C. Aers, L.B. Allard, B. Bryskiewicz, S. Charbonneau, D.C. Houghton, H.J. Labbé, J.P. McCaffrey, and A. Wang. Their contributions are gratefully acknowledged.

References

1. L. Esaki, *Highlights in Condensed Matter Physics and Future Prospects* (Plenum, New York, 1991).
2. S.S. Iyer, R.T. Collins, and L.T. Canham, *Light Emission from Silicon* (MRS, Pittsburgh, 1992).
3. P.M. Fauchet, C.C. Tsai, L.T. Canham, I. Shimizu, and Y. Aoyagi, *Microcrystalline Semiconductors: Materials Science & Devices* (MRS, Pittsburgh, 1993).
4. A. Uhlir, Bell Syst. Tech. J. **35**, 333 (1956); see also D.R. Turner, J. Electrochem. Soc. **105**, 402 (1958).
5. T. Unagami and M. Seki, J. Electro. Chem. Soc. **125**, 1339 (1978).
6. M.I.J. Beale, J.D. Benjamin, M.J. Uren, N.G. Chew, and A.G. Cullis, J. Cryst. Growth **73**, 622 (1985).
7. See, for example, G. Bomchil, A. Halimaoui, and R. Herino, Appl. Surf. Sci. **41/42**, 604 (1989).
8. M.I.J. Beale, N.G. Chew, M.J. Uren, A.G. Cullis, and J.D. Benjamin, Appl. Phys. Lett. **46**, 86 (1985).
9. K. Barla, R. Herino, G. Bomchil, J.C. Pfister, and A. Freund, J. Cryst. Growth **69**, 726 (1984).
10. Y. Watanabe, Y. Arita, T. Yokoyama, and Y. Igarashi, J. Electrochem. Soc. **122**, 1351 (1975).
11. K. Imai, Sol. State Electron. **24**, 159 (1981).
12. C. Pickering, M.I.J. Beale, D.J. Robbins, P.J. Pearson, and R. Greef, J. Phys. C **17**, 6535 (1984).
13. L.T. Canham, Phys. World **5**, No. 3, 41 (1992).
14. L.T. Canham. Appl. Phys. Lett. **57**, 1046 (1990).
15. See the news item in Electronics Times No. 590, 1 (11 April 1991).
16. A.G. Cullis and L.T. Canham, Nature **353**, 335 (1991).
17. V. Lehmann and U. Gösele, Appl. Phys. Lett. **58**, 856 (1991).
18. A. Bsiesy, J.C. Vial, F. Gaspard, R. Herino, M. Ligeon, F. Muller, R. Romestein, A. Wasiela, A. Halimaoui, and G. Bomchil, Surf. Sci. **254**, 195 (1991).
19. N. Koshida and H. Koyama, Jpn. J. Appl. Phys. **30**, L1221 (1991).
20. S. Gardelis, J.S. Rimmer, P. Dawson, B. Hamilton, R.A. Kubiak, T.E. Whall, and E.H.C. Parker, Appl. Phys. Lett. **59**, 2118 (1991).
21. A. Halimaoui, C. Oules, G. Bomchil, A. Bsiesy, F. Gaspard, R. Herino, M. Ligeon, and F. Muller, Appl. Phys. Lett. **59**, 304 (1991); but see also A. Gee, J. Electrochem. Soc. **107**, 787 (1960) for what may have been the earliest observation of electroluminescence in p-Si.
22. N. Koshida and H. Koyama, Appl. Phys. Lett. **60**, 347 (1992).

23. K.J. Nash, L.T. Canham, and A.G. Cullis, Europhys. News **23**, No. 10, 183 (1992).
24. S.M. Prokes, O.J. Glembocki, V.M. Burmudez, R. Kaplan, L.E. Friedersdorf, and P.C. Searson, Phys. Rev. B **45**, 13788 (1992).
25. C. Tsai, K.H. Li, D.S. Kinosky, R.Z. Qian, T.C. Hsu, J.T. Irby, S.K. Banerjee, B.K. Hance, and J.M. White, Appl. Phys. Lett. **60**, 1700 (1992).
26. S.M. Prokes, J. Appl. Phys. **73**, 407 (1993).
27. Y. Kato, T. Ito, and A. Hiraki, Appl. Surf. Sci. **41/42**, 614 (1989) and references therein.
28. M.S. Brandt, H.D. Fuchs, M. Stutzmann, J. Weber, and M. Cardona, Solid State Commun. **81**, 307 (1992); P. Deak, M. Rosenbauer, M. Stutzmann, J. Weber, and M.S. Brandt, Phys. Rev. Lett. **69**, 2531 (1992); M. Stutzmann, in these proceedings.
29. Y.H. Xie, W.L. Wilson, F.M. Ross, J.A. Mucha, E.A. Fitzgerald, J.M. Macaulay, and T.D. Harris, J. Appl. Phys. **71**, 2403 (1992).
30. V. Petrova-Koch, T. Muschik, A. Kux, B.K. Meyer, F. Koch, and V. Lehmann, Appl. Phys. Lett. **61**, 943 (1992).
31. S. Miyazaki, K. Shiba, K. Sakamoto, and M. Hirose, Optoelectronics **7**, 95 (1992).
32. L.T. Canham, M.R. Houlton, W.Y. Leong, C. Pickering, and J.M. Keen, J. Appl. Phys. **70**, 422 (1991).
33. S. Shih, C. Tsai, K.-H. Li, K.H. Jung, J.C. Campbell, and D.L. Kwong, Appl. Phys. Lett. **60**, 633 (1992).
34. A. Nakajima, T. Itakura, S. Watanabe, and N. Nakayama, Appl. Phys. Lett. **61**, 46 (1992).
35. K.H. Jung, S. Shih, T.Y. Hsieh, D.L. Kwong, and T.L. Lin, Appl. Phys. Lett. **59**, 3264 (1991).
36. C. Tsai, K.-H. Li, J. Sarathy, S. Shih, J.C. Campbell, B.K. Hance, and J.M. White, Appl. Phys. Lett. **59**, 2814 (1991).
37. M.A. Tischler, R.T. Collins, J.H. Stasthis, and J.C. Tsang, Appl. Phys. Lett. **60**, 639 (1992).
38. M. Yamada and K. Kondo, Jpn. J. Appl. Phys. **31**, L993 (1992).
39. S. Miyazaki, K. Shiba, K. Sakamoto, and M. Hirose, in Ref. [3], p. 269.
40. D.J. Lockwood, G.C. Aers, L.B. Allard, B. Bryskiewicz, S. Charbonneau, D.C. Houghton, J.P. McCaffrey, and A. Wang, Can. J. Phys. **70**, 1184 (1992).
41. G.D. Saunders and Y.-C. Chang, Phys. Rev. B **45**, 9202 (1992).
42. J.R. Proot, C. Delerue, and G. Allen, Appl. Phys. Lett. **61**, 1948 (1992).
43. M. Yamamoto, R. Hayashi, K. Tsunetomo, K. Ohno, and Y. Osaka, Jpn. J. Appl. Phys. **30**, 136 (1991).
44. M.S. Hybertsen, in Ref. [2], p. 179.
45. T. Takagahara and K. Takeda, Phys. Rev. B **46**, 15578 (1992).
46. A.J. Read, R.J. Needs, K.J. Nash, L.T. Canham, P.D.J. Calcott, and A. Qtiesh, Phys. Rev. Lett. **69**, 1232 (1992).
47. F. Buda, J. Kohanoff, and M. Parrinello, Phys. Rev. Lett. **69**, 1272 (1992).
48. T. Ohono, K. Shiraishi, and T. Ogawa, Phys. Rev. Lett. **69**, 2400 (1992).
49. M.S. Hybertsen and M. Needals, Phys. Rev. B, to be published.
50. T. van Buuren, Y. Gao, T. Tiedje, J.R. Dahn, and B.M. Way, Appl. Phys. Lett. **60**, 3013 (1992).
51. H. Takagi, H. Ogawa, Y. Yamazaki, A. Ishizaki, and T. Nakagiri, Appl. Phys. Lett. **56**, 2379 (1990).

52. D. Zhang, R.M. Kolbas, P. Mehta, A.K. Singh, D.J. Lichtenwalner, K.Y. Hsieh, and A.I. Kingon, in Ref. [2], p. 35.
53. See also S. Furukawa and T. Miyasato, Phys. Rev. B **38**, 5726 (1988), who observed quantum size effects on the optical band gap of nanocrystalline Si:H particles.
54. Takagi et al. [51] in their work on ultrafine Si particles found an inverse relationship between the PL peak energy and the nanocrystallite size, indicative of quantum size effects.
55. The research literature on the optical and other properties of p-Si is already quite extensive (see, for example, recent issues of Applied Physics Letters) and only some of the references involving key contributions to the subject can be cited here.
56. P.C. Searson, Appl. Phys. Lett. **59**, 832 (1991).
57. J.P. McCaffrey, Ultramicroscopy **38**, 149 (1991).
58. T. George, M.S. Anderson, W.T. Pike, T.L. Lin, R.W. Fathauer, K. H. Jung, and D.L. Kwong, Appl. Phys. Lett. **60**, 2359 (1992).
59. M.W. Cole, J.F. Harvey, R.A. Lux, D.W. Eckart, and R. Tsu, Appl. Phys. Lett. **60**, 2800 (1992).
60. S.R. Goodes, T.E. Jenkins, M.I.J. Beale, J.D. Benjamin, and C. Pickering, Semicond. Sci. Technol. **3**, 483 (1988).
61. R. Tsu, H. Shen, and M. Dutta, Appl. Phys. Lett. **60**, 112 (1992).
62. J.C. Tsang, M.A. Tischler, and R.T. Collins, Appl. Phys. Lett. **60**, 2279 (1992).
63. Z. Sui, P.P. Leong, I.P. Herman, G.S. Higashi, and H. Temkin, in Ref. [2], p. 13.
64. Y.-J. Wu, X.-S. Zhao, and P.D. Persans, in Ref. [2], p. 69.
65. G. Kanellis, J.F. Morhange, and M. Balkanski, Phys. Rev. B **21**, 1543 (1980).
66. H. Richter, Z.P. Wang, and L. Ley, Solid State Commun. **39**, 625 (1981).
67. R. Tsu, S.S. Chao, M. Ize, S.R. Ovshinsky, G.J. Jan, and F.H. Pollak, J. Phys. (Paris) **42**, C4-269 (1981).
68. Z. Iqbal and S. Veprek, J. Phys. C **15**, 377 (1982).
69. I.H. Campbell and P.M. Fauchet, Solid State Commun. **58**, 739 (1986).
70. J.C. Vial, A. Bsiesy, F. Gaspard, R. Hérino, M. Ligeon, F. Muller, R. Romestein, and R.M. Macfarlane, Phys. Rev. B **45**, 14171 (1992).
71. V. Vezin, P. Goudeau, N. Naudon, A. Halimaoui, and G. Bomchil, Appl. Phys. Lett. **60**, 2625 (1992).
72. J.E. Smith, Jr., M.H. Brodsky, B.L. Crowder, M.I. Nathan, and A. Pinczuk, Phys. Rev. Lett. **26**, 642 (1971).
73. S. Miyazaki, T. Yasaka, K. Okamoto, K. Shiba, K. Sokamoto, and M. Hirose, in Ref. [2], p. 185.
74. J.C. Barbour, D. Dimos, T.R. Guilinger, M.J. Kelly, and S.S. Tsao, Appl. Phys. Lett. **59**, 2088 (1991).
75. V.V. Doan and M.J. Sailor, Appl. Phys. Lett. **60**, 619 (1992).
76. J.C. Campbell, C. Tsai, K.-H. Li, J. Sarathy, P.R. Sharps, M.L. Timmons, R. Venkatasubramanian, and J.A. Hutchby, Appl. Phys. Lett. **60**, 889 (1992).
77. X.L. Zheng, W. Wang, and H.C. Chen, Appl. Phys. Lett. **60**, 986 (1992).
78. Z.Y. Xu, M. Gal, and M. Gross, Appl. Phys. Lett. **60**, 1375 (1992).
79. J. Sarathy, S. Shih, K.H. Jung, C. Tsai, K.-H. Li, D.L. Kwong, J.C. Campbell, S.-L. Yau, and A.J. Bard, Appl. Phys. Lett. **60**, 1532 (1992).
80. S. Shih, K.H. Jung, T.Y. Hsieh, J. Sarathy, J.C. Campbell, and D.L. Kwong, Appl. Phys. Lett. **60**, 1863 (1992).

81. L.E. Friedersdorf, P.C. Searson, S.M. Prokes, O.J. Glembocki, and J.M. Macaulay, Appl. Phys. Lett. **60**, 2285 (1992).
82. C.H. Perry, F. Lu, F. Namavar, N.M. Kalkhoran, and R.A. Soref, Appl. Phys. Lett. **60**, 3117 (1992).
83. S.M. Prokes, J.A. Freitas, Jr., and P.C. Searson, Appl. Phys. Lett. **60**, 3295 (1992).
84. X. Chen, B. Henderson, and K.P. O'Donnell, Appl. Phys. Lett. **60**, 2672 (1992).
85. M. Matsuura and T. Kamizato, Surf. Sci. **174**, 183 (1986).
86. R.P. Vasquez, R.W. Fathauer, T. George, A. Ksendzov, and T. Lin, Appl. Phys. Lett. **60**, 1004 (1992).
87. M. Stutzmann, in these proceedings.
88. M.A. Tischler and R.T. Collins, Solid State Commun. **84**, 819 (1992).

OPTICAL PROPERTIES OF SILOXENE AND SILOXENE DERIVATES

M. STUTZMANN, M.S. BRANDT, H.D. FUCHS, M. ROSENBAUER,
M.K. KELLY, P. DEAK, J. WEBER, AND S. FINKBEINER
Max-Planck-Institut für Festkörperforschung
Heisenbergstrasse 1
D-7000 Stuttgart 80
Germany

ABSTRACT. Structural and optical properties of siloxene ($Si_6O_3H_6$) and its derivates obtained by chemical substitution or annealing are reviewed. The preparation of siloxene is briefly described and results of x-ray diffraction and infrared absorption are shown. The equilibrium structures of stoichiometric siloxene and the electronic properties of the corresponding two-, one-, and zero-dimensional Si-clusters are obtained from quantum chemical calculations. Experimental results concerning luminescence, luminescence excitation, absorption coefficients, magnetic resonance, and stability are discussed.

1. Introduction

The integration of optoelectronic devices in standard silicon technology is one of the outstanding problems in modern semiconductor physics. Heteroepitaxy of GaAs on Si, strained layer Si-Ge superlattices and, more recently, anodically etched porous silicon are among the most promising approaches persued in the last years. In particular, the work on porous Si has triggered expectations that, after all, a Si-based optoelectronic may become reality [1,2]. However, a detailed microscopic understanding of the optical properties of porous Si is still missing. Most groups favour the quantum confinement approach, which explains the observation of strong optical transitions in porous silicon as a quantum size effect: carriers are confined in small Si "quantum-wires" or nanocrystals produced somehow during anodic oxidation with a sufficiently narrow distribution of sizes and shapes. Other researchers have invoked specific substances such as hydrogenated amorphous silicon, hydrogenated Si-surfaces, or polysilanes in order to explain the particular properties of porous Si. Based on a comparison of many structural and optical characteristics, our group has proposed that certain derivates of siloxene ($Si_6O_3H_6$) are the origin of the photo-luminescence detected in porous Si [3,4].

Irrespective of their role in porous Si, siloxene and structurally related substances are interesting materials on their own. Although discovered and first described more than a century ago by Wöhler [5], the electronic and optical properties of siloxene and its derivates remain basically unexplored. To our knowledge, so far only about 10–20 published studies exist, most of which date back to 1970 or even earlier. The purpose of this paper is to present an overview over some interesting aspects of siloxene, based partly on older work and partly on recent results of our group.

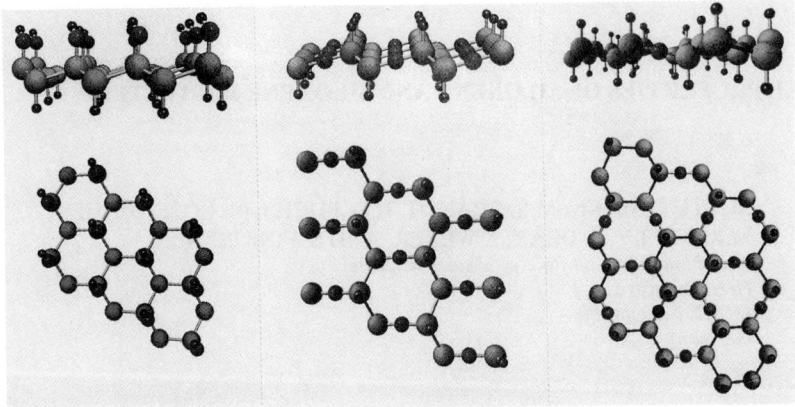

Figure 1. Equilibrium structure of different modifications of stoichimetric siloxene, $Si_6O_3H_6$: (a) Si planes, (b) Si chains, (c) sixfold Si rings. Light atoms symbolize Si, small dark atoms hydrogen, and larger dark atoms oxygen.

2. Basic Structural Properties of Siloxene

Siloxene belongs to the large class of silicon-based polymers or "silicon-backbone-materials" that are characterized by the fact that each silicon atom, in contrast to bulk crystalline or amorphous Si, has less than four silicon nearest neighbors [6]. The preparation of these materials occurs under conditions that are very different from those of crystalline Si growth or amorphous Si thin film deposition, so that investigations on these materials have basically remained within the domain of organosilicon chemistry. In the specific case of siloxene, the important chemical ingredients in addition to silicon are oxygen and hydrogen. Oxygen serves, as usual in organosilicon chemistry, as a link between two silicon atoms, allowing polymerization of the Si-clusters to macroscopic structures, whereas hydrogen serves as an efficient bond terminator which prevents the whole network from having unacceptably high defect densities. In Fig. 1(a) to (c), we show the three possible crystalline modifications of stoichiometric siloxene with the same sum formula, $Si_6O_3H_6$. The three modifications are characterized by a decreasing dimensionality of the Si-backbone structure. Figure 1(a) shows the Si-plane configuration, in which the Si atoms form corrugated planes very similar to the (111) surface of bulk crystalline Si. Each silicon atom is bonded to three other Si-atoms, and the remaining sp^3 hybrid which is not bonded to Si is saturated by OH and H radicals in such a way that ideally one side of the plane is terminated by H, and the other side by OH-groups. The Si planes are then stacked upon each other to form a three-dimensional crystal similar to graphite. The inter-plane bonding is provided by much smaller dipole forces. It is found both theoretically and experimentally that the Si-plane structure in Fig. 1(a) can lower its energy considerably by allowing the oxygen atoms to be inserted into the Si-plane in the form of Si–O–Si units. This oxidation of the silicon backbone can occur in two ordered ways. In Fig. 1(b) the oxygen atoms have been inserted in such a way as to isolate linear,

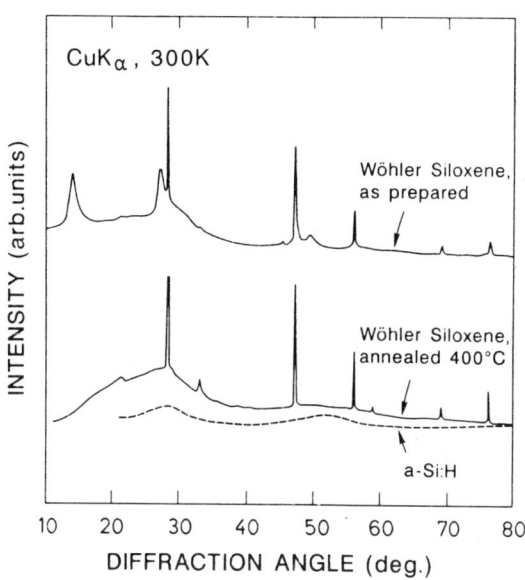

Figure 2. X-ray diffraction spectra of as-prepared Wöhler siloxene and after annealing at 400 °C. The spectrum of amorphous silicon is shown for comparison (dashed curve).

one-dimensional silicon chains, whereas in the case of Fig. 1(c) quasi-zero-dimensional Si_6-rings are formed. In Figs. 1(b) and (c), thus, one has the same planar bonding structure as in the case of Fig. 1(a), however, with a lower dimensionality of the Si backbone. Note that in the ideal case all three modifications can be entirely free of structural defects provided that the hydrogen and oxygen contents are as required by stoichiometry. For the structural characterization of pure siloxene, x-ray diffraction and infrared absorption are the most informative techniques [7,8]. Weiss et al. have performed a detailed study of the x-ray diffraction from siloxene crystals prepared under various conditions [7]. They find a Si-Si bonding distance within the planes shown in Fig. 1(a) which is almost identical to that of bulk crystalline Si, and report a spacing between adjacent Si-planes of ~ 6.3 Å. In Fig. 2, we show a powder diffraction pattern for as-prepared siloxene which clearly exhibits a peak at a diffraction angle (~ 14°) corresponding to the expected interlayer-spacing mentioned above. In addition, the same sample shows further peaks coinciding with or close to the powder diffraction pattern of randomly oriented Si crystals measured simultaneously with the siloxene sample (upper trace in Fig. 2). Annealing of the sample at 400 °C causes a complete disappearance of the peaks related to the Si-plane structure in Fig. 1(a), leaving only a broad background similar to the diffraction spectrum of amorphous silicon (lower two traces in Fig. 2). The amorphization caused by annealing is expected from the metastable nature of the plane structure in Fig. 1(a) against insertion of oxygen atoms, but indicates also that the oxygen insertion does not occur in a sufficiently ordered fashion to produce the structures shown in Figs. 1(b) and (c) selectively. Instead, a random mixture of all three phases is likely to result. Infrared (IR) absorption spectra of as-prepared siloxene are shown in the upper trace of Fig. 3. One clearly observes the well-known Si-H-related vibrational modes around 600–900 and 2100 cm^{-1}, the OH-related bands at 1600 and 3400 cm^{-1}, and the Si-O-Si stretching vibrations at 1000–

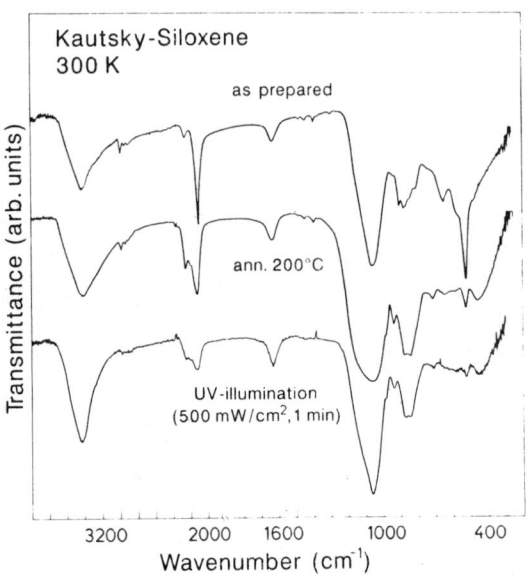

Figure 3. Room-temperature infrared transmittance of Kautsky siloxene as-prepared (upper trace), annealed at 200 °C (middle trace), and after UV illumination in air (lower spectrum).

1100 cm^{-1}. Characteristic for as-prepared siloxene are a sharp hydrogen stretch mode at 2100 cm^{-1} and a new Si-related mode at ~ 520 cm^{-1} of similar width. We have previously assigned this line to a phonon mode of the two-dimensional Si planes [4]. Upon annealing, the sharp Si-H vibrations and the 520 cm^{-1} Si-mode are quickly destroyed, as oxygen is inserted into the planes. The lower trace in Fig. 3 demonstrates that a similar irreversible destruction of the Si-planes can also be achieved by irradiation with ultraviolet (UV) light ($\hbar\omega$ = 3.5 eV) at room temperature in ambient atmosphere.

3. Preparation of Siloxene

Siloxene in the Si-plane modification is prepared by chemical transformation of $CaSi_2$. $CaSi_2$ is a layered silicide consisting of alternating Si and Ca planes [9]. For preparation of siloxene, the Ca-layers are removed by a topochemically reaction at low temperature and the remaining isolated Si-planes are stabilized by intercalation of H and OH bond terminators. The original recipe for this reaction was given by Wöhler [5] and consists simply of placing $CaSi_2$ into concentrated aqueous HCl at 0 °C for several hours. The resulting Wöhler siloxene is a bright yellow substance with a strong room-temperature luminescence in the green (~ 550 nm). Wöhler siloxene is stable for months under normal ambient conditions, but photooxidizes under UV illumination. A more careful preparation method for stoichiometric siloxene has been described by Kautsky [10]. Instead of using aqueous HCl, Kautsky siloxene is prepared with a mixture of HCl, water, and ethanol, again

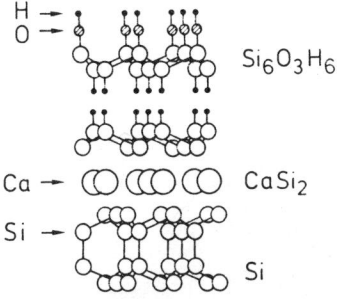

Figure 4. Comparison of the structure of crystalline Si, CaSi$_2$, and siloxene (Si$_6$O$_3$H$_6$) in the Si-plane modification.

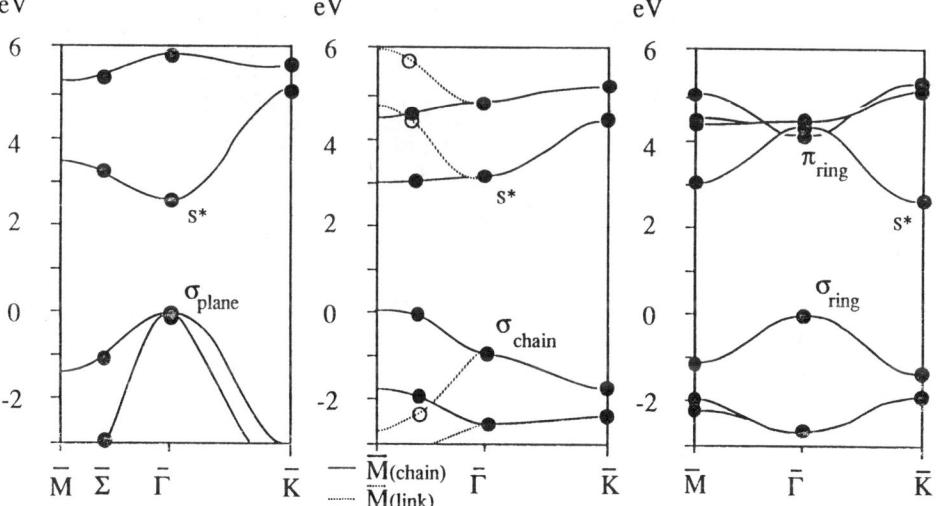

Figure 5. Calculated band structure of siloxene in the plane, chain, and ring modification. See text for details.

at 0 °C, but under exclusion of air and light. Kautsky siloxene is a whitish-grey substance which is very reactive and sensitive to UV light (cf. the lower trace in Fig. 3). It shows a blue photoluminescence under UV excitation. According to Kautsky, the yellow Wöhler-siloxene consists of a mixture of stoichiometric siloxene with derivates in various stages of oxidation.

An interesting aspect for the growth of siloxene on crystalline silicon is shown in Fig. 4: since the Si-Si distances in both CaSi$_2$ and siloxene (planar modification) agree with those of bulk crystalline Si to within better than 0.5% [7], it should be possible to grow siloxene epitaxially on Si(111) substrates. To this end, one first grows a thin epitaxial CaSi$_2$ layer [9], which is then transformed into siloxene by the topochemical reaction described above. First attempts to use this method for the preparation of thin luminescent siloxene layers on

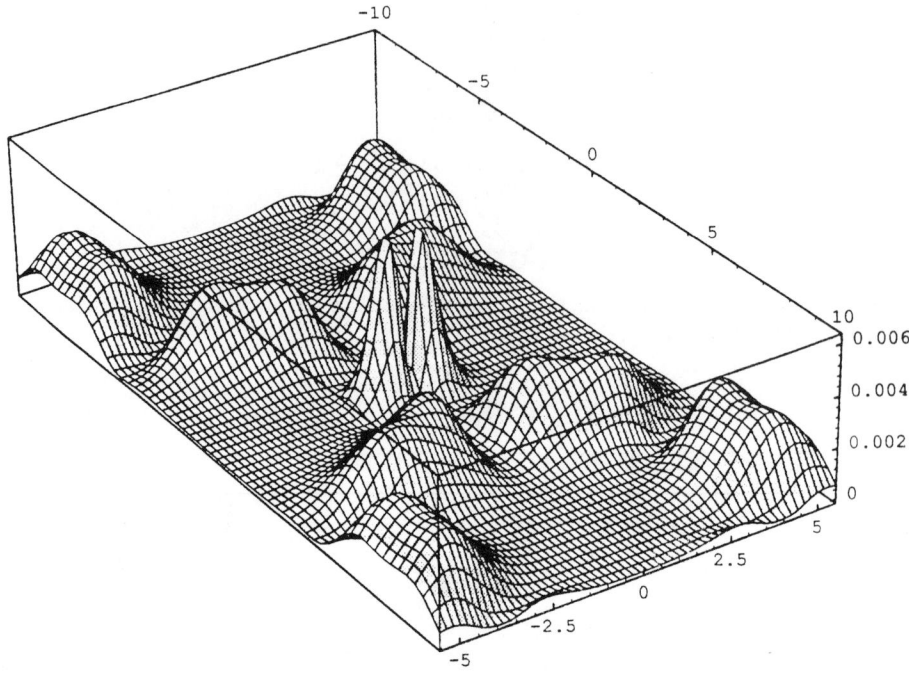

Figure 6. Contour plot of the electronic density for the highest valence band of siloxene in the Si_6 ring modification. The sharp peaks at the center are due to 2p orbitals of the oxygen bridge joining the two Si-rings. The positions of Si atoms are at the density minima.

crystalline Si are very promising [11].

4. Theoretical Modelling of Siloxene

Since up to now no theoretical studies of siloxene have been published, we have performed a number of calculations concerning electronic, structural, and vibrational properties of the three ideal siloxene modifications shown in Fig. 1. As a matter of fact, the structures in this figure are the result of minimizing the total energy of suitable cyclic clusters using semi-empirical quantum-chemical methods. Details of these calculations will be published elsewhere [12]. Here, we only mention briefly those results which are of interest for a better understanding of optical properties of siloxene. To begin, Fig. 5 summarizes our results for the electronic band structure of the plane, chain, and ring modifications of siloxene. Contrary to the indirect band structure of bulk silicon, the two-dimensional Si planes in siloxene give rise to a direct bandgap of ~ 2.7 eV at the center of the Brillouin zone. This is in reasonable agreement with calculations by Takeda and Shiraishi [13] for planar polysilane, $(SiH)_n$, where a direct gap of 2.6 eV at Γ and an indirect gap of 2.4 eV between

Γ and Σ was obtained (planar polysilane is obtained from siloxene by removing all oxygen atoms). In the Si-chain modification of siloxene (Fig. 1(b)) a somewhat larger direct gap of 3 eV is obtained for wavevectors parallel to the Si chains. The corresponding wavefunctions are essentially localized within the Si chains and are isolated from each other by the intermediate oxygen atoms. These act as effective potential barriers because of the larger bonding-antibonding splitting for Si–O compared to Si–Si bonds (dotted lines in Fig. 5). Indeed, the electronic structure of the chain-modification of siloxene is quite similar to that of a crystalline array of linear polysilane chains. There, the highest valence band and the lowest conduction band are found to be almost parallel and dispersionless over a large fraction of the Brillouin zone, with a direct gap of about 3.8 eV [13]. The band structure of siloxene in the Si_6-ring modification is shown on the right-hand side of Fig. 5. In this modification the direct gap is significantly larger than in the two other modifications (approximately 4–4.2 eV). The lowest transition between Γ and K is indirect, with an energy of about 2.8 eV.

It is interesting to look in more detail at the electronic properties of siloxene in the ring-modification, since in the relevant chemical literature the existence of such rings has been invoked in order to explain the remarkable colour and luminescence of certain siloxene derivatives [14–16]. Figure 6 shows a contour plot of the valence electron density for the ground state of stoichiometric siloxene in the Si_6-ring structure (Fig. 1(c)). The two narrow peaks in the center of Fig. 6 are due to electrons in the oxygen 2p orbitals. The two half-circles show the density of Si $3p_x$ and $3p_y$ valence electrons participating in the σ-bonding state giving rise to the highest valence band on the right side of Fig. 5. The positions of Si atoms coincide with the valleys in the ring structure. Note that there is essentially no density of Si 3p electrons at the Si–O–Si bridge connecting the two Si_6-rings. Thus, to a first approximation, adjacent Si_6-rings can be regarded as isolated entities with little interaction between their respective wavefunctions.

The main evidence for Si_6-rings acting as chromophores in siloxene comes from chemical substitution experiments. Looking again at Fig. 1(c), one notes that each Si atom belonging to a Si_6-ring is bonded to one hydrogen atom. It is now possible by more or less complicated chemical reactions to replace one hydrogen atom after the other with different monovalent radicals such as Cl, OH, NH_2, OCH_3, etc. Depending on the electro-negativity of the radicals, this substitution has pronounced effects on the optical properties (absorption and fluorescence) of the resulting siloxene derivatives. In particular, for a given radical, the substitution should occur in six distinct steps. Such a behavior has indeed been observed [14,15]. As a specific example, we have modelled the effect of OH substitution on the electronic properties of siloxene in the Si_6-ring modification. The different possible chemical compositions in this case are $Si_6O_3H_{6-n}(OH)_n$ with $0 \leq n \leq 6$, with the configuration n = 0 corresponding to the situation described in Figs. 5 and 6. With increasing degree of H versus OH substitution, $1 \leq n \leq 6$, we obtain from our calculations a significant decrease of both the direct and the indirect optical transition energies, which is in qualitative agreement with the available experimental evidence. A comparison between theoretical and experimental optical transition energies is given in Table 1.

5. Experimental Results and Discussion

The large spectral range which can be covered by the strong photoluminescence of siloxene

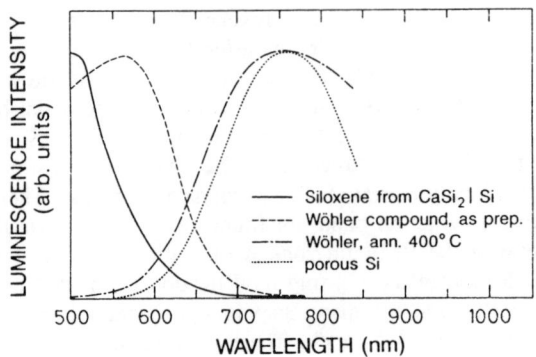

Figure 7. Photoluminescence spectra of siloxene prepared as a thin layer on crystalline silicon as well as of Wöhler siloxene in the as-prepared state and after annealing at 400 °C. The luminescence spectrum of anodically etched porous Si is shown for comparison. All spectra were taken at 300 K.

TABLE 1. Effect of H versus OH substitution on the optical properties of $Si_6O_3H_{6-n}(OH)_n$, $0 \leq n \leq 6$. $E(\Gamma)$ and $E(K)$ denote direct optical gaps at different points of the Brillouin zone; $E(K-\Gamma)$ is the calculated indirect gap for transitions between K and Γ. Experimental luminescence data are taken from the work of Kautsky and Herzberg [14].

n	$E(\Gamma)$ (eV)	$E(K)$ (eV)	$E(K-\Gamma)$ (eV)	E_{lum} (eV)
0	4.1	3.6	2.8	≥ 3.0
1	4.0	3.3	2.7	2.5
2	3.9	3.1	2.5	2.2
3	3.9	2.9	2.3	1.9
6	3.8	2.3	1.8	1.6

and siloxene derivates is summarized in Fig. 7. Shown is the blue luminescence observed for Kautsky siloxene with a peak around 500 nm, the green-to-yellow luminescence of Wöhler-siloxene, and the red luminescence of siloxene after annealing at 400 °C in ambient atmosphere. In the annealed state, the luminescence of siloxene resembles very much the strong visible luminescence observed in anodically etched porous silicon, suggesting a common origin of the luminescence in both materials [3,4].

The luminescence observed in as-prepared siloxene is, together with the photoluminescence obtained from porous Si, probably the most efficient radiative recombination process reported for Si-related materials at room-temperature. As demonstrated in Fig. 8, the external quantum efficiency for siloxene-luminescence at 300 K is at least two or three orders of magnitude higher than typical luminescence efficiencies observed for other Si-networks (crystalline Si, microcrystalline Si, amorphous Si, or disordered Si:O:H alloys). It increases by at least another order of magnitude upon cooling to 77 K. At this temperature, Wöhler siloxene probably reaches close to 100% internal

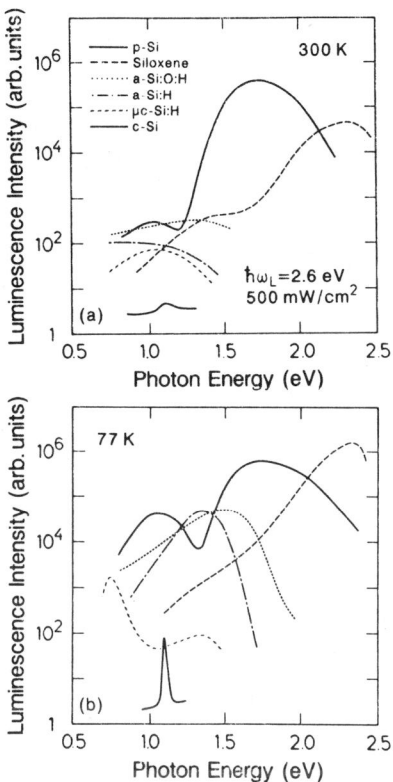

Figure 8. Comparison of the luminescence from porous Si, Wöhler siloxene, amorphous hydrogenated Si and Si-O alloys, microcrystalline silicon and crystalline Si. Note the logarithmic intensity scale. Upper figure, T = 300 K; lower figure, T = 77 K.

quantum efficiency for excitation in the blue spectral region.

Photoluminescence excitation spectra for Wöhler siloxene at room temperature are shown in Fig. 9, both with a linear and logarithmic intensity scale. As-prepared siloxene has its luminescence maximum at $\hbar\omega \approx 2.25$ eV and is most efficiently excited in a narrow peak centered around 2.5 eV. Excitation further into UV is less efficient by about a factor of two. If Wöhler siloxene is stored in air for several weeks the luminescence maximum and also the peak in the excitation spectrum shift to higher energies by about 0.1 eV. At the same time, the relative efficiency for excitation in the UV decreases by another factor of two. This is probably due to the formation of an oxidized surface layer which absorbs the incoming UV light. For both, as-deposited and aged Wöhler siloxene, the luminescence peak energy is almost independent of excitation energy above 2.6 eV. For excitation below 2.5 eV, the luminescence peak follows the excitation energy with a Stokes shift of about 0.25 eV. A similar behavior has been reported for siloxene at 4.2 K by Hirabayashi et al. [17].

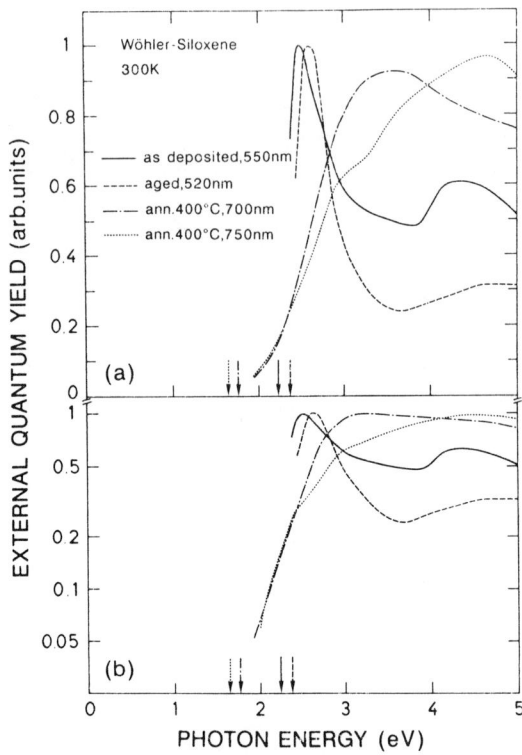

Figure 9. Luminescence excitation spectra of Wöhler siloxene at 300 K: (a) linear intensity scale, (b) logarithmic scale. Arrows denote the energy at which the luminescence was monitored.

As already mentioned in connection with Fig. 7, annealing of siloxene at temperatures above 200 °C leads to a change of the luminescence colour from blue or green to red. The influence of annealing temperature on various structural and optical properties is summarized in Table 2. The main observations are:

1. The Si-planes of the siloxene modification in Fig. 1(a) are quickly destroyed, as evidenced by the disappearance of the x-ray diffraction peak at $2\Theta \approx 14°$ (cf. Fig. 2). Simultaneously, the IR peak at 520 cm^{-1} disappears.
2. Above 200 °C hydrogen starts to evolve, accompanied by an increase of the ESR spin density associated with Si dangling bonds. The insertion of oxygen atoms into the silicon planes gives rise to a shift of the Si-H stretching frequency from 2100 cm^{-1} towards 2250 cm^{-1}.
3. The optical gap (defined as the energy at which the absorption reaches 50%) decreases from about 2.8 eV to 1.9 eV, which can be seen as a colour change from grey to red and eventually black.
4. A second luminescence band centered at about 700 nm (1.8 eV) slowly develops and reaches a pronounced intensity maximum after annealing at 400 °C for 10

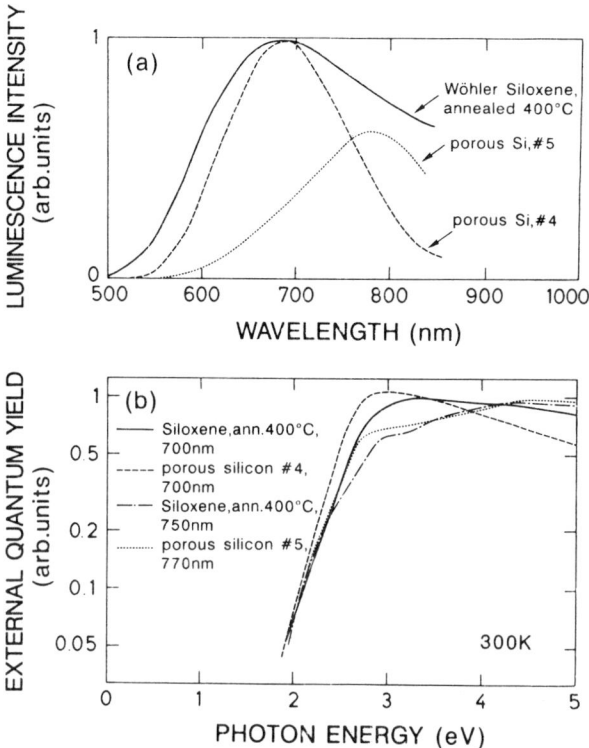

Figure 10. Comparison of (a) the photoluminescence and (b) the luminescence excitation in Wöhler siloxene annealed at 400 °C and in two porous Si samples.

min (see Fig. 7). The intensity of this luminescence is about a factor of 10 smaller than the green luminescence of as-prepared siloxene, which may be partly due to the larger defect density. Annealing at temperatures above 400 °C leads to a complete disappearance of the luminescence.

The luminescence excitation spectra of annealed Wöhler siloxene are also shown in Fig. 9. They are characterized by an almost flat response above 3 eV and a pronounced exponential tail for lower excitation identical to that of porous Si with a similar red luminescence (Fig. 10).

The optical gap of about 2.8 eV for as-prepared siloxene (Table 2) and the observation of a sharp, exciton-like peak in the luminescence excitation spectra of Fig. 9 are in good agreement with the optical properties predicted for the Si-plane modification on the basis of our quantum chemical calculations (Fig. 5). Further information about optical transitions in siloxene occurring at higher photon energies can be obtained from the energy dependence of the absorption coefficient determined by spectroscopic ellipsometry. In Fig. 11, we compare Wöhler siloxene with microcrystalline silicon. Starting from the direct gap at ~ 2.8 eV, siloxene exhibits a monotonically increasing absorption with a step at ~ 4.4 eV. This step can be assigned to direct transitions at the M-point of the Si-plane band

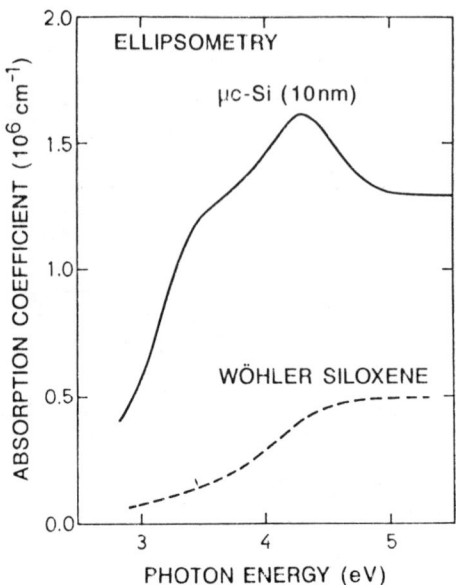

Figure 11. Absorption coefficient of Wöhler siloxene obtained from spectroscopic ellipsometry. Data for microcrystalline Si are shown for comparison.

TABLE 2. Changes in the structural and optical properties of Kautsky siloxene as a function of isochronal annealing in air. I(X) denotes the relative intensity of the X-ray diffraction peak corresponding to the interplane distance, I(IR) are the intensities of the infrared absorption peaks at the specified wavenumbers, and I(Lum) is the normalized luminescence intensity at 700 nm. N_S is the ESR spin density and E_{gap} denotes the optical gap (absorptance = 0.5).

T_A (°C)	I(X) @ d = 6.3 Å	I(IR) @ 520 cm^{-1}	I(IR) @ 2100 cm^{-1}	N_S (cm^{-3})	I(Lum) @ 700 nm	E_{gap}* (eV)	Colour
as-prepared	1	1	1	1×10^{16}	1	2.75	grey
100	-	0.5	0.9	1×10^{16}	5	2.8	grey
200	0.5*	0.3	0.8	1×10^{16}	5	2.9	grey
300	0*	0.1	0.7	1×10^{17}	10	2.8	grey
400	0	0	0.1	2×10^{18}	200	2.2	red
500	-	0	0	2×10^{19}	-	1.95	dark brown
600	-	0	0	2×10^{19}	-	1.9	black

*From Ubara et al. [8] for annealing in vacuum.

structure in Fig. 5. In crystalline Si, the absorption maximum at the same energy is due to direct transitions at the X-point. Note that the maximum absorption coefficient in siloxene is about a factor of three lower than in bulk silicon, as expected from the smaller Si–Si bond density in siloxene.

Finally, we mention a couple of other experimental results concerning the

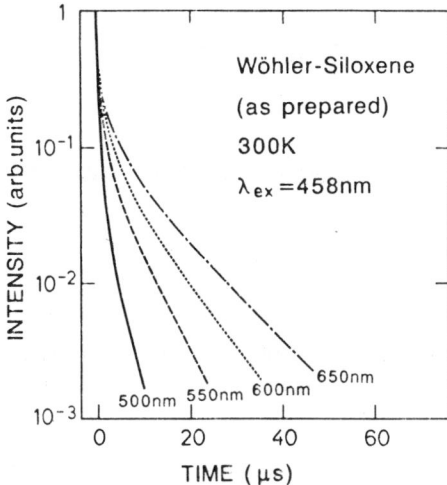

Figure 12. Luminescence decay in as-prepared siloxene following pulse excitation with a 458-nm laser at 300 K. Different transients correspond to different wavelengths at which the luminescence was measured.

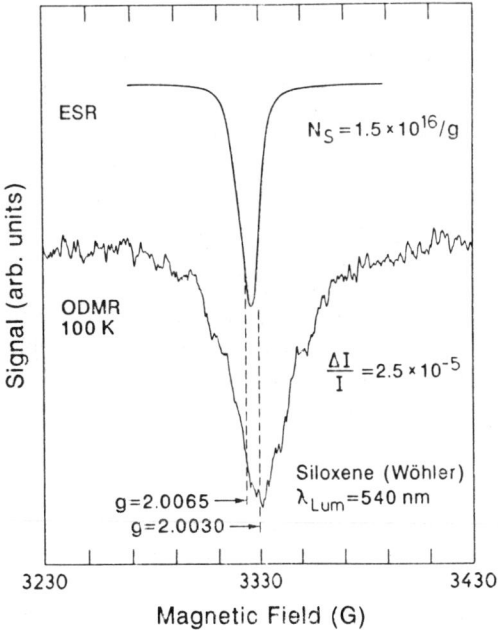

Figure 13. Electron spin resonance (ESR) and optically detected magnetic resonance spectra (ODMR) of Kautsky siloxene annealed at 400 °C with a luminescence at 660 nm.

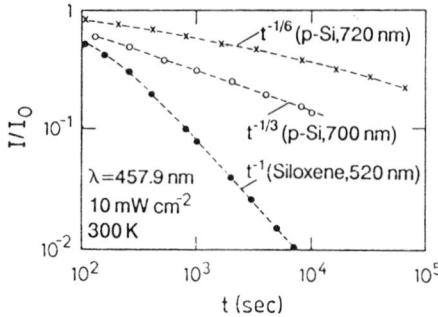

Figure 14. Normalized decay of the photoluminescence intensity in siloxene and porous Si during long illumination in air. Illumination was done at room-temperature with an intensity of 10mW/cm^2 at 458 nm.

photoluminescence of siloxene. First, Fig. 12 shows luminescence decays after pulse excitation with a blue laser line (λ_{ex} = 458 nm). The measurements were done at 300 K. The transients are not simple exponentials, but consist of a fast initial decay followed by a slower tail. Initial decay times are of the order of 1 µs or less, whereas the slower decay extends to time constants of approximately 10 µs, depending on the exact photoluminescence wavelength which is being monitored. It is interesting to compare the results in Fig. 12 to earlier time-resolved luminescence data at 4.2 K reported in Ref. [17]. There, almost the same behavior was observed, with an initial decay time constant of ~ 10 ns and a long transient extending to about 10 µs. This suggests that there is only a weak temperature dependence of the radiative lifetime in as-prepared siloxene.

In Fig. 13 we present optically detected magnetic resonance (ODMR) spectra for Kautsky siloxene annealed at 400 °C. The sample showed a red luminescence at about 660 nm and had a ESR spin density of 2×10^{18} spins/g. The g-value was 2.0050, indicating silicon dangling bond defects as the structural origin of the spins. At 100 K, the same spin signal is also observed in the ODMR as a quenching signal (decrease of the luminescence intensity at spin resonance). This demonstrates that the dangling bond defects act as nonradiative recombination centers in annealed siloxene.

An effect which is commonly observed in both porous silicon and in siloxene is a pronounced luminescence fatigue during long excitation with visible or UV light. Several examples are shown in Fig. 14, where excitation occurred at 300 K with a wavelength of 458 nm in air. The siloxene luminescence monitored at 520 nm decreases irreversibly by two orders of magnitude in about 2 hours. The decay follows a simple bimolecular reaction rate and is proportional to 1/t for long times. At the same time the form of the infrared absorption spectra changes as shown in Fig. 3, relating the luminescence decrease to a destruction of the siloxene Si-planes. Hirabayashi et al. observed a similar fatigue effect also at 4.2 K. They report that the fatigue is only partly reversible in as-prepared siloxene, whereas a sample annealed at 350 °C with a red luminescence exhibits a completely reversible luminescence fatigue [18]: after quenching the luminescence by prolonged illumination at 4.2 K to about one tenth of the initial intensity, annealing at room temperature produces a complete recovery of the luminescence. This behavior is reminiscent of the luminescence fatigue in hydrogenated amorphous silicon [19] and certainly deserves a detailed investigation in the future.

6. Summary

We have presented a brief review of structural and optical properties of siloxene and related chemical derivates. The basic structural units of stoichiometric siloxene are twodimensional silicon planes, linear Si chains, or sixfold Si rings connected by oxygen and terminated by H and OH radicals. Preparation of siloxene occurs via a topochemical reaction starting from $CaSi_2$. Quantum chemical calculations indicate band structures for the different modifications which are completely different from those of bulk crystalline Si and explain the interesting optical properties of siloxene. Siloxene shows a strong luminescence which can be tuned over the entire visible range. The green luminescence of as-prepared siloxene is most efficiently excited in a narrow band around 2.6 eV and exhibits decay times between 1 and 10 µs. Strong absorption ($\alpha = 5 \times 10^5$ cm^{-1}) in siloxene occurs for photon energies above 4.3 eV. The colour of the luminescence can be changed to red either by chemical substitution of hydrogen or by simple annealing in air at 400 °C. The latter material shows an exponential tail in the excitation spectra. ODMR measurements indicate that dangling bond defects in annealed siloxene act as nonradiative recombination centers.

Acknowledgments

We thank M. Cardona and H.-J. Queisser for support and helpful comments. Technical assistance by H. Hirt, P. Wurster, M. Siemers, and P. Andler is gratefully acknowledged. We especially thank A. Breitschwerdt for the IR measurements, and V. Lehmann and J. Köhler for performing the X-ray diffraction studies.

References

1. L.T. Canham, Appl. Phys. Lett. **57**, 1046 (1990).
2. A. Richter, P. Steiner, F. Kozlowski, and W. Lang, IEEE Electron Dev. Lett. **12**, 691 (1991).
3. M.S. Brandt, H.D. Fuchs, M. Stutzmann, J. Weber, and M. Cardona, Solid State Commun. **81**, 307 (1992).
4. M. Stutzmann, J. Weber, M.S. Brandt, H.D. Fuchs, M. Rosenbauer, P. Deak, A. Höpner, and A. Breitschwerdt, Adv. Solid State Phys. **32**, 179 (1992).
5. F. Wöhler, Lieb. Ann. **127**, 257 (1863).
6. N. Matsumoto, K. Takeda, H. Teramae, and M. Fujino, Adv. in Chem., Vol. 224 (Amer. Chem. Soc., Washington, 1990). p. 515.
7. A. Weiss, G. Beil, and H. Meyer, Z. Naturforschung **35b**, 25 (1980).
8. H. Ubara, T. Imura, A. Hiraki, I. Hirabayashi, and K. Morigaki, J. Non-Cryst. Solids **59&60**, 641 (1983).
9. J.F. Morar and M. Wittner, J. Vac. Sci. Technol. A **6**, 1340 (1988).
10. H. Kautsky, Z. Anorg. Chem. **117**, 209 (1921).
11. M.S. Brandt, A. Breitschwerdt, H.D. Fuchs, A. Höpner, M. Rosenbauer, M. Stutzmann,

and J. Weber, Appl. Phys. A **54**, 567 (1992).
12. P. Deak, M. Rosenbauer, M. Stutzmann, J. Weber, and M.S. Brandt, Phys. Rev. Lett. **69**, 2531 (1992).
13. K. Takeda and K. Shiraishi, Phys. Rev. B **39**, 11028 (1989).
14. H. Kautsky and G. Herzberg, Z. Anorg. Chem. **139**, 135 (1924).
15. E. Hengge, Chem. Ber. **95**, 648 (1962).
16. G. Schott, Z. Chem. (Leipzig) **3**, 41 (1963).
17. I. Hirabayashi, K. Morigaki, and S. Yamanaka, J. Non-Cryst. Solids **59&60**, 645 (1983).
18. I. Hirabayashi, K. Morigaki, and S. Yamanaka, J. Phys. Soc. Japan **52**, 671 (1983).
19. K. Morigaki, I. Hirabayashi, M. Nakayama, S. Nitta, and K. Shimakawa, Solid State Commun. **33**, 851 (1980).

Photograph Caption

Standing (rear): L.J. Sham, M.S. Skolnick, L. Eaves, K. von Klitzing, I. Bar-Joseph, S. Scandolo, M. Stutzman, J. Weiner, D.G. Steel.
Standing: D.J. Lockwood, F.M. Peeters, J.P. Kotthaus, A. Pinczuk, E.O. Göbel, J.F. Ryan, C. Kallin, C.L. Kane, P. Hawrylak, G. Abstreiter, C. Tejedor, J. Shah, D. Heitmann, E.I. Rashba, E. Burstein.
Squatting: R. Merlin, D. Gershoni, B. Deveaud, N.J. Pulsford, D.S. Chemla, J.M. Calleja, G.E.W. Bauer.

PARTICIPANTS

Professor Dr. G. Abstreiter
Walter Schottky Institut
Tech. Univ. München
Am Coulombwall
D-8046 Garching
Germany

Dr. D. Awschalom
Department of Physics
University of California
Santa Barbara, CA 93106
USA

Dr. I. Bar-Joseph
Institute of Physical Sciences
Weizmann Institute of Science
Rehovot 76100
Israel

Dr. G.E.W. Bauer
Department of Theoretical Physics
Delft University of Technology
Lorentzweg 1
2628 CJ Delft
The Netherlands

Professor E. Burstein
Department of Physics
University of Pennsylvania
Philadelphia, PA 19104
USA

Dr. J.M. Calleja
Dpto. Fisica Aplicada C-IV
Universidad Autonoma
E-28049 Cantoblanco, Madrid
Spain

Dr. D.S. Chemla, Director
Materials Sciences Division, MS 66
Lawrence Berkeley Laboratory
1 Cyclotron Road
Berkeley, CA 94720, USA

Dr. B. Deveaud
C.N.E.T., LAB/OCM
F-22301 Lannion Cedex
France

Professor L. Eaves
Physics Department
University of Nottingham
Nottingham NG7 2RD
UK

Dr. D. Gershoni
Department of Physics
Technion-Israel
Institute of Technology
Technion City
Haifa 32000
Israel

Professor E.O. Göbel
Fachbereich Physik
Philipps-Universität Marburg
Renthof 5
D-3550 Marburg
Germany

Professor B.I. Halperin
Lyman Lab. of Physics
Harvard University
Cambridge, MA 02138
USA

Dr. P. Hawrylak
Inst. for Microstructural Sciences
National Research Council of Canada
Ottawa, Ontario K1A 0R6
Canada

Professor Dr. D. Heitmann
Institut für Angewandte Physik
Universität Hamburg
Jungiusstrasse 11
D-2000 Hamburg 36
Germany

Professor C. Kallin
Department of Physics
McMaster University
1280 Main Street West
Hamilton, Ontario L8S 4M1
Canada

Professor C.L. Kane
Physics Department
University of Pennsylvania
Philadelphia, PA 19104-6396
USA

Professor Dr. K. von Klitzing
Max-Planck-Institut für
Festkörperforschung
Heisenbergstrasse 1
D-7000 Stuttgart 80
Germany

Professor Dr. J.P. Kotthaus
Sektion Physik
Universität München
Geschwister-Scholl-Platz 1
D-8000 München 22
Germany

Dr. D.J. Lockwood
Inst. for Microstructural Sciences
National Research Council of Canada
Ottawa, Ontario K1A 0R6
Canada

Dr. E.E. Mendez
IBM Research Division
T.J. Watson Research Center
P.O. Box 218
Yorktown Heights, NY 10598
USA

Professor R. Merlin
Physics Department, Randall Laboratories
University of Michigan
Ann Arbor, MI 48109
USA

Professor A.V. Nurmikko
Division of Engineering
Brown University, Box M
Providence, RI 02912
USA

Dr. F.M. Peeters
Department of Physics
University of Antwerp (U.I.A.)
Universiteitsplein 1
B-2610 Antwerp
Belgium

Dr. A. Pinczuk
AT&T Bell Laboratories
600 Mountain Avenue
Murray Hill, NJ 07974-0636
USA

Dr. A.S. Plaut
Department of Physics
Exeter University
Stocker Road
Exeter, EX4 4QL
UK

Dr. N. Pulsford
Philips Natuurkundig Laboratorium
gebouw WB1
Postbus 80000
5600 JA Eindhoven,
The Netherlands

Professor E.I. Rashba
Physics Department
University of Utah
Salt Lake City, Utah 84112
USA

Professor E.H. Rezayi
Physics & Astronomy Department
California State University
Los Angeles, CA 90032
USA

Dr. J.F. Ryan
Clarendon Laboratory
Parks Road
Oxford OX1 3PU
UK

Dr. S. Scandolo
Int. School for Advanced Studies
Via Beirut 2-4
I-34014 Trieste
Italy

Dr. J. Shah
AT&T Bell Laboratories, Rm. 4D415
Crawfords Corner Road
Holmdel, NJ 07733-3030
USA

Dr. L.J. Sham
Department of Physics
University of California, San Diego
La Jolla, CA 92093-0319
USA

Professor M.S. Skolnick
Department of Physics
Sheffield University
Sheffield S3 7RH
UK

Professor D.G. Steel
Dept. of Physics, Randall Laboratories
University of Michigan
Ann Arbor, MI 48109
U.S.A.

Dr. M. Stutzmann
Max-Planck-Institut für
Festkörperforschung
Heisenbergstrasse 1
D-7000 Stuttgart 80
Germany

Dr. C. Tejedor
Dept. Fisica Materia Condensada
Univ. Autonoma de Madrid
E-28049, Cantoblanco, Madrid
Spain

Dr. Rainer G. Ulbrich
IV. Physikalisches Institut
Universität Göttingen
Bunsenstrasse 11-15
D-3400 Göttingen
Germany

Dr. Joseph Weiner
AT&T Bell Laboratories
600 Mountain Avenue
Murray Hill, NJ 07974-0636
U.S.A.

Dr. C. Weisbuch
Thomson - CSF
Laboratoire Central de Recherches
Domaine de Corbeville
F-91404 Orsay Cedex
France

AUTHOR INDEX

Abstreiter, G., 327
Apal'kov, V.M., 63
Awschalom, D.D., 157

Bar-Ad, S., 173
Bar-Joseph, I., 173
Bauer, G.E.W., 91
Berg, A., 3
Böhm, G., 327
Brandt, M.S., 427
Brunner, K., 327
Buckle, P.D., 377, 387
Buhmann, H., 29
Burstein, E., 401

Calleja, J.M., 267
Carmel, O., 173
Chand, N., 337
Chemla, D.S., 101
Chen, W., 27
Clérot, F., 129
Cockburn, J.W., 377
Colas, E., 265
Cundiff, S.T., 187

Deak, P., 427
Dennis, B.S., 57, 267
Deveaud, B., 129
Devreese, J.T., 221
Dumas, C., 129

Eaves, L., 387
Evans, H.B., 387

Feldmann, J., 145
Finkbeiner, S., 427
Fisher, T.A., 387
Fitzgerald, E.A., 337

Florez, L.T., 265
Ford, R.A., 45
Foxon, C.T., 45
Fritze, M., 27
Fuchs, H.D., 427

Gershoni, D., 337
Göbel, E.O., 145
Goñi, A.R., 267
Gozdz, A.S., 265
Grahn, H.T., 239
Grambow, P., 265
Grey, R., 377

Halperin, B.I., 1
Harbison, J.P., 265
Harris, I.N., 45
Harris, J.J., 45
Haug, R.J., 29
Hawrylak, P., 295
Heitmann, D., 265, 351
Henini, M., 387
Hill, G., 377
Hirler, F., 327
Hwang, D.M., 265

Jiang, M.Y., 401

Kallin, C., 13
Kane, C.L., 365
Kapon, E., 265
Kash, K., 265
Katzer, D.S., 129
Kelly, M.K., 427
Klitzing, K. von, 3, 29, 265
Koch, M., 145
Köhler, K., 145
Kotthaus, J.P., 247

Kukushkin, I.V., 29
Kwok, S.H., 239

Lage, H., 265
Levinson, Y., 173
Lockwood, D.J., 409
Longo, J.P., 13

Martinez, G., 29
Meier, T., 145
Mendez, E.E., 373
Merlin, R., 239
Meurer, B., 351
Mowbray, D.J., 387

Nurmikko, A.V., 27

Pate, M.A., 377
Peeters, F.M., 221
Pfeiffer, L.N., 57, 267, 337
Pinczuk, A., 57, 267
Plaut, A.S., 265
Plessen, G. von, 145
Ploog, K., 29, 145, 239, 265, 351
Potemski, M., 29
Pulsford, N.J., 29

Rashba, E.I., 63
Rezayi, E.H., 79
Rodríguez, F.J., 281
Rosenbauer, M., 427
Ryan, J.F., 45

Samarth, N., 157
Scandolo, S., 213
Schmitt-Rink, S., 145
Schulze, A., 145
Sermage, B., 129
Shah, J., 117
Sham, L.J., 201
Shi, J.M., 221
Skolnick, M.S., 377, 387
Smyth, J.F., 157
Steel, D.G., 187
Strenz, R., 327
Stutzmann, M., 427

Tagg, W.I.E., 377
Tassone, F., 213
Tejedor, C., 281
Thomas, P., 145
Timofeev, V.B., 29
Turberfield, A.J., 45

Van der Gaag, B.P., 265

Wang, H., 187
Weber, J., 427
Weimann, G., 327
Weiner, J.S., 267, 337
Weisbuch, C., 311
West, K.W., 57, 267
White, C.R.H., 387
Whittaker, D.M., 377

SUBJECT INDEX

Absorption bistability, 213
Absorption spectroscopy, 1, 101, 265, 281, 351, 373, 401, 409, 427
Activation energy, 409
Alkali fullerides, 401
AlSb/InAsSb/GaSb, 213
Amorphous silicon, 409
Anodized silicon, 409
Antidots, *see* Quantum antidots
Asymmetric quantum wells, 213

Ballistic transport, 365
Band non-parabolicity, 213, 221, 247
Bandstructure calculations, 427
Bernstein modes, 351
Bi-excitons, 173, 187
Bir-Aronov-Pikus mechanism, 201
Bloch oscillations, 117
Bottleneck, *see* Relaxation bottleneck
Bound excitons, 91, 101, 267
Broken symmetry, 63

C_{60}, 401
Carrier relaxation, 327
$CaSi_2$, 429
Cathodoluminescence, 337
Chern-Simons gauge field, 1
Circular dichroism, 101
Coherent emission, 101, 311 (*see also* Stimulated emission)
Coherent nonlinear spectroscopy, 101, 117, 173, 187
Coherent optical excitations, 145
Collective excitations, 247, 267, 351, 365
Confined carriers, 157
Confined geometries, 157
Coulomb blockade, 365, 401
Coulomb charging energy, 351
Coulomb interaction, 13, 27, 101, 117, 187, 201, 247, 267, 281, 351, 365
Coupled quantum wells, 373

Cyclotron resonance, 13, 29, 57, 221, 247, 351

D-center, 221
Deformation potential model, 337
Delta doping, 29, 265
Density of states, 3, 101, 265, 295
Dephasing mechanism, 129, 187
Diluted magnetic semiconductors, 157
Dispersion relation, 13, 247, 265, 267, 281, 351
Dots, *see* Quantum dots
Double barrier resonant tunneling, 365, 377
Double barrier structure, 365, 377, 387
Double quantum wells, magnetically coupled, 157
D'yakonov-Perel mechanism, 201

Edge transport, 365
Effective mass approximation, 221
Electric field, 27, 117, 213, 239, 311, 373
Electrochemistry, 409
Electroluminescence, 377, 387
Electron condensation, 29
Electron correlations, 29, 45
Electron-electron interaction, 13, 29, 57, 63, 79, 247, 281, 351, 365
Electron gas, 295
Elektron-hole
  interaction, 27, 29, 79, 173, 281
  recombination, 201, 387
  scattering, 45
Electronic excitations, collective and single-particle, 101, 267, 365
Electron-phonon coupling, 221
Electron-phonon interaction, 221
Electron relaxation bottleneck, *see* Relaxation bottleneck
Electron spin relaxation, 201
Electron spin resonance, 3, 427
Electron transmission through barriers, 365
Electro-optical quantum devices, 373
Electro-optic anisotropy, 239

Electrostatic confinement, 247
Elliott-Yafet mechanism, 201
Energy gap, 1, 29, 57, 63, 79, 327, 409, 429
Exchange enhancement, 13, 57
Exchange interaction, 3, 201, 295, 401
Excited state populations, 377
Exciton binding energy, 409
Exciton-exciton interaction, 101
Exciton localization, 187
Excitons, 27, 63, 91, 101, 117, 129, 145, 157, 173, 187, 201, 213, 239, 265, 295, 311, 327, 401, 409
Exciton spin relaxation, 201
Exciton unbinding, 91
Extreme quantum limit, 29, 57

Fabry-Perot resonator, 311
Fano resonance, 27, 91
Far-infrared spectroscopy, 1, 247, 351 (*see also* Infrared spectroscopy)
Femtosecond optical spectroscopy, 101, 117, 145, 157
Fermi edge singularity, 27, 29, 45, 267, 281, 295
Fermi level, 5, 27, 91, 247, 265, 267, 281, 295, 327, 365
Fermions, 1
Fermi surface, 1
Field-effect devices, 247
Filling factor, 1, 3, 13, 29, 57, 63, 79, 365
Four-wave mixing, 101, 117, 145, 173, 187, 401
Fractional charges, 63
Fractional quantum Hall effect, 1, 13, 27, 29, 45, 57, 63, 79, 365
Franck-Condon mechanism, 401
Free carriers, 201
Free excitons, 129
Free polarization decay, 187
Frenkel excitons, 401
Fullerenes, 401

GaAs, 27, 45, 101, 117, 129, 187, 247, 267, 327, 365
GaAs/AlAs, 145, 239, 387
GaAs/AlGaAs, 3, 13, 29, 57, 91, 173, 221, 239, 265, 295, 311, 327, 337, 351, 373, 377
GaAs/InGaAs, 337
Gap excitations, 57, 63, 79
g-factor, 3
Ginzburg-Landau mechanism, 101
Green function method, 281

Hartree-Fock approximation, 13, 101, 351

Herzberg-Teller mechanism, 401
Heterostructures 3, 13, 27, 29, 91, 101, 157, 247, 295, 351, 373
Hidden symmetry, 63
Hole burning, 101, 187
Hole-phonon coupling, 387
Hole population inversion, 387
Hole spin relaxation, 201
Holographic lithography, 327

InAs/GaSb/AlSb, 373
Incompressible electron liquid states, *see* Incompressible quantum liquid
Incompressible quantum liquid, 45, 57, 63
Inelastic light scattering, 13, 57, 267 (*see also* Raman spectroscopy)
Infrared spectroscopy, 427 (*see also* Far-infrared spectroscopy)
InGaAs, 27, 337
InGaAs/InP, 311
Inhomogeneous broadening, 145, 187, 327
InSb, 247
Integral quantum Hall effect, 3, 13, 29, 57, 91
Interband excitations, 213, 267, 311, 373
Intersubband electronic excitations, 57, 257, 311, 373
Intersubband scattering rate, 377
Intraband excitations, 63, 247
Intrasubband electronic excitations, 257

Kohn's theorem, 247, 351
Korringa relaxation, 3

Landau level excitations, 1, 3, 13, 29, 57, 63, 101, 221, 295
Landau level filling factor, *see* Filling factor
Lateral confinement, 247, 337
Lateral superlattices, 247
Lifetime, 29, 129, 157, 173, 187, 201, 213, 373, 377, 401, 409
Linear electro-optic effect, 239
Lineshape, 29, 101, 145, 187, 213, 247, 327, 365, 409
Linewidth, 145, 187, 311, 327, 337, 351, 409
Localized excitons, 187, 409
Localized holes, 281, 295
Luminescence, *see* Photoluminescence
Luttinger liquid, 267, 365

Magnetic field, 1, 3, 13, 27, 29, 45, 57, 63, 79, 91,

101, 157, 173, 221, 247, 265, 295, 351, 365
Magnetic polarons, 157
Magnetic quantum wells, 157
Magnetic semiconductors, 157
Magneto-excitons, 63, 79, 101, 173
Magneto-optics, 29, 63, 79, 157, 221, 265
Magnetoplasmons, 13, 57, 247
Magnetorotons, 63
Magnetotransport, 373
Many-body effects, 101, 117
Many-electron states, 295
Mean-field approximation, 91, 101
Mesa etching, 247, 265
Mesoscopic systems, 247
Modulation doped, 45, 265, 267
Momentum scattering, 201
Motional narrowing, 201

Nanoparticles, 409, 427
Nanostructures, 117, 295, 365
Nonlinear optical properties, 101, 145, 187, 401
Nuclear spin relaxation, 3

One-dimensional electron gas, 101, 267, 281, 295
One-dimensional plasmon, 267
Optical anisotropy, electric field induced, 239
Optical bistability, 213
Optical Bloch equation, 117, 187
Optical gap energy, 409
Optical singularities, 267, 281
Optically detected magnetic resonance (ODMR), 427
Optically induced magnetization, 157
Optoelectronic devices, 311
Overhauser shift, 3

Parabolic quantum wells, 27, 267, 351
Phase space filling, 101
Phonons, 187, 201, 221, 311, 365, 387, 401, 409, 427
Photoexcited holes, 27, 29, 45, 79, 295, 327
Photogenerated electric field, 213
Photoholes, *see* Photoexcited holes
Photoluminescence, 27, 29, 45, 57, 63, 79, 91, 129, 157, 187, 201, 239, 265, 267, 281, 295, 311, 327, 337, 373, 401, 409, 427
Photoluminescence excitation, 57, 265, 267, 327, 337, 387, 401, 427
Photon echo, 101, 145, 173, 187
Photon quantization in microcavities, 311

Picosecond optical spectroscopy, 27, 187
p-i-n structures, 117, 337, 377, 387
Plasmons, 267
Pockels effect, 239
Polarization effects, 101, 117, 145, 157, 173, 187, 201, 239, 337
Polarons, 221, 401
Porous silicon, 409, 427

Quantized optoelectronic devices, 311
Quantum antidots, 247, 327
Quantum beats, 117, 145, 173
Quantum boxes, 311
Quantum chemical supercell calculations, 427
Quantum confinement, 101, 157, 187, 351, 409
Quantum dot atoms, 351
Quantum dots, 247, 265, 327, 351, 401, 409
Quantum Hall effect, 1, 3, 13, 27, 29, 45, 57, 63, 79, 91, 373
Quantum liquid, 45, 57 (*see also* Incompressible quantum liquid)
Quantum microcavities, 311
Quantum wells, 57, 91, 101, 117, 129, 145, 157, 173, 187, 201, 213, 221, 239, 267, 311, 327, 337, 373, 377, 387, 409
Quantum wires, 247, 265, 267, 281, 311, 327, 337, 409
Quasiholes, 79
Quasiparticles, 29, 45, 79

Rabi frequency, 101, 187, 311
Radiative recombination, 29, 129, 213, 295, 373, 401
Radiative trapping, 63
Raman spectroscopy, 1, 401, 409 (*see also* Inelastic light scattering)
Random-phase approximation, 267, 351
Relaxation bottleneck, 311, 327
Resonant excitation, 101, 129, 187, 401
Resonant injection of electrons and holes, 387
Resonant tunneling, 239, 365, 373, 377, 387

Shake-up processes, 63, 201, 295
Shallow etching, 327
Shallow impurities in superlattices, 221
Si, 409, 427
    amorphous, 409
    porous, 409, 427
Siloxene, 409, 427
Siloxene derivates, 427